MOLECULAR BIOLOGY OF THE CYTOSKELETON

MOLECULAR BIOLOGY OF THE CYTOSKELETON

Edited by

Gary G. Borisy
University of Wisconsin, Madison

Don W. Cleveland
Johns Hopkins University School of Medicine

Douglas B. Murphy
Johns Hopkins University School of Medicine

Cold Spring Harbor Laboratory
1984

MOLECULAR BIOLOGY OF THE CYTOSKELETON

Printed in the United States of America
Cover design by Emily Harste

Front cover: Distribution of microtubules in a 3T3 cell after microsurgery to sever them from the centrosome. The persistence of microtubules distal to the hole indicates that their stability does not depend on their being connected to the centrosome. (Courtesy of P. Kronebusch and G.G. Borisy, Laboratory of Molecular Biology, University of Wisconsin, Madison.)

Back cover: Construction of a conditionally lethal allele of the yeast actin gene by integrative replacement/disruption. A plasmid carrying a mutant, incomplete copy of the actin gene is integrated into the chromosomal locus resulting in one incomplete copy and one complete, yet mutant, copy. Excision of the plasmid leaves the mutation on the chromosome. (Courtesy of D. Shortle, Dept. of Microbiology, State University of New York at Stony Brook; P. Novick and D. Botstein, Dept. of Biology, Massachusetts Institute of Technology, Cambridge.)

Library of Congress Cataloging in Publication Data

Main entry under title:

Molecular biology of the cytoskeleton.

Bibliography: p.
Includes index.
1. Cytoskeleton. 2. Molecular biology.
I. Borisy, Gary G. II. Cleveland, Don W. III. Murphy, Douglas B.
QH603.C96M65 1984 574.87′5 84-17566
ISBN 0-87969-174-3

All Cold Spring Harbor Laboratory publications may be ordered directly from Cold Spring Harbor Laboratory, Box 100, Cold Spring Harbor, New York 11724. (Phone: 800-843-4388)

Conference Participants

Irene Abraham, *National Institutes of Health, Bethesda, Maryland*
Ueli Aebi, *Department of Cell Biology, Johns Hopkins University, Baltimore, Maryland*
Philip Anderson, *Department of Genetics, University of Wisconsin, Madison, Wisconsin*
Roberts Angerer, *Department of Biology, University of Rochester, Rochester, New York*
Joseph Ardizzi, *Department of Neurology, Baylor College of Medicine, Houston, Texas*
Ellen Baker, *Department of Biology, Yale University, New Haven, Connecticut*
Dorothy Baldwin, *Department of Biological Sciences, Carnegie Mellon University, Pittsburgh, Pennsylvania*
Ray Bandziulis, *Department of Biology, Yale University, New Haven, Connecticut*
Avri Ben-Ze'ev, *Department of Genetics, Weizmann Institute of Science, Rehovot, Israel*
Kerry Bloom, *Department of Biology, University of North Carolina, Chapel Hill, North Carolina*
Steven Blose, *Cold Spring Harbor Laboratory, Cold Spring Harbor, New York*
Miki Blumberg, *New York University Medical Center, New York, New York*
Gary G. Borisy, *Department of Molecular Biology, University of Wisconsin, Madison, Wisconsin*
David Botstein, *Department of Biology, Massachusetts Institute of Technology, Cambridge, Massachusetts*
Karen Brunke, *Zoecon Corporation, Palo Alto, California*
Fernando Cabral, *Department of Endocrinology, University of Texas Medical School, Houston, Texas*
Ming Chen, *Department of Molecular Biology, St. Jude Children's Research Hospital, Memphis, Tennessee*
Margaret Clarke, *Department of Molecular Biology, Albert Einstein College of Medicine, Bronx, New York*
Don W. Cleveland, *Department of Physiological Chemistry, Johns Hopkins University, Baltimore, Maryland*

Thomas A. Cooper, *Department of Anatomy, University of California, San Francisco, California*
Nicholas Cowan, *Department of Biochemistry, New York University Medical Center, New York, New York*
Susan Craig, *Department of Physiological Chemistry, Johns Hopkins Medical School, Baltimore, Maryland*
Julia Currie, *Department of Molecular Biology, New York State Institute for Basic Research and Developmental Disabilities, Staten Island, New York*
Arturo De Lozanne, *Department of Structural Biology, Stanford University School of Medicine, Stanford, California*
William Dove, *McArdle Laboratory, University of Wisconsin, Madison, Wisconsin*
David Drubin, *Department of Biochemistry, University of California, San Francisco, California*
Teresa Dunn, *Department of Physics and Chemistry, Johns Hopkins School of Medicine, Baltimore, Maryland*
Susan Dutcher, *Department of Molecular, Cellular and Developmental Biology, University of Colorado, Boulder, Colorado*
William Earnshaw, *Department of Cell Biology, Johns Hopkins University, Baltimore, Maryland*
Elizabeth Elliott, *Ontario Cancer Institute, Toronto, Canada*
Henry Epstein, *Department of Neurology, Baylor College of Medicine, Houston, Texas*
Ursula Euteneuer, *Department of Zoology, University of California, Berkeley, California*
Stephen Farmer, *Department of Biochemistry, Boston University Medical School, Boston, Massachusetts*
James Feramisco, *Cold Spring Harbor Laboratory, Cold Spring Harbor, New York*
Elaine Fuchs, *Department of Biochemistry, University of Chicago, Chicago, Illinois*
Chandler Fulton, *Department of Biology, Brandeis University, Waltham, Massachusetts*
David Gard, *Department of Biochemistry, University of California, San Francisco, California*
Irith Ginzburg, *Department of Neurobiology, Weizmann Institute of Science, Rehovot, Israel*
Michael Gottesman, *National Institutes of Health, Bethesda, Maryland*
Ruth Grebenau, *Department of Microbiology, Albert Einstein College of Medicine, Bronx, New York*
Karen Greer, *Department of Biology, Yale University, New Haven, Connecticut*
Keith Gull, *Biological Laboratory, University of Kent, England*
Peter Gunning, *Stanford Medical School, Palo Alto, California*
Radhey S. Gupta, *Department of Biochemistry, McMaster University, Hamilton, Ontario, Canada*
John Hammer, *NHLRI, National Institutes of Health, Bethesda, Maryland*
Jane Havercroft, *Department of Physiological Chemistry, Johns Hopkins Medical School, Baltimore, Maryland*
David Helfman, *Cold Spring Harbor Laboratory, Cold Spring Harbor, New York*
Howard Holtzer, *Department of Anatomy, University of Pennsylvania, Philadelphia, Pennsylvania*
Sandra Honda, *Department of Neurology, Baylor College of Medicine, Houston, Texas*
Tim Huffaker, *Department of Biology, Massachusetts Institute of Technology, Cambridge, Massachusetts*
Jonathan Izant, *Department of Genetics, Hutchinson Cancer Research Center, Seattle, Washington*
Jonathan Jarvik, *Department of Biological Sciences, Carnegie Mellon University, Pittsburgh, Pennsylvania*
José L. Jorcano, *Department of Cell and Tumor Biology, University of Heidelberg, Heidelberg, Federal Republic of Germany*
Gertrude Kasbekar, *Department of Cell Biology, National Science Foundation, Washington, D.C.*

Yuriko Kataoka, *Cold Spring Harbor Laboratory, Cold Spring Harbor, New York*
Larry Kedes, *VA Medical Center, Stanford University Medical Center, Palo Alto, California*
Dan Keihart, *Department of Cell Biology, Johns Hopkins Medical School, Baltimore, Maryland*
John Kilmartin, *Laboratory of Molecular Biology, Medical Research Council, Cambridge, England*
Marc Kirschner, *Department of Biochemistry, University of California, San Francisco, California*
Michael Krause, *Department of Biology, University of Colorado, Boulder, Colorado*
Thomas Kreis, *European Molecular Biological Laboratory, Heidelberg, Federal Republic of Germany*
Michael Kuchka, *Department of Biological Sciences, Carnegie Mellon Institute, Pittsburgh, Pennsylvania*
Steve W. L'Hernault, *Department of Embryology, Carnegie Institution of Washington, Baltimore, Maryland*
Elaine Lai, *Department of Biology, Brandeis University, Waltham, Massachusetts*
Scott Landfear, *Department of Tropical Public Health, Harvard School of Public Health, Boston, Massachusetts*
Birgitte Lane, *Imperial Cancer Research Fund, London, England*
Elias Lazarides, *Department of Biology, California Institute of Technology, Pasadena, California*
John Leavitt, *Department of Molecular Carcinogenesis, Linus Pauling Institute, Palo Alto, California*
Gloria Lee, *Department of Biochemistry and Biophysics, University of California, San Francisco, California*
M. Gwo-Shu Lee, *Department of Biochemistry, New York University Medical Center, New York, New York*
Paul A. Lefebvre, *Department of Genetics and Cell Biology, University of Minnesota, St. Paul, Minnesota*
Leslie A. Leinwand, *Department of Microbiology and Immunology, Albert Einstein College of Medicine, Bronx, New York*
Mindy Lewis, *Albert Einstein College of Medicine, Bronx, New York*
Ching Lin, *Department of Molecular Carcinogenesis, Linus Pauling Institute, Palo Alto, California*
Victor Ling, *Department of Medical Biophysics, Ontario Cancer Institute, Toronto, Canada*
Uriel Z. Littauer, *Department of Neurobiology, Weizmann Institute of Science, Rehovot, Israel*
Ricardo Maccioni, *Department of Biophysics and Genetics, University of Colorado, Denver, Colorado*
Alexander MacLeod, *Ludwig Institute for Cancer Research, Medical Research Council, Cambridge, England*
Hunter Martin, *Department of Biochemistry, University of Pittsburgh, Pittsburgh, Pennsylvania*
Hiroshi Maruta, *Department of Neurochemistry, Max-Planck-Institute, Martinsried, Federal Republic of Germany*
Fumio Matsumura, *Cold Spring Harbor Laboratory, Cold Spring Harbor, New York*
Shigeko Matsumura, *Cold Spring Harbor Laboratory, Cold Spring Harbor, New York*
Tim J. Mitchison, *Department of Biochemistry, University of California, San Francisco, California*
N. Ronald Morris, *Department of Pharmacology, Rutgers Medical School, Piscataway, New Jersey*
Douglas B. Murphy, *Department of Cell Biology, Johns Hopkins University, Baltimore, Maryland*
Peter Novick, *Department of Biology, Massachusetts Institute of Technology, Cambridge, Massachusetts*

Joanna Olmsted, *Department of Biology, University of Rochester, Rochester, New York*
Raymond Owens, *Imperial Cancer Research Fund, London, England*
Thoru Pederson, *Worcester Foundation, Shrewsbury, Massachusetts*
Shelden Penman, *Massachusetts Institute of Technology, Cambridge, Massachusetts*
Lorraine Pillus, *Massachusetts Institute of Technology, Cambridge, Massachusetts*
Gianni Piperno, *Rockefeller University, New York, New York*
Mark F. Pittenger, *Department of Physiological Chemistry, Johns Hopkins University School of Medicine, Baltimore, Maryland*
John R. Pringle, *Department of Biological Sciences, University of Michigan, Ann Arbor, Michigan*
Wim Quax, *Department of Biochemistry, University of Nijmegen, Nijmegan, The Netherlands*
Elizabeth Raff, *Department of Biology, Indiana University, Bloomington, Indiana*
Dennis R. Roop, *National Institutes of Health, Bethesda, Maryland*
Mark Rose, *Massachusetts Institute of Technology, Cambridge, Massachusetts*
Joel Rosenbaum, *Department of Biology, Yale University, New Haven, Connecticut*
Dan Sackett, *National Institutes of Health, Bethesda, Maryland*
Lino Saez, *Department of Microbiology, Albert Einstein College of Medicine, Bronx, New York*
Edward D. Salmon, *Department of Biology, University of North Carolina, Chapel Hill, North Carolina*
Peter Schatz, *Massachusetts Institute of Technology, Cambridge, Massachusetts*
Tim Schedl, *McArdle Laboratory, University of Wisconsin, Madison, Wisconsin*
Robert Schwartz, *Department of Cell Biology, Baylor College of Medicine, Houston, Texas*
Carolyn Silflow, *Department of Genetics and Cell Biology, University of Minnesota, St. Paul, Minnesota*
David Soifer, *Institute for Basic Research, Staten Island, New York*
Jim Spudich, *Department of Structural Biology, Stanford University School of Medicine, Stanford, California*
Kevin F. Sullivan, *Department of Physiological Chemistry, Johns Hopkins University, Baltimore, Maryland*
Kathy Talbot, *Ludwig Institute for Cancer Research, Medical Research Council, Cambridge, England*
William Theurkakuf, *Department of Biochemistry, Brandeis University, Waltham, Massachusetts*
James Thomas, *Department of Biology, Massachusetts Institute of Technology, Cambridge, Massachusetts*
Manuel M. Valdivia, *Department of Cell Biology, Baylor College of Medicine, Houston, Texas*
Richard Vallee, *Worcester Foundation, Shrewsbury, Massachusetts*
Dale Vandre, *Department of Molecular Biology, University of Wisconsin, Madison, Wisconsin*
Samuel Ward, *Department of Embryology, Carnegie Institution of Washington, Baltimore, Maryland*
James Weatherbee, *Department of Pharmacology, Rutgers Medical School, Piscataway, New Jersey*
Donald Weeks, *Zoecon Corporation, Palo Alto, California*
Robert Wiehing, *Worcester Foundation, Shrewsbury, Massachusetts*
William Welch, *Cold Spring Harbor Laboratory, Cold Spring Harbor, New York*
Carolyn Whitfield, *National Institutes of Health, Bethesda, Maryland*
Gerhard Wiche, *Department of Biochemistry, University of Vienna, Vienna, Austria*
Ben Williams, *Department of Biology, Yale University, New Haven, Connecticut*
Robin Wright, *Department of Biological Sciences, Carnegie Mellon University, Pittsburgh, Pennsylvania*
Zendra Zehner, *Department of Biochemistry, Medical College of Virginia, Richmond, Virginia*

First row: J. Spudich; G.G. Borisy, D.W. Cleveland, D.B. Murphy
Second row: T. Huffaker; S. Ward, P. Anderson; E. Raff
Third row: J. Rosenbaum, S.W. L'Hernault; W. Dove; E. Lazarides, M. Kirschner
Fourth row: Coffee Break; E. Lai, C. Fulton, N.R. Morris

Contents

Preface

During the past few years, exciting new findings have been reported on the structures and gene sequences of cytoskeleton proteins and on the mechanisms controlling their expression in development. To stimulate discussion among cell and molecular biologists involved in this new and expanding research field, we organized a meeting on the Molecular Biology of the Cytoskeleton at Cold Spring Harbor Laboratory in April 1984. The results of this meeting are summarized in this volume in 40 papers from some of the major contributors. We regret that due to space considerations, we could not accommodate contributions from all of the meeting participants.

The purpose of *Molecular Biology of the Cytoskeleton* is to present a synopsis of up-to-date information, to identify new trends in future research, and by including brief reviews of the respective disciplines, to serve as a reference for newcomers to the field. We hope it will also be useful as a textbook and guide to the original literature for those wishing to organize courses and seminars around the subject. The book is organized into three sections: (1) Cytoskeletal Proteins: Structures, Dynamics, and Isoforms, which describes the dynamics of tubulin, actin, myosin, and intermediate filaments as determined by in vitro and in vivo studies. Attention is directed to the isoforms, or variants, of tubulin, actin, myosin, and intermediate filaments. (2) Mutant Analysis of the Cytoskeleton, which reviews recent advances in mutant analysis as successfully applied to cytoskeleton proteins in several genetic systems, including nematodes, yeast, *Aspergillus, Physarum, Chlamydomonas,* and *Drosophila*, and a variety of vertebrate genomes. (3) Cytoskeletal Genes: Structure, Expression, and Regulation, which examines the analysis of cytoskeletal genes and pseudogenes and their organization within the genome and reviews recent approaches to the study of cytoskeletal gene regulation.

We would like to acknowledge and thank the staffs of several granting agencies for their efforts to provide support for the meeting. These include the National Science Foundation and the following divisions of the National Institutes of Health: the Fogarty International Center, the National Cancer Institute, and the National Institute of General Medical Sciences.

As organizers of the meeting and editors of these contributed papers, we would like to express our deep appreciation to Dr. James D. Watson and the Cold Spring Harbor staff for their enthusiasm and willingness to host and support the conference. Special thanks are due to Gladys Kist and Susan Schultz, whose organizational assistance helped the meeting run smoothly. We are also grateful to Nancy Ford, Dorothy Brown, and the Publications Office staff for their efforts in assembling and editing the manuscripts. Finally, we wish to thank the 120 meeting participants, whose preparation and interest made the meeting exciting and productive. Given their talents and enthusiasm, we anticipate there will be many exciting developments in the years ahead.

G.G.B.
D.W.C.
D.B.M.

CYTOSKELETAL PROTEINS: STRUCTURES, DYNAMICS, AND ISOFORMS

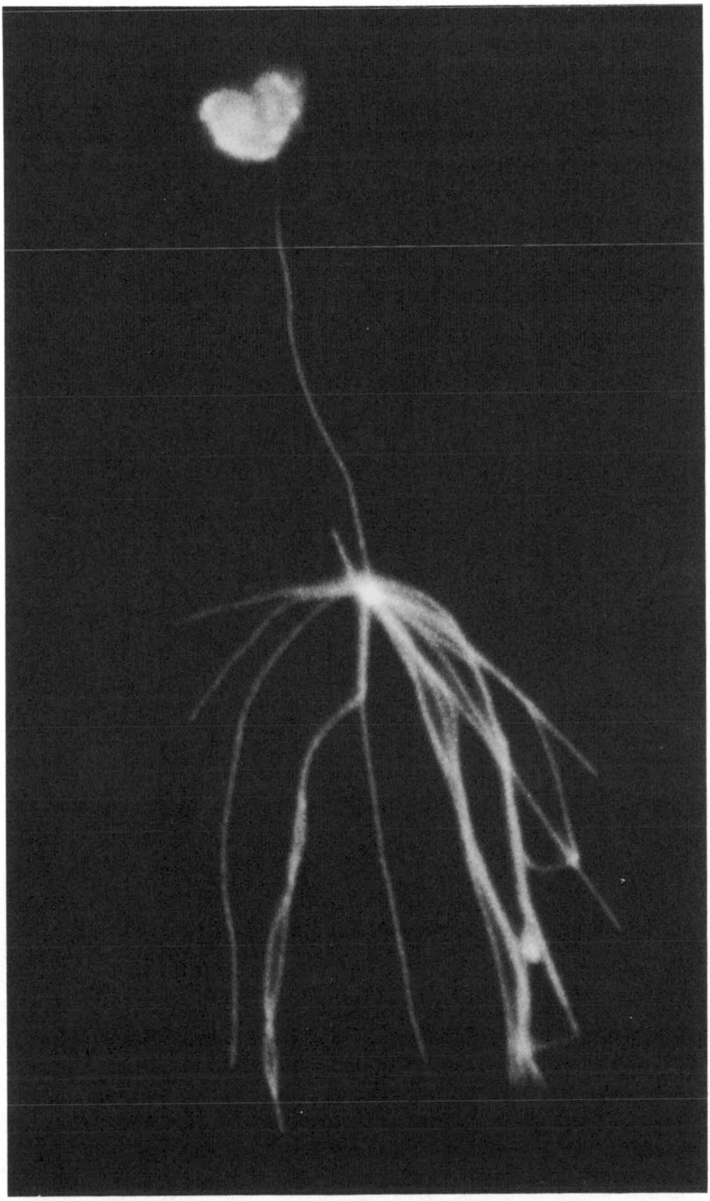

Recto: Microtubule capture by kinetochores occurs when CHO chromosomes are added to centrosomes containing microtubules. Purified interphase centrosomes were incubated with 25 μM tubulin at 37°C to produce microtubules. Microtubules and chromosome were visualized by immunofluorescence microscopy using tubulin antibody and Hoechst dye H23358 (see Mitchison and Kirschner, this volume). (Photo courtesy of T. Mitchison and M. Kirschner, University of California, San Francisco.)

Interphase-Mitosis Transition: Microtubule Rearrangements in Cultured Cells and Sea Urchin Eggs

Dale D. Vandre, Paul Kronebusch, and Gary G. Borisy

Laboratory of Molecular Biology
University of Wisconsin
Madison, Wisconsin 53706

In most eukaryotes, the transition occurring as cells enter mitosis, commonly referred to as the G_2/M-phase transition, is characterized by significant biochemical and morphological alterations. The most obvious of these transformations is a set of events involving the nucleus, including chromosome condensation, the disruption of the nuclear envelope, and the formation of the spindle apparatus. Mitosis is additionally characterized by specific processes that include the switching off of gene transcription (Paulson and Taylor 1982), cessation of intracellular transport and endocytosis (Berlin et al. 1978; Berlin and Oliver 1980; Warren et al. 1983), rearrangement of the endoplasmic reticulum and Golgi apparatus (Zeligs and Wollman 1979; Hiller and Weber 1982), modulation of Ca^{++} uptake (Holmes and Stewart 1977), and decrease in adenosine 3′:5′-monophosphate levels (Zeilig et al. 1976). Immunofluorescence studies have also documented rearrangements of the cytoskeleton, including the networks of microtubules (Weber et al. 1975; Brinkley et al. 1976; DeBrabander et al. 1979; Aubin et al. 1980a), microfilaments (Aubin et al. 1979), and intermediate filaments (Aubin et al. 1980b; Blose 1981). Of these, the most radical changes occur in the microtubule network. The interphase cytoplasmic array of microtubules disappears, and a new array, the mitotic spindle, forms. The mechanisms underlying this transformation are at present poorly understood. All of these systems return to their interphase configuration again during telophase at the M/G_1 transition.

It has been demonstrated that, upon fusion of an interphase cell and a mitotic cell, the interphase nuclear membrane breaks down and the chromatin condenses to form mitotic-appearing chromosomes (Johnson and Rao 1970). This implied not only that mitotic events were triggered by a distinct event, as opposed to a series of transitional steps, but also that a factor present in the mitotic cell was capable of prematurely inducing a mitotic state in the interphase cell. In a similar manner, amphibian oocytes that were arrested in late meiotic G_2 phase could be induced by microinjections of cytoplasm from maturing oocytes to undergo germinal vesicle breakdown and the subsequent events leading toward oocyte maturation. The substance mediating oocyte maturation is incompletely characterized and has been designated maturation-promoting factor (for review, see Masui and Clarke 1979; Maller and Krebs 1980). The appearance of related factors occurs also in mitotic cells during the G_2-to-prophase transition, as demonstrated by the induction of oocyte maturation by mitotic cell extracts (Sunkara et al. 1979; Nelkin et al. 1980).The widespread distribution of these factors in both meiotic and mitotic cells suggests that the activation of a mitosis-promoting factor may be a generalized phenomenon.

One possible mechanism for the triggering of mitosis might involve protein phosphorylation reactions. Protein phosphorylation is a widespread posttranslational regulatory mechanism (Cohen 1982), and it occurs at increased levels in oocytes after injection of maturation-promoting factor (Maller et al. 1977). Furthermore, partially purified preparations of maturation-promoting factor contain active protein kinases (Wu and Gerhart 1980), consistent with the idea that phosphorylation of target proteins might be key events in the interphase-mitosis transition.

Many proteins that change their degree of phosphorylation in a cell-cycle-dependent manner are components of the nucleus (Laskey 1983), and one major group of nuclear phosphoproteins are the histones. Histone H3 is unphosphorylated for the majority of the cell cycle but is selectively phosphorylated at serine residue 10 during mitosis (Paulson and Taylor 1982), whereas histone H1 increases its overall degree of phosphorylation during mitosis (Gurley et al. 1978). Experiments with temperature-sensitive cell lines that fail to phosphorylate histones H1 and H3 in a mitosis-specific manner indicate that histone phosphorylation may be correlated with chromosome condensation (Matsumoto et al. 1980; Yasuda et al. 1981). Histone phosphorylation alone, however, is not sufficient to cause chromosome condensation because it does not precede condensation in the expected manner (Gurley et al. 1978), and it can also occur without resulting condensation (Krystal and Poccia 1981). Other nonhistone proteins have been associated with chromosome condensation, including perichromin (McKeon et al. 1984) and a 35-kD phosphoprotein synthesized in temperature-sensitive mutant cells that undergo premature chromosome condensation at the elevated temperature (Yamashita et al. 1984).

In addition to the histones, the nuclear lamina is composed of three polypeptides that undergo cyclic phosphorylations. The lamina, composed of lamin A, B, and C, is associated with the nucleoplasmic surface of the inner nuclear membrane (Gerace et al. 1978). The lamins have a high level of associated phosphate during mitosis, and this phosphorylation has been proposed as being involved in the depolymerization of the nuclear envelope (Gerace and Blobel 1980). It has been postulated that maturation-promoting factor may be directly involved in these mitosis-specific lamin phosphorylations (Miake-Lye et al. 1983). Thus, major morphological events involving the nucleus during mitosis are associated with phosphorylations of component proteins, mediated perhaps by mitotic factors such as maturation-promoting factor.

In addition to the changes occurring in nuclear structure at mitosis, a number of structural rearrangements are observable in the different cytoskeletal networks. The organization of vimentin in intermediate filaments changes during cell division, altering from wavy filamentous bundles to a cagelike structure surrounding the nucleus (Aubin et al. 1980b; Zieve et al. 1980; Blose 1981). These alterations in filament structure during mitosis are correlated with increased phosphorylation of the vimentin (Evans and Fink 1982) at a minimum of two additional serine residues (Evans 1984).

Similar mitosis-specific posttranslational modifications of actin and tubulin have not been reported, although cytoskeletal systems composed of these proteins do undergo coordinate transformations during mitosis. An immunofluorescence study using antibodies directed against actin, myosin, and tropomyosin has shown that the well-developed microfilament bundles present in interphase and early prophase disappear near the time of nuclear-envelope breakdown. Microfilament bundles are not seen again in the cytoplasm until telophase (Aubin et al. 1979). The transformation of the cytoplasmic microtubule networks also occurs in tight synchrony with the nuclear mitotic events. The interphase network disappears when the spindle appears in prometaphase, and the spindle structure is lost in telophase as the interphase network of microtubules reemerges. The spindle microtubule arrangement is determined by the presence of two dominant microtubule-organizing centers: the kinetochores that function between prometaphase and telophase, and the centrosomes whose activity is greatly enhanced during this same period.

We discuss here evidence which suggests that the transformation of the microtubule network at mitosis is generated or triggered by a diffusible cellular component released from the nucleus or nuclear region and that the reorganization of the microtubules to form the mitotic spindle correlates with phosphorylation events at microtubule-organizing centers.

DISCUSSION

Microtubule Networks and Spindle Formation

The microtubule network in interphase and mitotic cells has been described by many investigators. However, the nature of the transition between these two states remains unclear. Is it a direct one where the microtubules of the interphase cell are reorganized into the spindle or is it an indirect one where the interphase network is first disassembled and then the spindle is constructed from subunits? Is the mitosis-to-interphase transition the reverse of the interphase-to-mitosis transition or do these transitions proceed by different routes? To answer these questions, we have reexamined the microtubule patterns during the transitions between interphase and mitosis in two different cell systems. We have examined glutaraldehyde-fixed samples of both sea urchin eggs and tissue-culture cells using anti-tubulin immunofluorescence to determine the typical microtubule patterns for each stage. All samples were reduced in 1 mg/ml $NaBH_4$ and then stained with a monoclonal antibody (clone YL1/2) raised against yeast tubulin (a gift from J.V. Kilmartin, this volume), followed by fluorescein-conjugated rabbit anti-rat IgG. Despite the differences in size, shape, and cell-cycle time that exist between the PtK_1 endothelial cell line and the developing sea urchin egg, there was a striking similarity in their microtubule patterns as they progressed through the cell cycle.

Sea Urchin Eggs

Samples of fertilized eggs from *Lytechinus pictus* were drawn from a beaker at 10-minute intervals, placed on polylysine-coated coverslips, allowed to adhere for 10 minutes, and then fixed and processed for immunofluorescence staining. The fixative consisted of 0.4 M Na-acetate (pH 6.0), 50 mM EGTA, 2% glutaraldehyde, and 1% Triton X-100. This procedure gave us excellent differential interference contrast images and better microtubule detail than we obtained using the usual methanol fixation (Harris et al. 1980; Bestor and Schatten 1981; Harris 1981).

Unfertilized eggs showed only diffuse cytoplasmic staining and no microtubule structures. Within 30 minutes after fertilization, the two pronuclei had fused and the monaster was present within the egg. The microtubules of the monaster began to break down from the egg center outward, forming a hollow sphere of radial microtubules. Two interphase asters arose from two spots positioned on opposite sides of the nucleus. The microtubules of the interphase asters ultimately extended out to the cell periphery so that most of the microtubules of the egg were focused on these two spots (Fig. 1a).

As the egg passed from interphase to prophase, the nuclear envelope broke down and short prophase asters appeared as brightly staining spheres with short microtubules extending from them. The cytoplasm outside the radius of these short asters was devoid of microtubules. Figure 1b shows what is apparently a transition stage from the long microtubules of the interphase aster to the short microtubules of prophase. The halo of microtubules near the egg surface indicates that the long microtubules may be lost from the center outward. Release of the chromosomes into the cytoplasm preceded formation of the kinetochore microtubule bundles. At metaphase, the two asters were connected by the kinetochore bundles as the chromosomes aligned on the spindle equator (Fig. 1c).

During anaphase, the kinetochore fibers shortened toward the poles and disappeared. During this stage, the metaphase asters grew outward toward the cell boundary. The formerly bright center became a hollow sphere devoid of fluorescence, and the microtubules of the aster appeared to increase in length (Fig. 1d). Occasionally, two spots were detected within the hollow sphere at the pole. In telophase, the volume of the cytoplasm was filled with microtubules that apparently were focused on a diffuse polar region. The spindle equator was often surprisingly free of microtubule staining (Fig. 1e).

Figure 1f shows an egg cleaved into two cells, each containing an interphase nucleus. Except for the microtubules of the midbody, the microtubules that fill the cytoplasm are focused on the region of the nuclear envelope. The configuration was similar to the monaster except for the asymmetry in the distribution of microtubules. The same microtubule patterns were repeated in the second cell

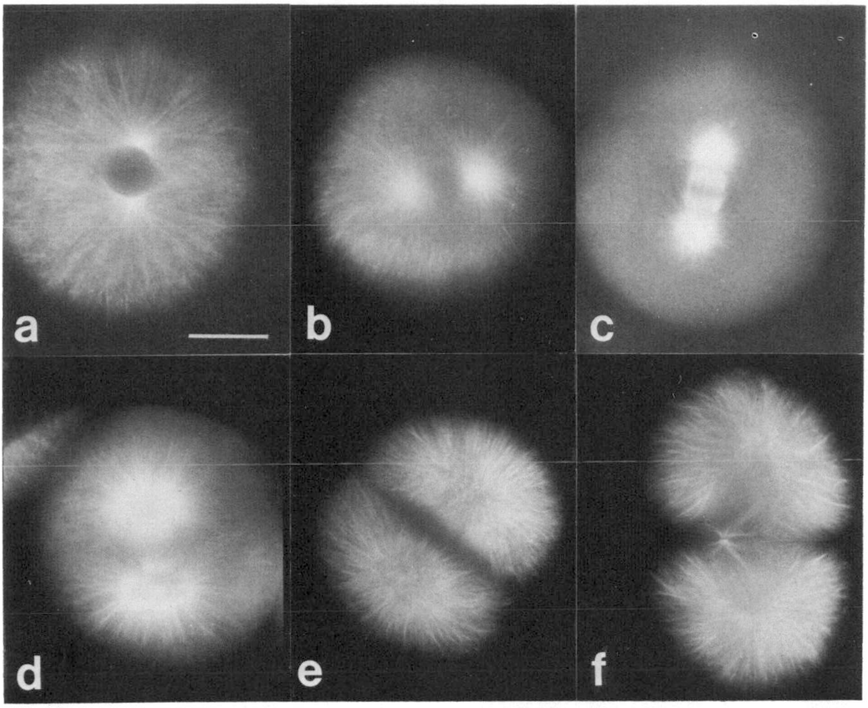

Figure 1
Immunofluorescent microtubule patterns during the transitions between interphase and mitosis in fertilized *Lytechinus pictus* eggs. (*a*) Egg containing an interphase nucleus and separated centrosomes that are the focus of the microtubule network; (*b*) transition stage between interphase and mitotic microtubule networks; (*c*) metaphase spindle; (*d*) anaphase spindle; (*e*) telophase prior to egg cleavage; (*f*) cleaved egg with daughter cells displaying an interphase microtubule pattern. Bar, 30 μm.

cycle. The diffuse monaster pattern was converted to two interphase asters. The long interphase centrosomal microtubules were converted to short prophase and metaphase asters. During anaphase and telophase, the asters enlarged to fill the cytoplasm with microtubules again.

Tissue-culture Cells

We have also examined the microtubule patterns during the transitions between interphase and mitosis in human fibroblasts, Chinese hamster ovary (CHO), and PtK_1 cells. Cells were fixed with 1% glutaraldehyde in phosphate-buffered saline, permeabilized in 0.1% Triton X-100, and processed for immunofluorescence. We discuss the pattern seen in the PtK_1 cells, since the microtubule distribution is not disturbed by the rounding up that occurs in mitosis in other cell lines.

Interphase cells contained an extensive array of microtubules that filled the cytoplasm. Often, there was a focus of microtubules on the centrosome. The prominence of this focus varied with cell type and from cell to cell within a population. In all of the cell types examined, we have seen a strong centrosomal focus of microtubules during prophase. Early prophase cells contained one focus, whereas cells in late prophase containing separated centrosomes showed two foci. Although the nucleus showed extensive chromatin condensation at this stage, the cytoplasmic array of microtubules remained very extensive and filled the cytoplasm as it does in interphase (Fig. 2a). CHO cells round up during mitosis, but were often found well spread at this stage of prophase with an extensive array of cytoplasmic microtubules. Cells fixed during the time of nuclear-envelope breakdown exhibited a transformed microtubule network. The centrosomal microtubules were more numerous and much shorter than at earlier stages of prophase. Many cells at this stage contained microtubule fragments distributed throughout the cytoplasm (Fig. 2b). Elongated prophase cells often contained many microtubules at the end of the cell extensions with no free microtubules near the nuclear area. Metaphase cells displayed short astral microtubules radiating out from the spindle poles, but no microtubules were present elsewhere in the cytoplasm (Fig. 2c).

Anaphase coincided with the reappearance of cytoplasmic microtubules. Cen-

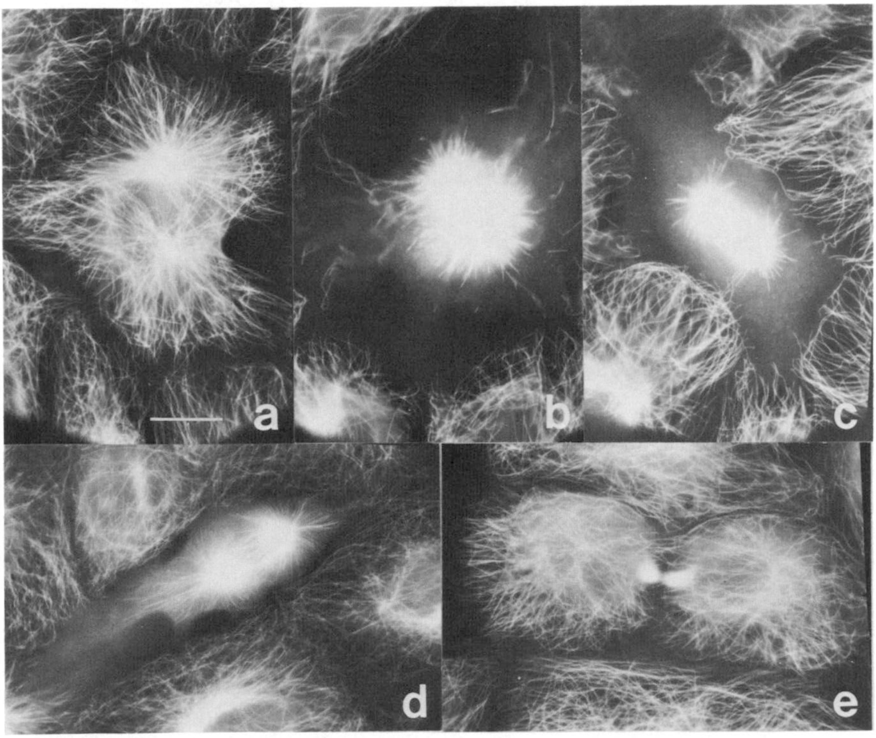

Figure 2
Immunofluorescent microtubule patterns in PtK_1 cells during the transitions between interphase and mitosis. (*a*) Prophase cell with separated centrosomes with an intact nuclear envelope. (*b*) Transition stage between interphase and mitotic microtubule networks. The nuclear envelope has broken down. (*c*) Metaphase spindle; (*d*) early anaphase; (*e*) early G_1 daughter cells after cleavage. Bar, 20 μm.

trosomal microtubules were longer than in metaphase and often there were non-centrosomal microtubules present as well (Fig. 2d). Late anaphase and telophase cells contained microtubule arrays that filled the entire cytoplasm. After cleavage, daughter cells displayed a microtubule pattern that was very much like that in interphase except for the presence of the midbody remnant (Fig. 2e).

Several key features of the microtubule cycle are common to both the sea urchin egg and the PtK_1 cells and thus may be considered general. In both systems, the interphase-to-mitosis transition appears to be indirect. The interphase microtubule network breaks down, and the spindle is formed from subunits rather than by a reorganization of the preexisting microtubules. However, the time frames for these two processes are overlapping, not separate. Asters grow from the prophase centrosomes while the interphase network still exists. A key event in the transition occurs about the time of nuclear-envelope breakdown or prometaphase. At this time, the cytoplasmic microtubules depolymerize and the astral tubules are trimmed to their mitotic size. The transition from mitosis to interphase in both systems proceeds by a different route. Instead of the breakdown of the old system and the formation of a new system, the process appears to be continuous. The astral tubules grow, extend into the cytoplasm, and differentiate into the interphase network.

Phosphorylation of Microtubule-organizing Centers

Monoclonal antibodies to phosphoproteins present in mitotic HeLa cells have recently been prepared by Davis and co-workers (1983). These antibodies were shown to react with a large number of proteins present in mitotic cell lysates from several different cell types. In addition, the amount of antigen present in any given cell correlated with its position in the cell cycle. Little or no antigenic material was found in interphase cells, but antigen quantity rose rapidly during G_2. Peak antigen levels were demonstrated at metaphase, and these amounts fell rapidly to interphase levels by the end of mitosis. Using one of these antibodies, MPM-2, we determined the immunofluorescence localization of reactive material to structural components of microtubule-organizing centers, namely, centrosomes, kinetochores, and midbodies of PtK_1 cells (Vandre et al. 1984). Similar distribution patterns were obtained with CHO cells as shown here. Monolayer cultures of cells on glass coverslips were partially synchronized with 2.5 mM thymidine, and 4 hours following release from thymidine, cells were treated with 0.04 μg nocodazole for 3 hours. Cells were fixed and stained with monoclonal antibody MPM-2 before and after release from the nocodazole block, followed by fluorescein-conjugated goat anti-mouse immunoglobulin as described recently (Vandre et al. 1984).

Immunoreactive material was restricted to faintly staining patches within the nucleus of interphase cells (Fig. 3a). The nucleus of prophase cells, on the other hand, was intensely reactive with the MPM-2 antibody. Like interphase cells, no cytoplasmic staining was evident with the exception of intense staining associated with the centrosomes (Fig. 3b). Once the nuclear envelope broke down at prometaphase, immunoreactive material was present throughout the cytoplasm (Fig. 3c). It appeared as if reactive material was released from the nucleus and spilled into the cytoplasm or, alternatively, that the kinase responsible for the

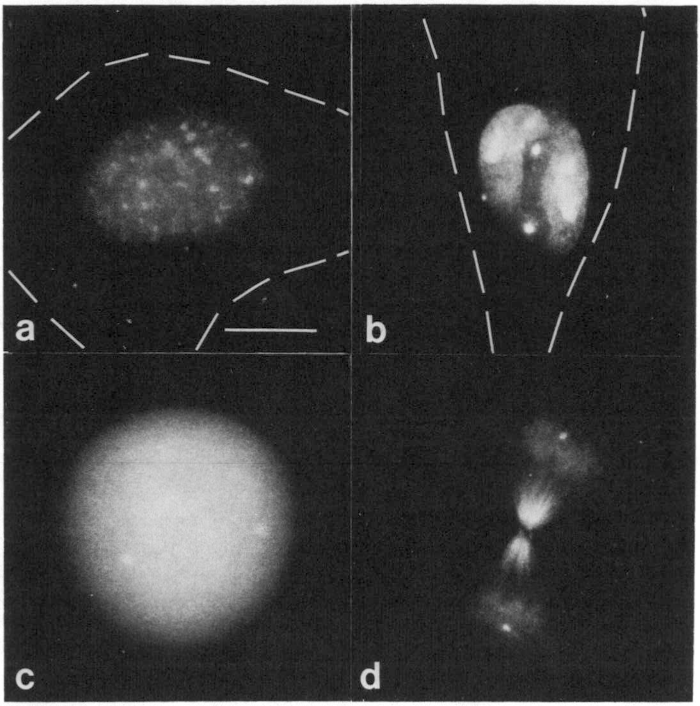

Figure 3
Monolayer CHO cells stained by indirect immunofluorescence with monoclonal antibody MPM-2. (*a*) Interphase cell with staining restricted to patches within the nucleus. The cell margins are indicated by the dashed line in *a* and *b*. (*b*) Prophase cell with intense staining within the intact nucleus and also at the duplicated centrosomes adjacent to the nuclear envelope. (*c*) Metaphase cell demonstrating antigenic material distributed throughout the cytoplasm and extending to the cell periphery after nuclear-envelope breakdown. (*d*) Telophase cell showing a reduction in overall fluorescence intensity with staining associated with the centrosomes, re-forming nucleus, and midbody. Bar, 10 μm.

mitosis-specific phosphorylations of cytoplasmic components was restricted within the nucleus until after nuclear-envelope breakdown. Cytoplasmic immunofluorescence was apparent from prometaphase through anaphase, but the greatest overall fluorescence was at metaphase (Fig. 3c). The staining intensity of the cytoplasmic immunoreactive material at these stages obscured the centrosome-specific localization clearly present at prophase. At cytokinesis, the majority of the cytoplasmic immunoreactive material was no longer apparent (Fig. 3d). Pronounced immunofluorescence was, however, detectable on either side of the phase-dense matrix of the midbody, the centrosomes, and the re-forming nucleus. The patchy staining within the nucleus did not become apparent again until after daughter cell separation.

Localization of MPM-2-reactive material could be improved if the cells were briefly extracted with detergent prior to fixation. This was apparently the result of removing soluble immunoreactive material from the cells during the extraction. In those cells extracted and fixed while in the nocodazole block, the centrosomes were detected as two intensely staining dots and were located centrally, with the chromosomes radiating away from them (Fig. 4a). After a release from nocodazole of 20 minutes, the majority of the cells had progressed into metaphase (Fig. 4b). The centrosomes were clearly evident at the poles of the spindle apparatus. Immunoreactive material was present also on the condensed chromosomes and perhaps the spindle fibers as well. Unlike PtK_1 cells, kinetochore staining was not clearly visible in the extracted but intact CHO cells; however, staining at presumptive kinetochore remnants was clearly visible in mitotic spindles isolated from the CHO cells. As in the PtK_1 cells, the association of MPM-2-reactive material with microtubule-organizing centers during mitosis suggests a functional role for phosphorylation in modulating the microtubule-organizing capacity of these organelles at mitosis.

Coordinate Control of Mitotic Events

There is a remarkable correlation between the change in the microtubule distribution and the appearance of MPM-2 staining during mitosis. This relationship holds even when extended to a comparison of different cell types such as the CHO and PtK_1 cells. During interphase, the microtubule network is only partially organized at the centrosome. At this time, MPM-2-reactive material is detected within the nucleus and only minimally at the centriole (Vandre et al. 1984). As the cell undergoes the transition to mitosis, the microtubule-initiating capacity of the centrosome increases appoximately fivefold (Kuriyama and Borisy 1981). Our tubulin immunofluorescence patterns show a dramatic increase in the number of centrosomal microtubules during prophase and prometaphase, which correlates

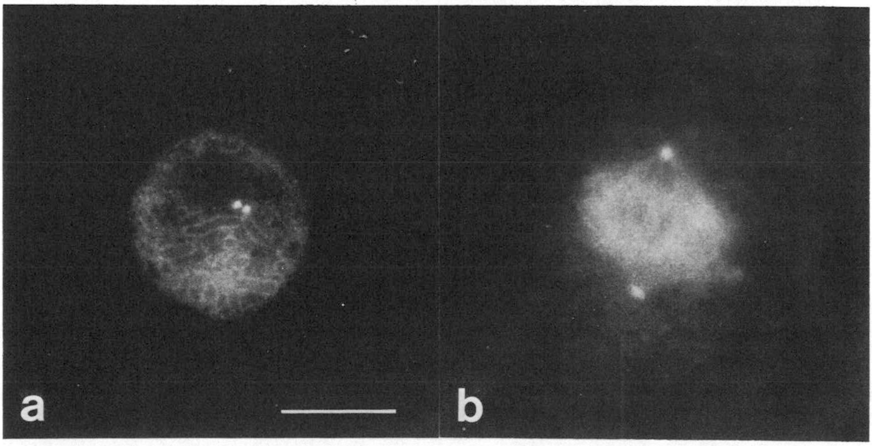

Figure 4
Immunofluorescence localization of MPM-2-reactive material in CHO cells following removal of soluble material by detergent extraction prior to fixation. (*a*) Mitotic cell after 4 hr of nocodazole (0.04 μg/ml) block; centrally oriented centrosomes remain immunoreactive during mitotic blockage. (*b*) After 20-min release from nocodazole, the metaphase spindle forms and staining is clearly present at the spindle poles and chromosomes in these extracted cells. Bar, 10 μm.

with the appearance of MPM-2 staining at the centrosome during these same stages. At the time of nuclear-envelope breakdown, the interphase cytoplasmic microtubule network is lost, which correlates with the appearance of MPM-2-reactive material in the cytoplasm. MPM-2 staining is then observed at the kinetochores, which are also capable of organizing microtubules. After the onset of anaphase, the microtubule-nucleating capacity of the centrosome decreases (Snyder et al. 1983), and we see a concomitant loss in its MPM-2 staining. Conversely, the cytoplasm once again fills with microtubules and MPM-2 staining disappears from the cell. We propose that the increased microtubule-nucleating activity of the centrosome is the result of centrosomal phosphorylation by mitotic factors, although no direct biochemical evidence linking these two events presently exists. In a similar manner, inactivation of the mitotic factors and activation of specific phosphatases would result in dephosphorylation of centrosomal components and a return to interphase levels of nucleating capacity. The complementary loss and reappearance of cytoplasmic microtubules may result from the phosphorylation of noncentrosomal, diffusible targets.

Tissue-culture cells can be held in M phase for several hours by treating them with microtubule-disrupting drugs like nocodazole. Although the mitotic condition is normally maintained for less than 1 hour, cells can be held in this state for several hours, and they will progress through a normal mitosis when the drug is washed out. We have observed that the mitotic MPM-2 staining of centrosomes and the cytoplasm is also maintained in these nocodazole-blocked cells (Fig. 4a). This suggests that spindle formation is a prerequisite for proper sequential activation of factors responsible for inactivation of mitosis-initiating factors.

One can propose a number of mechanisms in which mitosis-promoting factors could alter the microtubule distribution. If Ca^{++} levels control microtubule assembly as has been proposed (Hepler 1977; Harris 1981), then a mitotic factor may be released into the cytoplasm that alters Ca^{++} pumps so that Ca^{++} levels rise and cytoplasmic microtubules are depolymerized. Another mitotic factor localized at the centrosome may increase Ca^{++} sequestering activity in the spindle area, thereby allowing localized microtubule polymerization. Microtubule assembly could also be modulated by microtubule-associated proteins (MAPs), and it has been shown that the state of phosphorylation of certain MAPs affects their microtubule-promoting activity in vitro (Jameson and Caplow 1981; Murthy and Flavin 1983). The phosphorylation of one set of MAPs in the cytoplasm at mitosis may destabilize the interphase microtubule, whereas phosphorylation of another set of proteins at the centrosome may enhance localized nucleation of microtubules. The MPM-2 antibody recognizes a number of phosphoproteins that appear during mitosis (Davis et al. 1983). It is possible that a set of these proteins localized at the centrosome are responsible for the enhanced microtubule-initiating activity of these organelles at mitosis, whereas a different set is responsible for the disruption of the interphase microtubule network. These proteins would be dephosphorylated after anaphase, allowing microtubules to reappear throughout the cytoplasm.

As mentioned previously, when mitotic cells are fused with interphase cells, they induce premature chromosome condensation in the interphase cell (Johnson and Rao 1970). A time-lapse study of PtK_2–human cell hybrids has shown that premature condensation of the interphase chromatin does not begin until after nuclear-envelope breakdown of the prophase nucleus in the common cytoplasm (Peterson and Berns 1979). Thus, it appears that the factors responsible for premature chromsome condensation are released from the mitotic nucleus or are activated by a substance released from the nucleus. Again it appears that the presence and activity of these mitosis-inducing factors correspond with an event (nuclear-envelope breakdown) at which time we would expect MPM-2 reactivity to be high.

An important element in the interphase-mitosis transition may be a protein kinase with the function of directly modifying specific cellular targets or activating other kinase systems. This system would ultimately result in the phosphorylation of key cellular components required for chromosome condensation, nuclear-envelope breakdown, and cytoskeletal reorganization. These mitosis-specific processes may directly reflect posttranslational modification of lamins, histones, vimentin, or the components of microtubule-organizing centers. Alternatively, a more complex pathway involving modifications of ionic pumping or sequestering systems could generate environmental conditions favorable for microtubule assembly in the localized area surrounding the mitotic spindle while simultaneously maintaining conditions in the remainder of the cytoplasm that are unfavorable for microtubule assembly.

SUMMARY

Detailed observations of sea urchin eggs and PtK$_1$ cells indicate that the interphase-to-mitosis transition of microtubules is an indirect one, whereas the reverse transition appears continuous and to proceed by a different route. Microtubules are lost abruptly from the cytoplasm at the time of nuclear-envelope breakdown. The metaphase spindle then forms in a cytoplasm free of any other microtubules, but during anaphase and telophase, the cytoplasm once again fills with microtubules by growth of the asters. The formation of the spindle is preceded by a dramatic increase in the number of microtubules associated with the centrosome during prophase and prometaphase. After anaphase, the number of centrosomal microtubules returns to its interphase level.

Coincident with the microtubule rearrangements, we have shown the appearance of mitosis-specific phosphoproteins in association with microtubule-organizing centers, namely, centrosomes, kinetochores, and the midbody. The staining is initially detectable in the prophase nucleus, spreading throughout the cell when the nuclear envelope is broken down. Both centrosomal and cytoplasmic stainings disappear at telophase. We have proposed that the presence of these phosphoproteins may regulate both the enhanced microtubule-nucleating capacity of the centrosome and the decrease in cytoplasmic microtubules occurring during mitosis.

The interphase-mitosis transformation appears to be initiated by a distinct triggering event, altering the internal environment throughout the cell. This transformation may be generated by the activation of a single factor that then promotes a cascade of reactions, each specific for a different cellular target. One element in the transition system may be a protein kinase that directly modifies specific cellular targets or activates other kinase systems. This process would result in the phosphorylation of key cellular components required for chromosome condensation, nuclear-envelope breakdown, and cytoskeletal reorganization. The reorganization of the microtubule network is closely coupled with the appearance of the mitotic phase triggering event, and further investigation into this reorganization may help elucidate crucial elements underlying the molecular mechanisms of initiation and regulation of mitosis.

ACKNOWLEDGMENT

This work was supported by National Institutes of Health grant GM-30385 and by National Institutes of Health postdoctoral fellowship GM-009145-02 (to D.D.V.).

REFERENCES

Aubin, J.E., M. Osborn, and K. Weber. 1980a. Variations in the distribution and migration of centriole duplexes in mitotic PtK_2 cells studied by immunofluorescence microscopy. *J. Cell Sci.* **43:** 177.

Aubin, J.E., K. Weber, and M. Osborn. 1979. Analysis of actin and microfilament-associated proteins in the mitotic spindle and cleavage furrow of PtK_2 cells by immunofluorescence microscopy. *Exp. Cell Res.* **124:** 93.

Aubin, J.E., M. Osborn, W.F. Franke, and K. Weber. 1980b. Intermediate filaments of the vimentin-type and the cytokeratin-type are distributed differently during mitosis. *Exp. Cell Res.* **129:** 149.

Berlin, R.D. and J.M. Oliver. 1980. Surface functions during mitosis. II. Quantitation of pinocytosis and kinetic characterization of the mitotic cycle with a new fluorescence technique. *J. Cell Biol.* **85:** 660.

Berlin, R.D., J.M. Oliver, and R.J. Walter. 1978. Surface functions during mitosis. I. Phagocytosis, pinocytosis and mobility of surface-bound con A. *Cell* **15:** 327.

Bestor, T.H. and G. Schatten. 1981. Anti-tubulin immunofluorescence microscopy of microtubules present during the pronuclear movement of sea urchin fertilization. *Dev. Biol.* **88:** 80.

Blose, S.H. 1981. The distribution of 10 nm filaments and microtubules in endothelial cells during mitosis: Double-label immunofluorescence study. *Cell Motil.* **1:** 417.

Brinkely, B.R., G.M. Fuller, and D.P. Highfield. 1976. Tubulin antibodies as probes for microtubules in dividing and nondividing mammalian cells. *Cold Spring Harbor Conf. Cell Proliferation* **3:** 435.

Cohen, P. 1982. The role of protein phosphorylation in neural and hormonal control of cellular activity. *Nature* **296:** 613.

Davis, F.M., T.Y. Tsao, S.K. Fowler, and P.N. Rao. 1983. Monoclonal antibodies to mitotic cells. *Proc. Natl. Acad. Sci.* **80:** 2926.

DeBrabander, M., G. Geuens, J. DeMay, and M. Joniau. 1979. Light microscopic and ultrastructural distribution of immunoreactive tubulin in mitotic mammalian cells. *Biol. Cell.* **34:** 213.

Evans, R.M. 1984. Peptide mapping of phosphorylated vimentin: Evidence for a site-specific alteration in mitotic cells. *J. Biol. Chem.* **259:** 5372.

Evans, R.M. and L.M. Fink. 1982. An alteration in the phosphorylation of vimentin-type intermediate filaments is associated with mitosis in cultured mammalian cells. *Cell* **29:** 43.

Gerace, L. and G. Blobel. 1980. The nuclear envelope lamina is reversibly depolymerized during mitosis. *Cell* **19:** 277.

Gerace, L., A. Blum, and G. Blobel. 1978. Immunocytochemical localization of the major polypeptides of the nuclear pore complex-lamina fraction: Interphase and mitotic distribution. *J. Cell Biol.* **79:** 546.

Gurley, L.R., J.A. D'Anna, S.S. Barham, L.L. Deaven, and R.A. Tobey. 1978. Histone phosphorylation and chromatin structure during mitosis in Chinese hamster cells. *Eur. J. Biochem.* **84:** 1.

Harris, P.J. 1981. Calcium regulation of cell cycle events. In *Mitosis/cytokinesis* (ed. A.M. Zimmermand and A. Forer), p. 29. Academic Press, New York.

Harris, P., M. Osborn, and K. Weber. 1980. Distribution of tubulin-containing structures in the egg of the sea urchin *Strongylocentrotus purpuratus* from fertilization through first cleavage. *J. Cell Biol.* **84:** 668.

Hepler, P.K. 1977. Membranes in the spindle apparatus: Their possible role in the control of microtubule assembly. In *Mechanisms and control of cell division* (ed. T.L. Rost and E.M. Gifford, Jr.), p. 212. Dowden, Hutchinson, and Roos, Stroudsburg, Pennsylvania.

Hiller, G. and K. Weber. 1982. Golgi detection in mitotic and interphase cells by antibodies to secreted galactosyltransferase. *Exp. Cell Res.* **142:** 85.

Holmes, R.D. and P.R. Stewart. 1977. Calcium uptake during mitosis in the myxomycete *Physarum polycephalum. Nature* **269:** 592.

Jameson, L. and M. Caplow. 1981. Modification of microtubule steady-state dynamics by phosphorylation of the microtubule-associated proteins. *Proc. Natl. Acad. Sci.* **78:** 3413.

Johnson, R.T. and P.N. Rao. 1970. Mammalian cell fusion: Induction of premature chromosome condensation in interphase nuclei. *Nature* **226:** 717.

Krystal, G.W. and D.L. Poccia. 1981. Phosphorylation of cleavage stage histone H1 in mitotic and prematurely condensed chromosomes. *Exp. Cell Res.* **134:** 41.

Kuriyama, R. and G.G. Borisy. 1981. Microtubule nucleating activity of centrosomes in Chinese hamster ovary cells is independent of the centriole cycle but coupled to the mitotic cycle. *J. Cell Biol.* **91:** 822.

Laskey, R.A. 1983. Phosphorylation of nuclear proteins. *Philos. Trans. R. Soc. Lond. B* **302:** 143.

Maller, J.L. and E.G. Krebs. 1980. Regulation of oocyte maturation. *Curr. Top. Cell. Regul.* **16:** 271.

Maller, J., M. Wu, and J.C. Gerhart. 1977. Changes in protein phosphorylation accompanying maturation of *Xenopus laevis* oocytes. *Dev. Biol.* **58:** 295.

Masui, Y. and H.J. Clarke. 1979. Oocyte maturation. *Int. Rev. Cytol.* **57:** 185.

Matsumoto, Y., H. Yasuda, S. Mita, T. Marunouchi, and M. Yamada. 1980. Evidence for the involvement of H1 histone phosphorylation in chromosome condensation. *Nature* **284:** 181.

McKeon, F.D., D.L. Tuffanelli, S. Kobayashi, and M.W. Kirschner. 1984. The redistribution of a conserved nuclear envelope protein during the cell cycle suggests a pathway for chromosome condensation. *Cell* **36:** 83.

Miake-Lye, R., J. Newport, and M. Kirschner. 1983. Maturation-promoting factors induces nuclear envelope breakdown in cycloheximide-arrested embryos of *Xenopus laevis. J. Cell Biol.* **97:** 81.

Murthy, A.S.N. and M. Flavin. 1983. Microtubule assembly using the microtubule-associated protein MAP-2 prepared in defined states of phosphorylation with protein kinase and phosphatase. *Eur. J. Biochem.* **137:** 37.

Nelkin, B., C. Nichols, and B. Vogelstein. 1980. Protein factor(s) from mitotic CHO cells induce meiotic maturation in *Xenopus laevis* oocytes. *FEBS Lett.* **109:** 233.

Paulson, J.R. and S.S. Taylor. 1982. Phosphorylation of histones 1 and 3 and nonhistone high mobility group 14 by an endogenous kinase in HeLa metaphase chromosomes. *J. Biol. Chem.* **257:** 6064.

Peterson, S.P. and M.W. Berns. 1979. Mitosis in flat PtK_2-human hybrid cells. *Exp. Cell Res.* **120:** 223.

Snyder, J.A., B.T. Hamilton, and S.M. Mullins. 1983. Loss of mitotic centrosomal microtu-

bule initiation capacity at the metaphase-anaphase transition. *Eur. J. Cell. Biol.* **27:** 191.
Sunkara, P.S., D.A. Wright, and P.N. Rao. 1979. Mitotic factors from mammalian cells induce germinal vesicle breakdown and chromosome condensation in amphibian oocytes. *Proc. Natl. Acad. Sci.* **76:** 2799.
Vandre, D.D., F.M. Davis, P.N. Rao, and G.G. Borisy. 1984. Phosphoproteins are components of mitotic microtubule organizing centers. *Proc. Natl. Acad. Sci.* **81:** 4439.
Warren, G., C. Featherstone, G. Griffiths, and B. Burke. 1983. Newly synthesized G protein of vesicular stomatitis virus is not transported to the cell surface during mitosis. *J. Cell Biol.* **97:** 1623.
Weber, K., T. Bibring, and M. Osborn. 1975. Specific visualization of tubulin-containing structures in tissue culture cells. *Exp. Cell Res.* **95:** 111.
Wu, M. and J.C. Gerhart. 1980. Partial purification and characterization of the maturation-promoting factor from eggs of *Xenopus laevis. Dev. Biol.* **79:** 465.
Yamashita, K., T. Nishimoto, and M. Sekiguchi. 1984. Analysis of proteins associated with chromosome condensation in baby hamster kidney cells. *J. Biol. Chem.* **259:** 4667.
Yasuda, H., Y. Matsumoto, S. Mita, T. Marunouchi, and M. Yamada. 1981. A mouse temperature-sensitive mutant defective in H1 histone phosphorylation is defective in deoxyribonucleic acid synthesis and chromosome condensation. *Biochemistry* **20:** 4414.
Zeilig, C.E., R.A. Johnson, E.W. Sutherland, and D.L. Friedman. 1976. Adenosine 3′:5′ monophosphate content and actions in the division cycle of synchronized HeLa cells. *J. Cell Biol.* **71:** 515.
Zeligs, J.D. and S.H. Wollman. 1979. Mitosis in rat thyroid epithelial cells *in vivo*. I. Ultrastructural changes in cytoplasmic organelles during the mitotic cycle. *J. Ultrastruct. Res.* **66:** 53.
Zieve, G.W., S.R. Heidemann, and J.R. McIntosh. 1980. Isolation and partial characterization of a cage of filaments that surrounds the mammalian mitotic spindle. *J. Cell Biol.* **87:** 160.

Segregation of Isoactins in Skeletal-muscle Fibers

Susan W. Craig and Mark F. Pittenger
Johns Hopkins Medical School
Baltimore, Maryland 21205

José V. Pardo
Duke University Medical School
Durham, North Carolina 27705

Actins from vertebrate tissues fall into one of three electrophoretic classes: α (the most acidic), β, and γ. These electrophoretic classes correspond to six polypeptides that differ from each other by less than 5% of their amino acid sequence (Vandekerckhove and Weber 1978). Do these isoforms have a functional significance? Tissue-specific expression of actin isoforms, regulated expression of isoforms during muscle differentiation, and coexpression of multiple isoforms in a single cell suggest that isoactins might have specialized functions (Garrels and Gibson 1976; Whalen et al. 1976; Vandekerckhove and Weber 1978). But another possibility is that the isoforms are expressed in particular patterns only because their respective structural genes have become associated with specific regulatory elements (Fyrberg et al. 1981). The key questions are (1) Can isoactins substitute for each other in vitro and in vivo? and (2) Do isoactins substitute for each other in vivo?

Experiments to determine whether isoactins can substitute for each other in vitro have given a mixed answer. On the one hand, α-actin from vertebrate skeletal muscle and *Acanthamoeba* δ-actin copolymerize in vitro (Gordon et al. 1976). This result suggests that the even more closely homologous vertebrate actins are also likely to form copolymers, but this has not been demonstrated. Cytoplasmic β-actin and γ-actin are only slightly different from skeletal-muscle α-actin in polymerization properties (Gordon et al. 1977). On the other hand, the small number of studies with actin-binding proteins indicates that these proteins exhibit specificity for certain actins. Skeletal-muscle and *Acanthamoeba* actins both bind skeletal-muscle tropomyosin but under different ionic conditions and with opposite functional effects (Yang et al. 1977). Similarly, *Acanthamoeba* profilin is

5–10 times more effective in depolymerizing *Acanthamoeba* actin than vertebrate skeletal-muscle α-actin (Tseng and Pollard 1982). Unfortunately, there are no studies comparing the affinities and functions of actin-binding proteins with different vertebrate actin isoforms.

Experiments assessing whether isoactins can substitute for each other in vivo have been limited in number but uniform in result. Introduction of fluorescently labeled skeletal-muscle α-actin into fibroblasts or cytoplasmic actins into cultured cardiac cells resulted in incorporation of the actins into all actin-containing structures (Y.-L. Wang et al., in prep.). Introduction of cardiac α-actin into mouse L cells by gene transfection resulted in incorporation of cardiac actin into the Triton-insoluble cytoskeleton (Gunning et al. 1984). With the assumption that the introduced actin functions normally in the cell structures, these experiments suggest that isoactins can substitute for each other in vivo.

Whether isoactins *do* substitute for each other in vivo has been investigated by immunofluorescent localization of actin isoforms in cryostat sections of mature skeletal-muscle tissue. Antibody specific for cytoplasmic actin did not stain the I bands of sarcomeres but did stain the cytoplasmic surface of the sarcolemma, especially at the neuromuscular junction. The antibody also stained the muscle-fiber cytoplasm in a pattern corresponding to that of the sarcoplasmic reticulum (Lubit and Schwartz 1980; Hall et al. 1981). These results indicate that in vivo isoactins do not necessarily substitute for each other. In this paper, we summarize findings that extend the evidence for differential localization of isoactins in single muscle fibers.

DISCUSSION

Polyclonal Antibody to Vertebrate Actin Can be Used to Prepare Reagents That Discriminate between Isoactins

Two subpopulations of actin antibodies were isolated by affinity chromatography from a polyclonal antibody to gizzard actin (Fig. 1). One subpopulation recognizes γ-actins from smooth-muscle and nonmuscle cells but does not recognize skeletal-muscle α-actin. The other subpopulation recognizes determinants that are common to skeletal-muscle α-actin and the two γ-actin isoforms. Neither antibody recognizes cytoplasmic β-actin (Fig. 2A,B) (Pardo et al. 1983).

Anti-γ-actin Colocalizes with Mitochondria in Muscle Fibers and Does Not Bind to the Actin Filaments (I Bands) of Myofibrils

In transverse sections of adult mouse diaphragm stained with anti-γ, the actin is localized in punctate and ring-shaped structures (Fig. 3A, arrows, and Fig. 4A). The uneven distribution of γ-actin among individual fibers of the diaphragm is similar to the pattern seen in histochemical stain for mitochondria (Gauthier and Padykula 1966). In fact, the distribution of mitochondria, which is easily visible in the phase-contrast micrographs as dark dots (Figs. 3B and 4B), closely mirrors the anti-γ pattern. In contrast, similar sections stained with anti-α,γ (Fig. 3E)

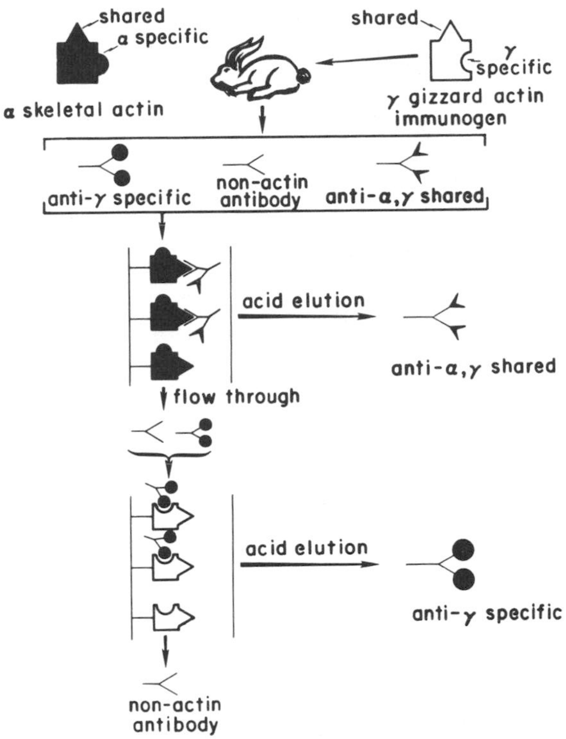

Figure 1
Schematic diagram illustrating affinity purification of isoactin-specific antibodies from polyclonal antisera. (Reprinted, with permission, from Pardo et al. 1983.)

show a pattern in which every muscle fiber is densely populated with brightly stained patches. This pattern represents staining of actin in myofibrils that have been sectioned transversely. In addition to the presence of γ-actin in muscle fibers, the vessels, myelin sheaths, and connective tissue are also stained intensely with anti-γ actin.

Longitudinal sections of muscle fibers stained with anti-γ show rows of bright dots running down the length of a fiber (Fig. 3C). The rows of fluorescent dots correspond to intermyofibrillar rows of phase-dense mitochondria (Pardo et al. 1983). Anti-γ does not stain detectably the periodic I bands of the muscle-cell sarcomeres (Fig. 3C), whereas in similar sections stained with anti-α,γ (Fig. 3F), the myofibrillar cross striations are evident. The anti-α,γ stain represents a superimposition of two actin patterns (mitochondrial and sarcomeric), but this cannot be seen in Figure 3F because of the difference in fluorescence intensity between the two structures.

Additional evidence that the anti-γ staining pattern reflects the cellular distribution of mitochondria was obtained by staining serial sections for succinic dehydrogenase, a mitochondrial marker enzyme. Fibers and regions of single fibers that stain for succinic dehydrogenase (Fig. 4C) also stain with the anti-γ

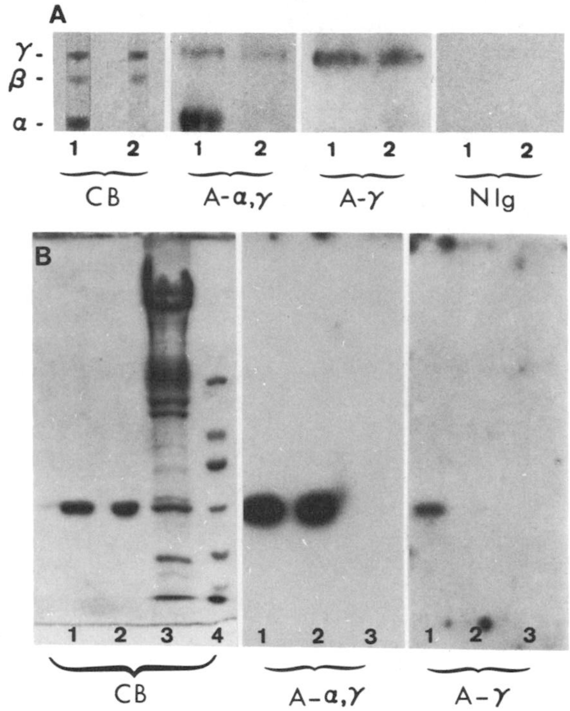

Figure 2
Immunoblots demonstrating specificity and selectivity of purified subpopulations of actin antibodies for different isoactins. (*A*) Isoelectric focusing gels. (Lanes *1*) A mixture of rabbit skeletal-muscle α-actin and cytoplasmic β-actin and γ-actin. The first panel is stained with Coomassie blue. The second, third, and fourth panels are immunoautoradiograms developed with anti-α,γ, anti-γ, and normal rabbit IgG, respectively. Gels in the second, third, and fourth panels were dried down before exposure to film so the bands are wider and slightly distorted compared with the Coomassie-blue-stained counterpart in the first panel. (*B*) Immunoblots of SDS gels. (Lanes *1*) Chicken gizzard γ-actin; (lanes *2*) rabbit skeletal-muscle α-actin; (lanes *3*) ghosts of human erythrocytes; (lane *4*) standards: α-actinin, 100K; bovine serum albumin, 68K; catalase, 60K; actin, 43K; aldolase, 40K; carbonic anhydrase, 29K. The first panel was stained with Coomassie blue. The second and third panels are immunoautoradiograms developed with anti-α,γ and anti-γ, respectively. (Reprinted, with permission, from Pardo et al. 1983.)

reagent (Fig. 4A). These regions are rich with mitochondria as observed by phase optics (Fig. 4B). Note, in particular, the mitochondrion-rich regions immediately underneath the sarcolemma of the red fibers.

The immunofluorescent patterns observed when muscle is stained with anti-γ actin are specific because they are abolished if the antibody is preabsorbed with homogeneous, γ-actin. Also, an affinity-purified keyhole limpet hemocyanin antibody does not stain sections of muscle tissue. This result shows that the affinity purification procedure does not cause IgG to stick to muscle tissue in the patterns we have observed with anti-γ actin (Pardo et al. 1983).

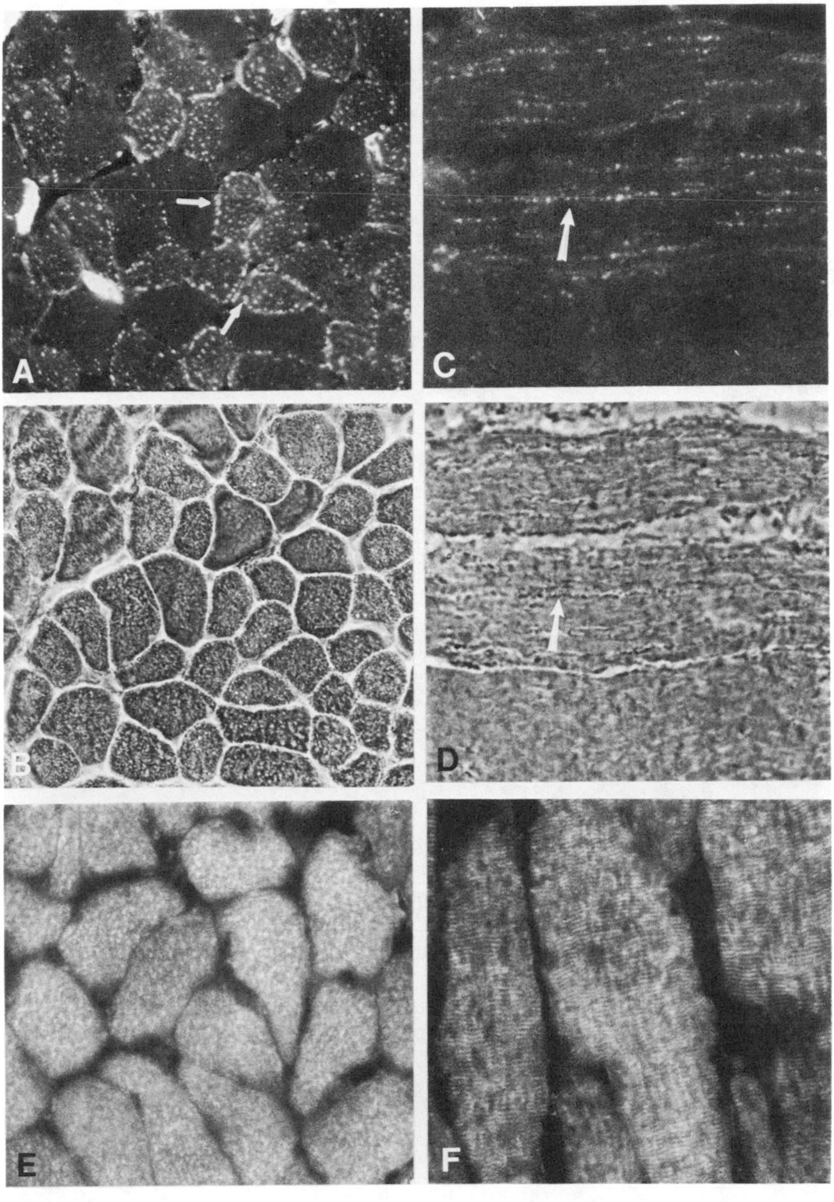

Figure 3

Localization of anti-γ-actin and anti-α,γ-actin in transverse and longitudinal cryostat sections of mouse diaphragm. (*A*) Transverse section stained with anti-β,γ-actin, fluorescence; (*B*) phase. (*C*) Longitudinal section stained with anti-γ-actin, fluorescence; (*D*) phase. (*E*) Transverse section stained with anti-α,γ-actin, fluorescence. (*F*) Longitudinal section stained with anti-α,γ, fluorescence. Arrows in micrographs point to mitochondria. Magnification: (*A,B*) 252 ×; (*C,D*) 621 ×; (*E,F*) 452 ×. (Reprinted, with permission, from Pardo et al. 1983.)

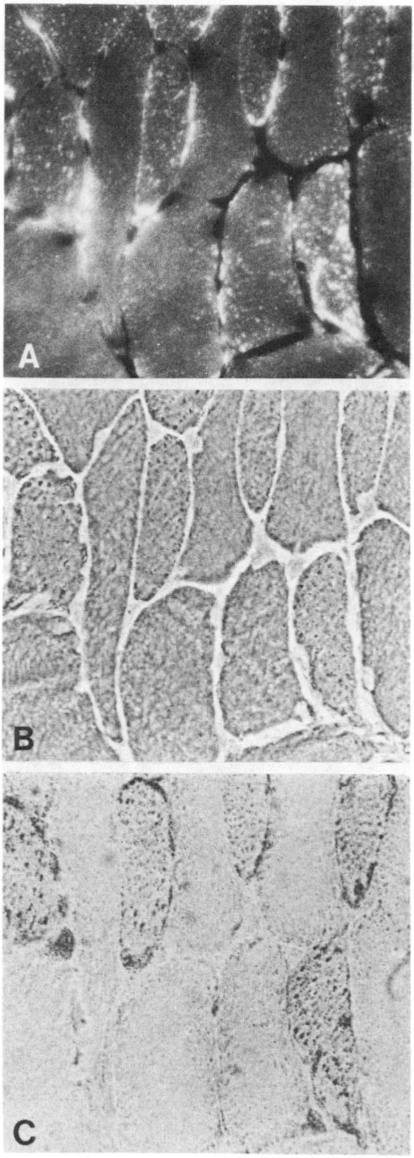

Figure 4
Colocalization of phase-dense mitochondria, γ-actin, and succinic dehydrogenase in muscle fibers of mouse diaphragm. (*A*) Transverse section stained with anti-γ-actin, fluorescence; (*B*) phase. (*C*) Serial section stained for succinate dehydrogenase, bright-field. Magnification, 424×. (Reprinted, with permission, from Pardo et al. 1983.)

γ-Actin Is Located in the Sarcolemmal Costameres and in the Z-disk Lattice Surrounding Individual Myofibrils

In skeletal muscles of the chicken, γ-actin is found associated with two other structures in addition to mitochondria. In transverse sections of anterior latissimus dorsi (ALD), posterior latissimus dorsi (PLD), and pectoralis muscles, the cortex of the fiber is intensely fluorescent, and inside the fiber there is a much fainter reticulum with occasional outlines of mitochondria (Fig. 5A,B). The inter-

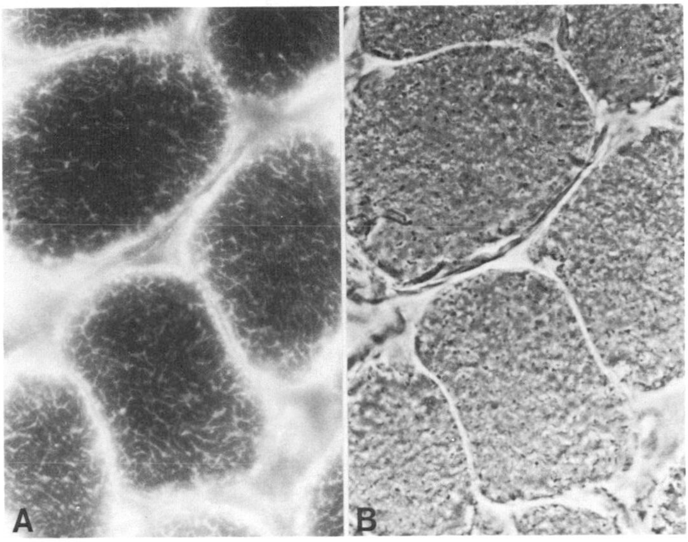

Figure 5
Immunofluorescence of γ-actin and α-actin plus γ-actin in transverse sections of chicken skeletal muscle. (*A*) Section of ALD stained for γ-actin. The cortex of the fibers is extremely bright and there is a dimmer internal reticulum that outlines the periphery of the myofibrils. (*B*) Phase micrograph of *A*. Magnification, 651×. (Reprinted, with permission, from Craig and Pardo 1983.)

nal reticulum resembles the honeycomb pattern that is seen when isolated Z-disk sheets are stained with anti-desmin (Granger and Lazarides 1978), suggesting that this pattern reflects the presence of γ-actin located at the periphery of the myofibrils. Localization of γ-actin in longitudinal sections confirms this assignment of position and in addition shows that γ-actin is organized at the sarcolemma in a two-dimensional lattice. In longitudinal sections, the cortex of the fibers is brightly stained (Fig. 6E); in addition, there is a faint internal cross-striated pattern (Fig. 6E). The width of these cross striations is clearly less than the width of the I band and closely approximates the width of the Z line (Fig. 7A,B). Combining the information from the transverse and longitudinal sections, it is clear that γ-actin is located in bands that encircle each myofibril at the level of the Z line or, in other words, γ-actin is present at the periphery of the Z disk.

The sarcolemma-associated fluorescence can be identified and separated from the Z-disk-associated fluorescence in longitudinal sections by the difference in fluorescence intensity (Fig. 6E). In sections that include the surface of the muscle fiber, it is apparent that the cortex-associated fluorescence is organized in an orthogonal lattice (Fig. 6A,C). The fluorescent bands that are transverse to the long axis of the fiber are in register with the I-Z-I region of the subjacent layer of myofibrils and are in the same region as the previously defined costameres or sites of sarcolemma to myofibril attachment (Craig and Pardo 1983; Pardo et al. 1983).

As in mouse diaphragm (Pardo et al. 1983), γ-actin is also associated with mitochondria in these chicken muscles (Fig. 6F,G, arrows). In mouse diaphragm,

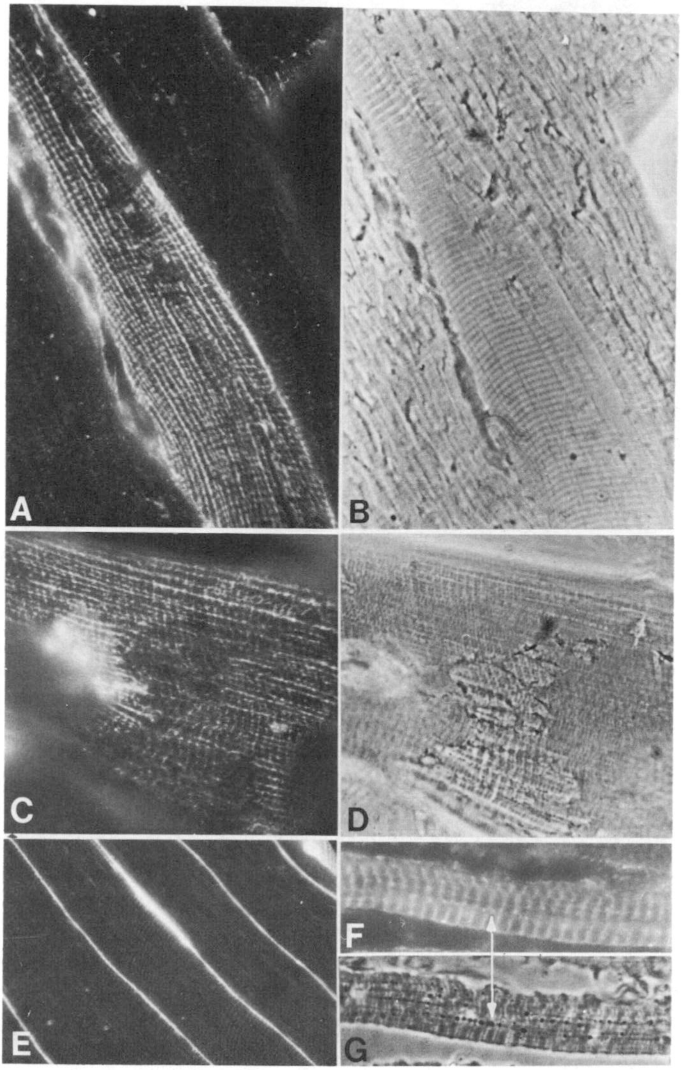

Figure 6

Localization of γ-actin in longitudinal sections of skeletal muscle. (*A*) Section of pectoralis showing organization of γ-actin in a two-dimensional lattice. (*B*) Phase micrograph of *A*. (*C*) Another example of the γ-actin lattice in a pectoralis fiber showing the costameres more clearly. (*D*) Phase micrograph of *C*, illustrating that the fluorescent costameres in *C* overlie the I bands of subjacent myofibrils. (*E*) Longitudinal section below the sarcolemma of a pectoralis fiber, illustrating the difference in fluorescence intensity between cortex-associated γ-actin and γ-actin at the Z lines of internal myofibrils. (*F*) Longitudinal section of PLD showing association of γ-actin with mitochondria (arrow). (*G*) Phase micrograph of *F*; mitochondria are labeled with arrow. Note also that the fluorescent bands in *F* are narrower than the I bands in *G*. Magnification: (*A–D*) 651 × ; (*E*) 350 × ; (*F*,*G*) 595 × . (Reprinted, with permission, from Craig and Pardo 1983.)

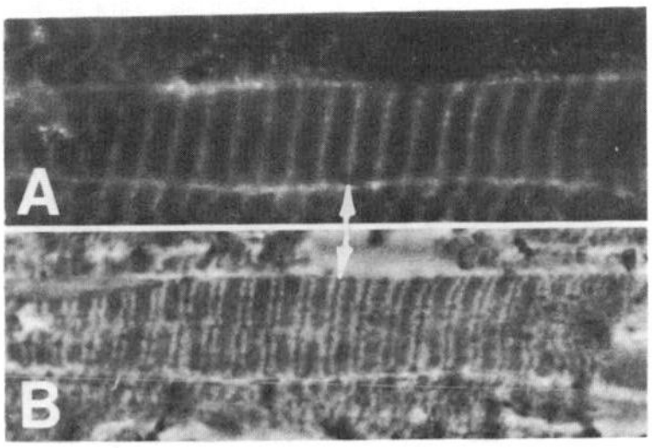

Figure 7
Myofibril-associated γ-actin overlies the Z line and not the entire I band of the myofibrils. Fluorescent (*A*) and phase (*B*) micrographs of an ALD stained with anti-γ-actin. The fluorescent bands of γ-actin in *A* (arrows) are narrower than the corresponding I bands shown in the phase micrograph (*B*). Magnification, 856×. (Reprinted, with permission, from Craig and Pardo 1983.)

the sarcolemma-associated γ-actin was not as intense as in the chicken muscles, and we could not detect myofibril-associated stain at all. The reason for this variation is not known.

CONCLUSIONS, CRITIQUE, AND SPECULATION

γ-Actin is not detected in sarcomeres but is found in association with mitochondria, at the periphery of Z disks, and in sarcolemmal costameres. Whether α-actin is also present in these structures is unknown because we do not have an antibody specific for skeletal-muscle α-actin. With the reservation that γ-actin might be present in sarcomeres, but arranged in such a way that the antigenic determinants are inaccessible to the anti-γ, the results show that during unperturbed in vivo development, actin isoforms are at least partially segregated in muscle fibers. The mechanism of this segregation is unknown.

The differential localization of isoactins could simply reflect differences in time of synthesis of the cytoplasmic actins relative to the time of sarcomerogenesis. If this explanation holds, the nonmuscle actins must have an extremely low turnover time because they persist for the life of the fiber. Another possibility is that the differential localization signifies that actin isoforms have specialized functions in addition to their common activities. These specialized functions might be expressed through differences in the sites for particular actin-binding proteins, resulting in a change in affinity or functional consequence of the actin:actin-binding protein interaction. From this viewpoint, the observed incorporation of transfected cardiac actin gene product into nonmuscle cytoskeleton (Gunning et al. 1984; N.M. McKenna et al., in prep.) might reflect only the copolymerization properties of these isoactins, rather than complete interchangeability of function in the cell.

ACKNOWLEDGMENT

This work was supported by grants from the National Institutes of Health (AI-13700) and the Muscular Dystrophy Association.

REFERENCES

Craig, S.W. and J.V. Pardo. 1983. Gamma actin, spectrin, and intermediate filament proteins colocalize with vinculin at costameres, myofibril to sarcolemmal attachment sites. *Cell Motil.* **3:** 449.

Fyrberg, E.A., B.J. Bond, N.D. Hershey, K.S. Mixter, and N. Davidson. 1981. The actin genes of *Drosophila*: Protein coding sequences are highly conserved but intron positions are not. *Cell* **24:** 107.

Garrels, J.I. and W. Gibson. 1976. Identification and characterization of multiple forms of actin. *Cell* **9:** 793.

Gauthier, G.F. and H.A. Padykula. 1966. Cytological studies of fiber types in skeletal muscle. A comparative study of mammalian diaphragm. *J. Cell Biol.* **28:** 333.

Gordon, D.J., J.L. Boyer, and E.D. Korn. 1977. Comparative biochemistry of non-muscle actins. *J. Biol. Chem.* **252:** 8300.

Gordon, D.J., Y.-Z. Yang, and E.D. Korn. 1976. Polymerization of *Acanthamoeba* actin. Kinetics, thermodynamics and co-polymerization with muscle actin. *J. Biol. Chem.* **251:** 7474.

Granger, B. and E. Lazarides. 1978. The existence of an insoluble Z disc scaffold in chicken skeletal muscle. *Cell* **15:** 1253.

Gunning, P., P. Ponte, L. Kedes, R.J. Hickey, and A.I. Skoultchi. 1984. Expression of human cardiac actin in mouse L cells: A sarcomeric actin associates with a nonmuscle cytoskeleton. *Cell* **36:** 709.

Hall, Z.W., B.W. Lubit, and J.H. Schwartz. 1981. Cytoplasmic actin in postsynaptic structures at the neuromuscular junction. *J. Cell Biol.* **90:** 789.

Lubit, B.W. and J.H. Schwartz. 1980. An anti-actin that distinguishes between cytoplasmic and skeletal muscle actins. *J. Cell Biol.* **86:** 891.

Pardo, J.V., M.F. Pittenger, and S.W. Craig. 1983. Subcellular sorting of isoactins: Selective association of gamma actin with skeletal muscle mitochondria. *Cell* **32:** 1093.

Tseng, P.C.-H. and T.D. Pollard. 1982. Mechanism of action of *Acanthamoeba* profilin: Demonstration of actin species specificity and regulation by micromolar concentrations of $MgCl_2$. *J. Cell Biol.* **94:** 213.

Vandekerckhove, J. and K. Weber. 1978. At least six different actins are expressed in higher mammals: An analysis based on the amino acid sequence of the amino terminal tryptic peptide. *J. Mol. Biol.* **126:** 783.

Whalen, R.G., G.S. Butler-Browne, and F. Gros. 1976. Protein synthesis and actin heterogeneity in calf muscle cells in culture. *Proc. Natl. Acad. Sci.* **73:** 2018.

Yang, Y., D.J. Gordon, E.D. Korn, and E. Eisenberg. 1977. Interaction between *Acanthamoeba* actin and rabbit skeletal tropomyosin. *J. Biol. Chem.* **252:** 3374.

Microtubule Dynamics and Cellular Morphogenesis

Tim Mitchison and Marc Kirschner
Department of Biochemistry and Biophysics, University of California San Francisco, California 94143

The problem of understanding morphogenesis on a cellular level is much more complicated than the problem of understanding the self-assembly of crystals, viruses, or even complex oligomeric structures. On a microscopic level, the individual components of the cell such as the membrane or cytoskeletal systems probably assemble by the usual rather restrictive rules of crystallographic symmetry, relaxed a little by quasi-equivalence (Caspar and Klug 1963). However, on a macroscopic level the assemblage of these structures must be extremely plastic. It must, for example, respond quickly to changes in growth conditions or to developmental stimuli. It must also be capable of being remodeled into many forms, representing different cell types and states while utilizing as major structural components the same elements. This problem is well exemplified by microtubules, whose surface lattice does not seem to vary from cell to cell (or in general from organism to organism) but whose organization varies profoundly during the cell cycle of a given cell. The fundamental question peculiar to the macroscopic organization of the cytoskeleton is, By what means does the cell localize polymerization in certain positions and along certain directions and at the same time prevent random polymerization of cytoskeletal elements?

There are several features of microtubule polymerization that may be related to the need for specific localization of assembly. Microtubules in cells polymerize from organizing centers, the most prominent of which is the centrosome, a region that contains the centriole pair. It is not known by what means these organizing centers specify that polymerization occurs in association with them, nor is it clear if and how these centers can influence assembly at great distances. Second, microtubules are known to possess a structural polarity that is reflected in a difference in the polymerization rates at their two ends. Thus, the cell can, in principle, differentially control assembly on the two microtubule ends. Third, microtubules bind and hydrolyze GTP during the course of assembly. This can lead to a thermodynamic as well as a kinetic difference between the two ends, further extending the scope for differential control of the two ends (see, e.g., Kirschner 1980; Hill and Kirschner 1982). The hydrolysis of GTP can also lead to other important transient features of polymerization first considered by Carlier and Pantaloni (1981), Hill and Carlier (1983), and Carlier et al. (1984), which together with our recent experiments form the basis of much of the analysis in this paper.

It is these unusual transient features of the polymerization reaction that can lead to unique properties of microtubules designed to allow for the macroscopic control of assembly. Knowledge of the nature of the nucleation reaction off centrosomes, coupled with an understanding of the non-steady-state behavior of microtubules, has provided important new clues as to how the cell can control overall organization of the microtubule cytoskeleton and how it can rapidly reorganize the cytoskeleton during morphogenetic processes. We will consider first the unique properties of microtubules themselves and then how these properties play a role in morphogenesis.

Role of GTP Hydrolysis in Microtubule Polymerization

The tubulin dimer has two nucleotide binding sites, one of which is nonexchangeable and invariably contains GTP in some structural role (Weisenberg et al. 1968; Spiegelman et al. 1977). The exchangeable site can bind both GTP and GDP, but the dimer will only polymerize readily into microtubules when liganded with GTP (Weisenberg 1972). When the tubulin dimer polymerizes, the GTP is hydrolyzed to GDP, although this hydrolysis is not necessary for assembly, since tubulin liganded with nonhydrolizable GTP analogs will also polymerize (Arai and Kaziro 1976; Penningroth et al. 1976). The release of free energy that accompanies GTP hydrolysis frees the microtubule from the thermodynamic constraints that apply to simple helical polymers (Oosawa and Asakura 1975) such as tobacco mosaic virus or hemoglobin S, since the polymer need not be in simple chemical equilibrium with a dimer at both ends. The polymerization reaction can be broken down into a series of rate constants for the partial reactions, and depending on the relative rates, different dynamic behavior is possible, of which one particular case would be "treadmilling" (Wegner 1976; Margolis and Wilson 1978; Hill and Kirschner 1982).

Carlier et al. (1984) first showed by turbidometric measurements that the depolymerization of microtubules is not simply the reverse of the polymerization reaction and, in particular, that depolymerization is much faster than expected by extrapolation of the polymerization rate at different tubulin concentrations. They hypothesized that growing microtubules have GTP subunits at their ends that have not yet hydrolyzed, forming a GTP-liganded cap. The rather small extrapolated off rate from polymerization studies reflects the dissociation of GTP-liganded subunits from such a cap, whereas during depolymerization, GDP-liganded subunits dissociate from bulk polymer at a much faster rate.

Using the method of Bergen and Borisy (1980), we have measured the individual off and on rates from each end of a microtubule by examining the growth off axoneme templates by electron microscopy and immunofluorescence. Using this independent method, we have confirmed the results of Hill and Carlier (1983) that there is a break in the plot of growth rate versus concentration. We find using purified tubulin that for both the plus and minus ends of the microtubule, the depolymerization rate during depolymerization is two to three orders of magnitude larger than that during polymerization, obtained by extrapolation (Mitchison and Kirschner 1984b).

However, during our studies of the elongation of microtubule nucleated by organelles (Mitchison and Kirschner 1984a,b), an important fact emerged that could not have been discovered by bulk measurements such as turbidometry. When individual microtubules were observed, they did not all behave in the same way; rather, two classes could be observed under a variety of conditions, a class of shrinking microtubules and a class of growing microtubules. From their behavior, we concluded that these two classes interconverted only rarely. An example of this is shown in Figure 1.

In this experiment, microtubules were assembled onto preformed seeds and then sheared. After attaining a steady state in polymer mass, determined by turbidometry, samples were withdrawn at various time intervals and fixed. The microtubules were then sedimented onto coverslips in the airfuge and then observed by immunofluorescence. By this assay, we were able to determine the length distribution and number concentration of microtubules as a function of time. Classical theory for assembly of equilibrium helical polymers predicts that length distribution and number concentration should only change very slowly once the polymer is at steady state, and the introduction of treadmilling should not change this (Hill 1980). However, we observed that mean polymer length increased rapidly and steadily after attainment of the steady state, and this was offset by a decrease in number concentration of microtubules (for a full account of this experiment, see Mitchison and Kirschner 1984a,b). To explain these data, we hypothesized that the microtubule ends existed in two different states: Growing microtubule ends have a cap of GTP-liganded subunits, whereas the shrinking ends have GDP-liganded subunits exposed as diagramed in Figure 2. To account for the persistence of microtubule in the growing or shrinking phase, growing and shrinking ends must interconvert rarely. We estimate that on average a shrinking microtubule will lose 20,000 subunits before it becomes recapped and starts to grow, at the steady-state dimer concentration. This precessive depolymerization will cause the disappearance of whole microtubules, and since pure tubulin has a low efficiency of spontaneous nucleation, the microtubule number concentration at the turbidity plateau will steadily decrease.

The data shown in Figure 1 also support the idea that a GTP cap is necessary to stabilize a microtubule end and that its loss leads to the initiation of depolymerization. When the growing microtubules were sheared, at the time indicated, a large, transient decrease in turbidity was observed. We interpret this as being due to the exposure of new GDP-liganded ends, which are exposed when the bulk polymer is broken by the shear. These new ends start to depolymerize rapidly, leading to an increase in dimeric tubulin; eventually, many of them become recapped with GTP tubulin and start to grow. The existence of the GTP cap, stabilizing the growing microtubules, depends on the fact that hydrolysis is a first-order reaction; once a subunit is incorporated into polymer, it has a fixed

LENGTH REDISTRIBUTION AT STEADY STATE

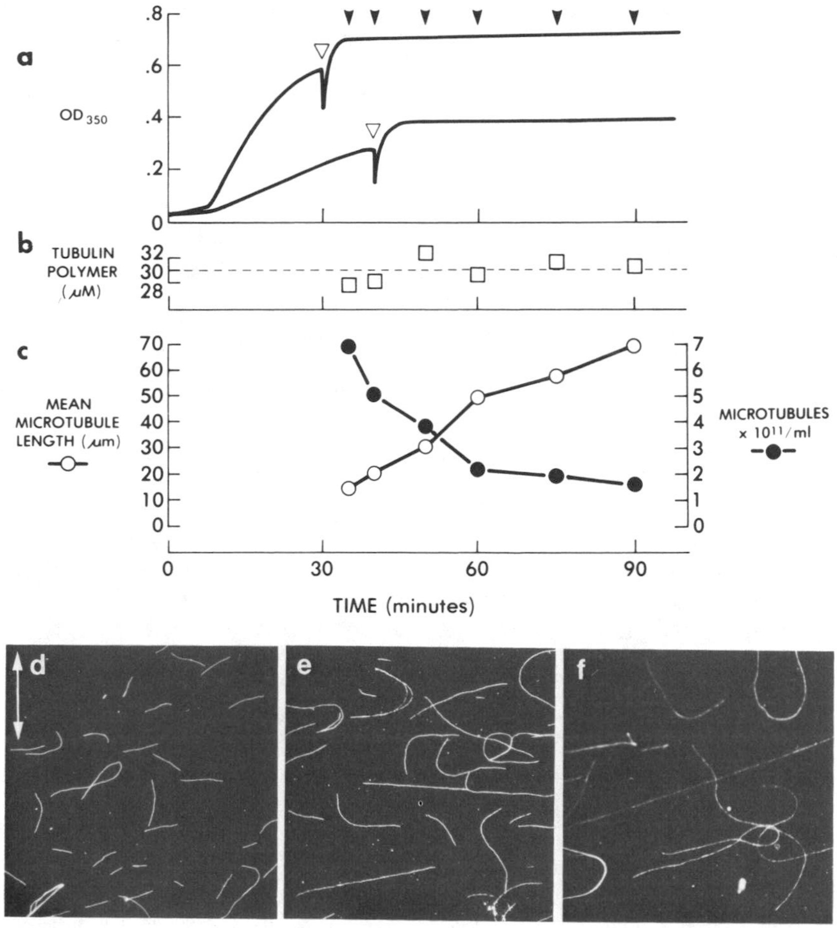

Figure 1

(*a*) Turbidometric assay of the assembly of phosphocellulose-purified tubulin into microtubules. Tubulin at a concentration of 59 μM dimer (upper trace) or 32 μM (lower trace) in assembly buffer (80 mM KPIPES, 1 mM EGTA, 1 mM $MgCl_2$, 1 mM GTP) plus a GTP-regenerating system (Terry and Purich 1979) was induced to assemble by addition of preformed microtubule seeds. At the time indicated by the open arrowhead, the solution was sheared by passage through a 21-gauge syringe needle. In the experiment shown in the upper trace, after a plateau in turbidity was reached, aliquots were removed at the times indicated by solid arrowheads. These aliquots were fixed in 1% glutaraldehyde at room temperature, diluted, and quantitatively sedimented onto 5-mm^2 polylysine-coated glass coverslips using the Beckman airfuge EM90 rotor at 90,000*g* for 15 min. The microtubules were visualized by immunofluorescence, photographed, measured, and counted. (*b*) Total tubulin polymer in each aliquot calculated by multiplying the mean length by the number concentration. (*c*) Mean length and number concentration as a function of time. Microtubules (500) were measured per point, and number concentration was determined by averaging 36 fields. (*d*,*e*,*f*) Representative fluorescent fields at 5, 15, and 60 min, respectively. Arrow, 30 μm.

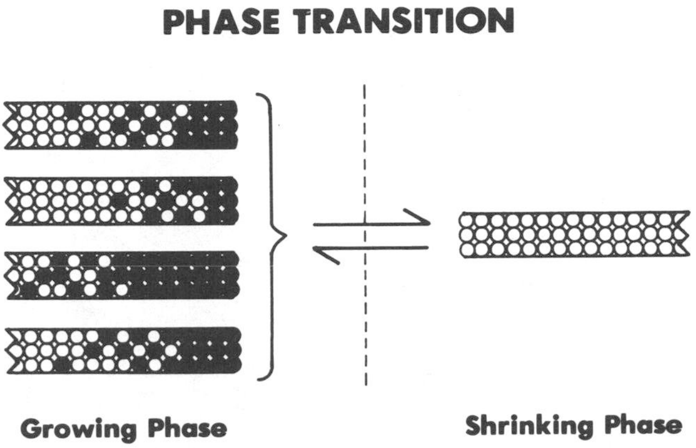

Figure 2
Model to account for coexistence of shrinking and growing microtubules. The majority of the microtubules in the population, shown on the left, are growing by addition of GTP-liganded subunits (●) to a GTP-liganded cap. Hydrolysis following polymerization results in a bulk of polymer consisting of GDP-liganded subunits (○). Rare fluctuations can lead to loss of the cap, leaving GDP-liganded subunits at the end of the microtubules. Such microtubules, shown on the right, start to depolymerize rapidly by loss of GDP-liganded subunits.

probability of hydrolysis per unit time. If the on rate, which is concentration-dependent, is fast compared to the hydrolysis rate, hydrolysis will lag behind polymerization, resulting in a GTP cap. Such an effect has been demonstrated by bulk measurements during rapid assembly by Carlier and Pantaloni (1981). The average length of such a cap thus depends on the dimer concentration, and the length of individual caps will fluctuate with time. At low tubulin concentration, when polymerization is relatively slow, the cap will only have a transient existence, and these individual microtubules will be unstable. We have seen such behavior following the dilution of microtubules initiated at high tubulin concentrations (Mitchison and Kirschner 1984a). The mechanism by which a rapidly shrinking GDP-liganded end can become recapped is less clear. It could reflect binding of GTP tubulin or exchange of nucleotide at the terminal subunit; experiments to distinguish these possibilities are in progress.

A schematic diagram of the polymerization rate as a function of concentration for one end of a microtubule is shown in Figure 3. The upper line illustrates the behavior of growing microtubules, where the slope gives approximately the second-order on rate (K_T), and the intercept, the off rate (K'_T). The lower line gives the depolymerization rate (K'_D), which may be independent of concentration. In principle, individual microtubules could exist in both states, at any concentration, depending on whether or not they have a GTP cap. The probability of the cap fluctuating to zero length, and a growing microtubule transiting to the shrinking phase, decreases with increasing dimer concentration. The probability of a shrinking end becoming recapped, and the microtubule transiting to the growing phase, is probably also concentration-dependent, increasing with increasing dimer concentration. Relative probabilities are indicated by arrows on the vertical transitions in Figure 3. A bulk measurement, such as turbidity, will average the behavior of many microtubules and give the heavy curve indicated, which trans-

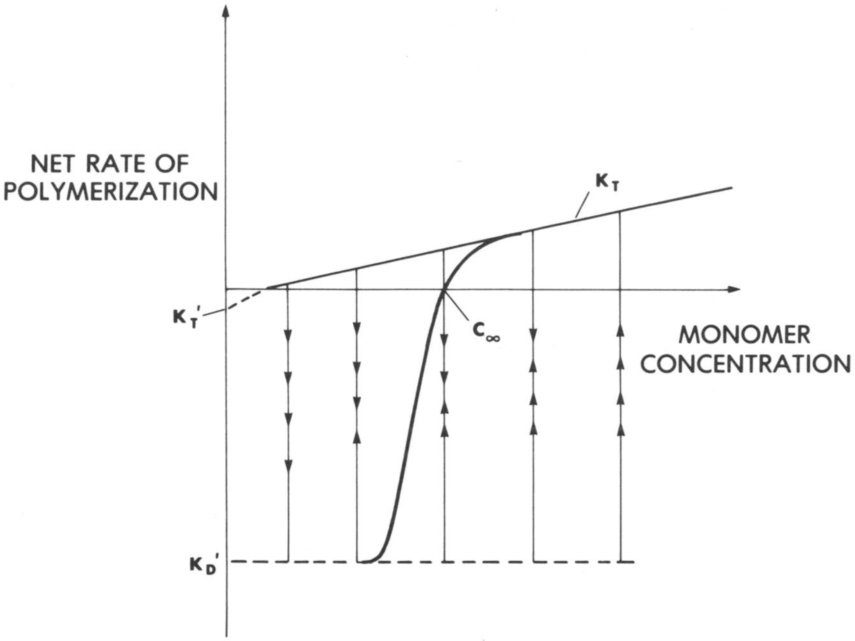

Figure 3
Schematic diagram showing polymerization rate as a function of concentration. The growing phase of microtubules has the kinetic parameters of the GTP cap, and behaves as the upper line, with on rate given by the slope K_T and off rate by the intercept, K'_T. The shrinking phase has an off rate, K'_D, the off rate of GDP-liganded subunits from a GDP lattice. Individual microtubules will generally remain in one phase, on either the upper or lower line, but occasional fluctuations will lead to transitions (vertical arrowheads). A bulk average will give the solid curve, which crosses the axis at C_∞, the steady-state concentration.

its smoothly between polymerization and depolymerization lines, crossing the axis at the steady-state concentration. In practice, the situation for free microtubules is complicated by the existence of two ends with different kinetics. Graphs similar to Figure 3 could be constructed independently for the plus and minus ends, and there is no reason why their steady-state concentration should be exactly the same; in this case, the bulk steady-state concentration would fall between that for plus and minus ends, and subunit uptake would occur preferentially at one end. If microtubules under some condition possessed high transition probabilities between the two phases, this would smooth out length fluctuations and could lead to treadmilling. Under the circumstances of the experiments here, the very low transition probabilities lead to extensive length fluctuations, which would swamp out treadmilling behavior.

Although the stabilization of microtubules by GTP caps can explain experimental data, at present, it remains hypothetical. Hill and Chen (1984) have constructed an explicit kinetic model for the elongation process, and using plausible rate constants, they have demonstrated that the coexistence of shrinking and growing phases is theoretically reasonable. Direct tests of the role of GTP hydrolysis could come from experiments with nonhydrolizable GTP analogs, and we have initiated such experiments. It is clear that microtubules formed from nonhydrolizable analogs depolymerize much more slowly than GDP-liganded microtubules, and there is probably no break in the rate-versus-concentration plot. However, the rather slow on and off rates found with the analogs make it difficult to absolutely rule out the coexistence of shrinking and growing microtubules.

In summary, the presence of a GTP cap can mean that a microtubule array is composed of two phases, growing and shrinking microtubules. Since these phases seem to convert infrequently, the population is extremely dynamic. In particular, at steady state, the population of microtubules consists of a majority growing slowly and a minority shrinking quickly. If this type of unusual dynamics, which we have termed dynamic instability, also occurs in cells, it will have many important implications. In particular, it provides a special role for microtubule-organizing centers (MTOC). We discuss below the biological implications of dynamic instability as well as experiments on microtubule assembly off organizing centers.

Biological Implications of Rapid Microtubule Depolymerization

During prophase, the interphase microtubule array must depolymerize, to be replaced by the mitotic array, and this process must occur very rapidly. In tissue-culture cells, microtubules ≈20-μm long must depolymerize in a few minutes, and the situation is more severe in eggs with large cytasters. Similar rapid depolymerization must occur during reorganization of the interphase network. Depolymerization rates based on the extrapolation of the polymerization line fall in the range of 1.5–20/sec (Johnson and Borisy 1977; Bergen and Borisy 1980). At these rates it would take a 20-μm microtubule, which contains 32,000 subunits, 27 minutes to 6 hours to depolymerize. However, using the off rate of GDP tubulin, we observe 340/sec for the plus end. The microtubule could disappear in 1.5 minutes, which is much more reasonable. The fastest depolymerization rates observed in vivo were those determined by E.D. Salmon (pers. comm. and this volume) for depolymerization of spindle microtubules following colchicine injection. He hypothesized that the observed off rate of ≈500/sec was too fast for endwise depolymerization, but we feel this number is not incompatible with our results.

These figures lead to the simple idea that the microtubule has evolved GTP hydrolysis primarily to facilitate *disassembly* as originally postulated by Weisenberg and Deery (1976). The favorable free energy of polymerization for GTP tubulin is decreased when GTP is hydrolyzed, presumably by a conformational change that weakens subunit interactions. The polymer effectively traps some of the free energy of triphosphate hydrolysis by weakening subunit interactions, and this portion of the free energy can only be released by depolymerization, which is restricted to the ends of the polymer. This leads to a polymer that can grow rapidly when assembled, but can be made to shrink rapidly, without the necessity for large changes in the intracellular environment. In this respect, the microtubule (and perhaps the microfilament) can be contrasted with structures such as bacterial flagella, phage tails, or intermediate filaments. The biological function of these helical polymers does not require rapid disassembly, and hence no chemical change is associated with polymerization.

Implications of Dynamic Instability for MTOCs

A more subtle role for the hydrolysis-induced polymer destabilization concerns the role of MTOCs. It has been known from early work with lysed cells (Snyder and McIntosh 1975; Brinkley et al. 1981) and mitotic lysates (McGill and Brinkley 1975; Gould and Borisy 1977; Summers and Kirschner 1979; Telzer and Rosenbaum 1979; Bergen et al. 1980; Ring et al. 1980) that centrosomes, and to a lesser extent kinetochores, are capable of nucleating microtubules in vitro. This means that the nucleated microtubules have a kinetic advantage over those assembling spontaneously, and analogous experiments can be performed in vivo by releasing cells from a drug-induced microtubule disassembly (Brinkley et al. 1976; Osborn and Weber 1976). However, it is not clear why in the long term (for which a kinetic advantage is insufficient explanation), microtubules should continue to arise from organizing centers. Models have been proposed to account for this long-term stability of nucleated arrays by postulating a thermodynamic advantage for MTOC-associated microtubules. Such an advantage could arise from differing critical concentrations at the two ends (Kirschner 1980) or local differences in critical concentration (DeBrabander et al. 1981), but neither idea has been proven experimentally.

To study the nucleation reaction in vitro, we have isolated in functional form two types of mammalian MTOCs, centrosomes and kinetochores. The centrosomes are highly purified and well characterized (Mitchison and Kirschner 1984a), whereas the kinetochores are present in an enriched preparation of metaphase chromosomes that is free of soluble components. This preparation was made by a modification of the aqueous polyamine method of Lewis and Laemmli (1982), from Chinese hamster ovary (CHO) cells synchronized with thymidine and then arrested in mitosis with a high concentration of vinblastine (10 μg/ml).

The organizing sites each show rather different behavior in vitro. The number of microtubules nucleated by each organelle as a function of tubulin concentration is shown in Figure 4. Considering the centrosome first, nucleation is first seen at 4 μM, and the curve has a sigmoid shape, plateauing around 20 μM. This result is quite surprising given that the steady-state concentration for this tubulin preparation is 14 μM, and thus the centrosome can nucleate microtubules at tubulin concentrations at which individual microtubules are unstable (Mitchison and Kirschner 1984a,b). This is not simply due to blocking of the minus ends of the microtubules, since subsequent experiments demonstrated that plus ends are unstable well above 4 μM. Such behavior can be best understood with reference to Figure 3, taking into account the two phases of polymerization-depolymerization and the persistence of the GTP cap. The centrosome must have a sufficiently efficient nucleating mechanism that a microtubule can be initiated rapidly, compared with GTP hydrolysis, by the incorporated tubulin subunits. The initiated microtubule starts to grow out with a GTP cap and is thus stabilized at least for the lifetime of the cap. Presumably, below 14 μM the microtubules are only transiently stable. Fluctuations of the cap size will eventually lead to the exposure of GDP subunits at the end of the microtubule, which will then start to shrink. How-

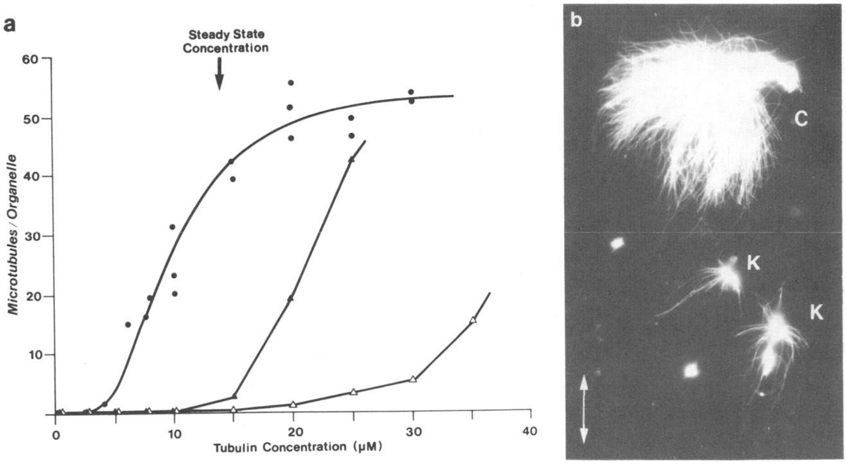

Figure 4
Microtubule nucleation in vitro. (*a*) The number of microtubules nucleated by purified CHO interphase centrosomes (●) or mitotic kinetochore pairs (△, ▲) is shown as a function of tubulin dimer concentration. The organelles were incubated with tubulin in assembly buffer at 37°C, fixed with glutaraldehyde, and sedimented onto electron microscopic grids (centrosomes) or coverslips (chromosomes) and visualized by shadowing or anti-tubulin immunofluorescence, respectively. The centrosomes were fixed at various times between 5 and 20 min, and the number of microtubules was constant with time. The kinetochores were regrown for 5 min (△) or 15 min (▲). One hundred organelles were averaged for each point. (*b*) Representative field from the kinetochore assay. This field shows two chromosomes, with the microtubules nucleated by the kinetochore pair (K) and also a mitotic centrosome (C), which contaminated the preparation. Note that the mitotic centrosome nucleates many more microtubules than do the interphase centrosomes. Also note that the microtubules nucleated by the kinetochores are much more heterogeneous in length, and generally shorter, than those nucleated by the centrosome. This field was from a preparation incubated for 15 min at 25 μM tubulin. Arrow, 15 μm.

ever, transient growth can be quite extensive, easily greater than the 20 μm, which can be visualized in the electron microscope. It is only at very low concentrations that limitations on the length of growth can be seen.

The kinetochore in contrast appears to be much less efficient at nucleation and cannot nucleate below the steady-state concentration. This organelle, unlike the centrosome, does not seem to be designed for nucleation, and perhaps nucleation is not normally an important part of kinetochore function. Nucleation by kinetochores has been demonstrated in vivo following release of a drug block (DeBrabander et al. 1981), but such experimental manipulation may lead to an artificially high tubulin dimer concentration following drug removal, which may drive nucleation as seen in vitro. Such a high concentration may not occur during a normal prophase, and the kinetochore may interact with microtubules principally by capturing polar microtubules (Reider and Borisy 1981; Pickett-Heaps et al. 1982; but see also Witt et al. 1980).

The centrosome and kinetochore differ in several other ways in their nucleation properties, and these are summarized in Table 1. These results can be summed up by saying that the centrosome is clearly designed to nucleate micro-

Table 1
Properties of Microtubule-organizing Centers In Vitro

	Centrosome	*Kinetochore*
Minimum tubulin concentration for nucleation	4 μM	15 μM
Saturable number of nucleation sites?	yes	no
Lag phase in nucleation?	no	yes
Uniform polarity of nucleation?	yes (plus end out)	no
Microtubule interactions other than nucleation	?	capture ?

tubules, with a uniform polarity, and can do so below the concentration where free microtubules are stable. Nucleation by the kinetochore, however, occurs only at higher concentrations and may reflect another type of interaction with tubulin or microtubules whose primary function is not nucleation, but may be capture.

The ability of the centrosome to nucleate below the steady-state concentration has important implications for the organization of the microtubule cytoskeleton, as diagramed in Figure 5. This figure illustrates the relative persistence of free and centrosomal microtubules. Here we diagram an experiment where a cell has just been released from a nocodazole-induced depolymerization (DeBrabander et al. 1980, 1981). Initially, the free-dimer concentration is high, so that although the centrosome will nucleate effectively, spontaneous polymers will also arise. However, as tubulin polymerizes, the free-dimer concentration decreases and the elongation rate slows. This leads to a shortening in the average GTP cap length, and both free and centrosomal microtubules start to transit into the depolymerizing phase. At this point an important distinction between free and centrosomal microtubules is apparent. Whereas the loss of a free microtubule is irreversible, since at low tubulin concentration, spontaneous nucleation is improbable, the depolymerization of a nucleated microtubule leaves an unoccupied nucleation site, which can grow a new microtubule. In the long term, only the nucleated polymer will be present. Each individual microtubule emanating from the centrosome is unstable, but a nucleated ensemble is always present. Microtubules are effectively restricted to the centrosome by its capacity for continuous nucleation.

In a closed system such as the cell, the relative stability of nucleated and free polymers will depend largely on the total amount of tubulin present and the number of nucleation sites. In a typical tissue-culture cell in interphase, the centrosome is small and may be saturated with microtubules. In this case, the free-dimer concentration may be at the steady-state level, and free and nucleated polymers may coexist (see, e.g., Karsenti et al. 1984a). However, when the cell enters mitosis, the centrosome becomes greatly augmented in its nucleation capacity (Telzer and Rosenbaum 1979; Kuriyama and Borisy 1981) as shown in

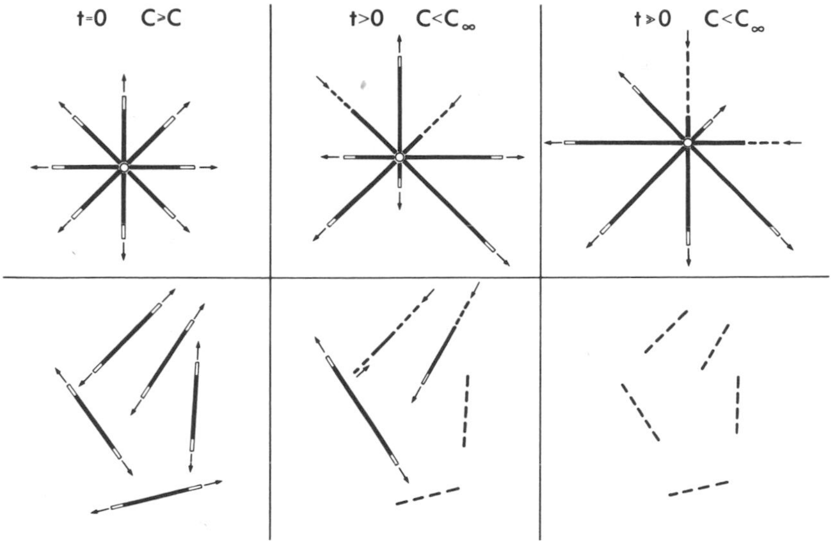

Figure 5
Persistence of centrosomal microtubules. (*Left*) This diagram shows schematically the changes in the microtubule cytoskeleton of a cell that has just been released from nocodazole-induced microtubule disassembly, interpreted in terms of the dynamic instability model. The open circle represents the centrosome, and the lines represent microtubules. Filled lines indicate GDP-liganded subunits, and the open tips represent the GTP-liganded cap. Arrow denotes the direction of growth of the microtubule. The dimer concentration is initially high. Both spontaneous and centrosomal microtubules are nucleated and grow out with GTP caps. (*Middle*) The decrease in dimer concentration leads to a decrease in growth rate and a destabilization of all microtubules, and some start to depolymerize. (*Right*) After a long time, the free microtubules have disappeared, whereas centrosomal microtubules continue to be nucleated.

vitro in Figure 4 a and b. These new, empty nucleation sites will start to absorb the free dimer into microtubules, bringing its concentration below the steady-state level. The average GTP cap size will decrease, and individual microtubules will become unstable. Eventually, the cell will contain only nucleated microtubules, in a highly dynamic array.

In summary, the half-life of a given microtubule in a closed system such as the cell, and thus the dynamics of the array, depends on the free-dimer concentration, which in turn depends on the number of microtubules present. Increased nucleation capacity will result in increased numbers of microtubules, decreased dimer concentration, and decreased stability of each microtubule.

This idea predicts that cells in mitosis should have fewer free cytoplasmic microtubules than in interphase and that the tubulin cytoskeleton is much more dynamic in mitosis than in interphase, both of which appear to be true (Karsenti et al. 1984b; Borisy; Salmon; both this volume). However, the situation is undoubtedly more complicated in living cells, due to the influence of nontubulin components that can alter tubulin polymerizability and the stability of microtubules.

Morphogenesis by Selection and Stabilization

Another interesting possibility that might arise from the dynamic behavior of microtubules we have described is diagramed in Figure 6. There are frequent instances in cellular morphogenesis when a cell has to direct microtubules toward a particular part of the cell, e.g., into an extending neurite or into a pseudopod during locomotion. The structure of the centrosome does not suggest any mechanism by which it would direct nucleated microtubules. Instead, we suggest that given the dynamic nature of microtubules, the centrosome may continuously nucleate microtubules in random directions. If proteins that could cap or stabilize the ends of microtubules become localized in a particular region of the cell as shown in Figure 6, then microtubules growing in that direction would become preferentially stabilized. Microtubules in other regions would continue to initiate depolymerization stochastically. Eventually, the majority of the cell's microtubules would be directed toward the stabilizing region and the cell would become polarized.

Microtubules that become capped, or otherwise stabilized, would differ from newly nucleated microtubules in their age. Conceivably, time-dependent processes such as lateral association of proteins or covalent modification could modify old microtubules, so that microtubules that are initially differentiated only by end interactions become distinguished along their lenghts. Such mechanisms could be invoked to explain the microtubule subclasses recently distinguished by antibody staining (Thompson et al. 1984). This time-dependent modification

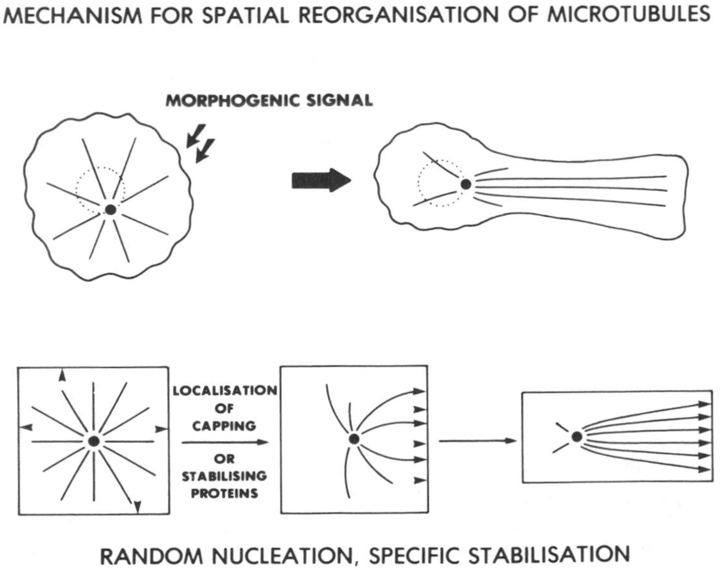

Figure 6
Morphogenesis by specific stabilization. Polarization of cells frequently involves redistribution of microtubules. This is postulated to occur by the local activation of capping or stabilizing factors in response to a morphogenetic signal. Microtubules are nucleated in random directions, but those that grow into a specific region of the cell become preferentially stabilized, leading to a polarization of the microtubule distribution.

of microtubules might have functional consequences, e.g., in intracellular transport, and this could exaggerate the regional differentiation caused by changes in local microtubule density.

Microtubule Capture by Kinetochores

A particularly important example of microtubule capture and stabilization may occur during prophase, as the kinetochore interacts with polar microtubules (Reider and Borisy 1981; Pickett-Heaps et al. 1982). Kinetochore microtubules are resistant to drug, pressure, and cold-induced depolymerization (Salmon et al. 1976), perhaps because the kinetochore blocks the plus end and the centrosome blocks the minus end, and together they prevent subunit dissociation.

We have begun to examine the question of microtubule capture by kinetochores in vitro, in experiments of the type shown in Figure 7. In this experiment, microtubule polymerization was first initiated off centrosomes and then chromosomes were added. This mimics the situation in prophase when the nuclear envelope breaks down, and naked kinetochores are exposed to the plus ends of

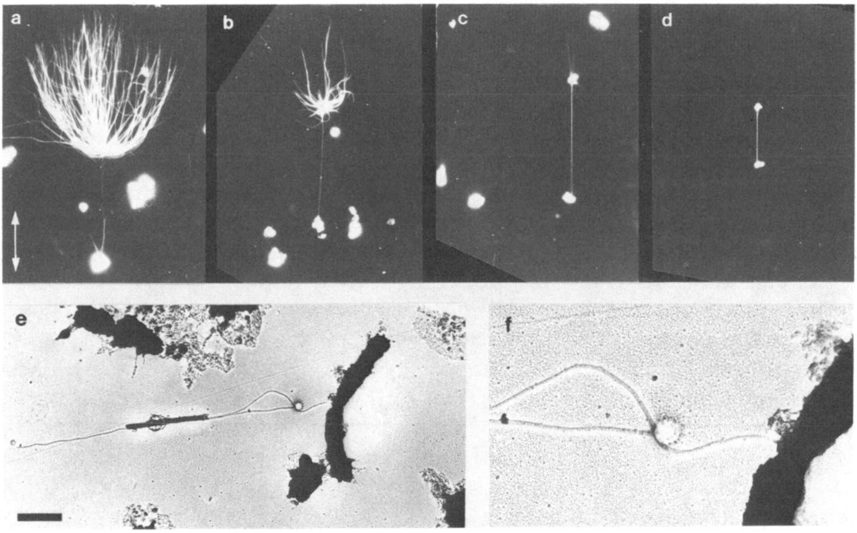

Figure 7
Microtubule capture by kinetochores. (*a–d*) Purified interphase centrosomes were incubated for 10 min with 25 μM tubulin at 37°C. CHO chromosomes were then added, and the mixture was diluted to 12 μM tubulin. After 10 min at 37°C, to allow interaction, the mixture was either fixed directly or diluted into 50 volumes of warm buffer and then fixed after various time intervals at 37°C. (*a*) Before dilution; (*b*) 30 sec, (*c*) 2 min, (*d*) 20 min after dilution. Complexes were sedimented onto coverslips through 30% glycerol cushions, and the microtubules were visualized by immunofluorescence. The chromosome, which is the lower body in each photograph, was visualized by Hoechst H23358 staining. Arrow, 15 μM. (*e,f*) A similar experiment with *Tetrahymena* axonemes used as the microtubule donor; complexes fixed 2 min after dilution, sedimented onto electron microscopic grids, and visualized by shadowing. Bar, 3 μm. (*f*) Higher magnification, showing the captured microtubule appearing to terminate in the kinetochore.

microtubules emerging from the poles. The mixture of chromosomes and microtubule asters was then incubated to allow interaction, and during this time the centrosomal microtubules continued to grow. The mixture can then be fixed immediately, or first diluted extensively with warm buffer and then fixed at various times to observe the effect of inducing microtubule depolymerization. When the mixture was fixed without dilution, we frequently observed complexes such as that shown in Figure 7a. Microtubules from the centrosome appear to contact, and terminate in, the kinetochore region of the chromosome. The kinetochore can be identified as the primary construction of the chromatin by Hoechst staining and as a bright pair of dots by anti-tubulin staining. Usually one microtubule forms the connection, but sometimes several microtubules seem to attach from one centrosomal microtubule array. Centrosomes that interact with more than one chromosome were also common. It is difficult to estimate the frequency with which these connections are made, but given the low concentration of both organelles, the formation of any complexes is quite impressive, and interaction of centrosomal microtubules with random pieces of debris or nonkinetochore regions of the chromosome was rare. In the cell, the geometry would presumably be much more favorable for making specific connnections.

If the connection is a specific one, it should have functional consequences for the captured microtubule. This specificity was tested by subjecting the complexes to depolymerizing conditions. A typical sequence following dilution into warm buffer is shown in Figure 7 b through d. The uncapped microtubules emanating from the centrosome rapidly start to depolymerize and are almost all gone after 1 minute. The centrosome-to-kinetochore microtubule, however, appears more stable, and complexes in which a characteristic bright dot (the centriole cylinder visualized by anti-tubulin immunofluorescence) are connected through a single (or sometimes multiple) microtubule to a kinetochore were seen for several minutes after dilution. At this point, there were virtually no other microtubules to be seen anywhere else on the coverslip, since only microtubules capped at both ends are stable under these conditions. Similar stable connections could be formed using *Tetrahymena* axonemes to initiate microtubule assembly, which cap one end of the resulting complex (Fig. 7e,f).

The experiment shown in Figure 7 is a preliminary result, and at present we have not determined the exact nature of the connection, nor whether all microtubule connections to the kinetochore completely block depolymerization. Our studies suggest that kinetochores may be able to interact with microtubules in several ways, and the possibility exists that it can act as a "sliding cap" allowing depolymerization only when coupled to translocation of the chromosome toward the centrosome. More work needs to be done on this question, which is particularly exciting in that it pertains to both the morphogenesis of the metaphase spindle and the mechanism of chromosome-to-pole movement in anaphase. We are confident, however, that the capture of microtubule ends by the kinetochore in vitro is a manifestation of an important intrinsic property of this organelle, which is probably important in the establishment of the mitotic spindle.

One of the most important questions about spindle structure concerns the origin of the polarity of spindle microtubules. Whether chromosome movement is driven by proteins associated with microtubules, e.g., dynienlike molecules, or by dynamics of the microtubules themselves, the direction of this movement is probably specified by the polarity of the microtubule lattice. The properties of microtubules, centrosomes, and kinetochores described in this paper can be combined to give a model, shown in Figure 8, that accounts for the uniform polarity found in the astral microtubules and kinetochore fibers at metaphase (Heidemann and McIntosh 1980; Euteneuer and McIntosh 1981). The data of E.D. Salmon et al. (pers. comm. and this volume) suggest that astral microtubules are highly dynamic in mitosis, and if their photobleaching data are explained by the dynamic instability model, then one can conclude that the average half-life of an astral microtubule in the sea urchin spindle is on the order of 10 seconds. In this model, microtubules are continually being nucleated by the pole with their plus ends out, losing the GTP cap, and then rapidly depolymerizing back to the centrosome. Our data strongly suggest that the centrosome is the most important nucleating center in mitosis and that the kinetochore has the ability to capture and cap microtubule ends. Thus, when the nuclear envelope breaks down (Fig. 8A), the kinetochore is exposed to transiently stable microtubule plus ends and can capture and cap them (Fig. 8B). Eventually, enough microtubules will be captured to form the kinetochore fiber, and the polarity of these microtubules will be plus ends toward the kinetochore, reflecting the absolute specificity of the centrosome for nucleating with plus ends out (Fig. 8C). In this model, the kinetochore need have little respect for polarity in its interaction with microtubules, and the polarity of the half-spindle is determined solely by the end specificity of centrosomal nucleation. Microtubules nucleated at the kinetochore appear first as randomly oriented fragments (DeBrabander et al. 1980, 1981) and may grow out with random polarity (Fig. 8B). However, the centrosome may not have the capacity to capture, so that microtubules nucleated at the kinetochore will quickly depolymerize. If the centrosome can capture, it is likely to be specific for the capture of minus ends, reflecting the end specificity of its nucleation capacity. The formation of the metaphase spindle, a paradigm in cellular morphogenesis, may thus be partially explicable with reference to the properties of its constitu-

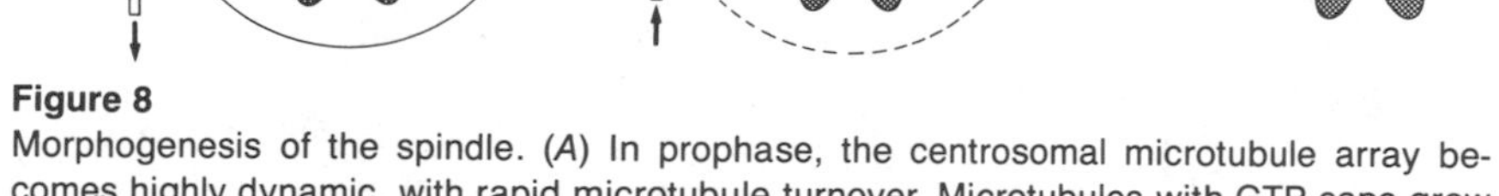

Figure 8

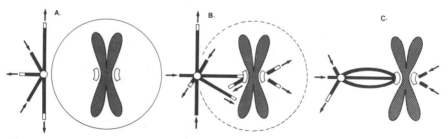

Morphogenesis of the spindle. (*A*) In prophase, the centrosomal microtubule array becomes highly dynamic, with rapid microtubule turnover. Microtubules with GTP caps grow out, lose the cap, and shrink. (*B*) Prometaphase. Microtubules with GTP caps grow toward, and interact with, the kinetochore. The kinetochore may also nucleate some microtubules. (*C*) Metaphase. Polar microtubules trapped by the kinetochore become stabilized and are removed from the dynamic pool. Sequential trapping of several microtubules leads to an ordered kinetochore fiber, whose polarity is defined by the centrosome. Microtubules nucleated at the kinetochore are not stabilized, and eventually disappear. Astral microtubules continue to turn over rapidly.

ents observed in vitro. The specificity of the centrosome in nucleating microtubules of uniform polarity defines the polarity of the half-spindle, the capture and cap reaction of the kinetochore stabilizes the kinetochore fibers, and the dynamic instability phenomenon implies that the final morphology of the spindle reflects morphogenesis by *stabilization* rather than by nucleation.

SUMMARY

In summary, the dynamic properties of tubulin are manifest most clearly by experiments that look at changes in the length of individual microtubules. These suggest that microtubule populations exist in two populations that interconvert infrequently. Detailed modeling studies by Hill and Chen (1984) suggest that this could be due to a cap of GTP subunits, whose size depends on the relative rates of assembly (a second-order process) and hydrolysis (a first-order process). Hence, the stability of the polymer is a function of the rate of polymerization and the tubulin concentration. Such a model can explain the transient but persistent growth of microtubules below the critical concentration and suggests that it is the ability of the centrosome to continually nucleate microtubules that allows it to produce an ensemble of microtubules well below the critical concentration of tubulin. Such a model can be used to explain the rapid reorganization of microtubules in cells, and particularly in the prophase-prometaphase transition. The models presented here emphasize the microtubule ends as controlling the stability, growth, or depolymerization of individual polymers. Our studies of centrosomes and kinetochores indicate that nucleation and capture are two specific interactions of the microtubules that lead to transient growth or specific stabilization of microtubules. It is expected that other factors that modify the ends of microtubules will be important in controlling the morphology of the microtubule arrays in cells.

ACKNOWLEDGMENTS

We thank T. Hill and U. Euteneuer for helpful discussion, S. Blose for the gift of anti-tubulin, and C. Cunningham-Hernandez for help in preparing the manuscript. This work was supported by grants from the National Institutes of Health and the American Cancer Society.

REFERENCES

Arai, T. and Y. Kaziro. 1976. Effect of guanine nucleotides on the assembly of brain microtubules: Ability of 5′-guanylyl imidodiphosphate to replace GTP in promoting polymerization of microtubules in vitro. *Biochem. Biophys. Res. Commun.* **69:** 369.

Bergen, L. and G.G. Borisy. 1980. Head to tail polymerization of microtubules in vitro. *J. Cell Biol.* **84:** 141.

Bergen, L., R. Kuriyama, and G.G. Borisy. 1980. Polarity of microtubules nucleated by centrosomes and chromosomes of CHO cells in vitro. *J. Cell Biol.* **84:** 151.

Brinkley, B.R., A.M. Fuller, and D.P. Highfield. 1976. Tubulin antibodies as probes for microtubules in dividing and nondividing mammalian cells. *Cold Spring Harbor Conf. Cell Proliferation* **3:** 435.

Brinkley, B.R., S.M. Cox, D.A. Pepper, L. Wible, S.C. Brenner, and R.L. Pardue. 1981. Tubulin assembly sites and the organization of cytoplasmic microtubules. *J. Cell Biol.* **90:** 554.

Carlier, M.-F. and D. Pantaloni. 1981. Kinetic analysis of GTP hydrolysis associated with tubulin polymerization. *Biochemistry* **20:** 1918.

Carlier, M.-F., T. Hill, and Y.-D. Chen. 1984. Interference of GTP hydrolysis in microtubule assembly: An experimental study. *Proc. Natl. Acad. Sci.* **81:** 771.

Caspar, D. and A. Klug. 1963. Physical principles of the construction of regular viruses. *Cold Spring Harbor Symp. Quant. Biol.* **27:** 271.

DeBrabander, M., G. Geuens, J. DeMey, G. Nuydens, and R. Willebrords. 1980. The microtubule nucleating and organizing activity of kinetochores and centrosomes in living PtK_2 cells. In *Microtubules and microtubule inhibitors* (ed. M. DeBrabander and J. DeMey), p. 255. Elsevier/North-Holland, Amsterdam.

DeBrabander, M., G. Geuens, R. Nuydens, R. Willebrords, and J. DeMey. 1981. Microtubule assembly in living cells after release from nocodazole block: The effects of metabolic inhibitors taxol and pH. *Cell Biol. Int. Rep.* **5:** 913.

Euteneuer, U. and J.R. McIntosh. 1981. Structural polarity of kinetochore microtubules in PtK_1 cells. *J. Cell Biol.* **89:** 338.

Gould, R.R. and G.G. Borisy. 1977. The pericentriolar material in CHO cells nucleates microtubule formation. *J. Cell Biol.* **73:** 601.

Heidemann, S.R. and J.R. McIntosh. 1980. Visualization of the structural polarity of microtubules. *Nature* **286:** 517.

Hill, T.L. 1980. Bioenergetic aspects and polymer length distribution in steady state head to tail polymerization of actin or microtubules. *Proc. Natl. Acad. Sci.* **77:** 4803.

Hill, T.L. and M.-F. Carlier. 1983. Steady state theory of the interference of GTP hydrolysis in microtubule assembly. *Proc. Natl. Acad. Sci.* **80:** 7234.

Hill, T.L. and Y.-D. Chen. 1984. Phase changes at the end of a microtubule with a GTP cap. *Proc. Natl. Acad. Sci.* (in press).

Hill, T.L. and M.W. Kirschner. 1982. Regulation of microtubule and actin filament assembly-disassembly. *Int. Rev. Cytol.* **84:** 185.

Johnson, K.A. and G.G. Borisy. 1977. Kinetic analysis of microtubule self assembly in vitro. *J. Mol. Biol.* **117:** 1.

Karsenti, E., S. Kobayashi, T. Mitchison, and M.W. Kirschner. 1984a. Role of the centrosome in organizing the interphase microtubule array: Properties of cytoplasts containing or lacking centrosomes. *J. Cell Biol.* **98:** 1763.

Karsenti, E., J. Newport, R. Hubble, and M.W. Kirschner. 1984b. The interconversion of metaphase and interphase microtubule arrays in *Xenopus* eggs. *J. Cell Biol.* **98:** 1730.

Kirschner, M.W. 1980. Implications of treadmilling for the stability and polarity of actin and tubulin polymers *in vivo*. *J. Cell Biol.* **86:** 330.

Kuriyama, R. and G.G. Borisy. 1981. Microtubule nucleating activity of centrosomes in CHO cells. *J. Cell Biol.* **91:** 822.

Lewis, C.D. and U.K. Laemmli. 1982. Higher order metaphase chromosome structure. *Cell* **29:** 171.

Margolis, R.L. and L. Wilson. 1978. Opposite end assembly disassembly of microtubules at steady state in vitro. *Cell* **13:** 1.

McGill, M. and B.R. Brinkley. 1975. Human chromosomes and centrioles act as microtubule nucleating sites. *J. Cell Biol.* **67:** 189.

Mitchison, T.J. and M.W. Kirschner. 1984a. Microtubule assembly nucleated by isolated centrosomes. *Nature* (in press).

———. 1984b. Dynamic instability of microtubule growth. *Nature* (in press).

Oosawa, F. and S. Asakura, eds. 1975. *Thermodynamics of the polymerization of protein.* Academic Press, New York.
Osborn, M. and K. Weber. 1976. Cytoplasmic microtubules grow from an organizing center to the plasma membrane. *Proc. Natl. Acad. Sci.* **73:** 867.
Penningroth, S.M., D.W. Cleveland, and M.W. Kirschner. 1976. In vitro studies of the regulation of microtubules. *Cold Spring Harbor Conf. Cell Proliferation* **3:** 1233.
Pickett-Heaps, J.D., D.H. Tippit, and K.R. Porter. 1982. Rethinking mitosis. *Cell* **29:** 729.
Reider, L.L. and G.G. Borisy. 1981. The attachment of kinetochores to the prometaphase spindle in PtK_1 cells. *Chromosoma* **82:** 693.
Ring, D., R. Hubble, D. Caput, and M.W. Kirschner. 1980. Isolation of microtubule organizing centers from mouse neuroblastoma cells. In *Microtubule inhibitors* (ed. M. De-Brabander and J. DeMay), p. 297. Elsevier/North-Holland, Amsterdam.
Salmon, E.D., D. Goode, T.K. Margol, and D.B. Bonar. 1976. Pressure-induced depolymerization of spindle microtubules. III. Differential stability in HeLa cells. *J. Cell Biol.* **69:** 443.
Snyder, J.A. and J.R. McIntosh. 1975. Initiation and growth of microtubules from mitotic centers in lysed mammalian cells. *J. Cell Biol.* **73:** 601.
Spiegelman, B.M., S.M. Penningroth, and M.W. Kirschner. 1977. Turnover of tubulin and the N site GTP in CHO cells. *Cell* **12:** 587.
Summers, K.R. and M.W. Kirschner. 1979. Characterization of the polar assembly and disassembly of microtubules observed *in vitro* by darkfield light microscopy. *J. Cell Biol.* **83:** 205.
Telzer, B.R. and J.L. Rosenbaum. 1979. Cell cycle dependent in vitro assembly of microtubules onto the pericentriolar material of HeLa cells. *J. Cell Biol.* **81:** 484.
Terry, B.J. and D.L. Purich. 1979. Nucleotide release from tubulin and NDP kinase action in microtubule assembly. *J. Biol. Chem.* **254:** 9469.
Thompson, W.C., D.J. Asai, and D.H. Carney. 1984. Heterogeneity among microtubules detected by a monoclonal antibody to alpha tubulin. *J. Cell Biol.* **98:** 1017.
Wegner, A. 1976. Head to tail polymerization of actin. *J. Mol. Biol.* **108:** 139.
Weisenberg, R.C. 1972. Microtubule formation in vitro in solutions containing low calcium concentrations. *Science* **177:** 1104.
Weisenberg, R.C. and W.J. Deery. 1976. Role of nucleotide hydrolysis in microtubule assembly. *Nature* **263:** 792.
Weisenberg, R.C., G.G. Borisy, and E.W. Taylor. 1968. The colchicine binding protein of mammalian brain and its relation to microtubules. *Biochemistry* **7:** 4466.
Witt, P.L., H. Ris, and G.G. Borisy. 1980. Origin of kinetochore microtubules in CHO cells. *Chromosoma* **81:** 483.

Dynamic Properties of Cytoskeletal Proteins in Focal Contacts

Thomas E. Kreis
European Molecular Biology Laboratory
D-6900 Heidelberg, Federal Republic of Germany

Zafrira Avnur, Joseph Schlessinger, and Benjamin Geiger
Department of Chemical Immunology
Weizmann Institute of Science
IL-76100 Rehovot, Israel

A number of physiological cellular processes, including cell motility and the control of anchorage-dependent cell growth and differentiation, depend on the formation of contacts between cells and extracellular substrata (Hay and Meier 1976; Folkman and Moscona 1978; Gospodarowicz et al. 1978; Ben Ze'ev et al. 1980). Most prominent, both in intact tissues and in cultured cells, are adherens-type junctions in which actin is attached to the inner side of the membrane through a vinculin-rich plaque (Geiger 1982). It has been proposed that the transmembrane anchorage in this area is triggered by local contacts, which lead to the reorganization (immobilization or clustering) of membrane receptors. This in turn induces local attachment of vinculin, and later of actin, to the membrane (Geiger 1983; Geiger et al. 1984). It is assumed that the assembly of the various junctional components, including proteins of the contact-specific cytoskeletal microdomains, plays a major role in the process of contact formation and stabilization and in the transmission of signals from the extracellular compartment through the plasma membrane to the cytoplasm.

As a model for such adherens-type junctions, we have chosen to study focal contacts (FC) of tissue-cultured cells. FC are the sites of the closest (10–20 nm) and possibly the strongest attachment of the cell membrane to the tissue-culture substrate (Abercrombie and Dunn 1975; Izzard and Lochner 1976, 1980; Grinnell 1978). Electron microscopic inspection revealed that microfilament bundles terminate in FC. Moreover, immunocytochemical analysis revealed that several microfilament-associated proteins, including actin, α-actinin, and vinculin, appear to be enriched near the endofacial surface of this particular membrane domain (Abercrombie et al. 1971; Heath and Dunn 1978; Geiger 1979, 1983; Kreis et al.

1979; Lazarides and Burridge 1979; Wehland et al. 1979). To this end, most of our knowledge about dynamic processes occurring within FC originated from studies utilizing drugs or immunocytochemical experiments performed on fixed cells.

Within the last few years, microneedle injection of fluorescently tagged proteins made it possible to visualize cytoskeletal structures within living cultured cells (for review, see Kreis and Birchmeier 1982). Recently, we have combined this microinjection technique with fluorescence photobleaching recovery (FPR) measurements (Schlessinger and Elson 1982). This combination enabled us to determine the dynamic properties of cytoskeletal structures such as microfilaments and microtubules within living cells (Kreis et al. 1982a,b; Scherson et al. 1984). These experiments indicated that a dynamic equilibrium exists between a soluble cytoplasmic pool of actin as well as of microtubule-associated proteins with their respective cytoskeleton-bound fractions.

In this study, we have applied FPR measurements on microinjected cells as well as in-situ-binding experiments to investigate in vivo the dynamic properties of the three major cytoskeletal components of FC, namely, actin, α-actinin, and vinculin. Our principal purpose was to analyze the mobilities of these different proteins within the junctional area and to shed light on the complexity of their assembly and exchange processes. It was anticipated that these data might help to clarify the dynamic organization of FC at a molecular level.

Distribution of Fluorescently Tagged Actin, α-Actinin, and Vinculin in Microinjected Living Chicken Gizzard Cells

Purified actin, α-actinin, and vinculin were labeled with rhodamine lissamine sulfonyl chloride (Fig. 1) and microinjected into living embryonic chicken gizzard cells. Microscopic analysis indicated that all three proteins became incorporated

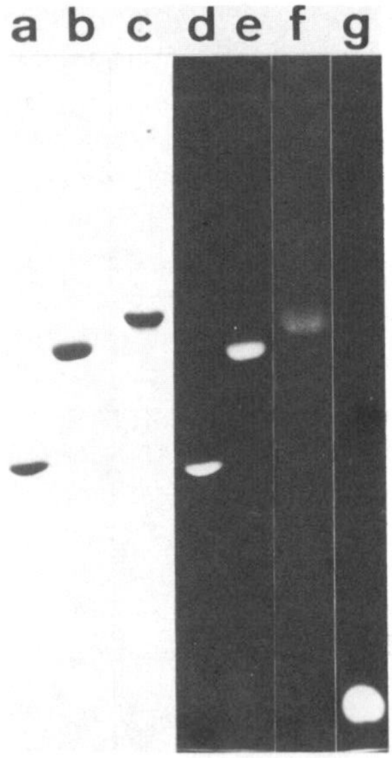

Figure 1
SDS-polyacrylamide gel electrophoresis of rhodamine-labeled proteins. (*a,d*) Actin isolated from rabbit skeletal muscle; (*b,e*) α-actinin; (*c,f*) vinculin from chicken gizzard; (*g*) free fluorochrome. Left three lanes are from the Coomassie-blue-stained gel; the others show the fluorescence of the corresponding unstained gel, as seen under UV light.

into specific cellular structures within 5–20 minutes after injection (Fig. 2). Actin was primarily associated with stress fibers that terminated in FC (Fig. 2a, arrowheads). α-Actinin was also associated with FC as well as with regularly spaced striations on stress fibers (Fig. 2b). Rhodamine-derivatized vinculin specifically associated with FC (Fig. 2c). Double-labeling experiments of injected cells with antibodies directed against the particular injected proteins coupled to different fluorophores indicated that the two fluorescence patterns were essentially identical, suggesting that the injected proteins became incorporated into the endogenous pool of the respective proteins and that the two were codistributed within the cells.

The amounts of microinjected proteins were quite low and did not exceed 5–10% of the respective endogenous protein. Time-lapse recording of recipient cells with a sensitive video-intensification system revealed that the microinjected cells were capable of mitosis (Fig. 3a,b) and normal motility (Figs. 3c–f and 4) and that the modified proteins participated also in the de novo formation of FC (Fig. 3f, arrowheads) and stress fibers (Fig. 4) throughout these processes. It seems that the microinjected proteins were stable within the recipient cells, as their typical patterns of distribution could usually still be visualized up to 50 hours after microinjection. From these observations, we concluded that microinjected actin, α-actinin, and vinculin readily incorporated into the cellular pool of endogenous protein and thus faithfully represent the endogenous pools of these proteins.

Using the optical system set for FPR measurements, we have tried to obtain an estimate for the packing density of the three proteins in or near FC. For this purpose, we have measured the net fluorescence intensity of each protein in FC areas and compared it to the concentration of a standard solution of the microinjected proteins. To calculate the local concentrations, we had to make a series of assumptions as to the exact volume of injected solution, the relative overall cytoplasmic concentrations of each protein, the fine geometry of FC, and the fraction of biologically active material in the injected solution. Since direct measurement of most of these properties on a single-cell basis is not available, the calculated values should be considered as a rough estimate only.

The highest packing density was found for actin with a calculated number of 1.1×10^6 molecules/μm^2 FC area. In view of the diameter of monomeric G-actin (~5.5 nm), this estimate implies that actin is forming a multilayered structure near FC, a notion that is certainly in agreement with the known structure of stress fibers. Based on similar measurements and on the same considerations, the approximate packing density of α-actinin and vinculin could be calculated. The values obtained were $3.5 \times 10^4/\mu m^2$ and $2 \times 10^4/\mu m^2$, respectively. Knowing the molecular dimensions of α-actinin and vinculin, this implies that the two proteins may be present in FC in several (2–3) layers. These numbers are greatly affected by the various assumptions mentioned above, and thus should be regarded as preliminary indications until more data are available.

Reorganization of Microinjected Proteins in Areas of FC of Living Cells

To determine microscopically the dynamic rearrangements of the cytoskeletal elements of FC in vivo, we have examined microinjected cells during cytokinesis

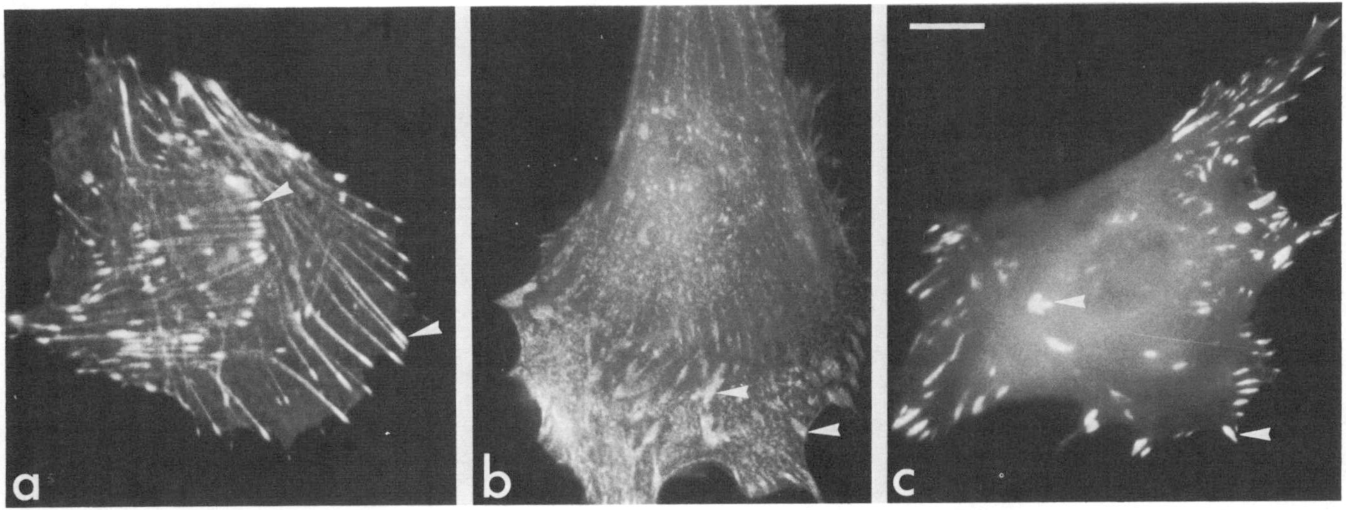

Figure 2
Distribution of rhodamine-labeled proteins in living chicken gizzard cells 60 min after microinjection. The cells were microinjected with rhodamine-actin (*a*), rhodamine-α-actinin (*b*), and rhodamine-vinculin (*c*). Arrowheads indicate focal contacts. Bar, 20 μm.

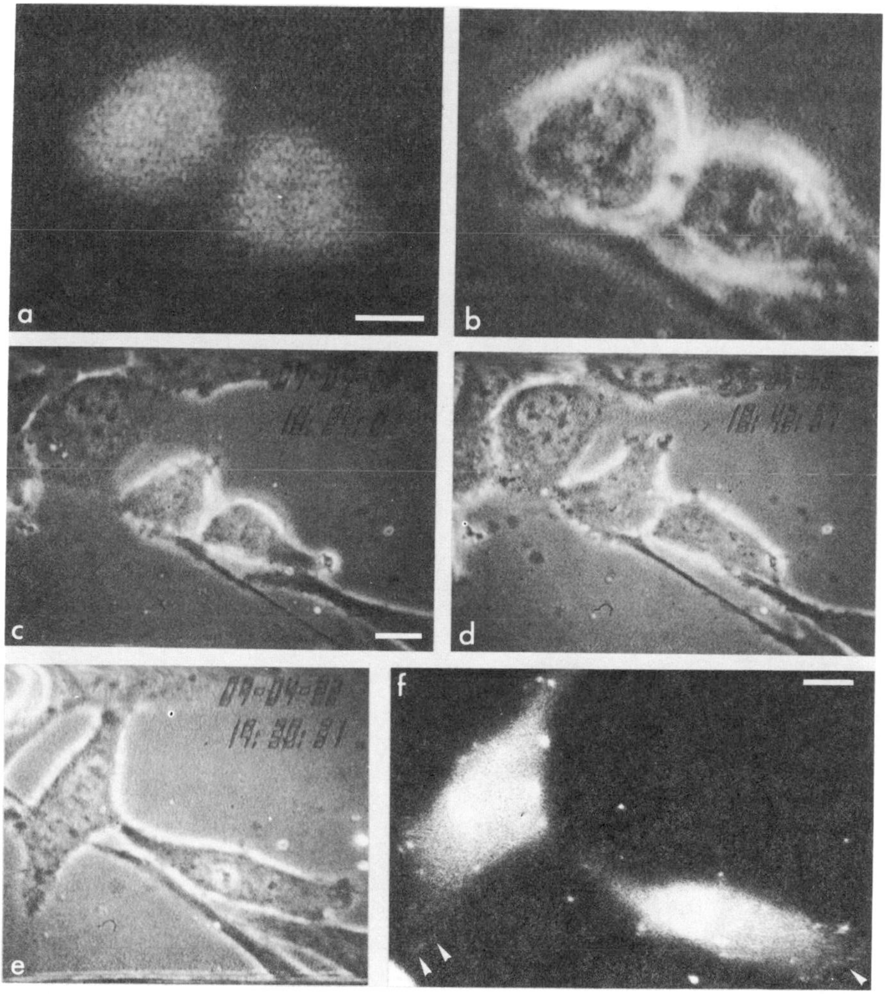

Figure 3
Video-intensified time-lapse cinematography of cytokinesis of a cell microinjected 3 hr prior to mitosis with rhodamine-labeled vinculin. The two daughter cells at the beginning of cytokinesis (*a,b*) and 14 (*c*), 28 (*d*), and 61 (*e,f*) min later. (*b–e*) Cells visualized by phase-contrast microscopy; (*a,f*) distribution of rhodamine-vinculin as seen by fluorescence microscopy. Arrowheads indicate focal contacts. Bar, 20 μm.

and locomotion by video-intensified time-lapse cinematography. Figure 3 shows a pair of chicken gizzard daughter cells derived from a parental cell that was microinjected with rhodamine-labeled vinculin 3 hours before it entered mitosis. As the daughter cells started to respread on the substrate, many small vinculin-rich plaques were noticed in these cells, probably corresponding to newly formed FC (Fig. 3f, arrowheads). This indicated that the injected vinculin was readily incorporated into the nascent FC.

Another example of FC reorganization was found in motile cells microinjected

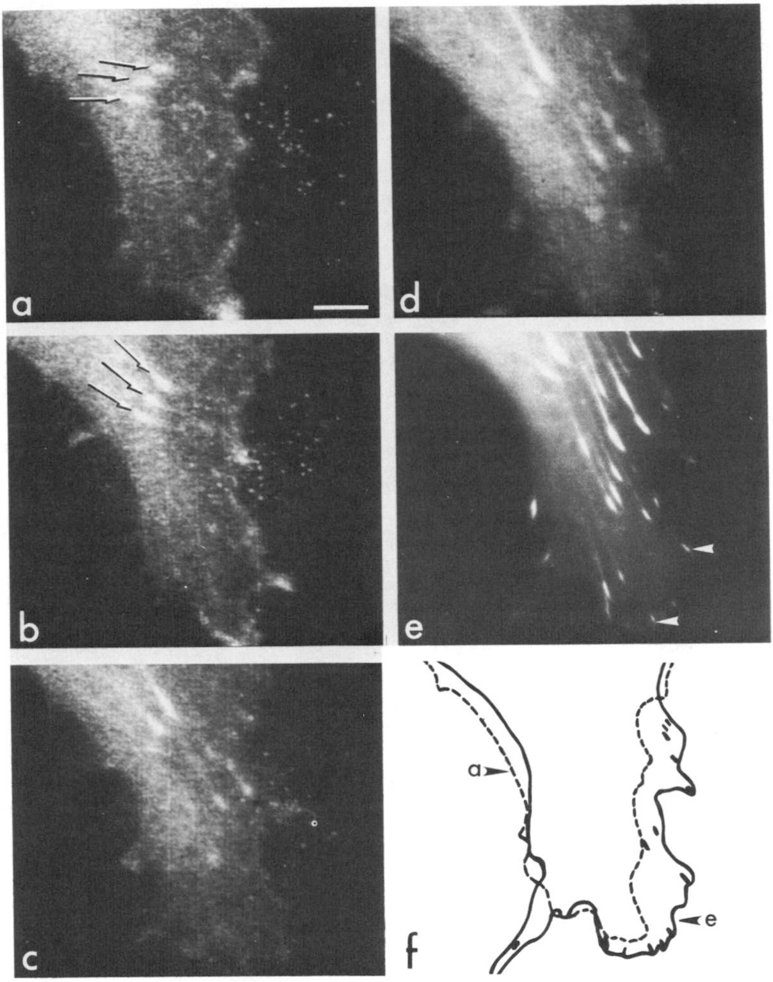

Figure 4
Video-intensified time-lapse cinematography of a motile chicken gizzard cell microinjected with rhodamine-labeled actin. (*a–d*) Fluorescence-intensified images photographed from the TV monitor with a polaroid camera. Patterns of rhodamine-actin 1 hr after microinjection at the beginning of the recording (*a*) and 15 (*b*), 35 (*c*), and 45 (*d*) min later. The same cell was fixed 60 min after initiation of the experiment and photographed in the photomicroscope without intensification (*e*). (*a,b*) Group of three FC marked with arrows to indicate rearrangement of FC; (*f*) borders of the cell shown in *a* (dashed line) and *e* (solid line) superimposed to illustrate locomotion of the cell and formation of new FC (indicated by arrowheads in *e*). Bar, 10 μm.

with rhodamine-labeled actin and recorded over a period of about 60 minutes (Fig. 4a–d). The cells were then fixed and photographed directly on film to obtain a higher level of resolution and contrast than that obtained by video intensification (Fig. 4e). Using marker points on the substrate, we could show that the

Table 1
Mobility of Rhodamine-labeled Actin, α-Actinin, and Vinculin in Microinjected Chicken Gizzard Fibroblasts as Measured by Fluorescence Photobleaching Recovery

Protein	*Cellular domain*	*Diffusion coefficient ($D[cm^2/sec] \times 10^{-9}$)*	*Fractional recovery (%)*	*Rate of slow recovery ($\tau/2$;min)*[a]
Actin	FC area	3.1 ± 1.1 (30)[b]	18 ± 7	4.1 ± 2.8 (23)
Actin	interfibrillary	3.2 ± 1.2 (35)	65 ± 13	n.d.
α-Actinin	FC area	2.8 ± 1.0 (10)	40 ± 10	2.7 ± 1.2 (6)
α-Actinin	interfibrillary	2.5 ± 1.4 (12)	76 ± 5	n.d.
Vinculin	FC area	3.5 ± 1.2 (30)	43 ± 8	2.1 ± 0.9 (15)
Vinculin	interfibrillary	2.9 ± 1.1 (30)	>80	n.d.
Bovine serum albumin	perinuclear	6.3 ± 0.9 (7)	>90	
Goat immunoglobulin	perinuclear	6.6 ± 0.7 (12)	>90	

[a]n.d. indicates not determined.
[b]Number of determinations.

leading lamella of this cell was clearly advancing during that period (Fig. 4f). Two significant observations were made in these experiments: (1) Several new FC formed at the ventral surfaces of the advancing lamella (Fig. 4e, arrowheads). (2) Several FC clearly changed their orientations (Fig. 4a,b, arrows). This latter observation was somewhat surprising, since the common view is that FC are rather static relative to the substrate. We propose that reorientation of FC is a manifestation of coordinated and directional assembly and disassembly of FC, namely, growth in one region of a FC and decrease in size in a different region. This may lead to an apparent "turning" or "sliding" of the region itself.

Dynamic Properties of Actin, α-Actinin, and Vinculin in FC as Measured by FPR Experiments

To analyze the dynamic properties of FC-associated proteins within living cells, we have applied the technique of FPR to cells microinjected with fluorescently tagged actin, α-actinin, and vinculin. Details concerning the FPR measurements and interpretation of the FPR results were performed as described and discussed in more detail elsewhere (Kreis et al. 1982a, Geiger et al. 1984).

Recovery of fluorescence after bleaching of labeled actin, α-actinin, and vinculin in FC occurred in two distinct phases: An initial fast recovery which reached a plateau about 10 seconds after bleaching, and a slow phase of fluorescence recovery with a halftime ($\tau/2$) on the order of minutes. The rates of the fast and slow recoveries measured for the three FC-bound proteins are summarized in Table 1. We have found that the fast-recovery fraction exhibited a similar diffusion coefficient for all three proteins studied on the order of $D \cong 3 \times 10^{-9}$ cm^2/sec. Nevertheless, significant differences were detected in the rates of slow recovery. Actin in FC exhibited the slowest recovery with an average $\tau/2$ of 4.1 minutes, which was considerably larger than that obtained for α-actinin and vinculin.

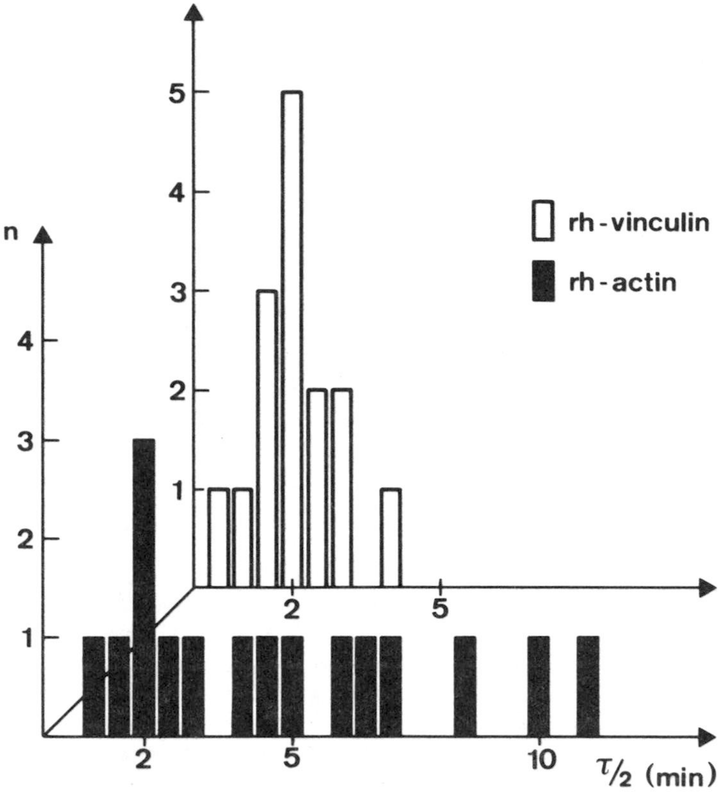

Figure 5
Histogram of the distribution of halftimes (τ/2) of slow recovery of fluorescence after photobleaching, measured for rhodamine-actin and vinculin in FC of living chicken gizzard cells. *n* Indicates the number of values, determined in separate measurements, with identical τ/2.

These FPR experiments indicated that each of the three proteins exists in two major forms: a diffusible fraction, which seems to be uniformly distributed throughout the cells, and an "immobile" fraction associated with the FC (actin and α-actinin are also associated with other cellular domains as previously shown) (Kreis et al. 1982a; Geiger et al. 1984). The quantitative relationships between the two pools cannot easily be assessed, since we do not know what fraction within the diffusible pool is biologically active. The FPR experiments indicated that there is a constant exchange of vinculin, α-actinin, and actin between the soluble pool and the FC-bound fraction. This conclusion was based on the finding that there was a distinct slow recovery of fluorescence for each of the three labeled proteins. The rate of this recovery was not affected by alteration of the size of the bleaching beam, indicating that this recovery reflected an exchange process between the diffuse pool of the protein and the FC. Similar results were obtained for actin in stress fibers (Kreis et al. 1982a) and microtubule-associated proteins in microtubules (Kreis et al. 1982b; Scherson et al. 1984). Interestingly, the variability in the rate of slow recovery of actin measured in different FC was significantly larger than that of vinculin (Fig. 5). This finding suggested that FC-bound actin is highly heterogeneous in its kinetic properties, in contrast to the more homogeneous pattern measured for vinculin. The molecular basis for this heterogeneity is not clear. Moreover, there was no correlation between the size of a FC or its labeling intensity and the kinetics of slow recovery.

Binding of Vinculin to FC of Isolated Substrate-attached Membranes

To characterize the requirements for the binding of soluble vinculin to FC further, we have used in situ "decoration" experiments similar to those described previously (Avnur et al. 1983). Cells of various types, derived from a variety of species, were treated with $ZnCl_2$, and their intact ventral membranes were isolated as described previously (Avnur and Geiger 1981; Avnur et al. 1983). The membranes were incubated with rhodamine- or fluorescein-labeled vinculin and then fixed and immunolabeled for a variety of other cytoskeletal proteins using a different fluorophore. Examples for such experiments are shown in Figure 6. Here, ventral membranes of cultured NRK cells were incubated with fluorescein-labeled vinculin (Fig. 6a) and counterlabeled with rhodamine-labeled phalloidin (Fig. 6c). As seen, the labeled vinculin bound specifically to FC at the termini of actin bundles.Those areas were positively identified as FC by interference-reflection optics (Fig. 6e). The binding of labeled vinculin to FC occurred under a large variety of environmental conditions, including broad pH range (5.8–7.8), ionic strength (20–300 mM KCl), $CaCl_2$ (0–10 mM), and $MgCl_2$ (0–20 mM).

An interesting and surprising observation was made when we tried to compete out the incorporation of labeled vinculin using an excess of unlabeled protein. Unlike the results obtained with α-actinin, addition of unlabeled vinculin in concentrations of up to 1 mg/ml had only a limited effect on the binding of the fluorescently labeled proteins (the latter were added in concentrations of 20–50 μg/ml). Nevertheless, it should be emphasized that this binding to membranes or to detergent-permeabilized cells (Avnur et al. 1983) was highly spatially specific. A possible interpretation of the inhibition results might be that binding of vinculin to FC is *not* a simple saturable process and that vinculin might undergo self-aggregation in the contact area.

Another relevant aspect approached by the decoration experiments was related to vinculin in the contact areas of transformed cells. It has been previously shown that, upon transformation of cells with Rous sarcoma virus (RSV), most FC deteriorate and that the cytoskeletal proteins of the plaque are reorganized into clusters of dots or "rosettes" (David-Pfeuty and Singer 1980; Rohrschneider 1980; Rohrschneider et al. 1982). The possibility was raised that the phosphorylation of vinculin on tyrosyl residues might alter its mode of binding to FC (Sefton et al. 1981). To examine directly the binding of vinculin to the rosettes in RSV-transformed cells, we have isolated ventral membranes by the $ZnCl_2$ method from a transformed rat cell line NB77 (derived from NRK, infected with RSV). As shown in Figure 6b,d,f, the binding of the normal chick protein occurred in the rosettes that are typical of the transformed state. Furthermore, the binding of the exogenous protein to the modified contacts showed the same characteristics as

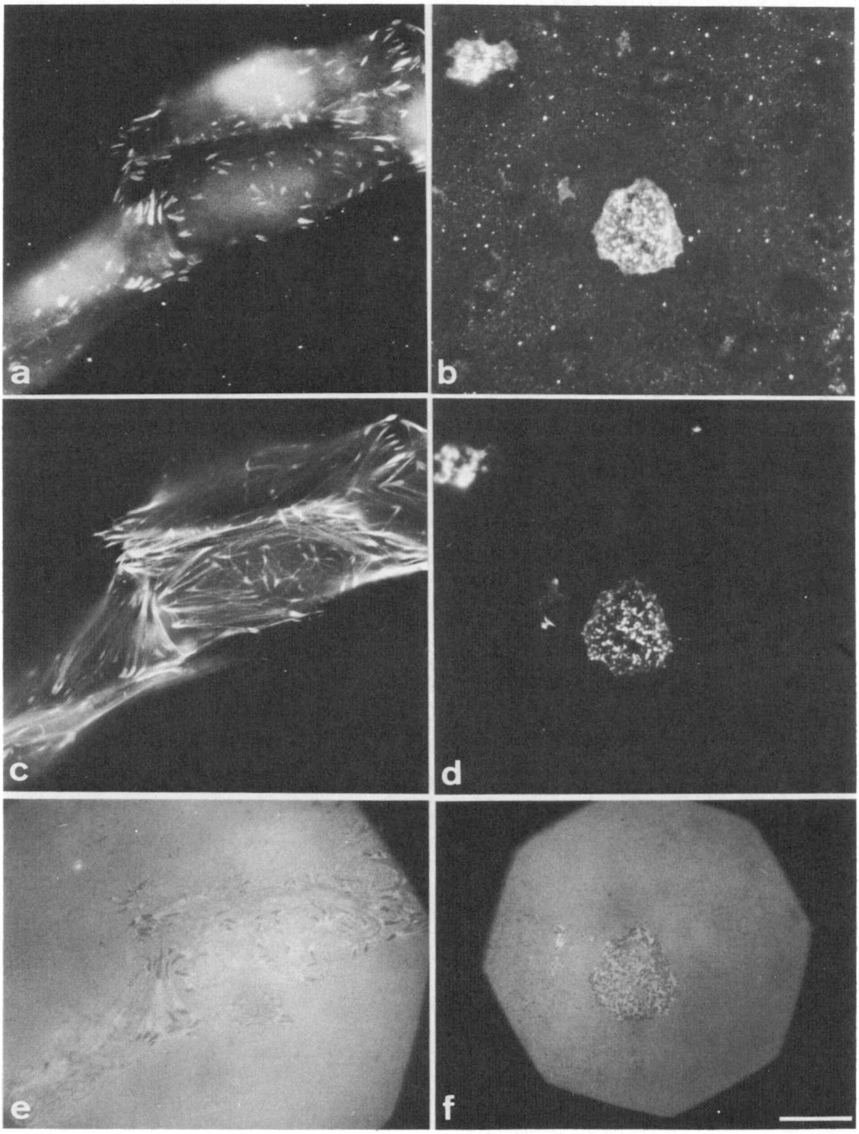

Figure 6

Binding of vinculin to FC of isolated substrate-attached membranes. Intact ventral membranes of $ZnCl_2$-treated NRK cells (*a,c,e*) or NB77, an RSV-transformed NRK cell line (*b,d,f*), were incubated with fluorescein-labeled vinculin (*a,b*) and counterlabeled with rhodamine-phalloidin (*c,d*). FC of the labeled ventral membranes were visualized by interference-reflection microscopy (*e,f*). Bar, 20 μm.

the binding to authentic FC: It occurred also at low temperatures (0–4°C) and did not require exogenous ATP. We cannot completely rule out the possibility that the exogenous protein was phosphorylated by membrane-bound protein kinase, but in view of the insensitivity of the binding under a variety of buffer conditions, including low temperature or ATP- and Ca^{++}-free buffers, this possibility seems unlikely.

CONCLUSIONS

The experiments described here elucidate some of the dynamic molecular features of FC-bound cytoskeletal components. Our observations were based on two complementary approaches: microinjection of fluorescently labeled proteins in combination with FPR measurements and in-situ-binding assays to isolated membranes.

FPR measurements on microinjected cells provide a useful approach for the determination of the dynamic properties of specific molecules in defined microdomains within the cytoplasm. Obviously, the reliability of the conclusions drawn from such experiments depends on the validity of several important assumptions: (1) The microinjected, fluorescently labeled, proteins should have properties identical to those of their native counterparts; (2) microinjection itself should not affect the native organization and the dynamic properties of the various proteins or the cytoplasmic structures with which they are associated; and (3) the laser beam used for fluorescence photobleaching should not damage the labeled cells nor alter their behavior. Evidence for the validity of these assumptions, in addition to those presented here, has been discussed previously (Kreis and Birchmeier 1980, 1982; Kreis et al. 1982a; Blose et al. 1984; Geiger et al. 1984).

The amounts of the various proteins injected into the cells were relatively small and did not greatly affect the amounts of the respective endogenous proteins. The experiments could be performed a long time after injections were administered (more than a day), after which an equilibrium was reached between the injected protein and the cellular pools. Moreover, the capacity of cells to move or divide was not significantly affected.

Although possible limited photodamage to the cells due to photobleaching cannot be ruled out at present, many precautions were taken to render this possibility unlikely. We have limited the intensity of the beam and have performed all experiments under low levels of illumination using the video-intensified camera.

On the basis of FPR experiments, several conclusions can be proposed: (1) All three proteins tested, namely, actin, α-actinin, and vinculin, maintain an exchange between a diffusible cytoplasmic pool and a FC-bound immobile fraction. (2) Although the rate of diffusion in the cytoplasm is similar for all three proteins, the rate of exchange (slow recovery) is different: Actin usually showed a significantly slower recovery than α-actinin and vinculin. (3) From these differences and from the results of previous experiments, it appears that the exchange of α-actinin and vinculin does not depend on the mobility of actin nor is it restricted by it. (4) The relatively rapid exchange indicates that the shape and size of FC may be controlled by the exchange between the soluble and junctional pools. Thus, increased association will lead to overall growth or recession of the FC. A localized change in the rate of exchange in different regions of a single FC can lead to alterations in its shape and orientation and to an apparent turning or sliding movement as discussed above. (5) The decoration experiments with isolated membranes indicated that vinculin-specific binding sites exist in FC and that the binding is not readily saturable. This finding is compatible with the suggestion that vinculin may undergo self-aggregation and form a multilayered plaque at the cytoplasmic faces of the membrane. Previous results (Avnur et al. 1983) indicated that the binding of vinculin to the plaque was largely actin-independent. (6) The specific association of vinculin with contacts was found in membranes prepared from a large variety of cells including epithelial and fibroblastic cells, normal and transformed, and in cells derived from a variety of species (chick, rat, mouse, bovine, and human). This has suggested that the vinculin-binding sites in the contact are evolutionarily conserved. It might be added that most antibodies prepared against vinculin exhibited a much narrower species specificity.

A major question that is still unresolved concerns the nature of the vinculin-binding sites in the plaque or in the junctional membrane. Identification of such components, determination of their relationships to "contact receptors" specific for FC, and their manipulations using approaches such as those described here may shed light on the molecular steps involved in the biogenesis of cell contacts.

ACKNOWLEDGMENTS

We thank Dr. T. Wieland for the generous gift of rhodamine-phalloidin and Dr. S. Pfeiffer for critically reading the manuscript. T.E.K. was a recipient of an EMBO long-term fellowship. This work was supported by grants from the Muscular Dystrophy Association (B.G.), from the National Institutes of Health (J.S.), and from the U.S.-Israel Binational Science Foundation (J.S.). B.G. is an incumbent of the C. Revson chair in biology.

REFERENCES

Abercrombie, M.J. and G.A. Dunn. 1975. Adhesions of fibroblasts to substratum during contact inhibition observed by interference reflection microscopy. *Exp. Cell Res.* **92:** 57.

Abercrombie, M.J., E.M. Heaysman, and S.M. Pegrum. 1971. The locomotion of fibroblasts in culture. IV. Electron microscopy of the leading lamella. *Exp. Cell Res.* **67:** 359.

Avnur, Z. and B. Geiger. 1981. Substrate-attached membranes of cultured cells: Isolation and characterization of ventral cell membranes and the associated cytoskeleton. *J. Mol. Biol.* **153:** 361.

Avnur, Z., J.V. Small, and B. Geiger. 1983. Actin independent association of vinculin with the cytoplasmic aspects of the plasma membrane in cell-substrate contacts. *J. Cell Biol.* **96:** 1622.

Ben Ze'ev, A., S.R. Farmer, and S. Penman. 1980. Protein synthesis requires cell-surface contact while nuclear events respond to cell-shape in anchorage-dependent fibroblasts. *Cell* **21:** 365.

Blose, S.H., D.I. Meltzer, and J.R. Feramisco. 1984. 10 nm filaments are induced to collapse in living cells microinjected with monoclonal and polyclonal antibodies against tubulin. *J. Cell Biol.* **98:** 847.

David-Pfeuty, T. and S.J. Singer. 1980. Altered distributions of the cytoskeletal proteins, vinculin, and α-actinin in cultured fibroblasts transformed by Rous sarcoma virus. *Proc. Natl. Acad. Sci.* **77:** 6687.

Folkman, J. and A. Moscona. 1978. Role of cell shape in growth control. *Nature* **273:** 345.

Geiger, B. 1979. A 130K protein from chicken gizzard: Its localization at the termini of microfilament bundles in cultured chicken cells. *Cell* **18:** 193.

———. 1982. Involvement of vinculin in contact-induced cytoskeletal interactions. *Cold Spring Harbor Symp. Quant. Biol.* **46:** 671.

———. 1983. Membrane-cytoskeleton interaction. *Biochim. Biophys. Acta* **737:** 305.

Geiger, B., Z. Avnur, T.E. Kreis, and J. Schlessinger. 1984. The dynamics of cytoskeletal organization in areas of cell contact. *Cell and muscle motility* (ed. J.W. Shay), vol. 5, p. 195. Plenum Press, New York.

Gospodarowicz, D., G. Green, and C.R. Birdwell. 1978. Determination of cellular shape by the extracellular matrix and its correlation with the control of cellular growth. *Cancer Res.* **38:** 4155.

Grinnell, F. 1978. Cellular adhesiveness and extracellular substrata. *Int. Rev. Cytol.* **53:** 65.

Hay, E.D. and S. Meier. 1976. Stimulation of corneal differentiation by interaction between cell surface and extracellular matrix. II. Further studies on the nature and site of transfilter "induction." *Dev. Biol.* **52:** 141.

Heath, J.P. and G.A. Dunn. 1978. Cell to substrate contacts of chicken fibroblasts and their relation to the microfilament system. A correlated interference-reflexion and high-voltage electron-microscopy study. *J. Cell Sci.* **29:** 197.

Izzard, C.S. and L.R. Lochner. 1976. Cell-to-substrate contact in living fibroblasts. An interference-reflection study with an evaluation of the technique. *J. Cell Sci.* **27:** 129.

———. 1980. Formation of cell-to-substrate contacts during fibroblast motility: An interference-reflection study. *J. Cell Sci.* **41:** 81.

Kreis, T.E. and W. Birchmeier. 1980. Stress fiber sarcomeres are contractile. *Cell* **22:** 555.

———. 1982. Microinjection of fluorescently labeled proteins into living cells, with emphasis on cytoskeletal proteins. *Int. Rev. Cytol.* **75:** 209.

Kreis, T.E., B. Geiger, and J. Schlessinger. 1982a. Mobility of microinjected rhodamine actin within living chicken gizzard cells determined by fluorescence photobleaching recovery. *Cell* **29:** 835.

Kreis, T.E., K.H. Winterhalter, and W. Birchmeier. 1979. In vivo distribution and turnover of fluorescently labeled actin microinjected into human fibroblasts. *Proc. Natl. Acad. Sci.* **76:** 3814.

Kreis, T.E., T. Scherson, U.Z. Littauer, J. Schlessinger, and B. Geiger. 1982b. Dynamics of microtubules in living cultured cells. *J. Cell Biol.* **9512:** 348a.

Lazarides, E. and K. Burridge. 1979. α-Actinin: Immunofluorescent localization of a muscle structural protein in nonmuscle cells. *Cell* **6:** 289.

Rohrschneider, L.R. 1980. Adhesion plaques of Rous sarcoma virus-transformed cells contain the *src* gene product. *Proc. Natl. Acad. Sci.* **77:** 3514.

Rohrschneider, L., M. Rosok, and K. Shriver. 1982. Mechanism of transformation by Rous sarcoma virus: Events within adhesion plaques. *Cold Spring Harbor Symp. Quant. Biol.* **46:** 953.

Sefton, B.M., T. Hunter, E.H. Ball, and S.J. Singer. 1981. Vinculin: A cytoskeletal target of the transforming protein of Rous sarcoma virus. *Cell* **24:** 165.

Scherson, T., T.E. Kreis, J. Schlessinger, U.Z. Littauer, G.G. Borisy, and B. Geiger. 1984. Dynamic interaction of fluorescently labeled microtubule associated proteins in living cells. *J. Cell Biol.* (in press).

Schlessinger, J. and E.L. Elson. 1982. Fluorescence methods for studying membrane dynamics. In *Methods of experimental physics* (ed. H. Ehrenstein and H. Lecar), vol. 20, p. 197. Academic Press, New York.

Wehland, J., M. Osborn, and K. Weber. 1979. Cell-to-substratum contacts in living cells: A direct correlation between interference-reflexion and indirect immunofluorescence microscopy using antibodies against actin and α-actinin. *J. Cell Sci.* **37:** 257.

Expression of a Unique β-Tubulin Variant in Chicken Red-cell Development

Douglas B. Murphy, Kathleen T. Wallis, and William A. Grasser
Department of Cell Biology and Anatomy
Johns Hopkins University School of Medicine
Baltimore, Maryland 21205

Our understanding of the dynamics of microtubule function in complex processes such as mitosis and cell motility has benefited considerably from the biochemical analysis of tubulin and microtubule-associated proteins (MAPs) and the mechanisms that govern their self-assembly in vitro. To understand the mechanisms that control the regulation of microtubule assembly, much interest has been focused on posttranslational modifications of tubulin, such as α-tubulin tyrosinolation (Kumar and Flavin 1982) and flagellar α-tubulin modification (L'Hernault and Rosenbaum 1983), and on MAPs and their phosphorylation (Vallee 1980; see also Vallee et al., this volume). However, with the discovery that each tubulin subunit is encoded by multiple, unique genes in a wide variety of organisms including vertebrates, attention is now being focused on the significance of tubulin gene expression in development (for review, see Cleveland 1983). Raff and co-workers (Kemphues et al. 1982; Raff et al. 1982; see also this volume) demonstrated through mutant analysis in *Drosophila* that different tubulin genes are expressed in somatic, embryonic, and meiotic tissues; Havercroft and Cleveland (see Sullivan et al., this volume) have shown that the expression of the four prominent β-tubulin genes of the chicken are also expressed in a tissue-specific manner. Although the significance of the differential expression of tubulin genes in development is not yet clear, a major goal of our studies, and the idea discussed in this paper, is to determine whether tubulins with different primary structures may assemble and function differentially in the cell.

Fulton and Simpson (1976) and Stephens (1975) were among the first to investigate the idea that different tubulin isoforms may perform different functions in the cell, an idea Fulton termed the "multitubulin hypothesis." Although there have been numerous attempts to demonstrate the unique identities and functions for tubulins from different sources, the differences that have been observed have been slight and their significance in determining microtubule function remains unclear. In this paper, we describe a new β-tubulin variant (tubulin isoform) from chicken erythrocytes that exhibits differences in primary structure and assembly properties in a dramatic way (Murphy and Wallis 1983a,b). The tubulin in these blood cells forms a microtubule bundle adjacent to the plasma membrane, which is thought to be important for the generation and maintenance of asymmetry in cell shape. Marginal bands are characteristic of the nucleated erythrocytes and thrombocytes of most vertebrates, including those of some mammals (Behnke 1970). We purified tubulin from chicken erythrocytes and brain tissue and showed that the β-tubulin subunits have different chemical and physical properties, different peptide maps, and different assembly properties during in vitro polymerization (Murphy and Wallis 1983a,b). The extreme specificity of β-tubulin expression in erythrocytes is indicated by two observations: (1) biochemical differences exist between the β subunits, not between the α subunits, and (2) the variant β subunit is only contained in erythrocytes and thrombocytes of the blood. The

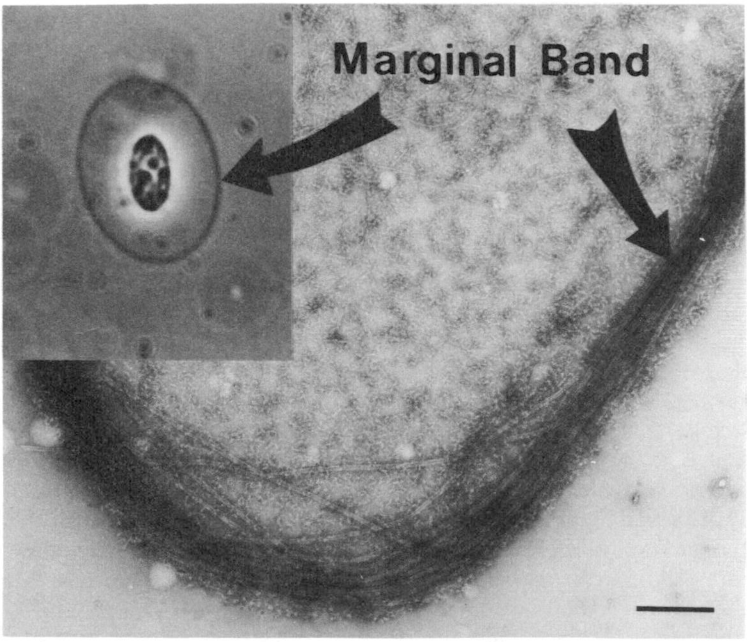

Figure 1
Microtubule bundle (marginal band) in a chicken erythrocyte ghost. The bundle in cross section contains 5–12 microtubules and may consist of just a few microtubules coiled many times around the cell. Erythrocytes were lysed in saline containing 5 mM Mg and 0.5% Triton X-100 and negatively stained with 1% uranyl acetate. Bar, 0.5 μm; magnification, 21,038×. (*Inset*) Erythrocyte ghost with nucleus and marginal band as seen by phase-contrast light microscopy. The cell diameter is 15 μm.

specificity of expression and the differences in peptide maps and amino acid composition suggest that these tubulin isoforms represent different gene products.

In this paper, we show that brain and erythrocyte tubulins display marked differences in their self-assembly in vitro, a fact that may be related to their functions in vivo. We also show that the onset of erythrocyte tubulin synthesis in developing chicken bone marrow is closely tied to the processes of erythropoiesis and cell-shape differentiation. Although not yet proved, it appears likely that the function of the erythrocyte tubulin is related to special requirements for microtubule assembly and function during erythrocyte differentiation. Since both brain and erythrocyte tubulins can be isolated in gram quantities, it should be possible to determine their structures and functions through comparative biochemical analysis and to relate these differences to their tissue-specific expression in development.

DISCUSSION

Chicken erythrocytes contain a marginal band of microtubules that is thought to generate and maintain their flattened shape (Fig. 1). The microtubules in erythrocyte marginal bands are distinguished from microtubules in other cells by their extreme length and stability, about which more will be said below. In chicken erythrocytes, virtually all of the tubulin can be shown by peptide mapping to contain the unique β-tubulin variant (Murphy and Wallis 1983a). By gel analysis and radioimmunoassay, we determined that tubulin comprises 3% of the nonhemoglobin protein (0.6% of the total cell protein) and that it is unequally distributed in monomer (26%) and polymer (74%) fractions in the cell, as determined by a cell-extraction procedure using Triton X-100. Both of these fractions, however, contain the unique β-tubulin variant.

Physicochemical Properties

Electrophoretic Mobility

When fractionated by SDS-polyacrylamide gel electrophoresis (Laemmli gel, pH 8.8), the α and β subunits of brain tubulin nearly coelectrophorese, whereas those of erythrocyte tubulin are widely separated (Fig. 2a). However, at pH 9.1, the subunits of both samples are clearly separated with $R_f\beta/R_f\alpha = 1.15$ (Murphy and Wallis 1983a).

Isoelectric Point

On two-dimensional O'Farrell gels (Fig. 2b), the pI values of both α-tubulins are similar but not identical (pI 5.2). The β-tubulin subunits from brain and erythrocytes have pI values of 5.0 and 5.4, respectively. Unlike brain tubulin and most other tubulins, the β subunit from erythrocytes is considerably more alkaline relative to the α subunit. It is estimated that a difference of at least 3–4 charged amino acids would be necessary to account for the large shift in pI.

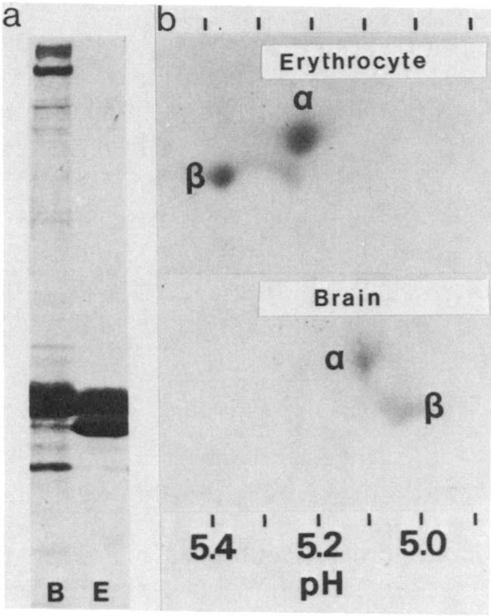

Figure 2
Electrophoretic properties of tubulin. (*a*) Electrophoretic mobility on SDS-polyacrylamide gels. Microtubule proteins (tubulin and associated proteins) were isolated by three cycles in vitro assembly and disassembly from erythrocytes (E) and brain tissue (B). Tubulin and high-molecular-weight MAPs (seen only in the brain sample [B]) are clearly visible. τ Factors (50–70 kD) are detected in both samples by immunoblotting but are not easily distinguished on the gel. The Laemmli SDS gel at pH 8.8 separates α and β subunits of the erythrocyte sample (E) but not of the brain sample (B). (*b*) Determination of pI. Samples of phosphocellulose-purified, urea-denatured tubulin (purity >99%) were fractionated by isoelectric focusing in the first dimension followed by electrophoresis at pH 9.1 in the second dimension. In erythrocytes, the β subunit is more alkaline than the α subunit.

Peptide Maps

The α and β subunits from brain tissue and erythrocytes were isolated and digested with α-chymotrypsin and fractionated by a two-dimensional peptide-mapping procedure on thin-layer plates (Fig. 3). By labeling the amino groups of the peptides with fluorescamine, all of the peptides in the digest can be visualized (Stephens 1978). From analysis of maps on which electrophoresis was performed at pH 6.5, the β subunits were determined to share only 60% of 90 resolvable peptides, indicating major differences in primary structure between the two subunits. The α subunits, in contrast, were nearly identical. Excluding differences in spot intensities, each α subunit would have at most only one or two unique peptides. These results confirm our earlier reports of β-chain differences, in which we examined the distribution of iodinated chymotryptic peptides by autoradiography after mapping on cellulose thin-layer plates (Murphy and Wallis 1983a). The extreme divergence in the peptide maps of β-tubulin suggests several differences in the primary sequences of the two subunits. A selective post-translational modification of the β subunit is unlikely, since amino acid analysis did not indicate the presence of modified residues but did reveal large differences in the relative abundance of several amino acids (see below).

Amino Acid Composition

The composition of tubulin subunits fractionated to homogeneity on SDS-hydroxyapatite is shown in Table 1. The table summarizes the composition and differences in the number of amino acids contained in the α and β subunits of brain and erythrocyte tubulins. Both α and β subunits showed differences in their compositions, but the differences were greater in both magnitude and extent for

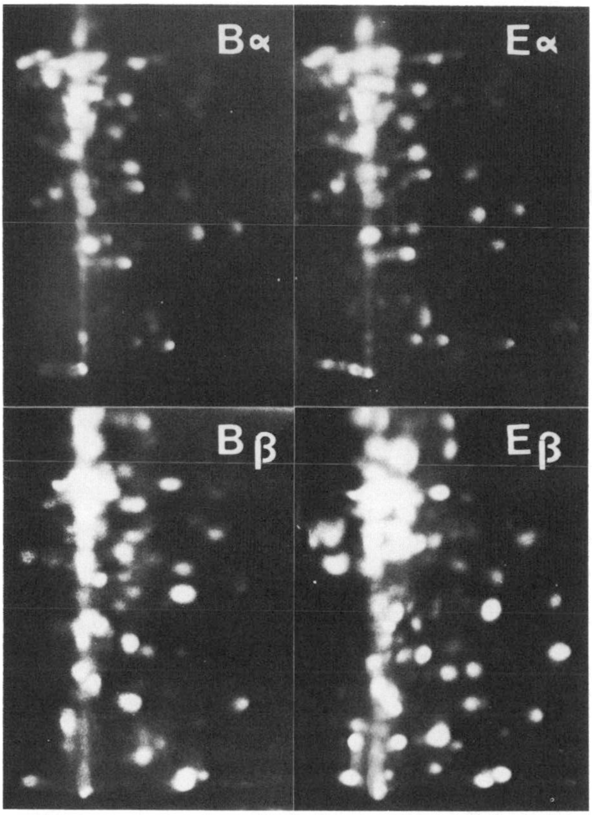

Figure 3
Peptide maps of α- and β-tubulin after treatment with α-chymotrypsin. α- and β-tubulin subunits were isolated on SDS-hydroxyapatite according to the method of Little (1979) using phosphocellulose tubulin that had been reduced and alkylated (Crestfield et al. 1963). The peptide maps were prepared on Silica-gel TLC plates, sprayed with fluorescamine, and photographed exactly as described by Stephens (1978). Ascending chromatography in chloroform-methanol-ammonium hydroxide at pH 12.3 (bottom to top) followed by electrophoresis (left to right) in pyridine-acetic acid, pH 3.5. Origins are near the bottom left corners of the maps.

the β subunit. For proline, basic, aromatic, and sulfur-containing amino acids, few differences were seen; however, there was a net loss of 15 neutral, hydroxylic, and carboxylic amino acids and a net gain of 10 bulky aliphatic amino acids, indicating that the erythrocyte β subunit had relatively increased hydrophobic properties. Taken together, the extensive differences in peptide maps and amino acid compositions indicate widely divergent primary structures for the two β-tubulin subunits. Since it is unlikely from amino acid analysis that these differences are due to posttranslational modifications, we propose that these proteins also contain different amino acid sequences and that they are the products of different β-tubulin genes.

Table 1
Amino Acid Composition of Tubulin Subunits

	α-Tubulin			*β-Tubulin*		
Amino acid	*brain*	*erythrocyte*	*difference*[a]	*brain*	*erythrocyte*	*difference*[a]
Asx	45.6	48.2	+3	52.6	54.2	+2
Thr	28.9	29.4	+1	29.8	24.9	−5*
Ser	23.2	25.4	+2	31.4	30.5	−1
Glx	57.8	58.2		65.9	60.2	−6
Pro	21.1	20.7		19.3	19.2	
Gly	39.2	41.1	+2	39.9	35.8	−4*
Ala	38.1	41.0	+3	32.7	31.8	−1
Val	28.9	28.5		27.2	27.2	
Met	8.8	8.8		15.4	13.0	−2*
Ile	21.8	19.4	−2*	13.8	19.8	+6****
Leu	33.5	34.0	+1	33.1	37.4	+4*
Tyr	18.3	18.2		16.3	16.1	
Phe	20.1	20.1		21.6	21.6	
His	11.9	14.5	+3**	10.4	9.2	−1*
Arg	22.2	22.5		22.8	26.0	+3*
Lys	20.6	18.0	−3*	16.3	12.9	−3**
Cys	10.9	10.9		8.3	9.8	+2*

Number of amino acid residues determined as moles/50,000, which is based on the molecular weight determined by nucleotide sequencing (Valenzuela et al. 1981).

[a]Differences: Erythrocyte minus brain, expressed to the nearest whole residue. Asterisks indicate deviations compared to brain subunits, with each asterisk representing a deviation of 10%; the minimum deviation thought to be significant by performing each determination in triplicate is 5%. Time-dependent degradation of serine and threonine was negligible. Tryptophan was not determined, but is known to number 4 per final chain weight of chicken brain tubulin by nucleotide sequencing (Valenzuela et al. 1981).

Tubulin Variants Exhibit Different Assembly Properties In Vitro

Erythrocyte tubulin distinguishes itself from brain tubulin in being significantly slower to self-assemble in vitro while exhibiting a critical concentration lower than that of brain tubulin. It is possible that these differences are related to differences in protein function in the cell. Tubulin from both tissue sources is readily isolated as microtubule protein (tubulin plus MAPs) by cycles of temperature-dependent assembly and disassembly (Murphy and Wallis 1983b). The brain protein contains high-molecular-weight MAPs plus some 50–70-kD τ factors; the erythrocyte protein contains, in addition to tubulin, only τ factors and some hemoglobin (Fig. 2a). However, unlike Sloboda and Dickersin (1980), who reported the presence of high-molecular-weight MAPs in the marginal bands of chickens and amphibia, we have not observed high-molecular-weight MAPs in the erythrocyte tubulin preparations.

When cold-depolymerized samples are incubated at 37°C, the erythrocyte protein polymerizes at a slow rate after a 5–7-minute lag period in assembly buffer containing 0.1 M Na-PIPES and 1 mM Mg-GTP and 5% glycerol at pH 6.9 (Fig. 4). As shown in the inset in Figure 4, however, subunits are able to immediately assemble upon the addition of exogenous microtubule seeds, demonstrating that the subunits are fully competent to assemble during the lag period. Hence, the long lag time and slow rate of self-assembly appear to be due to a

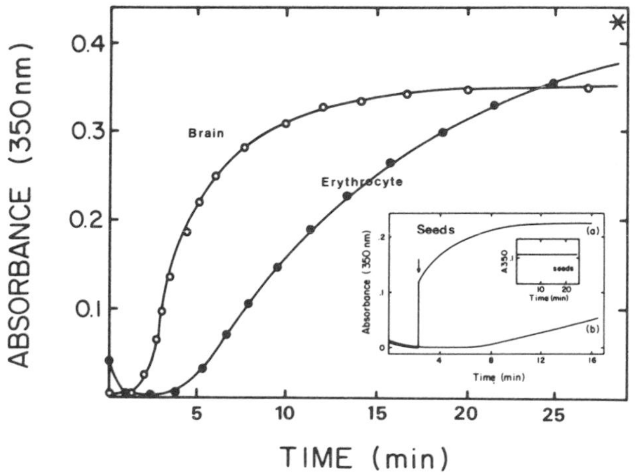

Figure 4
Polymerization of brain and erythrocyte tubulins. Microtubule protein (C_2S, 2.0 mg/ml) from brain tissue (○) and erythrocytes (●) was prepared in assembly buffer and monitored at 37°C at 350 nm with a recording spectrophotometer. (*Inset*) Addition of erythrocyte tubulin subunits onto microtubule seeds. Microtubule protein (C_2S, 2.0 mg/ml) in assembly buffer at 5°C was polymerized in a cuvette at 37°C (tracing *b*) or was prewarmed for 1 min, after which a small aliquot of sheared microtubule seeds was added (tracing *a*). Microtubule elongation observed immediately after seed addition demonstrates that the subunits are competent to assemble during the lag period observed for self-assembly (tracing *b*).

reduced rate of nucleation, rather than to inefficient microtubule elongation (for a general discussion of these parameters of microtubule assembly, see Kirschner 1978). The effect of the slow rate of self-assembly on the length of erythrocyte tubulin polymers is striking. Erythrocyte microtubules are 2–3 times longer than brain microtubules assembled in vitro at the same protein concentration and under the same conditions. Part of the explanation for this phenomenon may lie in the reduced number of MAPs, since nucleation of phosphocellulose-purified tubulin from erythrocytes can be dramatically enhanced by the addition of erythrocyte τ, brain MAPs, or even by DEAE-Dextran (not shown). However, as described below, the complete explanation appears even more complex, requiring consideration of the rate-retarding effect on microtubule growth by tubulin oligomers in the erythrocyte tubulin preparation.

Role of Tubulin Oligomers in Erythrocyte Tubulin Assembly

When microtubule seeds are added to preparations of depolymerized microtubule protein under assembly conditions, the rate of growth for brain protein is greater than that for erythrocyte protein, confirming the above observations that brain protein polymerizes faster than erythrocyte protein (Fig. 5A). However, the opposite results are obtained if microtubule seeds are added to preparations of phosphocellulose-purified tubulin; in this case, the initial rate for erythrocyte tubulin is double that for brain tubulin (Fig. 5B). This is all the more remarkable because the stimulatory factors (MAPs, τ) have been removed and because the

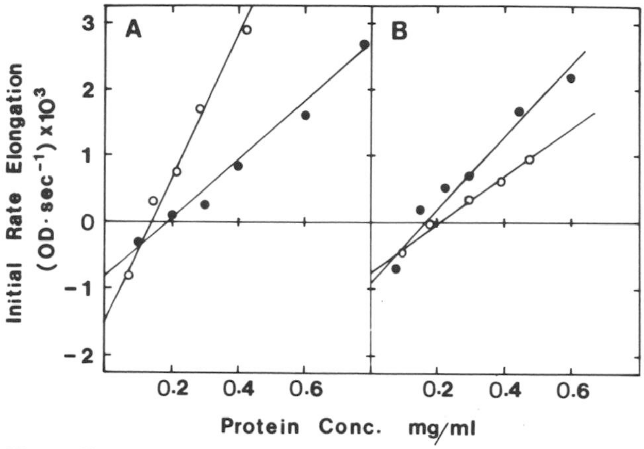

Figure 5
Comparisons of initial rates of microtubule elongation for four protein preparations. The relative rate constants for assembly and disassembly and the critical concentrations for assembly were determined by measuring the initial rates of elongation of microtubule seeds at different subunit concentrations. The procedures used for measuring and evaluating initial rates as defined by Johnson and Borisy have been described previously (Murphy et al. 1977). Brain microtubule seeds (15 μl, 5 mg/ml, 0.2-μm length) were added to a cuvette containing depolymerized tubulin subunits that were prewarmed for 45 sec at 30°C, and the initial change in turbidity was recorded at 350 nm. In these plots, the slope, *y* intercept, and *x* intercept are proportional to the assembly and disassembly rate constants, and the critical concentration. Data in *A* and *B* and Fig. 6 are separate experiments based on different seed preparations. Accordingly, the values of slopes and *y* intercepts can only be used in a relative sense within a given figure and may not be compared with the values in the other figures. (*A*) Microtubule protein (tubulin plus MAPs) containing rings from brain (○) and erythrocytes (●). (*B*) Phosphocellulose-purified tubulin without MAPs or ring oligomers from brain (○) and erythrocytes (●).

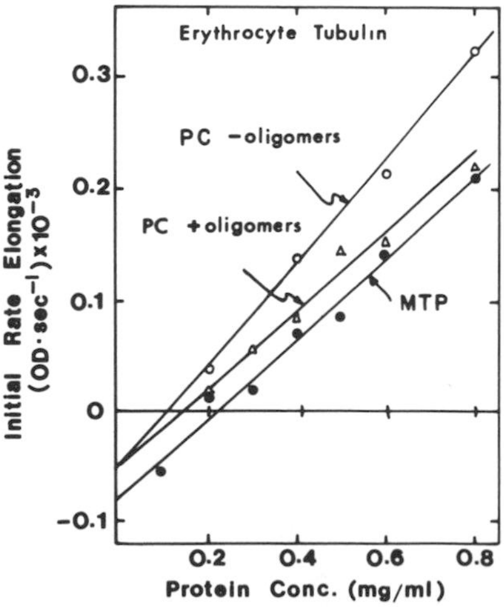

Figure 6
Tubulin ring oligmers inhibit microtubule elongation in vitro. Initial rate measurements at various tubulin concentrations were determined as described in Fig. 5 for three different preparations: (●) Microtubule protein (H_3P) containing MAPs and ring oligomers; (○) phosphocellulose-purified tubulin without MAPs or rings; (△) phosphocellulose-purified tubulin without MAPs but treated to contain rings.

preparations are known not to contain any inhibitory factors. The increased rate of elongation appears to be due to the effect of phosphocellulose in dispersing tubulin ring oligomers. We reported previously that erythrocyte tubulin at 1–2 mg/ml aggregates extensively at 5°C, forming rings of fully functional protein that readily settle out of solution and that account for up to half the protein. During chromatography on the phosphocellulose, the tubulin sample goes from cloudy to completely clear, and ring oligomers are dispersed as determined by electron microscopy. One can directly demonstrate the inhibitory effect of rings on the rate of elongation by comparing preparations of phosphocellulose-purified tubulin subunits with and without rings. Rings can be induced to re-form for this purpose by simply polymerizing and then depolymerizing an aliquot of the purified tubulin; after depolymerization, the preparation becomes turbid and can be demonstrated by electron microscopy to contain rings. When seeds are added to the ring-containing preparation, the slow assembly rate is again observed (Fig. 6). This oligomer effect on assembly has only been observed for the erythrocyte tubulin and has not been observed for brain tubulin.

At the present time, we do not know whether the tubulin oligomers observed in vitro also occur in vivo or, if they do, whether their presence also influences the rate of microtubule assembly in the cell. However, we do know from previous measurements of soluble and polymeric tubulin in erythrocytes that approximately 75% is in a "polymeric form," being nonextractable from Triton-permeabilized erythrocyte ghosts. However, from the microtubule number count and the size of the marginal band, we can only account for half of the tubulin polymer as being in the marginal band and speculate that the remainder may be in the form of oligomers, perhaps rings, that cannot be readily extracted from the cell. It is possible that for marginal-band formation, the majority of the tubulin pool (2 mg/ml) is maintained in the form of rings to assure a low concentration of 6S tubulin dimers. These conditions would reduce the rates of nucleation and elongation and allow the formation of long microtubule polymers such as are found in the marginal band. It remains to be demonstrated, however, that rings in this cell, or in any cell, are significant for the regulation of the rate of microtubule assembly in vivo.

Tubulin Expression in Developing Cells of the Bone Marrow

We examined chicken bone marrow cells with rabbit antibodies to the erythrocyte β subunit by immunofluorescence microscopy to determine the time and pattern of β-tubulin expression during erythrocyte differentiation. We confirmed that normal chicken bone marrow contains 70% red-cell precursors and that this percentage can be boosted to 85% during phenylhydrazine-induced anemia. Erythroblasts are readily distinguished from all other bone marrow cells by the presence of a large spherical nucleus, each containing a single large nucleolus. Most white-cell precursors have indented or irregularly shaped nuclei and indistinct nucleoli. When examined by immunofluorescence microscopy after labeling with affinity-purified antibodies to erythrocyte β-tubulin (Fig. 7a), approximately 50% of the total marrow cells in phenylhydrazine-treated chicks were determined to contain the erythrocyte-specific subunit, and a large number of early-stage erythroblasts were identified (Fig. 7b,c, arrowheads) that contained little, if any,

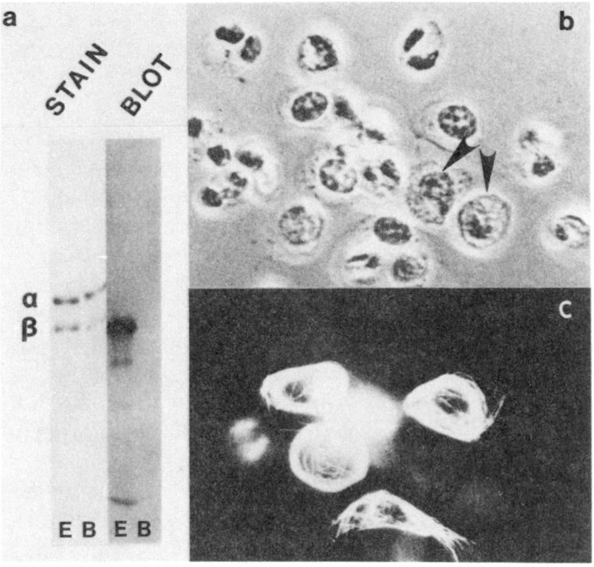

Figure 7
Immunofluorescence microscopy of chicken bone marrow cells stained with erythrocyte β-tubulin antibody. (*a*) Specificity of rabbit antibody to erythrocyte β-tubulin. Microtubule proteins from chicken brain (B) and erythrocytes (E) were fractionated by SDS-gel electrophoresis at pH 9.1 to resolve the α and β subunits in both preparations (Coomassie stain and immunoblot after incubation in the presence of affinity-purified antibody and iodinated protein A). The erythrocyte β subunit is specifically labeled. (*b*,*c*) Immunofluorescence microscopy. Bone marrow from the leg bone of a 10-day-old chick made anemic with phenylhydrazine (Wright and Van Alstyne 1931) was streaked on a coverglass and prepared for immunofluorescence microscopy using the polyethylene glycol stabilization procedure in the absence of aldehyde fixatives as described by Osborn and Weber (1982).

of the erythrocyte-specific subunit. Thus, undifferentiated progenitor cells (early erythroblasts and proerythroblasts) do not appear to contain significant amounts of erythrocyte β-tubulin, whereas late-stage erythroblasts and more mature subsequent stages do. In contrast, all marrow cells become labeled when stained with rabbit antibodies to sea urchin tubulin and which are known to bind to tubulin from both red and white blood cells. Examination of the patterns of antibody labeling indicates that β-tubulin synthesis is most active during the period of hemoglobin synthesis and reveals early erythroid precursors with little or no specific β-tubulin. It is not yet possible to distinguish whether the sudden appearance of the β-tubulin subunit in later-stage cells is due to a change in the rate of tubulin synthesis, a switch in gene expression, or a combination of these processes. Regardless of the exact details, a tubulin gene switching event appears to be involved. It will be interesting to correlate this event at a molecular level with other molecular aspects of erythropoiesis and to determine how the erythrocyte tubulin variant functions to fulfill uniquely tubulin assembly requirements during erythrocyte differentiation.

SUMMARY

Chicken erythrocytes contain a β-tubulin variant (tubulin isoform) that is distinguishable from chicken brain tubulin by its electrophoretic properties, amino acid composition, and peptide maps. The two tubulins are considered to be the products of different β-tubulin genes. The respective α-tubulin subunits appear to be similar, if not identical, polypeptides.

When polymerized in vitro, erythrocyte tubulin self-assembles after a long lag period and with a lower critical concentration than brain tubulin. The long lag time and slow assembly are due in part to the presence of tubulin ring oligomers that reduce rates of nucleation and tubulin subunit elongation. The formation of tubulin oligomers is a conspicuous feature of the erythrocyte tubulin assembly and may be related to the requirements of marginal-band formation in developing erythroblasts.

Immunofluorescence microscopy of developing bone marrow cells reveals a large number of erythroblast precursors that do not label with affinity-purified antibodies to erythrocyte β-tubulin but that are labeled by an antibody to sea urchin tubulin. The observations indicate a switch in tubulin gene expression that occurs at a time in development corresponding to the onset of hemoglobin synthesis but preceding marginal-band formation and cell morphogenesis.

ACKNOWLEDGMENT

We thank Dr. Joel Shaper (Department of Oncology, Johns Hopkins Medical School) for performing the amino acid determinations and for providing advice on sample preparation.

REFERENCES

Behnke, O. 1970. Microtubules in disc-shaped blood cells. *Int. Rev. Exp. Pathol.* **9:** 1.

Cleveland, D.W. 1983. The tubulins: From DNA to RNA to protein and back again. *Cell* **34:** 330.

Crestfield, A.M., S. Moore, and W.H. Stein. 1963. The preparation and enzymatic hydrolysis of reduced and *S*-carboxymethylated proteins. *J. Biol. Chem.* **238:** 622.

Fulton, C. and P.A. Simpson. 1976. Selective synthesis and utilization of flagellar tubulin. The multitubulin hypothesis. *Cold Spring Harbor Conf. Cell Proliferation* **3:** 987.

Kemphues, K.J., T.C. Kaufman, R.A. Raff, and E.C. Raff. 1982. The testis-specific β-tubulin subunit in *Drosophila melanogaster* has multiple functions in spermatogenesis. *Cell* **31:** 655.

Kirschner, M.W. 1978. Microtubule assembly and nucleation. *Int. Rev. Cytol.* **54:** 1.

Kumar, N. and M. Flavin. 1982. Modulation of some parameters of assembly of microtubules in vitro by tyrosinolation of tubulin. *Eur. J. Biochem.* **128:** 215.

L'Hernault, S.W. and J.L. Rosenbaum. 1983. *Chlamydomonas* α-tubulin is post-translationally modified in the flagella during flagellar assembly. *J. Cell Biol.* **97:** 258.

Little, M. 1979. Identification of a second beta chain in pig brain tubulin. *FEBS Lett.* **108:** 283.

Murphy, D.B. and K.T. Wallis. 1983a. Brain and erythrocyte microtubules contain different β-tubulin polypeptides. *J. Biol. Chem.* **258:** 7870.

———. 1983b. Isolation of microtubule protein from chicken erythrocytes and determination of the critical concentration for tubulin polymerization in vitro and in vivo. *J. Biol. Chem.* **258:** 8357.

Murphy, D.B., K.A. Johnson, and G.G. Borisy. 1977. Role of tubulin-associated proteins in microtubule nucleation and elongation. *J. Mol. Biol.* **117:** 33.

Osborn, M. and K. Weber. 1982. Immunofluorescence and immunocytochemical procedures with affinity purified antibodies: Tubulin-containing structures. *Methods Cell Biol.* **24A:** 98.

Raff, E.C., M.T. Fuller, T.C. Kaufman, K.J. Kemphues, J.E. Rudolph, and R.A. Raff. 1982. Regulation of tubulin gene expression during embryogenesis in *Drosophila melanogaster. Cell* **28:** 33.

Sloboda, R.D. and K. Dickersin. 1980. Structure and composition of the cytoskeleton of nucleated erythrocytes. I. The presence of microtubule-associated protein 2 in the marginal bands. *J. Cell Biol.* **87:** 170.

Stephens, R.E. 1975. Structural chemistry of the axoneme: Evidence for chemically and functionally unique tubulin dimers in outer fibers. In *Molecules and cell movement* (ed. S. Inoue and R.E. Stephens), p. 181. Raven Press, New York.

———. 1978. Fluorescent thin-layer peptide mapping for protein identification and comparison in the subnanomole range. *Anal. Biochem.* **84:** 116.

Valenzuela, P., M. Quiroga, J. Zaldivar, W.J. Rutter, M.W. Kirschner, and D.W. Cleveland. 1981. Nucleotide and corresponding amino acid sequences encoded by alpha and beta tubulin mRNAs. *Nature* **289:** 650.

Vallee, R. 1980. Structure and phosphorylation of MAP-2. *Proc. Natl. Acad. Sci.* **77:** 3206.

Wright, G.P. and M. Van Alstyne. 1931. The development of primitive avian red corpuscles on incubation in vitro. *Folia Haematol.* **46:** 26.

Kinetochore Antigens: Complexity, Synthesis, and Evolutionary Conservation

John V. Cox and Joanna B. Olmsted
Department of Biology
University of Rochester
Rochester, New York 14627

The discovery that sera from patients having the CREST variant of scleroderma stain the centromere of mitotic chromosomes and discrete spots within interphase nuclei (Fritzler and Kinsella 1980; Moroi et al. 1980; Tan et al. 1980) has made it possible to study the organization and composition of this chromosomal region in more detail. The number of labeled spots in interphase nuclei has been shown to correspond to the chromosome number (Moroi et al. 1980; Brenner et al. 1981) and to double during late G_2 of the cell cycle (Brenner et al. 1981). The retention of typical immunofluorescent labeling patterns in nuclear (Cox et al. 1983) and chromosomal (Cox et al. 1983; Earnshaw et al. 1984) matrix preparations, or in preparations in which isolated chromosomes have been treated with *Alu*I (Lica and Hamkalo 1983) or additional agents (Valdivia and Brinkley, this volume), suggests that the antigens are an integral component of the chromosomes in both interphase and mitotic cells. Using immunoprecipitation and immunoblotting, our studies (Cox et al. 1983) have demonstrated that a discrete set of antigens (34,000 daltons [34K], 23K, 20K, and 14K) react with the anticentromere sera. As assessed by absorption of staining patterns (Cox et al. 1983), or using antibodies affinity purified from any of these antigens immobilized on blots (J.V. Cox, unpubl.), all of these protein species were identified as being present in the centromere of interphase or mitotic chromosomes. This paper summarizes details of the complexity, phosphorylation, and evolutionary conservation of these antigens.

DISCUSSION

Complexity of the Antigenic Species

Immunoprecipitates from ^{35}S-labeled nuclei, chromosomes, and matrices were analyzed by two-dimensional gel electrophoresis (O'Farrell et al. 1977). As shown in Figure 1, the proteins precipitated by the centromere serum were undetectable

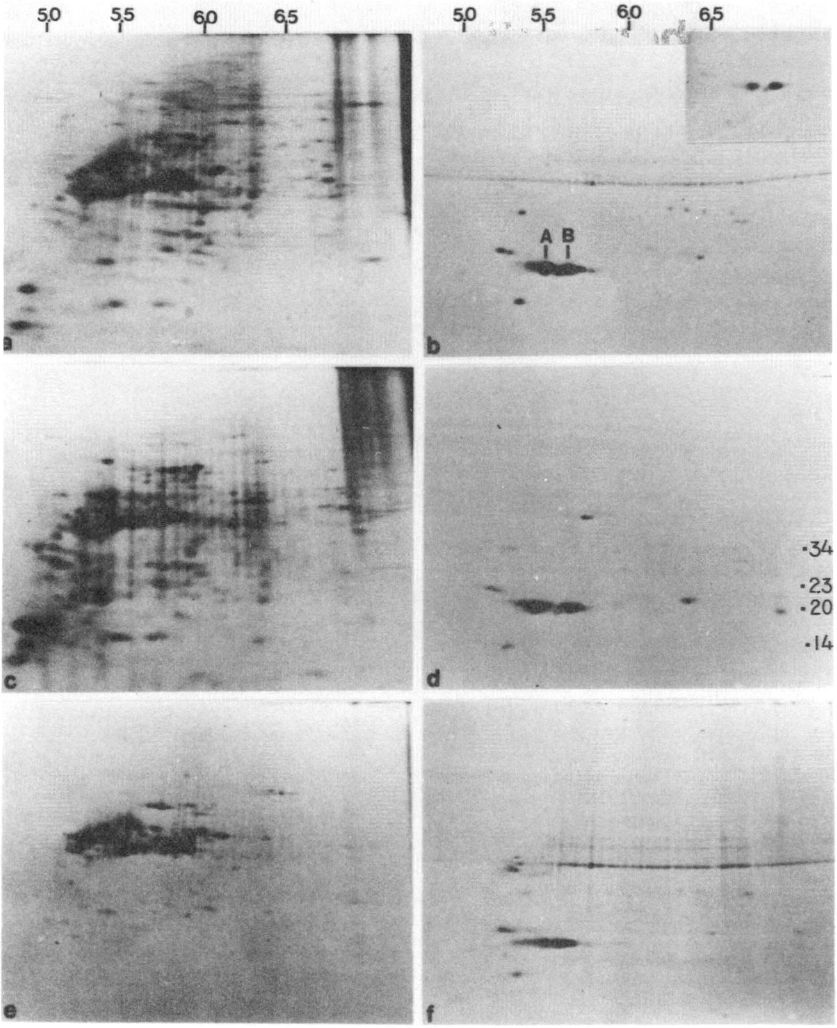

Figure 1
Two-dimensional O'Farrell gels of total ^{35}S-labeled proteins (*a,c,e*) and immunoprecipitates obtained with anticentromere serum (*b,d,f*) from isolated HeLa nuclei (*a,b*), chromosomes (*c,d*), and interphase matrices (*e,f*). Note that the immunoprecipitated proteins are not detectable in the total protein fractions. Two species at 20K, referred to as 20A (acidic) and 20B (basic), are designated in panel *b*. A shorter exposure of this region (*b, insert*) shows an additional protein of slightly lower molecular weight between 20A and 20B.

in any of the protein gels, indicating that the kinetochore antigens represented a very small percentage of the protein in each sample. The immunoprecipitates from each fraction were very similar in protein composition, although the abundance of individual proteins varied. For example, the 34K protein was present in reduced amounts in chromosomes and nuclear matrices as compared to interphase nuclei.

The main cluster of proteins in each immunoprecipitate ranged in pI from 5.2 to 5.9. The 14K and 34K proteins had pI values of 5.4, whereas several distinct isoelectric variants in the 20K and 23K regions were resolved. Two protein species, with pI values of 5.2 and 5.3, were identified at 23K; an additional protein of 23K focused at 6.4 and was most evident in chromosome immunoprecipitates (Fig. 1d). The proteins at 20K ranged in pI from 5.4 to 5.9, with two major spots at 20K (20A and 20B) apparent in immunoprecipitates from nuclei (Fig. 1b) and chromosomes (Fig. 1d).

To determine whether any of these protein species were variants arising from phosphorylation, immunoprecipitates from ^{32}P-labeled cell fractions were also analyzed on two-dimensional gels (Fig. 2). These gels demonstrate that the only phosphorylated species are at 20K and 23K and that these immunoprecipitated proteins are also a minor component of the phosphorylated protein in each fraction. The correspondence between these phosphorylated antigens and those identified from ^{35}S-labeled material was examined further by analyzing the effect of alkaline phosphatase digestion on the pI of the immunoprecipitated species. One-dimensional gel analyses had demonstrated that treatment of immunoprecipitates with 50 units/ml of alkaline phosphatase for 2 hours was sufficient to remove all ^{32}P label from the proteins (Cox 1983). An immunoprecipitate obtained from ^{35}S-labeled nuclei and treated with alkaline phosphatase prior to gel analysis was compared with immunoprecipitates from ^{32}P-labeled nuclei (Fig. 3a) and ^{35}S-labeled nuclei (Fig. 3b) that were untreated. As shown in Figure 3, alkaline phosphatase treatment (panel c) resulted in the loss of the most acidic protein at 23K (marked by arrow). These data suggested that this spot arises from the phosphorylation of a single unmodified species with a pI of 5.3. The pattern in the 20K region was more complex. Comparison of the alkaline-phosphatase-treated immunoprecipitate (panel c) with the untreated ^{32}P-labeled (panel a) and ^{35}S-labeled (panel b) immunoprecipitates shows that the phosphorylated species at 20K comprises a small percentage of antigens at this molecular weight. Alkaline phosphatase treatment causes the loss of the 20K protein between pI 5.4 and pI 5.6 (Fig. 3c, bracket). It can also be seen (panel c) that 20A is resolved into two molecular-weight species (Fig. 3c, dashes) following alkaline phosphatase treatment. The origin of this split is unknown, although the various levels of phosphorylation of the higher-molecular-weight species (Fig. 3c, upper dash) probably account for the ^{32}P labeling seen in this region. Pulse-chase experiments with [^{35}S]methionine have indicated that the acidic species at 20K and 23K appear within 90 minutes, and these data corroborate the phosphatase experiments that suggest these spots are phosphorylated variants of the more basic 20K and 23K species. Although the proteins in the 20K region are all antigens precipitated by the centromere serum, it remains to be determined whether these represent distinct polypeptides or are polypeptides that are related to one another, perhaps by other types of modification.

Cell-cycle Expression of Antigens

The biosynthesis of the centromere antigens through the cell cycle was examined by pulse labeling synchronized HeLa cells at various times and analyzing

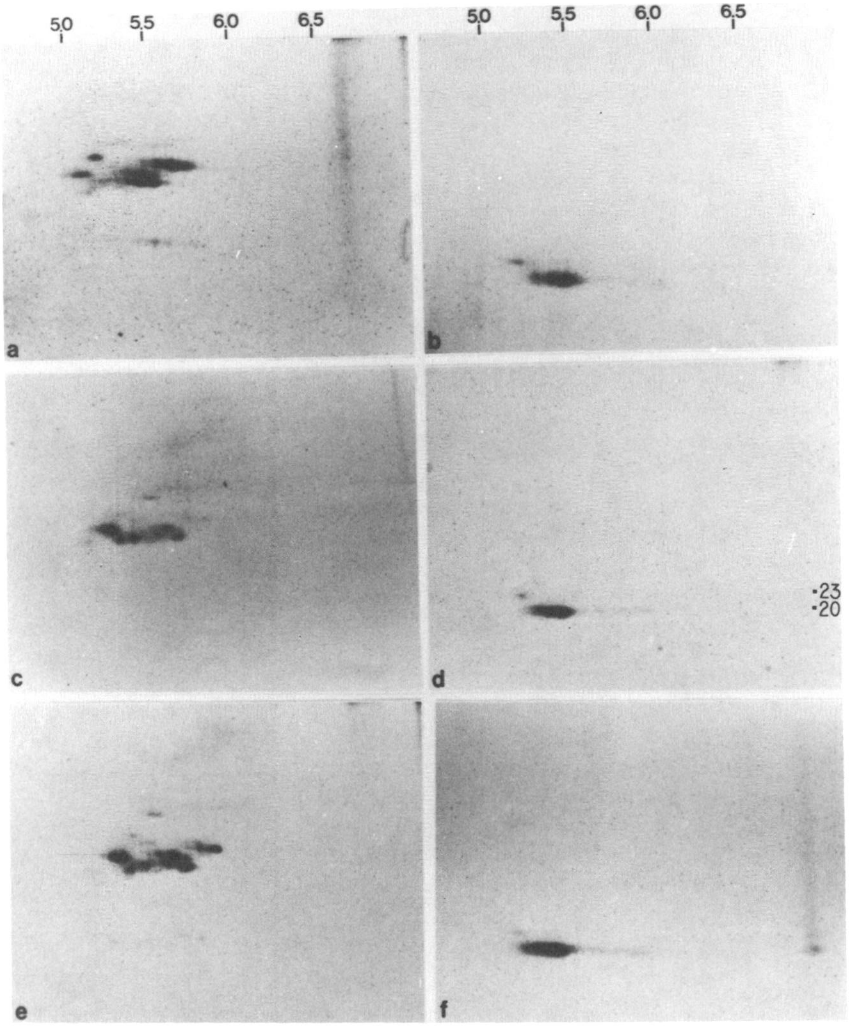

Figure 2
Two-dimensional O'Farrell gels of total ^{32}P-labeled proteins (*a,c,e*) and immunoprecipitates obtained with anticentromere serum (*b,d,f*) from isolated HeLa nuclei (*a,b*), chromosomes (*c,d*), and interphase matrices (*e,f*). Proteins of 20K and 23K are present in the immunoprecipitate from each fraction, and there appear to be three discrete phosphorylated species at 20K (evident in panel *b*).

the immunoprecipitated material on two-dimensional gels (Cox 1983). The level of synthesis of both the 20K and 23K proteins was similar at all stages of the cell cycle. The 34K protein was preferentially synthesized during G_2 and M and was present in reduced amounts at all other stages. The 14K protein was found in constant amounts through S and M but was entirely absent at phases corresponding to G_1 through early S. These data indicate that the synthesis of at least

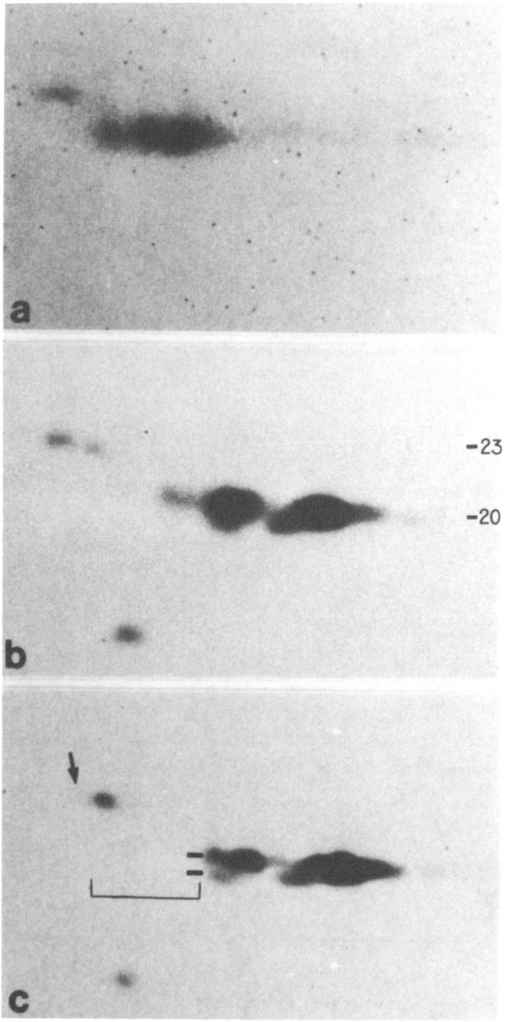

Figure 3

Two-dimensional O'Farrell gels of alkaline-phosphatase-treated immunoprecipitates from HeLa nuclei. An immunoprecipitate obtained with anticentromere serum from ^{35}S-labeled HeLa nuclei was immobilized on protein A–Sepharose beads; digested with DNase (250 μg/ml), RNase (250 μg/ml), and alkaline phosphatase (50 units/ml) for 2 hr at 0°C; and then run on a two-dimensional gel (panel *c*). Immunoprecipitates from ^{32}P-labeled nuclei (*a*) and ^{35}S-labeled nuclei that were treated with DNase and RNase alone are also shown. Note that the alkaline phosphatase treatment results in the loss of the 20K proteins with pI values between 5.4 and 5.6 (bracket).

the 14K and 34K proteins is modulated during the cell cycle. Similar experiments in which synchronized cells were pulsed with ^{32}P showed no significant changes in the phosphorylation of the 20K and 23K species at any stage (J.V. Cox, unpubl.).

Evolutionary Conservation of Antigens

Our previous results demonstrated that the 14K, 20K, and 23K antigens were also present in immunoprecipitates from L, 3T3, PtK_1, neuroblastoma, and Indian muntjac cells (see Cox et al. 1983) and that typical nuclear staining was observed in all of these cell lines (Cox 1983). The existence of antigens reacting with the antiserum in more evolutionary diverse eukaryotic organisms was examined using immunofluorescent and immunoblotting techniques. Cell lines derived from frog (*Rana pipiens*, haploid line) or *Drosophila* (embryonic Kc) showed the same speckled pattern of nuclear fluorescence observed in mammalian cell lines (Cox 1983). As shown in Figure 4, immunoblotting analysis of nuclei isolated from a frog cell line (lane 2), *Drosophila* tissue-culture cells (lane 3), yeast nuclei (lane 4), and *Tetrahymena* macronuclei (lane 5) indicates that all of these organisms contain an antigen of 14K. In some organisms, the antiserum reacts with other proteins; some of these are similar, but not identical, in molecular weight to the proteins recognized in HeLa nuclei (lane 1). These results suggest that at least the 14K protein may be a conserved component of the centromere region in a wide variety of eukaryotic organisms.

CONCLUSIONS

We have identified four major antigens localized at the centromere; the 20K and 23K proteins show varying states of phosphorylation, and the synthesis of the 14K and 34K species changes during the cell cycle. Although the structure of the kinetochore varies widely among organisms, the 14K protein appears to be conserved in a number of evolutionarily distant species. It is yet to be determined whether this protein may be associated with the distinctive properties of centro-

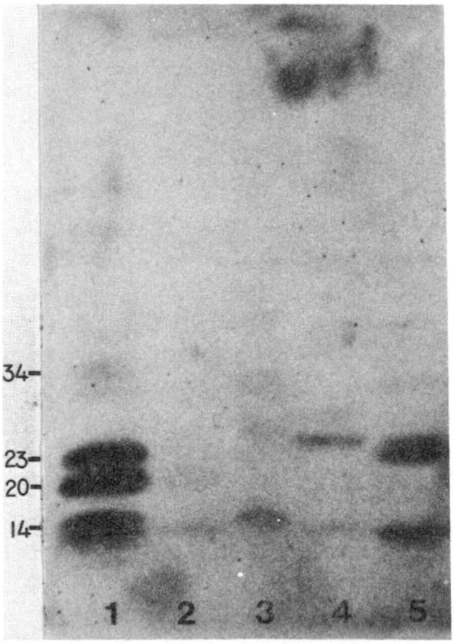

Figure 4
Immunoblot of nuclei from a variety of eukaryotic organisms probed with anticentromere serum. Nuclei isolated from HeLa cells (*1*), *Rana pipiens* haploid cell line (*2*), *Drosophila* embryonic Kc cells (*3*), *Saccharomyces cerevisiae* (*4*), and *Tetrahymena pyriformis* (macronuclei) (*5*) were run on a 10–25% gradient gel and blotted. Binding of anticentromere serum to the blot was detected with ^{125}I-labeled protein A. Note that the species at 14K is present in all samples.

mere chromatin that have been defined from studies on yeast plasmids (Bloom et al., this volume). We have also demonstrated that the anticentromere sera block microtubule assembly at the kinetochore in lysed cell models (Cox et al. 1983). Given all of these data, it seems likely that regulation of the amount and/or phosphorylation of these minor proteins will be important in centromere organization and in the generation of a functional kinetochore at the onset of mitosis.

ACKNOWLEDGMENT

This work was supported by National Institutes of Health grant GM-22214 awarded to J.B.O.

REFERENCES

Brenner, S., D. Pepper, M.W. Berns, E. Tan, and B.R. Brinkley. 1981. Kinetochore structure, duplication, and distribution in mammalian cells: Analysis by human autoantibodies from scleroderma patients. *J. Cell Biol.* **91:** 95.

Cox, J.V. 1983. "Human anticentromere antibodies: Distribution, characterization of antigens, and effect on microtubule organization." Ph.D. thesis, University of Rochester, Rochester, New York.

Cox, J.V., E.A. Schenk, and J.B. Olmsted. 1983. Human anticentromere antibodies: Distribution, characterization of antigens, and effect on microtubule organization. *Cell* **35:** 331.

Earnshaw, W.C., N. Halligan, C. Cooke, and N. Rothfield. 1984. The kinetochore is part of the metaphase chromosome scaffold. *J. Cell Biol.* **98:** 352.

Fritzler, M.J. and T.D. Kinsella. 1980. The CREST syndrome: A distinct serologic entity with anticentromere antibodies. *Am. J. Med.* **69:** 520.

Lica, L. and B. Hamkalo. 1983. Preparation of centromeric heterochromatin by restriction endonuclease digestion of mouse L929 cells. *Chromosoma* **88:** 42.

Moroi, Y., C. Peebles, M. Fritzler, J. Steigerwald, and E.M. Tan. 1980. Autoantibody to centromere (kinetochore) in scleroderma sera. *Proc. Natl. Acad. Sci.* **77:** 1627.

O'Farrell, P.Z., H.M. Goodman, and P.H. O'Farrell. 1977. High resolution two-dimensional electrophoresis of basic as well as acidic proteins. *Cell* **12:** 1133.

Tan, E.M., G.P. Rodnan, I. Garcia, Y. Moroi, M. Fritzler, and C. Peebles. 1980. Diversity of antinuclear antibodies in progressive systemic sclerosis. Anti-centromere antibody and its relationship to CREST syndrome. *Arthritis Rheum.* **23:** 617.

Biochemical Studies of the Kinetochore/Centromere of Mammalian Chromosomes

Manuel M. Valdivia and Bill R. Brinkley
Department of Cell Biology
Baylor College of Medicine
Houston, Texas 77030

The centromere region located at the primary constriction of metaphase chromosomes serves as the site for the attachment of spindle fibers needed for chromosome segregation during mitosis and meiosis. In most eukaryotic cells, the specific binding site for centromere-associated microtubules consists of a structurally differentiated region called the kinetochore, which forms a structural interface between spindle microtubules and centromeric chromatin. Although the kinetochore has been well characterized at the light and electron microscopic level, relatively little is known about the molecular components of this site. An understanding of the molecular basis for spindle microtubule initiation and attachment at the kinetochores, as well as the mechanism of how this structure facilitates chromosome segregation, will require a detailed knowledge of the centromere, including the DNA, proteins, and other chromatin components specifically associated with the centromere region.

The discovery of human autoantibodies directed against centromere components (Moroi et al. 1980; Brenner et al. 1981) has provided a promising new strategy for dissecting the molecular components of the kinetochore. In this report, we describe a protocol for the initial isolation and partial purification of the centromere/kinetochore of mammalian metaphase chromosomes using scleroderma CREST antiserum as an immunofluorescent and immunoelectron microscopic and biochemical probe. The final kinetochore pellet contained disklike structures resembling intact kinetochores of the metaphase chromosome, which were stained by immunoperoxidase. Two major groups of polypeptides were enriched in the fraction as indicated by two-dimensional gel electrophoresis, and a protein of 17K–18K was reactive by immunoblotting.

Strategy and Protocol for Kinetochore Isolation

The centromere region of metaphase chromosomes is highly resistant to nuclease digestion as compared with the chromosome arms (Rattner et al. 1978). The chromosomal structures that remain after nuclease digestion and after dehistonization, called the scaffold (Laemmli et al. 1978), maintain reactivity with the CREST antiserum (Earnshaw et al. 1984). These findings along with the availability of CREST antiserum, which reacted specifically with the kinetochore plates of metaphase chromosomes (Brenner et al. 1981), have enabled us to develop a procedure to purify the centromere/kinetochore region of mammalian metaphase chromosomes.

Our procedure is begun by obtaining a highly purified fraction of metaphase chromosomes from HeLa or Chinese hamster ovary (CHO) cells isolated by the hexylene glycol procedure of Wray and Stubblefield (1970) and including a final purification step involving a glycerol gradient (Gooderham and Jeppesen 1983). The fraction shown in Figure 1 is free of chromosome aggregates, nuclei, or interphase cells but contains an occasional centriole. When stained with CREST antiserum, the kinetochore appeared as two fluorescent spots as seen on chromosomes in intact cells (Fig. 1A). Staining with immunoperoxidase showed the antibodies to be localized in a narrow zone on the lateral sides of the centromere (Fig. 1B).

The isolated chromosomes were subjected to extensive digestion of DNA with micrococcal nuclease (200 units/ml) and complete dehistonization with heparin (1 mg/ml). After digestion, the residual chromosome contained less than 1% of the total chromosomal DNA and no detectable histones, as indicated by SDS-gel electrophoresis. With regard to remaining DNA and protein components, the residual chromosome consists largely of scaffold (Adolph et al. 1977). As shown in Figure 2, a and b, the kinetochore component remained with the digested chromosomes as indicated by CREST staining and immunoperoxidase. The scaffold material contains mostly nonhistone proteins and can be dissolved by treatments such as phenol, urea, and trypsin. Therefore, the heparin-resistant scaffolds were exposed to detergent (NP-40 plus cholate) to release some nonhistone proteins, followed by 4 M urea to dissolve the remaining structure. It should be pointed out that the residual chromosomal material retained the capacity to bind CREST antiserum, indicating the resistance of the kinetochore component to denaturation with urea (Fig. 2c,d). Finally, a density gradient (metrizamide) was used to obtain an enriched fraction of stained material (kinetochores). As shown in Figure 3, a and b, the enriched fraction stained brightly with CREST antiserum, attesting to the unexpected stability of the kinetochore component to harsh treatments such as 4 M urea. When sections of the metrizamide fraction were examined after immunoperoxidase staining, the remaining material existed either as electron-dense masses or as disklike structures, much like those associated with chromosomes of intact metaphase cells (Fig. 3b, arrowheads). Several fractions of the metrizamide gradient appeared to be enriched in stained material with very little contamination present. Our findings suggest that the kinetochore region is much denser and more resistant than the rest of the scaffold material in the chromosome, thus enabling us to purify it by density gradient. It is obvious, however, that the harsh effects of urea alter kinetochore structure somewhat as indicated by the various morphological forms seen in Figure 3b. At this time we do

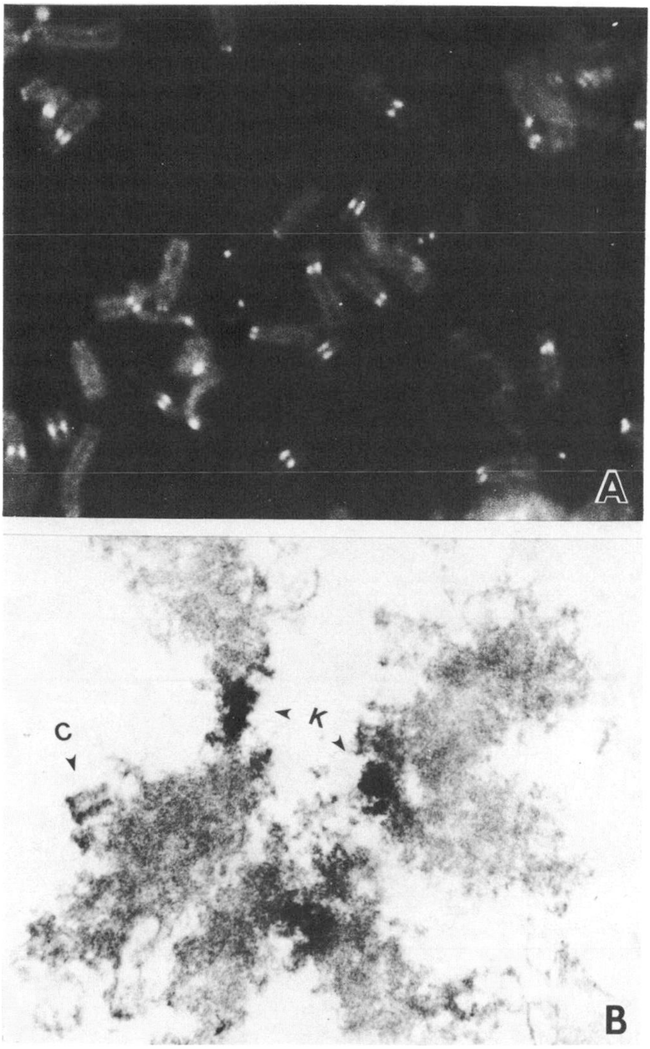

Figure 1
Fluorescence light microscopy and electron micrographs of CHO chromosomes. In both cases, human antiserum specific for kinetochore was used as first antibody, followed by anti-human fluorescein (*A*) or anti-human peroxidase (*B*). Arrowheads point to kinetochore (K) and centriole (C). Magnification: (*A*) 2,065 ×; (*B*) 35,910 ×.

not know whether or not these isolated kinetochores can interact with microtubules. Such functional tests will be performed in the future.

Initial Biochemical Studies

The fraction highly enriched in kinetochores described above has proved to be very instructive for biochemical studies. By quantitative analysis, we have deter-

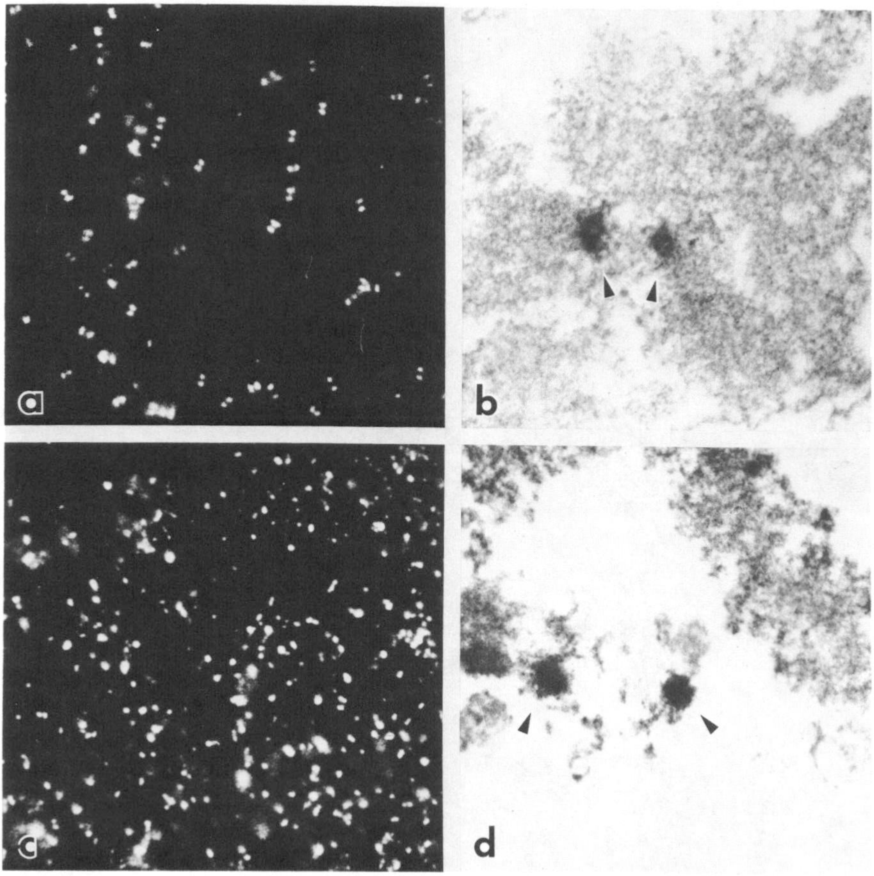

Figure 2
CHO metaphase chromosomes digested with micrococcal nuclease and heparin to produce scaffold structure. Immunofluorescence (*a*,*c*) and electron micrographs (*b*,*d*) of scaffold structure before (*a*,*b*) and after (*c*,*d*) detergent (NP-40 plus cholate) and urea treatment. Note the kinetochore pairs in scaffold structure (*a*) but not after urea treatment (*c*). Arrowheads point to kinetochore. Magnification: (*a*) 1,617 ×; (*b*) 17,780 ×; (*c*) 1,750 ×; (*d*) 21,315 ×.

mined that the amount of protein present in our most purified fractions represents less than 8% of the total chromosomal protein, and the DNA that copurified with the kinetochore fraction represented less than 1% of the total chromosomal DNA. Also, a small amount of RNA is present in the final fraction. Analysis by standard one-dimensional SDS-gel electrophoresis showed that the major difference in protein composition of whole chromosomes and the kinetochore fraction was primarily the absence of histones in the latter. Although this could be an indication that histones are absent in the kinetochore, it is more likely that histones are not necessary for the staining of kinetochores by CREST antiserum. It may also be related to previous reports that the nucleosome structure of chro-

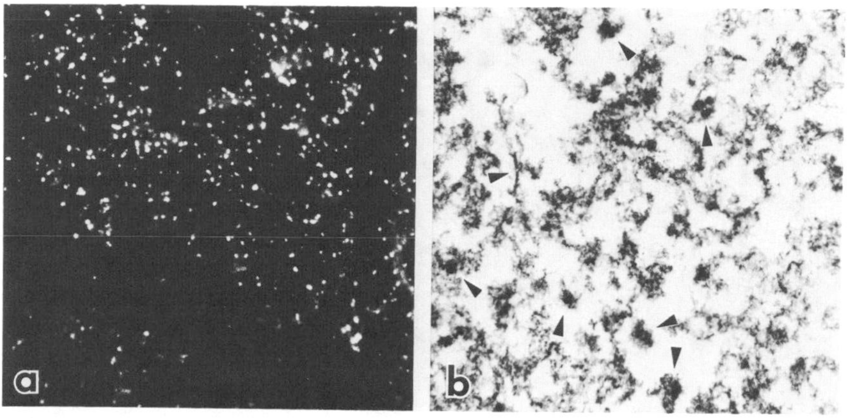

Figure 3
Isolated CHO kinetochore region. (*a*) Fluorescence light microscopy of chromosomes treated with MNase, heparin, detergent, and urea and finally loaded in a metrizamide gradient to enrich stained (kinetochore) material. (*b*) Electron micrographs of similar fraction seen by staining with immunoperoxidase. Arrowheads indicate putative kinetochores. Magnification: (*a*) 1,155×; (*b*) 7,350×.

matin in the centromere region presents differences with respect to the rest of the chromosome (Musich et al. 1977). Through the improved resolution afforded by two-dimensional silver-stained gels, we were able to show that the kinetochore/centromere fraction was enriched in at least ten proteins. Figure 4A shows the entire protein composition of HeLa chromosomes, and Figure 4B shows the proteins of the final kinetochore fraction. The major protein in both is actin (42K), which has been previously reported in whole chromosomes (Sanger 1975; Gooderham and Jeppesen 1983). Other workers have shown that it is difficult to completely eliminate actin from highly purified chromosome and chromosome scaffold preparations (Lewis and Laemmli 1982). Nevertheless, it is difficult to rule out the possibility that actin is a contaminating protein from the cytoplasm. From the amount that is routinely retained in our kinetochore/centromere fractions, we suggest that actin may be a structural component of the kinetochore/centromere.

Two major groups of polypeptides with molecular weights of 34K–40K and 50K–55K were enriched in our purified fractions when compared with bulk chromosome proteins (Fig. 4). A very similar set of proteins with essentially identical distribution on two-dimensional gels was found in a different cell line (CHO cells, not shown).

Using the Western transfer technique of Towbin et al. (1979), we have identified the antigen in kinetochore/centromere preparations that is recognized by the CREST serum used in these studies. A protein of 17K–18K was consistently reactive with three different CREST sera using two different mammalian cell lines. Serum from one patient cross-reacted with an additional 39K protein (Fig. 5). As shown in Figure 5, the protein fraction transferred to the nitrocellulose paper (Towbin et al. 1979) corresponded to the kinetochore/centromere fraction lacking histones. We were unable to demonstrate a specific immunoblot when histones were present in the transferred proteins. Using the technique of Olmsted (1981),

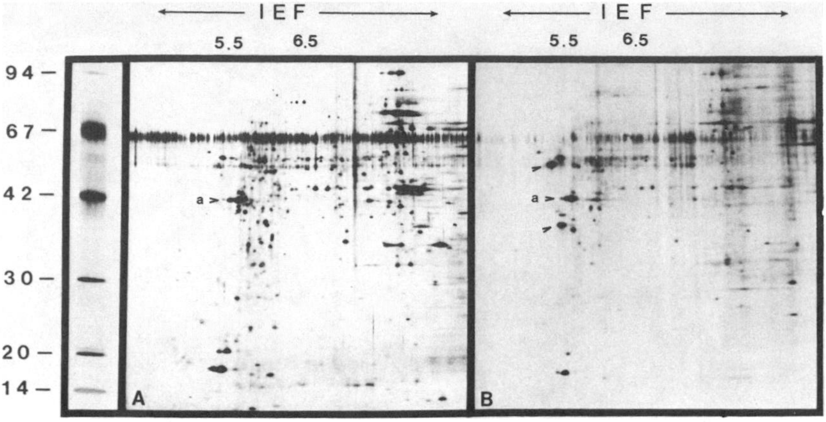

Figure 4
Two-dimensional polyacrylamide gel electrophoresis analysis of kinetochore fraction. (*A*) HeLa whole metaphase chromosomes. (*B*) HeLa kinetochore fraction. The position of enriched polypeptides in kinetochore fraction is shown by arrowheads in *B*. Actin, which is a major protein in both samples, is indicated by a. Standard marker proteins are phosphorylase b, 94K; albumin, 67K; ovalbumin, 42K; carbonic anhydrase, 30K; trypsin inhibitor, 20K; α lactalbumin, 14K.

we have succeeded in staining kinetochores using antibodies eluted from the 17K–18K band after immunoblotting.

A protein of similar molecular weight to our 17K–18K protein was also found by Earnshaw et al. (1984) along with several other proteins. There are, however, some interesting discrepancies between the findings of Cox et al. (1983) and those of Earnshaw et al. (1984) that may be explained by differences in various scleroderma CREST antisera. Thus, different sera could recognize different antigens even though the staining reaction appeared identical. Alternately, differences in the titer of the antibodies could explain discrepancies in Western blots. Since the CREST serum is not monospecific, it is also possible that some of the bands shown by Western blots are for antigens other than those localized in the kinetochore. Obviously, more studies are necessary before these interesting possibilities can be determined.

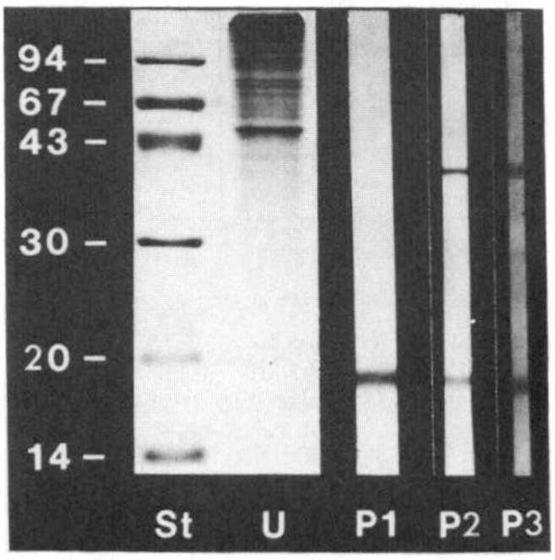

Figure 5
Western blot of isolated HeLa kinetochore fraction, probed with human anti-kinetochore. Proteins from isolated HeLa kinetochores were run in a 15–5% polyacrylamide gradient gel, electrophoretically transferred to nitrocellulose paper, and probed with three different anti-kinetochore sera (P1, P2, P3). Standard marker proteins are the same as in Fig. 4.

SUMMARY AND CONCLUSIONS

The association of chromosomes with spindle fibers is an essential feature of chromosome segregation in mitosis and meiosis. To understand the mechanism of chromosome segregation in higher eukaryotic organisms, it is essential to understand the molecular basis of spindle attachment to chromosomes via the kinetochore. An initial procedure is described in this paper that permits the purification of the kinetochores from mammalian metaphase chromosomes. A preliminary biochemical analysis of this fraction has served to identify several polypeptides that may be candidates for specific components of the kinetochore/centromere. Of special interest is a protein of approximately 18K associated with the kinetochore and recognized by immunoblotting using human autoantibodies from scleroderma CREST patients.

Several centromere DNA sequences have been isolated from yeast chromosomes (Clarke and Carbon 1980). Some of these clones have been used as affinity probes to identify specific centromere-binding proteins (Bloom et al., this volume). By using human autoimmune antibodies specific for kinetochores, several polypeptides have now been identified as possible components of the kinetochore/centromere from this work as well as that of Cox and Olmsted (this volume) and Earnshaw et al. (1984). Together, these experiments provide considerable promise that the molecular basis of the kinetochore–spindle fiber association may soon be elucidated.

ACKNOWLEDGMENTS

We thank Linda Wible for technical assistance, Bill Mollon for electron microscopy assistance, Bonnie Dunbar and Donnie Bundman for performing two-dimensional gels, and Pat Williams for typing the manuscript. This research was supported by grant CA-23022 awarded to B.R.B. M.M.V. was supported by a research fellowship of Fundacion Juan March of Spain.

REFERENCES

Adolph, K.W., S.M. Cheng, J.R. Paulson, and U.K. Laemmli. 1977. Isolation of a protein scaffold from mitotic HeLa cell chromosomes. *Proc. Natl. Acad. Sci.* **74:** 4937.

Brenner, S., D. Pepper, M.W. Berns, E. Tan, and B.R. Brinkley. 1981. Kinetochore structure, duplication and distribution in mammalian cells: Analysis by human autoantibodies from scleroderma patients. *J. Cell Biol.* **91:** 95.

Clarke, L. and J. Carbon. 1980. Isolation of a yeast centromere and construction of functional small circular minichromosomes. *Nature* **287:** 504.

Cox, J.V., E.A. Schenk, and J.B. Olmsted. 1983. Human anti-centromere antibodies: Distribution, characterization of antigens, and effect on microtubule organization. *Cell* **35:** 331.

Earnshaw, W.C., N. Halligan, C. Cooke, and N. Rothfield. 1984. The kinetochore is part of the metaphase chromosome scaffold. *J. Cell Biol.* **98(1):** 352.

Gooderham, K. and P. Jeppesen. 1983. Chinese hamster metaphase chromosomes isolated under physiological conditions. A partial characterization of associated non-histone proteins and protein cores. *Exp. Cell Res.* **144:** 1.

Laemmli, U.K., S.M. Cheng, K.W. Adolph, J.R. Paulson, J.R. Brown, and W.R. Baumbach.

1978. Metaphase chromosome structure: The role of non-histone proteins. *Cold Spring Harbor Symp. Quant. Biol.* **42:** 351.

Lewis, C.D. and U.K. Laemmli. 1982. Higher-order metaphase chromosome structure: Evidence for metalloprotein interactions. *Cell* **29:** 171.

Moroi, Y., C. Peebles, M.J. Fritzler, J. Steigerwald, and E.M. Tan. 1980. Autoantibody to centromere (kinetochore) in scleroderma sera. *Proc. Natl. Acad. Sci.* **77:** 1627.

Musich, P.R., F.L. Brown, and J.J. Maio. 1977. Subunit structure of chromatin and the organization of eukaryotic highly repetitive DNA: Nucleosomal proteins associated with a highly repetitive mammalian DNA. *Proc. Natl. Acad. Sci.* **74:** 3297.

Olmsted, J.B. 1981. Affinity purification of antibodies from diazotized paper blots of heterogeneous protein samples. *J. Biol. Chem.* **256(23):** 11955.

Rattner, J.B., G. Krystal, and B.A. Hamkalo. 1978. Selective digestion of mouse metaphase chromosomes. *Chromosoma* **66:** 259.

Sanger, J.W. 1975. Presence of actin during chromosomal movement. *Proc. Natl. Acad. Sci.* **72(6):** 2451.

Towbin, H., T. Staehelin, and J. Gordon. 1979. Electrophoretic transfer of proteins from polyacrylamide gels to nitrocellulose sheets: Procedure and some applications. *Proc. Natl. Acad. Sci.* **76:** 4350.

Wray, W. and E. Stubblefield. 1970. A new method for the isolation of chromosomes, mitotic apparatus or nuclei from mammalian fibroblast at near neutral pH. *Exp. Cell Res.* **59:** 469.

Nuclear Matrix–Intermediate Filament Scaffold: Subfraction of the Skeletal Framework That Retains Proteins Specific to Cell Type

Edward G. Fey, Katherine M. Wan, and Sheldon Penman
Department of Biology
Massachusetts Institute of Technology
Cambridge, Massachusetts 02139

The skeletal framework is the complex structure underlying cell morphology and internal organization. It is composed of both the major filament systems (microfilaments, intermediate filaments, and microtubules) and many interconnecting structural elements. Together, these form dense anastomosing three-dimensional networks. The skeletal framework is revealed after lipids are extracted and soluble proteins are released from intact cells with Triton X-100 in a suitably buffered medium (Brown et al. 1976; Lenk et al. 1977; Webster et al. 1978; Schliwa and Van Blerkom 1981). The skeletal framework contains the nucleus and cytoplasmic protein filaments arranged in a stable, self-supporting network of protein filaments and is covered by a protein layer or plasma lamina composed of plasma membrane proteins remaining after lipid extraction (Ben-Ze'ev et al. 1979). The framework has been shown to retain quantitatively the polyribosomes in all cell types examined (Lenk et al. 1977; Fulton et al. 1980; Cervera et al. 1981). Free ribosomes and considerable mRNA are found in the soluble fraction, but these do not engage in translation. These studies suggest (but have not yet proven) that binding of polyribosomes to the skeletal framework may be obligatory for protein synthesis.

The skeletal framework can be further dissected into distinct structural subfractions, each with a unique protein composition and morphology. Of particular significance is the core structure composed of the nuclear matrix and intermediate filaments, which we have termed the nuclear matrix–intermediate filament (NM-IF) scaffold. These studies of cell structure and associated biochemistry utilize two recently developed techniques. One is the sequential extraction procedure that selectively divides the cellular constituents into four specific fractions. The full utility of this extraction procedure is realized when the sequentially prepared cellular structures are examined morphologically, utilizing embedment-free microscopy. The sequential extractions yield cell structures that retain cellular morphology in an apparently native configuration.

The specificity of the fractionation protocol is shown by the unique protein composition of each cellular substructure. The localization of populations of proteins in individual fractions has been demonstrated using two-dimensional gel electrophoresis and immunological identification of proteins. We have shown several examples of specific proteins that comprise the skeletal framework localized in the NM-IF scaffold (Fey et al. 1984). In this paper, we examine both the biochemical and morphological distributions of four proteins during fractionation of HeLa cells into soluble and skeletal subfractions. Using a monoclonal antibody raised to a major 42-kD cytoskeletal protein, we show that, unlike actin, this protein is exclusively retained in the skeletal framework.

Further fractionation of the skeletal framework using nuclease digestion and elution with ammonium sulfate generates three distinct protein subfractions of the skeletal framework: the salt-labile cytoskeleton (CSK), the nuclease/salt-released chromatin (Chrom), and the salt/nuclease-resistant NM-IF fractions. Of these, the NM-IF fraction is of particular interest. The NM-IF contains only 5% of total cellular protein, yet it faithfully retains many aspects of the morphology of intact cells. In addition, the NM-IF displays large compositional changes characteristic of individual cell types, much more so than any of the other fractions. These characteristics suggest that the NM-IF may be important in determination and retention of differentiated phenotype.

Composition of the Skeletal Framework

Extraction of intact cells with Triton X-100 solubilizes more than 95% of the cellular phospholipid. The removal of the plasma membrane barrier allows the soluble proteins of the cell to diffuse into the extraction buffer. Depending on cell type, the soluble proteins comprise from 55% to 65% of the total cellular protein (Lenk et al. 1977; Lenk and Penman 1979; Fey et al. 1984; Staufenbiel and Deppert 1984). When monolayer cells are extracted with Triton X-100, the membrane-depleted skeletal framework remains in association with the substrate. This skeletal framework has been shown to retain, with remarkable fidelity, many of the morphological characteristics of differentiated cell structure, including the proteinaceous plasma lamina that retains the configuration of the plasma membrane (Ben-Ze've et al. 1979).

We have previously shown that the skeletal framework is composed of many hundreds of different proteins. This skeletal framework has now been subfractionated into three distinct protein populations, each representing a unique biological and morphological domain of the cell (Capco et al. 1982; Fey et al. 1984). The CSK is solubilized by extracting the skeletal framework with 0.25 M ammonium sulfate. The proteins of the CSK comprise 20–30% of the total cellular protein and are, with some significant exceptions (e.g., actin), different from the soluble proteins. The remaining structure contains the nucleus anchored to the substrate by an intermediate filament network. This structure was further fractionated by digestion with DNase and RNase, followed by a second elution with 0.25 M $(NH_4)_2SO_4$. This procedure removes all proteins that require nucleic acid integrity (i.e., the chromatin fraction) for their association with the nucleus. The structure that remains after these fractionation steps is composed of a protein nuclear matrix associated with a dense web of intermediate filaments and is called the NM-IF scaffold (Fey et al. 1984). The two fractions have very different protein compositions with few polypeptides in common.

In this paper, we examine some protein components of the skeletal framework that are identified by polyclonal and monoclonal antibodies. The monoclonal antibodies were obtained from mice immunized with antigens from specific subfractions of the skeletal framework. Two monoclonal antibodies were used in the experiments described here; a 42-kD protein is shown to be specifically retained in the cytoskeleton fraction, and a 52-kD protein is completely localized in the NM-IF scaffold. Neither protein displays any detectable soluble form.

The proteins of the NM-IF scaffold vary considerably from one cell type to another. The cytokeratins, a family of proteins known to be principally expressed in cells of epithelial origin (Tseng et al. 1982; Schmid et al. 1983), are quantitatively retained in the NM-IF scaffold and are observed to vary in size and quantity in different epithelial cell types. These differences are among those observed when comparing the NM-IF scaffold proteins from epithelial HeLa cells with those obtained from fibroblastic 3T3 cells.

Localization of Cytoskeletal and NM-IF Antigens

Monoclonal antibodies were obtained using both CSK and NM-IF protein fractions as the source of antigen. Two of the antibody-producing lines that have been cloned and characterized are described here. These antibodies are specific to a 42-kD cytoskeletal protein and a 52-kD nuclear protein. Each of these proteins showed a specific localization among the four fractions produced in the fractionation of HeLa cells using the protocol described above. Polyclonal antibodies to two other major structural antigens were obtained; actin antibody was purchased from Transformation Research (Framingham, Massachusetts), and antibody to the nuclear lamins was a generous gift from L. Gerace (Johns Hopkins Medical School).

Figure 1 shows the results of an experiment where the antigens to each of these monoclonal antibodies were localized in the four structural fractions. Equivalent amounts of protein from each fraction were electrophoretically separated on 10% polyacrylamide gels (Laemmli 1970). The proteins were transferred to strips of nitrocellulose which were then incubated with antibodies to actin (Fig. 1a), the 42-kD protein (Fig. 1b), the 52-kD protein (Fig. 1c), and the nuclear lamin proteins (Gerace et al. 1978) (Fig. 1d). An appropriate second antibody, conjugated to horseradish peroxidase, was then incubated with each strip, and the antibodies were visualized after reaction with 4-chloro-1-naphthol and H_2O_2 (Hawkes et al. 1982). As observed previously (Fey et al. 1984), actin is present in all four subcellular fractions. The actin present in the soluble fraction probably consists largely of G-actin, whereas some of the cytoskeletal actin is F-actin in the form of microfilaments (see below) (Simpson and Spudich 1980). In

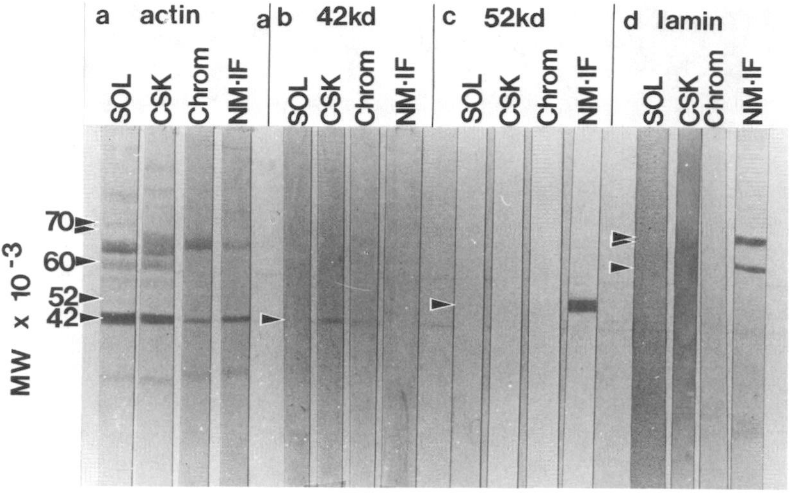

Figure 1
Identification of actin, a 42-kD cytoskeleton protein, the lamin proteins, and a 52-kD nuclear matrix protein in individual fractions from HeLa cells. Monolayers of HeLa cells were fractionated to produce soluble (SOL), cytoskeleton (CSK), chromatin (Chrom), and nuclear matrix-intermediate filament (NM-IF) components. Protein from 8×10^6 cells was separated by polyacrylamide gel electrophoresis and transferred to nitrocellulose, and specific antigens were identified in each fraction using a second antibody coupled to horseradish peroxidase as described in the text. Actin appeared in all four fractions (*a*), whereas the 42-kD protein identified by monoclonal antibody analysis was observed almost entirely in the CSK fraction (*b*). Some of the 42-kD protein is released into the chromatin but there is no detectable 42-kD protein present in the soluble fraction. The 60-kD, 67-kD, and 70-kD lamin proteins (*d*) as well as the 52-kD nuclear matrix protein (*c*) are observed entirely in the NM-IF complex with no detectable staining in any other fraction.

contrast, the 42-kD protein identified by the monoclonal antibody (Fig. 1b) is localized entirely in the cytoskeleton fraction. This protein is present as a major component of the cytoskeleton (pI 5.6). The monoclonal antibody to this protein shows no cross-reactivity with actin as determined by immunoblot analysis on two-dimensional polyacrylamide gels (data not shown). Localization of the 42-kD protein entirely on the skeletal framework, even after brief pulse labeling, suggests that the protein has little, if any, soluble phase.

When the 52-kD protein is localized in immunoblots, it is found entirely in the NM-IF fraction. This result, using HeLa cell protein, is in complete agreement with the observation in the epithelial MDCK cell line (Fey et al. 1984). The 60-kD, 67-kD, and 70-kD lamin proteins are the major components of the pore-lamin complex (Gerace et al. 1978) and are also a major component of the nuclear matrix (Staufenbiel and Deppert 1984). Figure 1d shows that these proteins are localized entirely in the NM-IF fraction. Each of these components of the skeletal framework—the 42-kD cytoskeleton protein, the 52-kD nuclear matrix protein, and the lamins—is quantitatively retained in the skeletal framework. None of these proteins are observed at any time in the soluble protein fraction, supporting the previous suggestion that many components of the skeletal framework are organized into the structural phase from the time of synthesis (Fulton and Wan 1983). Detailed analysis of the synthesis and assembly of these proteins is under way.

Spatial Localization of Proteins in the Skeletal Framework and the NM-IF Scaffold

HeLa cells were grown on glass coverslips and used to examine the distribution of the 42-kD protein, actin, the 52-kD protein, and the lamins on intact cells, skeletal frameworks, and NM-IF scaffolds (Fig. 2). Intact cells (Fig. 2a,d,g,j) were fixed and permeabilized in cold methanol (−20°C) for 10 minutes and postfixed in 3.7% formaldehyde. Skeletal frameworks (Fig. 2b,e,h,k) and NM-IF scaffolds (Fig. 2c,f,i,l) were prepared as described previously (Fey et al. 1984) and fixed in 3.7% formaldehyde. Fixed specimens were washed extensively in phosphate-buffered saline (PBS). Antibodies were applied at 1:40 dilutions in PBS and incubated for 30 minutes at 37°C. The appropriate second antibody, labeled with tetramethyl-rhodamine isothiocyanate, was applied, and the cells were examined using epifluorescence.

Actin is observed as a major protein component of all four biochemical fractions in every cell type examined thus far (Capco et al. 1982; Fey et al. 1984). The distribution of actin fluorescence (Fig. 2a–c) varied considerably during the course of this fractionation. In intact, permeabilized cells (Fig. 2a), actin was observed in a diffuse staining pattern evenly distributed over the entire cell. When phospholipid and soluble proteins were removed, the resulting skeletal framework (Fig. 2b) displayed a filamentous actin fluorescence in the cytoplasmic region of the cell and an intense nuclear fluorescence (Fig. 2b). In the NM-IF scaffold (Fig. 2c), actin fluorescence is observed in perinuclear filaments as well as in the nuclear matrix itself. The 42-kD protein identified by the monoclonal antibody used in Figure 1b stains both cytoplasmic filaments and the nucleus. This pattern is observed both in intact cells (Fig. 2d) and in skeletal framework preparations (Fig. 2e). The 42-kD protein is absent in the NM-IF scaffold preparation (Fig. 2f). This observation agrees with the immunoblot analysis in Figure 1b, where no detectable 42-kD antigen is observed in the NM-IF fraction.

The 52-kD antigens and the nuclear lamin proteins are observed entirely in the NM-IF scaffold in immunoblot analysis (Fig. 1c,d). Immunofluorescent localization of these antigens shows a constant nuclear organization that appears unaltered during fractionation (Fig. 2g–l). The lamin proteins (Fig. 2j–l) displayed a uniform nuclear fluorescence, whereas the 52-kD antigen was observed as in a punctate nuclear pattern (Fig. 2g–i). These results suggest that the lamins and the 52-kD protein are retained in a highly stable nuclear form. This salt-resistant nuclear matrix is a biochemically unique and morphologically stable substructure of most cells. In all cases, the spatial localization of antigens in cellular structures agrees closely with the distribution of the antigens in the immunoblot analysis described in Figure 1.

Proteins of the NM-IF Scaffold Vary with Cell Type

To establish the universality of the protocol described here, a number of different cell types have been fractionated, and the protein fractions have been compared

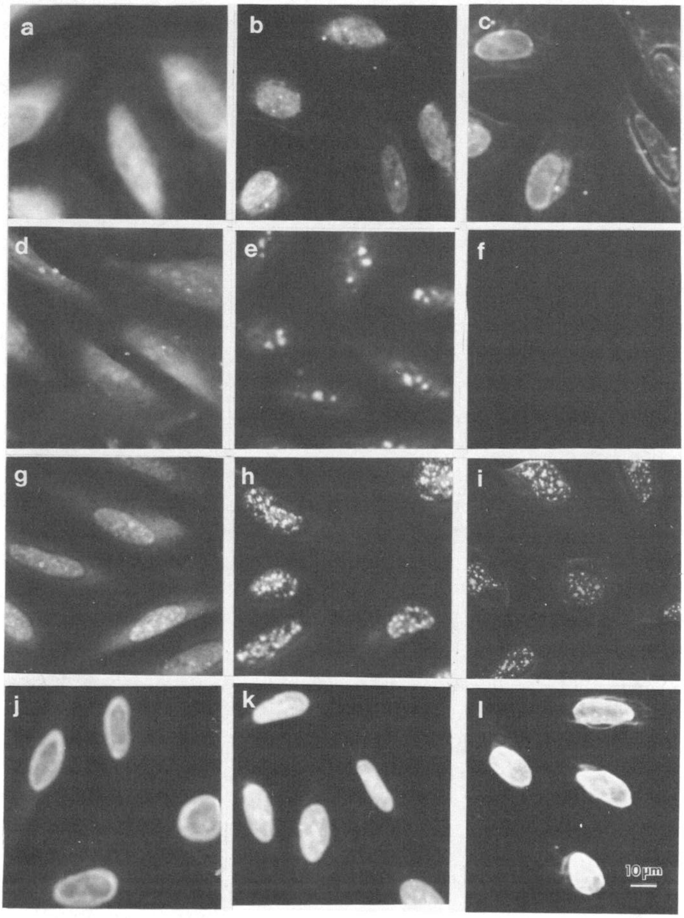

Figure 2
Localization of actin, a 42-kD cytoskeletal protein, the lamin proteins, and a 52-kD nuclear matrix protein in intact HeLa cells, skeletal frameworks, and NM-IF scaffold preparations. HeLa cells were grown on glass coverslips, and intact cell preparations (*a,d,g,j*) were fixed and permeabilized in cold methanol (−20°C, 10 min) and postfixed in 3.7% formaldehyde in PBS (0°C, 30 min). Skeletal frameworks (*b,e,h,k*) were obtained after extraction in cytoskeletal buffer containing 0.5% Triton X-100 (Fey et al. 1984). Soluble proteins were removed and the frameworks were fixed in 3.7% formaldehyde (0°C, 30 min). NM-IF scaffolds (*c,f,i,l*) were isolated by sequential extraction with 0.25 M $(NH_4)_2SO_4$, digestion with 100 μg/ml DNase I and 100 μg/ml RNase A followed by a further salt extraction. Scaffold preparations were washed in digestion buffer and fixed in formaldehyde (Fey et al. 1984). Antigens were localized in these preparations after incubation with the first antibody and a subsequent incubation with an appropriate second antibody labeled with tetramethyl-rhodamine isothiocyanate. Distribution of antigens was visualized using immunofluorescence microscopy. Actin (*a–c*) distribution is shown to be diffuse in intact cells (*a*), localized on filaments or in the nucleus of skeletal framework (*b*), or entirely on cytoplasmic filaments and the nuclear matrix in the NM-IF preparation (*c*). The 42-kD protein is observed in the intact cell and in the skeletal framework localized both on cytoplasmic filaments and in the nucleus (*d,e*). All 42-kD protein fluorescence is removed in the NM-IF scaffold (*f*). The lamins uniformly stain the nuclei (*j–l*), whereas the 52-kD protein (*g–i*) is observed with punctate nuclear staining. The localization of these proteins is unaltered during the entire course of fractionation.

using both one- and two-dimensional polyacrylamide gel analysis. Figure 3 shows a fluorogram of a polyacrylamide gel that compares proteins obtained from the epithelial HeLa cell line with those obtained from the fibroblastic 3T3 line. These cells were labeled for 12 hours with [^{35}S]methionine prior to fractionation, and equivalent cell numbers were loaded for each cell type in a particular fraction. Major differences in protein composition between analogous fractions are indicated by black dots. From this analysis, it is clear that the overall protein composition of each fraction of these two very different cell types is remarkably similar. Of the differences detectable in this analysis, the majority are observed in the NM-IF scaffold. As this fraction retains all the cytokeratin and vimentin intermediate filaments, proteins known to be differentially expressed in diverse cell types (Tseng et al. 1982; Schmid et al. 1983), the alterations in composition between 42-kD and 52-kD are probably due to differences in these proteins. There are a number of other compositional changes in the NM-IF that are most easily observed in two-dimensional analysis (Fig. 5). The proteins of the NM-IF scaffold may be involved in the establishment and retention of differentiated cell morphology (Fey et al. 1984).

Immunofluorescence microscopy was used to compare the distribution of specific proteins in the NM-IF scaffolds of HeLa and 3T3 cells. HeLa and 3T3 cells were grown on glass coverslips and fractionated as described above to yield the stable NM-IF scaffold structure. These preparations were stained using a double-antibody technique where the first antibody was directed against the lamins (Fig. 4a,b), vimentin (Fig. 4c,d), and the cytokeratins (Fig. 4e,f). These experiments illustrate the cell-type specificity of intermediate filament protein networks in epithelial cells. Both cell types express the lamins and vimentin, but only the epithelial HeLa cells display cytokeratin fluorescence. These results suggest that some of the cell-type-specific proteins are involved in cellular architecture responsible for the establishment of differentiated cell phenotype.

The proteins present in the NM-IF scaffolds of HeLa and 3T3 cells were compared by two-dimensional gel electrophoresis (Fig. 5). The lamins (L), vimentin (V), cytokeratins (C), and actin (A) were identified by immunoblot analysis on separate two-dimensional gels (data not shown), and the positions of these proteins are indicated by arrowheads. The specificity of cytokeratin proteins to the epithelial HeLa cell type is confirmed by this analysis. In addition to the cytokeratins, there are a number of major proteins in these NM-IF preparations that are specifically expressed in HeLa or in 3T3. The identification of these proteins and the correlation of their presence in tissues and animal species are the object of continued research. Utilization of the fractionation protocol described here allows a rapid purification of this cellular fraction that is enriched in cell-type-specific proteins.

Structure of the NM-IF Scaffold Revealed by Embedment-free Sections

Examination of NM-IF scaffold preparations using embedment-free sections in transmission electron microscopy provides clear images of protein filament net-

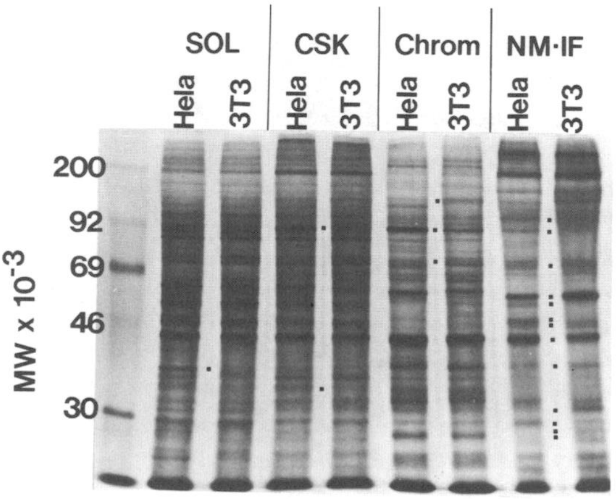

Figure 3
Polyacrylamide gel profiles comparing protein fractions from HeLa cells with those obtained after fractionation of 3T3 fibroblasts. HeLa and 3T3 cells were grown in monolayers, and proteins in both cell types were labeled in [^{35}S]methionine for 12 hr. Cells were fractionated as in Fig. 1 and analyzed after separation on a 10% polyacrylamide gel. Equal numbers of cells from each cell type were compared in each individual fraction. The number of radioactive counts loaded for each fraction was adjusted to facilitate comparison of all four fractions on one gel. (●) Major differences between HeLa and 3T3 protein bands within a fraction. From this analysis, it is apparent that the protein fractions obtained from two different cell types are remarkably similar and that the majority of protein differences are observed in the NM-IF scaffold.

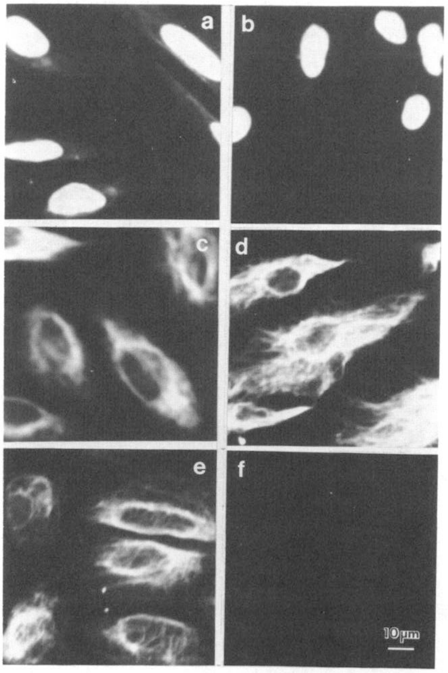

Figure 4
Immunofluorescent localization of lamins, vimentin, and cytokeratins in the NM-IF scaffolds of HeLa and 3T3 cells. NM-IF scaffolds were prepared from HeLa (*a,c,e*) and 3T3 (*b,d,f*) monolayers, and immunofluorescent localization of proteins was performed as described in Fig. 2. Both the lamin proteins (*a,b*) and vimentin (*c,d*) have a remarkably similar distribution in the NM-IF scaffolds of both cell types. The cytokeratins (*e,f*), although abundant in the epithelial HeLa NM-IF scaffold (*e*), are entirely absent in the fibroblastic 3T3 NM-IF scaffold (*f*). The cytokeratins are intermediate filament proteins that are characteristic of epithelial cells and can be used to identify cell type in a diagnostic assay (Tseng et al. 1982; Schmid et al. 1983). The NM-IF scaffold is therefore a cellular substructure that faithfully retains histological localization of a major class of proteins involved in cell and tissue differentiation.

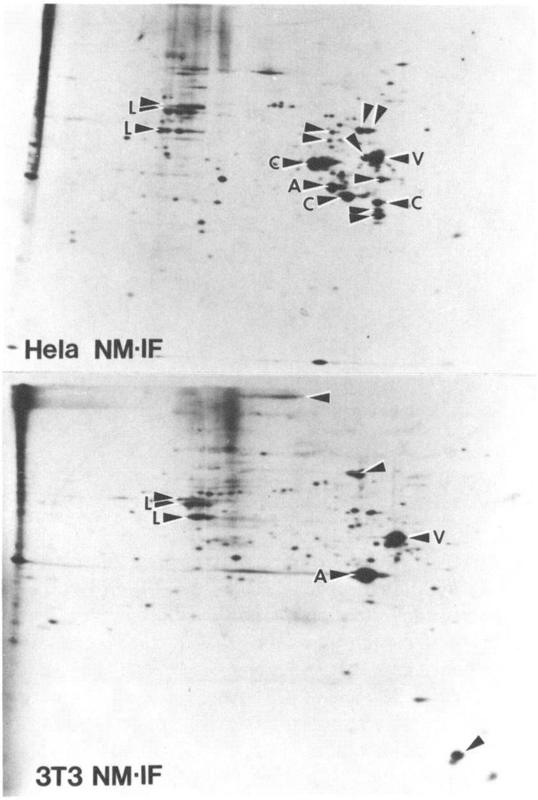

Figure 5
Two-dimensional polyacrylamide gel comparison of the proteins in the NM-IF scaffolds of HeLa and 3T3 cells. HeLa and 3T3 cell monolayers were labeled with [^{35}S]methionine and fractionated as described in Fig. 3. The proteins of the NM-IF scaffold from each of these cell types were analyzed by two-dimensional gel electrophoresis according to the method of O'Farrell (1975). Localization of the cytokeratins (C), lamins (L), vimentin (V), and actin (A) was determined by immunolocalization on two-dimensional gels transferred to nitrocellulose (data not shown). The position of these proteins is indicated in each gel. Arrowheads indicate proteins other than the cytokeratins that vary from 3T3 cells to HeLa cells. From this analysis, it is clear that a number of proteins retained in the NM-IF scaffold appear to be specific to individual cell types.

works (Capco et al. 1982; Fey et al. 1984). Figure 6a shows an embedment-free whole mount of an NM-IF scaffold preparation from HeLa cells. Cytoplasmic filaments appear to be connected with the lamina of the nuclear matrix. This image precludes examination of the interior of the nuclear matrix. Utilizing a recently developed sectioning technique (Capco et al. 1984), it is possible to analyze the interior structures of unembedded sections from nuclear matrix preparations. Figure 6b shows an embedment-free section of the cytoplasmic filaments and nuclear matrix interior of a HeLa NM-IF scaffold.

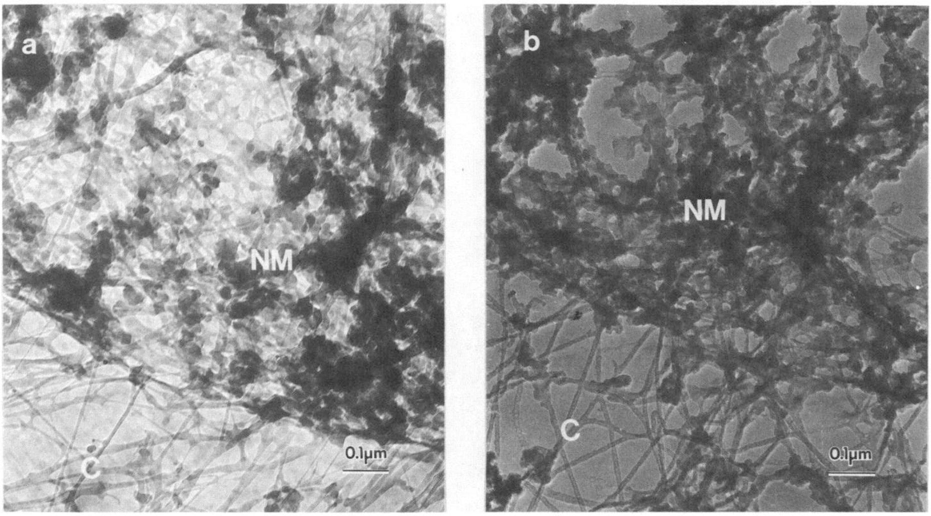

Figure 6
NM-IF scaffolds of HeLa cells viewed in whole mount or embedment-free section in transmission electron micrographs. NM-IF scaffold structures were prepared from HeLa cells and viewed in transmission electron microscopy. A whole mount (*a*) of the nuclear matrix (NM) displays the direct interaction of cytoplasmic filaments (C) with the lamina of the nuclear matrix. When NM-IF scaffold preparations were sectioned using a recently developed technique of embedment-free sectioning (Capco et al. 1984), the fibrillar interior of the nuclear matrix was observed (*b*). This structure is essentially devoid of DNA, RNA, and histones, yet retains the morphology of an intact nucleus.

DISCUSSION

The results described here briefly show some of the insights afforded by the sequential extraction procedure combined with embedment-free electron microscopy. It is worth emphasizing that the procedures yield essentially identical results with nearly every cell type tested so far. The method is insensitive to small variations in procedure, since the cell fractions are separated according to radically different criteria, i.e., release by Triton X-100 (soluble fraction), solubilization by 0.25 M $NH_4(SO_4)_2$ (cytoskeleton fraction), etc. The degree of fraction separation is remarkable as judged from the absence of common proteins. Nevertheless, the procedures are sufficiently gentle to preserve architectural integrity and spatial morphology. The structural networks are obtained free of soluble proteins and can be examined by embedment-free electron microscopy employing either the entire structure as whole mounts or resinless sections. These structural networks are masked in the conventional embedded section by the embedding resin.

Perhaps the most important aspects of these procedures are the many important and sometimes surprising insights afforded into cell-structure biochemistry. Also, several kinds of analyses are made possible by these techniques. The well-separated fractions offer unique sources of antigens for generating monoclonal antibodies to structural elements at high efficiency. For example, the 52-kD nuclear matrix antibody was obtained from a fraction that amounts to only 3–5% of total cell protein. The generation of cytoskeletal monoclonal antibodies is also facilitated by the removal of the irrelevant proteins of the soluble and other fractions. One such monoclonal antibody directed against the 42-kD acidic skeleton protein is shown here. Antibodies directed against the chromatin proteins are also accessible.

An important contribution to our understanding of cell structure has been the isolation of the nuclear matrix together with the associated intermediate filaments in a clearly visualized and biochemically well-defined preparation that we have termed the nuclear matrix–intermediate filament scaffold or NM-IF. The physical properties of this structure are quite different from the rest of cell architecture in that it is resistant to quite high ionic strength. The extraction conditions described here represent the most gentle conditions that permit a complete removal of chromatin and cytoskeleton. As will be reported elsewhere, the nuclear matrix as prepared here behaves metabolically as a distinct set of proteins, suggesting that a physiologically distinct cellular substructure has been isolated. Most notable is the retention of important aspects of cell and tissue morphology by the NM-IF even though this fraction amounts to only a small percentage of total cell protein.

The presumptive importance of the nuclear matrix for cell functions has been noted by a number of laboratories (Berezney 1979; Vogelstein et al. 1980; Brasch 1982; Setterfield et al. 1983). Of particular interest here is its probable role in organizing chromatin architecture. The separation of a well-defined nuclear matrix fraction has afforded a remarkable result that may be of considerable significance. Comparing cells of radically different types such as the epithelial HeLa cells and the fibroblastic 3T3 line shows a surprising similarity of the composition of the protein fractions with the exception of the NM-IF. This is a consistent result from a number of cell types, suggesting that a major protein correlate of phenotype is to be found in the NM-IF. This result is not entirely unprecedented, since the keratins that vary markedly in epithelial cells of different origin belong to the NM-IF as prepared here. We have shown that many other proteins of this substructure, including presumably those of the internal nuclear matrix, also vary with cell type. In contrast, the chromatin fraction shows very little change. The NM-IF scaffold may very well play a fundamental role in determining cell specialization and differentiation.

REFERENCES

Ben-Ze'ev, A., A. Duerr, F. Solomon, and S. Penman. 1979. The outer boundary of the cytoskeleton: A lamina derived from plasma membrane proteins. *Cell* **17:** 859.

Berezney, R. 1979. Dynamic properties of the nuclear matrix. In *The cell nucleus* (ed. H. Busch), vol. VII, p. 413. Academic Press, New York.

Brasch, K. 1982. Fine structure and localization of the nuclear matrix in situ. *Exp. Cell Res.* **140:** 161.

Brown, S., W. Levinson, and J. Spudich. 1976. Cytoskeletal elements of chicken embryo fibroblasts revealed by detergent-extraction. *J. Supramol. Struct.* **5:** 119.

Capco, D.G., G. Krochmalnic, and S. Penman. 1984. A new method of preparing embedment-free sections for transmission electron microscopy: Applications to the cytoskeletal framework and other three-dimensional networks. *J. Cell Biol.* **98:** 1878.

Capco, D.G., K.M. Wan, and S. Penman. 1982. The nuclear matrix: Three-dimensional architecture and protein composition. *Cell* **29:** 847.

Cervera, M., G. Dreyfuss, and S. Penman.1981. Messenger RNA is translated when associated with the cytoskeletal framework in normal and VSV-infected HeLa cells. *Cell* **23:** 113.

Fey, E.G., K.M. Wan, and S. Penman. 1984. The epithelial cytoskeletal framework and nuclear matrix-intermediate filament scaffold: Three-dimensional organization and protein composition. *J. Cell Biol.* **98:** 1973.

Fulton, A.B. and K.M. Wan. 1983. Many cytoskeletal proteins associate with the HeLa cytoskeleton during translation in vitro. *Cell* **32:** 619.

Fulton, A.B., K.M. Wan, and S. Penman. 1980. The spatial distribution of polyribosomes in 3T3 cells and the associated assembly of proteins into the skeletal framework. *Cell* **20:** 849.

Gerace, L., A. Blum, and G. Blobel. 1978. Immunocytochemical localization of the major polypeptides of the nuclear pore complex-lamina fraction. *J. Cell Biol.* **79:** 546.

Hawkes, R., E. Niday, and J. Gordon. 1982. A dot-immunobinding assay for monoclonal and other antibodies. *Anal. Biochem.* **119:** 142.

Laemmli, U.D. 1970. Cleavage of structural proteins during the assembly of the head of bacteriophage T4. *Nature* **227:** 680.

Lenk, R. and S. Penman. 1979. The cytoskeletal framework and poliovirus metabolism. *Cell* **16:** 289.

Lenk, R., L. Ransom, Y. Kaufmann, and S. Penman. 1977. A cytoskeletal structure with associated polyribosomes obtained from HeLa cells. *Cell* **10:** 67.

O'Farrell, P.H. 1975. High-resolution two-dimensional electrophoresis of proteins. *J. Biol. Chem.* **250:** 4007.

Schliwa, M. and J. Van Blerkom. 1981. Structural interaction of cytoskeletal components. *J. Cell Biol.* **90:** 222.

Schmid, E., D.L. Schiller, C. Grund, J. Stadler, and W.W. Franke. 1983. Tissue type-specific expression of intermediate filament proteins in a cultured epithelial cell line from bovine mammary gland. *J. Cell Biol.* **96:** 37.

Setterfield, G., R. Hall, T. Bladon, J.E. Little, and J.G. Kaplan. 1983. Changes in structure and composition of lymphocyte nuclei during mitogenic stimulation. *J. Ultrastruct. Res.* **82:** 264.

Simpson, P.A. and J.A. Spudich. 1980. ATP-driven steady-state exchange of monomeric and filamentous actin from *Dictyostelium discoideum. Proc. Natl. Acad. Sci.* **77:** 4610.

Staufenbiel, M. and W. Deppert. 1984. Preparation of nuclear matrices from cultured cells: Subfractionation of nuclei in situ. *J. Cell Biol.* **98:** 1886.

Tseng, S.C.G., M.J. Jarvinen, W.G. Nelson, J.-W. Huang, J. Woocock-Mitchell, and T.-T. Sun. 1982. Correlation of specific keratins with different types of epithelial differentiation; monoclonal antibody studies. *Cell* **30:** 361.

Vogelstein, B., D.M. Pardoll, and D.S. Coffey. 1980. Supercoiled loops and eukaryotic DNA replication. *Cell* **22:** 79.

Webster, R.S., D. Henderson, M. Osborn, and K. Weber. 1978. Three-dimensional electron microscopical visualization of the cytoskeleton of animal cells: Immunoferritin identification of actin- and tubulin-containing structures. *Proc. Natl. Acad. Sci.* **75:** 5511.

Tubulin Dynamics in Microtubules of the Mitotic Spindle

Edward D. Salmon
Department of Biology
University of North Carolina
Chapel Hill, North Carolina 27514

A critical factor in understanding the mechanisms that move chromosomes and regulate the assembly of mitotic spindles is identifying the pathways by which tubulin subunits exchange with microtubules in mitotic spindle fibers. The spindle is composed of a bipolar arrangement of microtubules, as diagramed in Figure 1A for a metaphase spindle in a first-division sea urchin embryo. Mitotic microtubules that extend away from the region of the centrosome and spindle poles in vivo all have "plus" distal structural polarity (Fig. 1A), similar to the structural polarity of microtubules that have been nucleated from centrosomes in vitro (Telzer and Haimo 1981; McIntosh and Euteneuer 1984). Four possible pathways for the exchange of tubulin with spindle microtubules in vivo have been proposed, and these are summarized diagrammatically in Figure 1B. All but one of the proposed pathways (see no. 4 in Fig. 1B) (Inoué and Sato 1967; Bajer and Molé-Bajer 1975; Inoué and Ritter 1975) are based on end-dependent subunit exchange models, the mechanism by which microtubules apparently assemble in vitro (Hill and Kirschner 1982; Mitchison and Kirschner, this volume).

In this paper, I discuss recent findings from two experimental approaches that are directed toward elucidating the pathways of tubulin exchange with spindle and cytoplasmic microtubules in living cells. In the first approach, the rate of tubulin dissociation from nonkinetochore spindle microtubules was measured and compared to rates of tubulin association expected from the characteristic kinetic properties of end-dependent microtubule polymerization in vitro (Salmon et al. 1984a). The second approach involves measurements of microtubule dynamics in living cells using tubulin fluorescent analog cytochemistry and fluorescence redistribution after photobleaching (FRAP). This work has been done in collabo-

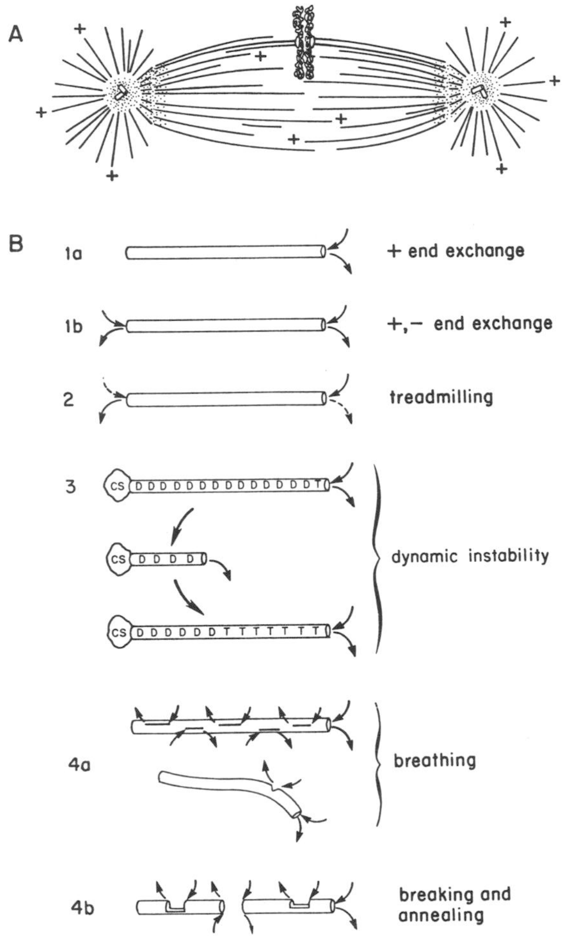

Figure 1
(*A*) Sketch of the current concept of microtubule arrangements in a metaphase first-division sea urchin spindle. Such a cartoon is based on detailed structural studies but must still be considered preliminary and merely indicative of the true state of affairs. The structure of the spindle poles, in particular, is poorly understood. Only one chromosome pair is drawn for clarity. Kinetochore-microtubules are defined as microtubules attached to kinetochores. (*B*) Four possible pathways of tubulin exchange with spindle microtubules: (1) subunit exchange by reversible association-dissociation events at one (*a*) or both (*b*) ends of a microtubule; (2) treadmilling or head-to-tail polymerization as a consequence of GTP hydrolysis; (3) dynamic instability as a consequence of the loss of tubulin-GTP "caps" at the plus ends; (4) subunit exchange all along the apparent length of a microtubule either by insertion or deletion of tubulin within the microtubule wall (*a*) or by rapid breaking and reannealing (*b*). (Reprinted, with permission, from Salmon et al. 1984b.)

ration with J.R. McIntosh's laboratory (University of Colorado). Initial results for microtubule arrays in sea urchin and mammalian culture cells were described in detail recently (Salmon et al. 1984b; Saxton et al. 1984). I focus here on nonkinetochore microtubule dynamics in sea urchin spindles. Kinetochore fiber microtubules are differentially stable, and their assembly is more complex (see Salmon and Begg 1980; Rieder 1982; Mitchison and Kirschner, this volume).

Investigating Tubulin Dissociation Rates In Vivo by Microinjecting Colchicine and Tubulin-Colchicine Complex into Living Cells

When microtubule polymerization is abruptly blocked at metaphase or early anaphase, the intrinsic rate of tubulin dissociation from microtubules is revealed. In experiments that are described in detail elsewhere (Salmon et al. 1984a), colchicine, Colcemid, or nocodazole at high intracellular concentrations was used to bind tubulin rapidly, thus blocking polymerization. Nonkinetochore microtubules depolymerized within a characteristic time of 20 seconds in sea urchin embryos (Fig. 2) and 10–15 seconds in PtK_1 tissue-culture cells; these values are based on measurements of changes in birefringence retardation of spindle fibers.

Colchicinelike drugs bind to the tubulin dimer at a site that is not exposed when the dimer is assembled into microtubules (Wilson and Meza 1973; Margolis and Wilson 1977; Deery and Weisenberg 1981; Schiff and Horwitz 1981). Tubulin dimer bound with colchicine (T-C) is the active intermediate for the rapid depolymerization of spindle microtubules following the injection of colchicine into sea urchin cells. P. Wadsworth has found in preliminary studies conducted in my laboratory that microinjecting mitotic cells with >2 μM intracellular concentration of T-C complex (made from porcine brain tubulin; Engelborghs and Lambeir 1980) produces loss of spindle birefringence retardation within 20 seconds, the same characteristic time as for spindle disassembly caused by injected colchicine at high concentrations. Colchicine binds to and dissociates from tubulin very slowly (Borisy and Taylor 1967); 2 μM free intracellular colchicine has little effect on spindle assembly during mitosis.

In the sea urchin embryo, following microinjection of 0.1–5.0 mM intracellular colchicine, the birefringence retardation of nonkinetochore microtubules dimin-

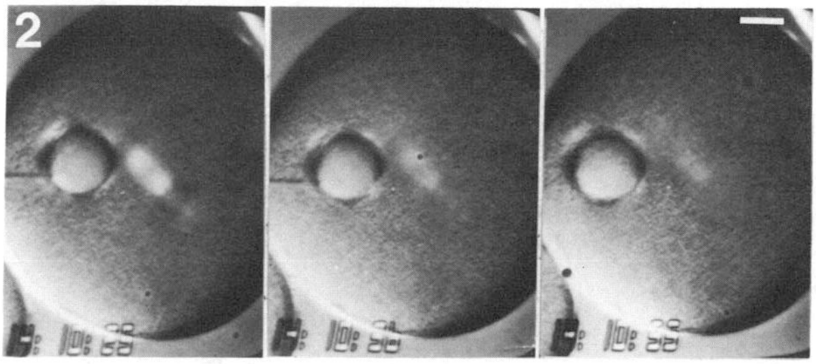

Figure 2
Polarization micrographs showing the rapid loss of spindle birefringence following the microinjection of 0.5 mM intracellular colchicine (for details, see Salmon et al. 1984a). Time is given in hr:min:sec on each frame. By 20 sec, most nonkinetochore fiber birefringence retardation has disappeared. Magnification, 202×; scale, 20 μm.

ished uniformly throughout the spindle (Fig. 2). The birefringence-retardation decay followed exponential kinetics with a first-order rate constant of $k = 0.11 \pm 0.023$ sec^{-1} and a corresponding half-time of $t_{1/2} = 6.5 \pm 1.1$ sec. The initial rate of tubulin dissociation from nonkinetochore spindle microtubules is in the range of $k_{d(spindle)}$ = 180–992 dimers/sec/microtubule (Salmon et al. 1984a). This range is based on k = 11% of the initial polymer/sec and the average length of a half-spindle microtubule, which is estimated from electron micrographs to be between 1 μm and 5.5 μm. If all half-spindle microtubules have one end in the centrosome or pole region, the so-called "minus" end, then the 5.5-μm estimate, corresponding to $k_{d(spindle)}$ = 992 dimers/sec/microtubule, appears to be most likely for microtubules in the sea urchin half-spindle, which is typically 11–12 μm long.

This estimate of non-steady-state $k_{d(spindle)}$ appears to approximate the steady-state rate of tubulin exchange with the nonkinetochore spindle microtubules. Several birefringence studies of spindle reassembly or augmentation (for review, see Salmon et al. 1984a) demonstrate that the rate of microtubule polymerization in marine embryos is at least 30% as fast as the rate of disassembly observed after injection of high intracellular colchicine. The differences in the rates of disassembly and assembly, however, may be due in part to delays caused by initiation events during assembly.

The Rate of Tubulin Dissociation Is Much Faster Than Expected from In Vitro End-dependent Association Parameters

How does the value of $k_{d(spindle)}$ compare with rates of tubulin association expected from kinetic parameters measured for end-dependent polymerization in vitro? Values for the bimolecular rate constant for tubulin association with microtubules in vitro range from 1×10^6 to 20×10^6 M^{-1}sec^{-1} (summarized in Salmon et al. 1984a). The association rate constant in vivo will be smaller than the in vitro rate constant because the diffusion rate of tubulin in sea urchin cytoplasm is eight to ten times slower than in vitro (Salmon et al. 1984c). The expected association rate constant for tubulin in the sea urchin embryo is $k_{a(embryo)} = 0.1 \times 10^6$ to 2.5×10^6 M^{-1}sec^{-1}. This range of values predicts significantly slower steady-state microtubule polymerization-depolymerization than seen in vivo if the critical concentration for spindle tubulin assembly in vivo is about 2 μM, the value measured for in vitro reassembly buffers (Keller and Rebhun 1982): $(0.1–2.5) \times 10^6$ M^{-1}sec^{-1} $\times$ 2 μM = 0.2–5.0 dimers/sec. Even for the highest estimated concentration of tubulin in the egg (range 5–27 μM; Pfeffer et al. 1976), the rate of tubulin association-dissociation does not balance if microtubules assemble only by addition of subunits at the ends.

Increasing the number of exchange sites along the apparent length of microtubules solves this problem (see no. 4 in Fig. 1B). If microtubules are constantly breaking and reannealing, then not all microtubule "minus" ends are located near the spindle poles or in the centrosome. On the other hand, the concentration and mobility of tubulin in the spindle region may be significantly different than in the surrounding cytoplasm.

Investigating Microtubule Dynamics with DTAF-Tubulin and FRAP Techniques

A powerful approach for analyzing microtubule dynamics in living cells is the combination of fluorescent analog cytochemistry (Kreis and Birchmeier 1982; Wang et al. 1983; Taylor et al. 1984) and measurements of FRAP (Jacobson et al. 1983). In our approach, tubulins labeled with dichlorotriazinyl amino fluorescein (DTAF-tubulin) were microinjected into cells where they mimicked the behavior of endogenous tubulin, thus serving as tracers of the cellular tubulin pool during the assembly and disassembly of microtubules. FRAP techniques are well established for measuring the mobility of fluorescent molecules in cells (Jacobson et al. 1983). With laser microbeaming, fluorescence can be bleached in a restricted, well-defined region of a cell or organelle. Because photobleaching is permanent, recovery of fluorescence in a bleached region requires redistribution of unbleached, labeled tubulin subunits into the irradiated area. The instrumentation procedures we have used for laser photobleaching and quantifying temporal changes in the spatial distribution of fluorescence using low-light-level video cameras and photographic procedures (video FRAP) are described in detail elsewhere, along with a description of measurements of the rate of tubulin diffusion in the cytoplasm of sea urchin embryos (Salmon et al. 1984c).

Procedures developed in McIntosh's laboratory for purifying bovine or porcine neurotubulin labeled with DTAF (DTAF-to-protein ratio ~1.0) and the in vitro assembly characteristics of DTAF-tubulin are described by Leslie et al. (1984). DTAF-tubulin has assembly-disassembly characteristics in reassembly buffers in vitro that are typical of unlabeled tubulin. Normal microtubules grow from DTAF-tubulin nucleated from the ends of axonemal fragments and purified centrosomes, and DTAF-tubulin does not appear to bind to the walls of microtubules in vitro.

When DTAF-tubulin was injected at low concentrations into living sea urchin embryos (Fig. 3), or mammalian tissue-culture cells, fibrous patterns of fluorescence formed, similar to anti-tubulin immunofluorescence and polarization im-

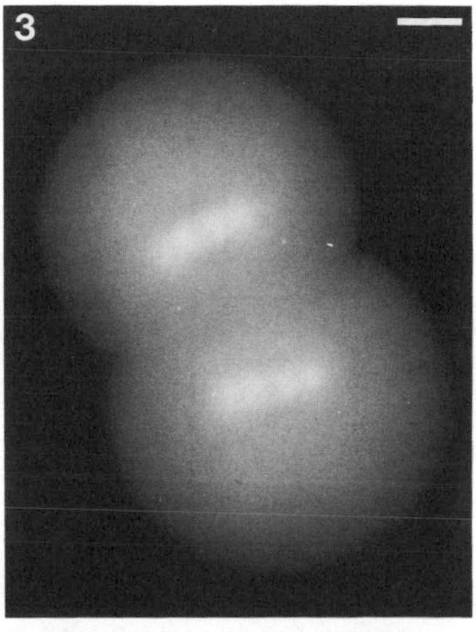

Figure 3
Micrograph of the DTAF-tubulin distribution in metaphase mitotic cells of a second-division embryo of *Lytechinus variegatus*. The embryo was injected with DTAF-tubulin about 30 min before first mitosis. The pattern of DTAF-tubulin fluorescence in the spindle-aster complex is similar to birefringence images seen in polarization microscopy, as described by Salmon et al. (1984b). Magnification, 310×; scale, 20 μm.

ages of microtubule arrays at similar stages in the cell cycle (for photographs and details, see Salmon et al. 1984b and Saxton et al. 1984). Changes in fluorescence intensity of microtubule arrays in living cells injected with DTAF-tubulin followed the normal fluctuations in assembly of cytoplasmic and mitotic spindle microtubules throughout the cell cycle, and fluorescence disappeared upon application of nocodazole, as happens with normal microtubules in vivo.

Spindle FRAP

In the sea urchin spindle experiments, cells were injected with DTAF-tubulin before mitosis, so the DTAF-tubulin equilibrated with the endogenous tubulin pool and the labeled tubulin dimers were incorporated uniformly into the microtubules that formed the mitotic spindle. After about 50% of the initial fluorescence was bleached, Salmon et al. (1984b) found that the recovery of fluorescence in either large or small bleached regions of the sea urchin mitotic spindle had four striking characteristics: It was exponential, rapid, nearly complete, and uniform (Fig. 4).

As measured by the video FRAP methods, the recovery of fluorescence in the bleached region was an exponential function of time. The time for 50% recovery ($t_{1/2}$) was 19.4 ± 6.2 seconds. Maximum recovery, about 93% initial fluorescence intensity, was reached within 60–90 seconds. Fluorescence recovered uniformly throughout the spindle, but no distinct geometrical pattern could be resolved.

More recently, Wadsworth and Salmon have analyzed the recovery of fluorescence using photon-counting, photomultipler recording methods that provide better sensitivity and temporal resolution than the video recording methods used previously (Salmon et al. 1984b). The recovery of fluorescence in the bleached region of a half-spindle was seen to involve fast and slow exponential phases (Fig. 5a,b). The fast phase is due to diffusion of tubulin within the bleached region at rates similar to the diffusion of tubulin in the cytoplasm (cf. Fig. 5c). The second phase is due to the incorporation of unbleached DTAF-tubulin into microtubules of the spindle fibers. We have not analyzed these kinds of data in detail, but the half-times for the second phase of recovery of fluorescence in the

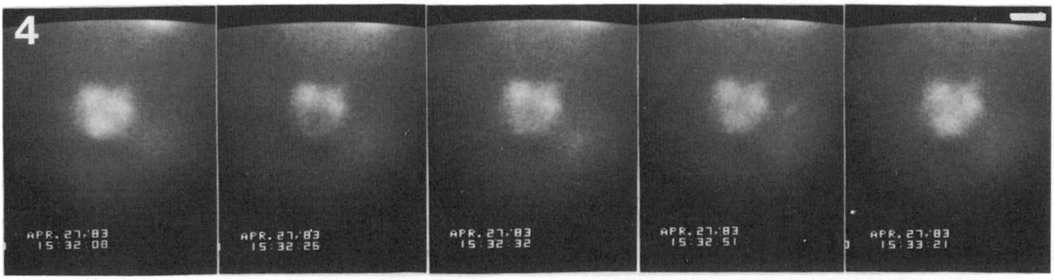

Figure 4
Fluorescence recovery after photobleaching the lower half-spindle of a tripolar metaphase spindle in a first-division embryo of *Lytechinus variegatus*. Diameter of the bleaching beam was about 4 μm. Video images were recorded as described by Salmon et al. (1984b,c). Magnification, 175×; scale, 20 μm.

bleached region are in the range of 14–20 seconds, similar to the half-times measured by the video FRAP method. Again, recovery of fluorescence was nearly

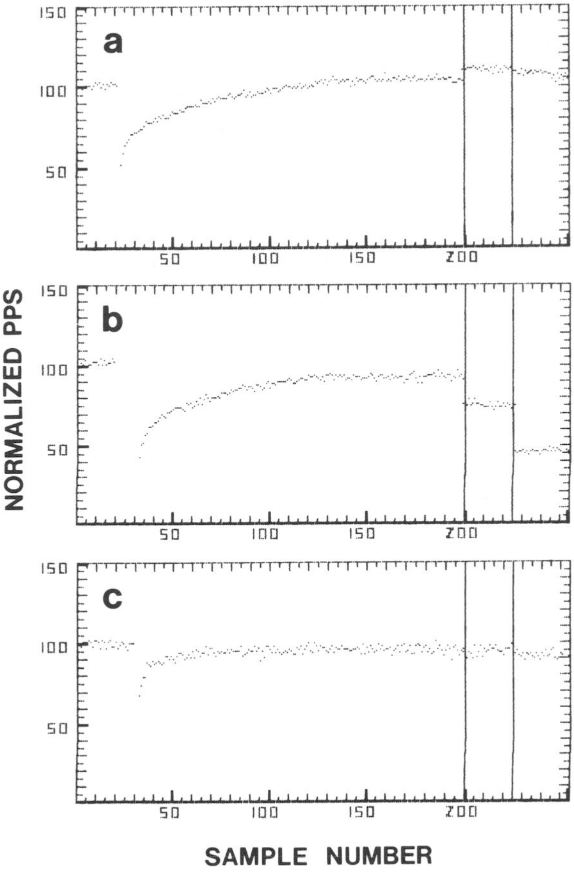

Figure 5
Computer records of FRAP of the central half-spindle region of metaphase spindles (*a,b*) and a cytoplasmic region adjacent to the spindle in *b* for first-division embryos of the sea urchin *Lytechinus variegatus* (*c*). Fluorescence excitation and bleaching were produced by a laser microbeam ($\lambda = 488$ nm and 4.2 μm dia). During the measurement phase, the laser microbeam intensity was attenuated by a factor of about 5×10^{-4} compared with the bleaching intensity. Bleaching period was about 300 msec. The number of photon counts was recorded every 0.5 sec using an EMI 9863A photomultiplier, C-10 photon counter (EMI) and an Apple computer to sample and store the data. Initial fluorescent intensities before bleaching were about 5000 pulses per second (pps) for the central half-spindle region (*a,b*) and 2000 pps for the adjacent cytoplasm (*c*). A total of 750 samples were taken in each experiment. The first 200 samples were plotted, then samples 450–475 (between the first and second vertical bars), and then samples 720–750 (second vertical bar to end). Note that in *a*, the spindle goes into anaphase about 2 min after bleaching. In b, the spindle goes into anaphase shortly after bleaching, and the normal decrease in microtubule assembly during anaphase is seen in later stages of recovery. In *c*, cell cleavage occurs during the late stages of recovery, and the optical measurement path through the cell is reduced.

complete (>90%) within 60 seconds. The exact extent of recovery is difficult to measure in sea urchin spindles because of the normal changes in assembly that occur as the spindle progresses through mitosis (see Fig. 5b).

Does the rapid rate of fluorescence recovery that we observed represent steady-state microtubule dynamics or reassembly of microtubules following depolymerization induced by the high-intensity photobleaching light? The 14–20-second half-time for recovery of fluorescence is similar to the reassembly rate following depolymerization of spindles by various agents (for review, see Salmon et al. 1984a). However, evidence thus far shows that microtubules in vivo (Salmon et al. 1984b; Saxton et al. 1984) and in vitro (Leslie et al. 1984) do not depolymerize significantly when fluorescence is photobleached. Photobleaching can cross-link microtubule proteins in vitro under some conditions, and the possibility that photobleaching does submicroscopic damage to the tubulin lattice has not yet been ruled out (Leslie et al. 1984). Nevertheless, the available evidence is positive that the FRAP results reflect steady-state microtubule dynamics.

This conclusion is also supported by the difference in FRAP between spindle and cytoplasmic microtubules. For PtK_1 and BSC cultured cells, Saxton et al. (1984) have measured half-times of recovery of fluorescence in bleached regions of metaphase spindles of about 12 ± 7 seconds, similar to the forward incorporation rate of DTAF-tubulin into the spindle fibers when injected at metaphase (Saxton et al. 1984), to the rate of incorporation of tubulin labeled with DTAF into sea urchin astral fibers (Wadsworth and Sloboda 1983), and to the sea urchin spindle FRAP rates. Bleached regions of cytoplasmic microtubule arrays in interphase cells also recover fluorescence exponentially with time, but the half-time of the nondiffusional recovery is 235 ± 80 seconds, or about 19 times slower. The difference in FRAP rates between spindle and cytoplasmic microtubules resembles the different rates of depolymerization for nonkinetochore spindle microtubules and interphase cytoplasmic microtubule arrays when assembly is blocked by nocodazole (M. de Brabander, pers. comm.; E.D. Salmon et al., unpubl.).

Tubulin Exchange Is Not a Simple Equilibrium at Microtubule Ends

The observed 20-second half-time for turnover of tubulin within microtubules of sea urchin spindle fibers in our FRAP studies clearly cannot be accounted for by simple equilibrium exchange of tubulin at the ends of microtubules (see no. 1 in Fig. 1B). At steady state, the average number of tubulin dimers, m, replaced inward from one end of a microtubule as a function of time, t, and dissociation constant, k_d, is given by (Zeeberg et al. 1980; Hill and Kirschner 1982): $m = 2\,(k_d t/\pi)^{1/2}$. If $k_d = 992$ dimers/sec, corresponding to an estimated average length of 5.5 μm for a nonkinetochore half-spindle microtubule, then $m = 159$ dimers replaced within 20 seconds at each end, which corresponds to 2% of the number of tubulin dimers in a 5.5-μm microtubule (Salmon et al. 1984b). Thus, simple equilibrium exchange of subunits at the ends of spindle microtubules should lead to very slow FRAP rates, but this was not what we observed in living spindles. In contrast, the simple equilibrium model does predict the extremely slow recovery of fluorescence that we observed in vitro after photobleaching astral microtubules that were generated by nucleating DTAF-tubulin assembly from purified centrosomes (Leslie et al. 1984).

Analysis of Other Proposed Pathways

High-efficiency treadmilling (see no. 2 in Fig. 1B) could, in principle, exchange all the tubulin molecules in the spindle fiber microtubules with a half-time of 20 seconds (Salmon et al. 1984b). However, if treadmilling does occur and microtubules are organized by the centrosomes and the spindle poles, as diagramed in Fig. 1A, then we should see the bleached region transported poleward during recovery, as predicted by the model of Margolis and Wilson (1981). So far, the evidence for poleward transport during fluorescence recovery is negative.

The spindle FRAP results are consistent with either the exchange of tubulin subunits at sites all along the length of spindle fiber microtubules (see no. 4 in Fig. 1B) or rapid, continuous depolymerization and repolymerization of spindle fiber microtubules as proposed by Mitchison and Kirschner (this volume) (see no. 3 in Fig. 1b). Some form of microtubule "breathing" and tubulin intercalation or rapid breaking and reannealing of microtubules would produce the uniform, exponential recovery of fluorescence in spindle fibers observed after photobleaching. However, there is no direct evidence yet that tubulin subunits can exchange at sites along a microtubule other than at its ends. Perhaps unknown proteins sever microtubules or disrupt the tubulin lattice in a manner analogous to the action of some proteins on actin filaments (Bonder and Mooseker 1983).

Mitchison and Kirschner (this volume) propose that whole microtubules can be turning over rapidly at steady-state assembly equilibrium, a so-called "dynamic instability" (see no. 3 in Fig. 1B). They postulate that the hydrolysis of GTP, bound to tubulin at the ends of microtubules, regulates rapid, catastrophic depolymerization of entire microtubules (Carlier et al. 1984). If microtubules are continuously depolymerizing, then regrowing from the centrosome, our FRAP results indicate that the whole process must occur rapidly, but nonsynchronously, among the population of nonkinetochore spindle microtubules with a half-time of about 20 seconds. The colchicine injection studies described above show that nonkinetochore spindle microtubules do depolymerize rapidly (as their scheme requires) when elongation ceases. The major reservation with this scheme is the same puzzle raised from the colchicine injection experiments. The rate of end-dependent microtubule elongation, based on the highest estimated tubulin concentration in the egg and the largest rate constant measured for end-dependent association of tubulin (corrected for the rate of tubulin diffusion in the cytoplasm), is too slow to polymerize a 5-μm microtubule within a half-time of 20 seconds (e.g., 27 μM $\times$ 2 $\times$ 10^6 M^{-1}sec^{-1} $\times$ 20 sec/1635 dimers μm^{-1} = 0.66 μm).

SUMMARY

Rates of tubulin dissociation and flux with microtubules in nonkinetochore spindle fibers have been measured in living cells. The spindle FRAP rates are clearly inconsistent with simple equilibrium exchange of tubulin subunits at microtubule ends (see no. 1 in Fig. 1B). The uniform pattern of fluorescence recovery that we observed is inconsistent with the treadmilling model proposed by Margolis and Wilson (1977) (see no. 2 in Fig. 1B). The FRAP data are, in principle, consistent with the dynamic instability model of Mitchison and Kirschner (see no. 3 in Fig. 1B), but, based on in vitro kinetic parameters and the rate of tubulin diffusion in the cytoplasm, end-dependent microtubule elongation does not appear to be fast enough. The dynamic equilibrium exchange of tubulin subunits at sites all along microtubules (see no. 4 in Fig. 1B; Inoué and Sato 1967; Inoué and Ritter 1975), however, is consistent with the current in vivo rate data, although the mechanism(s) underlying this process cannot yet be explained.

ACKNOWLEDGMENTS

I thank Dick McIntosh and his laboratory colleagues for a very productive and exciting sabbatical year in Boulder, Colorado, where many of the tubulin fluorescence studies were initiated. Thanks also to Pat Wadsworth, Tom Hays, Mary Porter, Ken Jacobson, Lans Taylor, Tim Mitchison, and Marc Kirschner for stimulating discussions and to Nancy Salmon for editing and typing the manuscript. This work was supported by grant GM-24364 from the National Institutes of Health.

REFERENCES

Bajer, A.S. and J. Molé-Bajer. 1975. Lateral movements in the spindle and the mechanism of mitosis. In *Molecules and cell movement* (ed. S. Inoué and R.E. Stephens), p. 77. Raven Press, New York.

Borisy, G.G. and E.W. Taylor. 1967. The mechanism of action of colchicine. Colchicine binding to sea urchin eggs and the mitotic apparatus. *J. Cell Biol.* **34:** 535.

Bonder, E.M. and M.S. Mooseker. 1983. Direct electron microscopic visualization of barbed end capping and filament cutting by intestinal microvillar 95-Kdalton protein (villin): A new actin assembly assay using *Limulus* acrosomal process. *J. Cell Biol.* **96:** 1097.

Carlier, M.-F., T.L. Hill, and Y. Chen. 1984. Interference of GTP hydrolysis in the mechanism of microtubule assembly: An experimental study. *Proc. Natl. Acad. Sci.* **81:** 771.

Deery, W.J. and R.C. Weisenberg. 1981. Kinetic and steady-state analysis of microtubules in the presence of colchicine. *Biochemistry* **20:** 2316.

Engelborghs, Y. and A. Lambeir. 1980. The physical chemistry of tubulin-colchicine interaction. In *Microtubules and microtubule inhibitors* (ed. M. De Brabander and J. DeMey), p. 133. Elsevier/North-Holland, Amsterdam.

Hill, T. and M.W. Kirschner. 1982. Bioenergetics and kinetics of microtubule and actin filament assembly-disassembly. *Int. Rev. Cytol.* **78:** 1.

Inoué, S. and H. Ritter, Jr. 1975. Dynamics of mitotic spindle organization and function. In *Molecules and cell movement* (ed. S. Inoué and R.E. Stephens), p. 3. Raven Press, New York.

Inoué, S. and H. Sato. 1967. Cell motility by labile association of molecules. *J. Gen. Physiol.* **50:** 259.

Jacobson, K., E. Elson, D. Koppel, and W. Webb. 1983. International workshop on the application of fluorescence photobleaching techniques to problems in cell biology. *Fed. Proc.* **42:** 72.

Keller, R.C.S., III and L.I. Rebhun. 1982. *Strongylocentrotus purpuratus* spindle tubulin. I. Characteristics of its polymerization and depolymerization in vitro. *J. Cell Biol.* **93:** 788.

Kreis, T. and W. Birchmeier. 1982. Microinjection of fluorescently labeled proteins into living cells with emphasis on cytoskeletal proteins. *Int. Rev. Cytol.* **75:** 209.

Leslie, R.J., W.M. Saxton, T.J. Mitchison, B. Neighbors, E.D. Salmon, and J.R. McIntosh. 1984. Assembly properties of fluorescein labeled tubulin in vitro before and after fluorescence bleaching. *J. Cell Biol.* **99:**(in press).

Margolis, R.L. and L. Wilson. 1977. Addition of colchicine-tubulin complex to microtubule ends: The mechanism of substoichiometric colchicine poisoning. *Proc. Natl. Acad. Sci.* **74:** 3466.

———. 1981. Microtubule treadmills—Possible molecular machinery. *Nature* **293:** 705.

McIntosh, J.R. and U. Euteneuer. 1984. Tubulin hooks as probes for microtubule polarity: An analysis of the method and an evaluation of data on microtubule polarity in the mitotic spindle. *J. Cell Biol.* **98:** 525.

Pfeffer, T.A., C.F. Asnes, and L. Wilson. 1976. Properties of tubulin in unfertilized sea urchin eggs. *J. Cell Biol.* **69:** 599.

Rieder, C.L. 1982. The formation, structure, and composition of the mammalian kinetochore and kinetochore fiber. *Int. Rev. Cytol.* **79:** 1.

Salmon, E.D. and D.A. Begg. 1980. Functional implications of cold-stable microtubules in kinetochore fibers of insect spermatocytes during anaphase. *J. Cell Biol.* **85:** 853.

Salmon, E.D., M. McKeel, and T. Hays. 1984a. The rapid rate of tubulin dissociation from microtubules in the mitotic spindle in vivo measured by blocking polymerization with colchicine. *J. Cell Biol.* **99:** 1067.

Salmon, E.D., R.J. Leslie, W.M. Saxon, M.L. Karow, and J.R. McIntosh. 1984b. Spindle microtubule dynamics in sea urchin embryos. Analysis using a fluorescein-labeled tubulin and measurements of fluorescence redistribution after laser photobleaching. *J. Cell Biol.* **99:** (in press).

Salmon, E.D., W.M. Saxton, R.J. Leslie, M.L. Karow, and J.R. McIntosh. 1984c. Diffusion coefficient of fluorescein-labeled tubulin in the cytoplasm of embryonic cells of a sea urchin: Measurement by video image processing of fluorescence redistribution after photobleaching. *J. Cell Biol.* **99:** (in press).

Saxton, W.M., D.L. Stemple, R.J. Leslie, E.D. Salmon, and J.R. McIntosh. 1984. Tubulin dynamics in cultured mammalian cells. *J. Cell Biol.* **99:** (in press).

Schiff, P.B. and S.B. Horwitz. 1981. Taxol assembled tubulin in the absence of exogenous guanosine 5-triphosphate or microtubule-associated proteins. *Biochemistry* **20:** 3247.

Taylor, D.L., P.A. Amato, K. Luby-Phelps, and P. McNeil. 1984. Fluorescent analog cytochemistry. *Trends Biochem. Sci.* **9:** 88.

Telzer, P.B. and L.T. Haimo. 1981. Decoration of spindle microtubules with dynein: Evidence for uniform polarity. *J. Cell Biol.* **89:** 373.

Wadsworth, P. and R.D. Sloboda. 1983. Microinjection of fluorescent tubulin into dividing sea urchin eggs. *J. Cell Biol.* **97:** 1249.

Wang, K.J., R. Feramisco, and J.F. Ash. 1983. Fluorescent localization of contractile proteins in tissue culture cells. *Methods Enzymol.* **85:** 514.

Wilson, L. and I. Meza. 1973. The mechanism of action of colchicine. Colchicine binding properties of sea urchin tail outer doublet tubulin. *J. Cell Biol.* **58:** 709.

Zeeberg, B., R. Reid, and M. Caplow. 1980. Incorporation of radioactive tubulin into microtubules at steady-state: Experimental and theoretical analysis of diffusional and directional flux. *J. Biol. Chem.* **255:** 9891.

Differential Cellular and Subcellular Distribution of Microtubule-associated Proteins

Richard B. Vallee, George S. Bloom, and Francis C. Luca

Cell Biology Group, The Worcester Foundation for Experimental Biology Shrewsbury, Massachusetts 01545

Cytoplasmic microtubules have been implicated in a wide variety of cellular functions. Among the more well established and noteworthy of these functions are mitosis and cellular morphogenesis. In addition, considerable evidence exists indicating that microtubules provide a framework to which other organelles, including intermediate filaments and a variety of membranous organelles, are attached. Microtubules have also been strongly implicated in transporting membranous organelles through the cell. How microtubules manage to participate in such a wide variety of apparently unrelated processes is not clear. Are all cellular microtubules alike or do microtubules vary in composition and structure to accomplish different tasks?

Cytoplasmic microtubules are composed of two classes of proteins. The microtubule wall is made up of tubulin. A variety of additional proteins, which have been termed microtubule-associated proteins (MAPs), have been found to copurify with microtubules. At least some of the MAPs have the appearance of arms on the surfaces of isolated microtubules. For this reason, it is believed that these proteins serve to mediate the interaction of microtubules with other cellular organelles. They are not absolutely required for the formation of microtubules, since they can bind to preformed microtubules and can be dissociated without apparent alteration of the polymer structure. They promote the assembly of tubulin into microtubules quite dramatically, although tubulin can be induced to self-assemble. Those polymers that are formed from tubulin alone generally exhibit numerous structural errors, something that does not seem to occur in the presence of the MAPs. All of these observations suggest that the MAPs act as accessory factors in microtubule formation and function.

As discussed elsewhere in this volume, numerous isoforms of tubulin have been identified. It appears that to some extent, the functional diversity of microtubules will prove to be due to the existence of tubulin isoforms or to posttranslational modifications of the tubulin molecule. Nonetheless, the tubulins are most remarkable in the degree to which their properties have been conserved, as judged both from conservation of primary sequence and from the virtually identical structural properties of microtubules in widely divergent species.

As we will discuss here, this is in contrast to the MAPs, which show considerable variation in properties from cell type to cell type and appear to be far less conserved in evolution than tubulin. This conclusion has come in part from the biochemical isolation of MAPs from several tissue and cellular sources. We review this work briefly. More recently, we have prepared polyclonal and monoclonal antibodies to MAPs isolated from sea urchin eggs and from brain tissue. The latter work, in particular, has revealed a degree of complexity in the molecular properties and the distribution of the MAPs that was previously unexpected. We discuss this work in some detail below.

COMPARISON OF MAPS FROM DIFFERENT SOURCES

Most work on the biochemistry of cytoplasmic microtubules has been performed with microtubules isolated from vertebrate brain tissue. This is because neuronal tissue is a particularly rich source of microtubules. Several MAPs have been found to copurify with the brain microtubules (Fig. 1; Table 1). Most prominent of the MAPs are the high-molecular-weight proteins traditionally known as MAP 1 and MAP 2. In addition, brain microtubules contain a tetrad of electrophoretic species known as the τ MAPs. A pair of low-molecular-weight species termed the LMW MAPs have been found to be associated with MAP 1 (Vallee and Davis

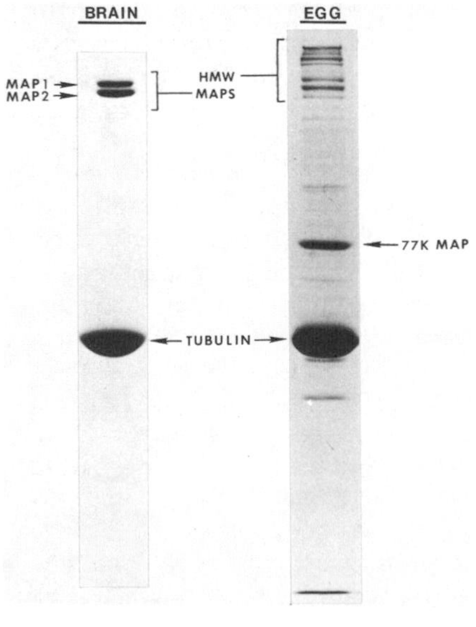

Figure 1
Comparison of brain and sea urchin microtubules on 9% polyacrylamide electrophoretic gels (Laemmli 1970). Sources for the microtubules were calf cerebral cortex (from Vallee 1982) and unfertilized eggs of *L. variegatus* (from Vallee and Bloom 1983), both prepared using taxol. Scale for the two lanes is slightly different.

Table 1
Well-characterized Microtubule-associated Proteins

Principal source	*Name*	M_r	*Properties*	*References*
Brain tissue	MAP 1 (A,B,C)	300,000–350,000	projection on microtubule surface; promotes assembly	Sloboda et al. (1975); Vallee and Davis (1983); Bloom et al. (1984b)
	MAP 2 (A,B)	270,000–285,000	projection on microtubule surface; promotes assembly	Murphy et al. (1977); Herzog and Weber (1978); Kim et al. (1979)
		70,000	protein associated with MAP 2; function unknown	Vallee et al. (1981)
	τ	55,000–62,000	asymmetric; promotes assembly	Cleveland et al. (1977)
	type II cAMP-dependent protein kinase	54,000; 39,000	enzyme associated with MAP 2	Vallee et al. (1981); Theurkauf and Vallee (1982)
	LMW MAPs	28,000; 30,000	light chains of MAP 1	Vallee and Davis (1983)
Cultured cells		255,000 210,000–220,000 125,000	asymmetric asymmetric; promotes assembly asymmetric; promotes assembly	Bulinski and Borisy (1979); Weatherbee et al. (1980); Olmsted and Lyon (1981); Duerr et al. (1981)
		80,000; 69,000	unknown	Duerr et al. (1981)
Sea urchin eggs		200,000–350,000 150,000 100,000 77,000 37,000	includes surface projections and microtubule cross-linkers	Vallee and Bloom (1983)

Modified from Vallee and Bloom (1984).

1983). In addition, we have identified three polypeptide components associated with MAP 2, two of which we identified as the subunits of a type II cAMP-dependent protein kinase (Vallee et al. 1981; Theurkauf and Vallee 1982). We find numerous other bands at low levels in brain microtubule preparations. It has not yet been determined whether these proteins actually copurify with the microtubules or, instead, represent unrelated contaminants.

How representative is the brain microtubule preparation of cytoplasmic microtubules in general? Several other sources of microtubules have been examined with varying degrees of thoroughness (Table 1). HeLa cells are the most extensively studied cultured cell line. They represent a relatively undifferentiated source of microtubules and provide a useful comparison with brain, in which most cells are differentiated and highly specialized as well. Sea urchin eggs represent a source in which the major function of microtubules appears to be in mitosis. The subunit proteins of mitotic microtubules are stockpiled in abundance in the unfertilized egg, providing an excellent source for biochemical analysis. This contrasts with the situation in cultured cells, where only a small fraction of cells in any given culture preparation are in mitosis and most microtubules that are isolated are involved in interphase functions.

As is indicated in Table 1, HeLa cells contain an assortment of MAPs that seems to be entirely different from that found in brain tissue. A band of 210,000 daltons is the most prominent MAP in HeLa cells, followed in abundance by a band of 125,000 daltons. Examination of other cell lines has revealed a picture more nearly similar to that obtained with HeLa than with brain. The content of MAPs in brain and cultured cells is not mutually exclusive, however, as revealed by the finding of the MAP-1 polypeptides in cultured cells in general (see below) (Bloom et al. 1984a) and by the identification of what may be the 210,000-dalton cultured cell MAP in brain tissue (Vallee 1982). MAP 2 has also been identified in HeLa cells at very low levels (Weatherbee et al. 1982) and in some other cell lines as well. However, in relative abundance, this MAP does, indeed, show striking differences in cellular and even subcellular distribution, as discussed below.

Isolation of microtubules from the sea urchin egg proved to be quite difficult using traditional procedures that rely on the self-assembly of microtubules from cytosolic extracts (as has been found now for most systems other than brain tissue and one or two cultured cell lines). To overcome this problem, we used a new procedure developed in this laboratory involving the use of taxol to promote microtubule assembly (Vallee 1982). Using this procedure, we identified yet a different pattern of MAPs from that seen in brain and cultured cells (Fig. 1) (Vallee and Bloom 1983). The apparent sea urchin MAPs included a large number of proteins, which varied widely in both size and abundance. Could the MAPs of the mitotic spindle be so complex or were some of these proteins adventitious contaminants or fragments of other protein species? To answer these questions, we initiated a program of monoclonal antibody production, and we have used the antibodies to establish whether the proteins identified in our biochemical preparations are, indeed, components of the spindle. To date, we have found that antibodies to five distinct proteins in the sea-urchin-egg microtubule preparation react with the mitotic spindle, as judged from immunofluorescence microscopy (Bloom et al. 1983; Vallee and Bloom 1983; G.S. Bloom et al., in prep.). This suggests that the pattern of bands observed in the purified sea urchin microtubule preparations is, indeed, a faithful representation of the composition of mitotic spindle microtubules. In hindsight, this may not be surprising in view of the complexity of another microtubular organelle, the axoneme. We now are inclined to believe that the spindle may, similarly, be composed of a highly complex network of interconnecting elements (see Vallee and Bloom 1983), which, as in the axoneme, may in turn represent complex, multisubunit structures.

Among the sea urchin MAPs were proteins equivalent in size to most of the MAPs identified in brain tissue, HeLa cells, and other cultured cell lines. Whether the sea urchin MAPs are structurally and functionally equivalent to the MAPs described in vertebrate tissues and cells is not certain, although it seems reasonable to expect that those MAPs that are functional components of the mitotic spindle would tend to be conserved throughout evolution.

ISOFORMS OF HIGH-MOLECULAR-WEIGHT BRAIN MAPS

The high-molecular-weight MAPs found in brain tissue have been the most extensively studied of all the MAPs. This is due in part to the relatively large amount of these proteins that may be isolated. (An average preparation of microtubules from calf brain in our laboratory yields approximately 300 mg of purified microtubules, which contains about 75 mg of high-molecular-weight MAPs.) In addition, it was observed several years ago that the presence of these proteins in microtubule preparations correlated with the observations of fine projections on the microtubule surface (Dentler et al. 1975; Murphy and Borisy 1975). This suggested that the high-molecular-weight MAPs in particular might play a role in the interaction of microtubules with other cellular organelles and were, therefore, of considerable importance in understanding how microtubules function. Of equal interest was the observation that, as has been found for MAPs in general, these proteins had the effect of promoting microtubule assembly in vitro (Murphy et al. 1977).

Identification of Five High-molecular-weight MAP Isoforms

As noted earlier, the high-molecular-weight MAPs generally appear as two electrophoretic components on high-percent polyacrylamide gels. These components were termed MAP 1 and MAP 2 by Sloboda et al. (1975). On low-percent polyacrylamide gels, these proteins resolve into multiple bands. We have observed that the multiplicity of the high-molecular-weight MAP bands is most striking in microtubules prepared from white-matter regions of the brain (Figs. 2–4). However, the same bands may be detected in gray matter or whole brain, both of which have been used traditionally as starting material in the preparation of brain microtubules.

Figures 2 through 4 show microtubules purified from bovine white matter analyzed on a low-percent polyacrylamide gel. Some gray matter was included in the preparation to provide sufficient MAP 2 (see below) for purposes of comparison with the MAP-1 electrophoretic species. Five bands are readily distinguishable in these preparations. Two of these correspond to the MAP-2 band observed on high-percent polyacrylamide gels, and we have therefore named them MAP 2A and MAP 2B. This designation is also based on the biochemical similarity of the two species. Both are stable at elevated temperature (Kim et al. 1979), a very unusual property for proteins of this size. The two have not yet proved to be

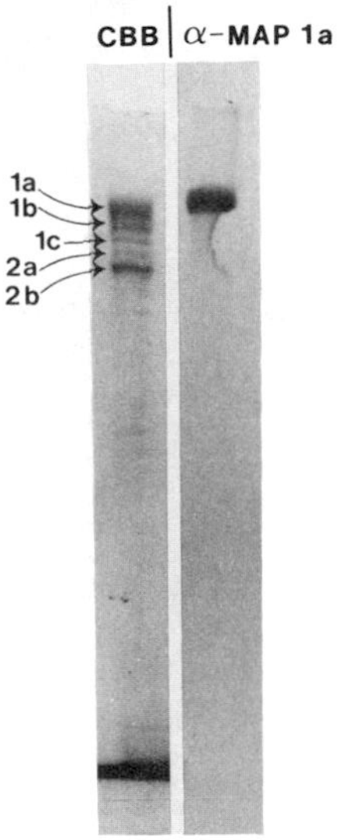

Figure 2
Immunoblot analysis using a monoclonal antibody to MAP 1A. Microtubules were prepared using taxol from calf brain white matter. The microtubles were subjected to electrophoresis using 4% polyacrylamide (Laemmli 1970) and examined by Coomassie brilliant blue staining and immunoblot analysis. (Reprinted, with permission, from Bloom et al. 1984b.)

separable by chromatographic means (Kim et al. 1979; Vallee et al. 1981) and show quite similar peptide MAPs (G.S. Bloom and R.B. Vallee, unpubl.). We have found the same bands to be cross-reactive using a monoclonal antibody (Fig. 3).

The existing evidence indicates, furthermore, that the smaller polypeptide does not derive from the larger by proteolysis. This is indicated by studies showing that both bands are present in tissues and cells dissolved directly in electrophoresis sample buffer (De Camilli et al. 1984). In addition, the ratio of the two MAP-2 bands varies with species of origin (compare pig brain microtubules shown in Murphy et al. [1977] with calf brain microtubules shown in Fig. 3), brain region (Caceres et al. 1984; DeCamilli et al. 1984), and stage of development (Burgoyne and Cumming 1984). The two MAP-2 species do not appear to represent different posttranslational states of the same polypeptide, because phosphorylation of MAP 2 by its associated protein kinase or dephosphorylation does not change the observed electrophoretic pattern (Theurkauf and Vallee 1983 and unpubl.). For these reasons, we believe that MAP 2A and MAP 2B are distinct proteins. It should be mentioned that as yet there is no evidence to suggest that they perform distinct functions.

We refer to the remaining high-molecular-weight MAP polypeptides as MAP 1 on the basis of their greater size and distinguish them in order of increasing electrophoretic mobility as MAP 1A, MAP 1B, and MAP 1C. However, our existing evidence indicates that these species are not as closely related as are MAP 2A and MAP 2B. For example, MAP 1C appears to be quite different from all of the other MAPs. It is relatively resistant to proteolysis, whereas the other high-molecular-weight MAPS are unusually sensitive to proteolysis and are digested at similar rates (Bloom et al. 1984b). Peptide maps of the other two MAP-1 polypeptides have also proved to be quite distinct (G.S. Bloom et al., in prep.).

Further evidence that the several MAP isoforms differ in primary structure has come from immunological analysis. Figure 2 shows the result of immunoblot

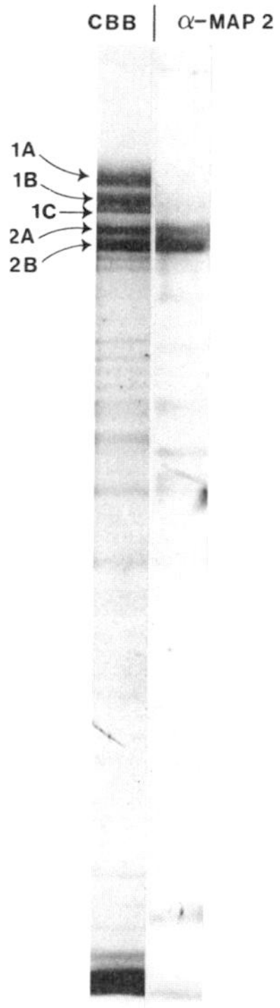

Figure 3
Immunoblot analysis using a monoclonal antibody to MAP 2A and MAP 2B. Microtubules were prepared using taxol from calf brain white matter containing a small amount of gray matter to provide sufficient MAP 2 for comparative purposes. Electrophoresis was performed using 4% polyacrylamide following the procedure of Laemmli (1970) but with the omission of SDS from the stacking and separating gel and the inclusion of 2 M urea in the separating gel.

analysis of the high-molecular-weight MAPs with a monoclonal antibody raised against electrophoretically purified MAP 1A (Bloom et al. 1984b). Strong reaction was seen with MAP 1A alone. Figure 4 shows the result of immunoblot analysis performed with a monoclonal antibody raised against electrophoretically purified MAP 1B. This antibody represents one of five that we have recently raised against this polypeptide (G.S. Bloom et al., in prep.). Four of the five antibodies show no reactivity with MAP 1A, as seen for the antibody described in Figure 4. Some reaction may be observed with electrophoretic bands of lower molecular weight than MAP 1B. However, we believe that the observed reactivity is with fragments of MAP 1B rather than with the MAP-2 polypeptides or with MAP 1C. The antibody did not react with purified MAP 2. In addition, exposure of the MAPs to chymotrypsin under conditions designed to selectively preserve MAP 1C (Bloom et al. 1984b) abolished immunoreactivity.

One of the anti-MAP-1B antibodies does react with a species of the approximate electrophoretic mobility of MAP 1A. However, reaction with this band is weaker than with MAP 1B, despite the somewhat higher level of MAP 1A in our preparations. For this reason, it seems unlikely that the antibody recognizes identical epitopes in MAP 1A and MAP 1B, although the full explanation for this result is not clear.

Taken together, these results indicate that MAP 1B is unlikely to be a fragment of MAP 1A. Whether the two polypeptides may be partially related in sequence remains to be determined.

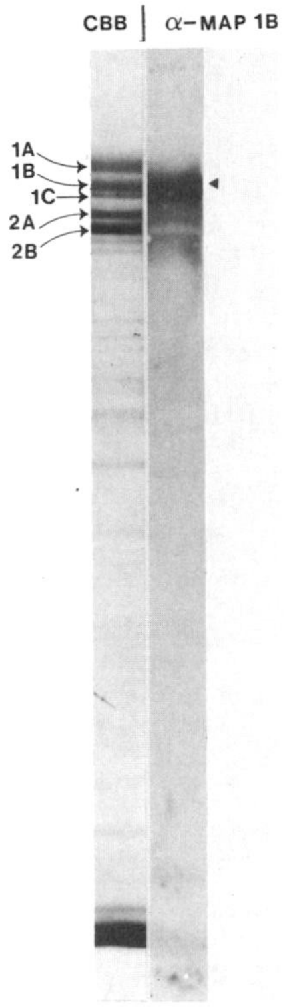

Figure 4
Immunoblot analysis using one of five monoclonal antibodies to MAP 1B. Sample preparation and electrophoresis were carried out as described for Fig. 3.

Biochemical Separation and Proteolytic Dissection of MAP 1 and MAP 2

Our ability to fractionate the high-molecular-weight MAPs has not kept pace with our identification of the MAP isoforms. The high-molecular-weight MAPs can be separated biochemically into MAP-1 and MAP-2 preparations containing the respective MAP-1 and MAP-2 polypeptides. Further work will be needed to subfractionate these preparations. However, it seems worthwhile to review the findings made with the existing MAP preparations.

As noted above, the MAP-2 polypeptides copurify under a variety of conditions. Exposure of microtubules to elevated temperature leaves MAP 2 and the τ MAPs soluble, and the MAP-2 polypeptides can then be separated from the τ polypeptides by gel-filtration chromatography (Herzog and Weber 1978; Kim et al. 1979). Alternatively, the MAP-2 polypeptides can be purified under less-harsh conditions, taking advantage of the tendency of the MAP-1 polypeptides to form high-molecular-weight aggregates (Vallee et al. 1981). This leaves most of the MAP 2 in a nonaggregated form, which may be further purified quite readily. MAP 2 prepared by this method was found to contain accessory proteins that are destroyed when the protein is prepared by exposure to elevated temperature (Vallee et al. 1981; Theurkauf and Vallee 1982).

The MAP-1 polypeptides were much more difficult to isolate, as the result of the tendency to form high-molecular-weight aggregates noted above. These aggregates contain contaminating proteins, including MAP 2. Complete elimination of MAP 2 from MAP 1 had proved to be impossible using traditional chromatographic procedures (Kuznetsov et al. 1981a; Vallee and Davis 1983). To overcome this problem, we took advantage of our observation that the amount of MAP 2 in microtubules prepared from calf brain white matter was quite low (Vallee 1982; Vallee and Davis 1983).

MAP 1 purified from white matter contained both MAP 1A and MAP 1B. The MAP-1 and MAP-2 preparations shared two important properties. When recombined with tubulin, both had the appearance of periodically arranged filamentous projections, or arms, on the microtubule surface (Herzog and Weber 1978; Kim et al. 1979; Vallee and Davis 1983). The dimensions of the arms were similar, although the average distance between arms differed somewhat. Both preparations also promoted microtubule assembly (Murphy et al. 1977; Herzog and Weber 1978; Kim et al. 1979; Kuznetsov et al. 1981b; Vallee and Davis 1983).

Proteolytic digestion studies have provided evidence for both common and distinct features of the MAP-1 and MAP-2 polypeptides. All of the high-molecular-weight MAP polypeptides are unusually sensitive to proteolytic degradation under nondenaturing conditions, with the exception of MAP 1C (Bloom et al. 1984b). In fact, the time course of the disappearance of the MAP-1 and MAP-2 electrophoretic species during tryptic digestion appeared to be identical (Vallee and Borisy 1977). We have shown that the primary site for both tryptic and chymotryptic cleavage of MAP 2 is at the base of the projecting portion of the molecule (Vallee and Borisy 1977; Vallee 1980). We believe that MAP 1A and MAP 1B may be structured similarly on the basis of the similar digestion kinetics for MAP 1 and MAP 2 and the concomitant disappearance of what appeared to be all of the projections from microtubules containing both classes of MAP molecules (Vallee and Borisy 1977). In addition, with the use of our anti-MAP-1A antibody (see below), we have identified a large ($M_r > 200,000$) fragment of MAP 1A that is released from the microtubule during chymotryptic digestion (G.S. Bloom and R.B. Vallee, unpubl.). This may be equivalent to the projection fragments produced from MAP 2 that are of similar size and have similarly lost the ability to bind to microtubules (Vallee and Borisy 1977; Vallee 1980).

The results of peptide mapping conducted under nonnative conditions have also been contradictory. This may be due to the complexity of high-molecular-weight MAP polypeptides, as discussed in this paper. Kuznetsov et al. (1981a) found that MAP 1 and MAP 2 differed markedly in their one-dimensional peptide maps. However, Wiche and co-workers found MAP 1 and MAP 2 to be quite similar (Herrmann et al. 1984). As noted earlier, our laboratory has found that peptide maps of MAP 1A and MAP 1B are quite distinct (G.S. Bloom et al., in prep.). In addition, we observed that peptide maps of MAP 2A and MAP 2B were quite similar, although very different from maps of both of the MAP-1 species. The basis for the conflicting results obtained by the three laboratories is not certain, although, in view of our own data, we suspect that differences in the isoform content of the different preparations that were used could be involved.

Some striking biochemical differences between MAP 1 and MAP 2 have been identified. It has been known for several years that when microtubules are exposed to elevated temperatures, MAP 2 remains soluble, whereas MAP 1 precipitates (Herzog and Weber 1978; Kim et al. 1979). This suggests that the two proteins are markedly different in their structural properties. However, it is also possible that it is a small portion of MAP 1 that is responsible for coagulation under conditions of elevated temperature and that the molecule is for the most part structurally similar to MAP 2. The latter view is supported by evidence that both classes of MAPs survive other solution extremes (8 M urea and 6 M guanidine) (Vallee and Davis 1983).

Another striking biochemical difference between MAP 1 and MAP 2 is in their content of low-molecular-weight subunits. MAP 2 purified under nondenaturing conditions was found to contain three smaller polypeptides at low (~1:20) stoichiometry (Vallee et al. 1981; Theurkauf and Vallee 1982). Two of the polypeptides were subsequently found to be the subunits of a cAMP-dependent protein kinase that phosphorylates MAP 2 itself (Theurkauf and Vallee 1982). Despite the fractional stoichiometry of the complex, the associated kinase represented a substantial portion of the total cAMP-dependent protein kinase in brain tissue (from 30% to 35%, as assayed by several independent means). MAP 2 is, so far, unique among the MAPs in having an associated enzyme. This may be one indication that MAP 2 is a highly specialized protein. Further evidence to this effect is discussed below.

Our biochemical preparation of MAP 1 isolated from bovine white matter was also found to contain low-molecular-weight subunits (Vallee and Davis 1983). These polypeptides differed in molecular weight from the polypeptides associated with MAP 2 (M_r = 28,000 and 30,000 for the MAP-1-associated polypeptides; M_r = 39,000 and 54,000 for the subunits of the MAP-2-associated protein kinase; M_r = 70,000 for a MAP-2-associated species of unknown function). In contrast to MAP 2, the MAP-1-associated species, which we refer to as light chains, were present in the complex at close to 1 mole of each per mole of high-molecular-weight polypeptide.

Combined with our more recent information on the respective isoforms of MAP 1 and MAP 2, it seems clear that the MAP-1 and MAP-2 groups are quite distinctive, although they do share some important properties. As we have discussed, each of the MAP-1 polypeptides has distinctive properties. We expect that the same will prove to be true for the MAP-2 polypeptides, although the differences between MAP 2A and MAP 2B are likely to be more subtle. These two polypeptides differ in apparent size by approximately 15,000 daltons (Murphy et al. 1977). If this, indeed, represents a difference in amino acid content, such a difference would be substantial and should eventually be detectable in the biochemical properties of the proteins.

COMPARISON OF THE DISTRIBUTION OF MAP-1 AND MAP-2 POLYPEPTIDES

Differential Distribution of High-molecular-weight MAP Isoforms

Our first indication that the MAP-1 and MAP-2 polypeptides might be differentially distributed in cells came from a comparison of microtubules prepared from bovine brain gray-matter and white-matter regions (Vallee 1982). This distinction between gray and white matter results from the disposition of neuronal cells in nervous tissue. White matter contains only axons and the supporting glial cells. Gray matter contains neuronal cell bodies and dendritic processes as well as axons and glia. In an earlier report, Matus and co-workers (1981) had found using a polyclonal antibody prepared against the entire complement of high-molecular-weight MAPs that immunoreactivity was restricted to the dendritic processes and cell bodies of neurons. No staining was detected in axons or in glial cells. This contrasted with a considerable body of ultrastructural evidence for the existence of abundant cross-linking elements between microtubules and other organelles in axons (see, e.g., Smith et al. 1975; Hirokawa 1982). Since there have been increasing indications that axonal microtubules are involved in the intracellular transport of these organelles (Berlinrood et al. 1973; Hayden et al. 1983), the identification of the cross-linking elements became of even greater interest.

In gray matter, the source commonly used in the preparation of microtubules, the MAP-1 and MAP-2 bands observed on low-percent polyacrylamide gels were both found to be prominent (Vallee 1982). In white matter, however, we found that the level of MAP 2 was considerably lower. MAP 1, on the other hand, was present in the same ratio to tubulin as in gray matter. This suggested that MAP 2 was restricted in its distribution, whereas MAP 1 was more widespread, and was the first indication that the two classes of high-molecular-weight MAPs might be differentially distributed.

Closer examination of the high-molecular-weight MAP polypeptides on more highly resolving polyacrylamide gels revealed that the composition of MAP 1 in the two brain regions also differed, although the total amount of MAP was constant. In gray matter, MAP 1A was by far the most abundant species, whereas MAP 1B and MAP 1C were present in considerably lower amount. Although MAP 1A was still the most abundant MAP-1 component in white matter, MAP 1B and MAP 1C were considerably more abundant (Bloom et al. 1984b; G.S. Bloom et al., in prep.).

To examine the distribution of the MAPs more closely, we have conducted an extensive survey of brain and spinal cord regions using immunofluorescence microscopy of tissue sections. We have used a polyclonal antibody to MAP 2 (Miller et al. 1982; Theurkauf and Vallee 1982; Bloom et al. 1984b; De Camilli et al. 1984) and monoclonal antibodies to MAP 1A (Bloom et al. 1984a, 1984b) and MAP 1B (this paper) in this work. These studies have borne out our biochemical analysis of gray- and white-matter microtubules and have revealed striking differences in the distribution of the high-molecular-weight MAPs.

MAP 2 was almost entirely restricted in its distribution to gray-matter regions (Miller et al. 1982; Bloom et al. 1984b; De Camilli et al. 1984). Within gray matter, MAP 2 was found exclusively in the dendritic processes and, at lower levels, the cell bodies of neurons. This indicated a remarkably restricted distribution for MAP 2, particularly in view of the fact that it is the most abundant of the MAPs in the brain. It should be mentioned that occasional, weak axonal staining has been observed in peripheral nerve roots (Papasozomenos et al. 1982) and in some white-matter tracts in the central nervous system (Caceres et al. 1984), suggesting that there could be a low level of MAP 2 in some axons. This result is consistent with our biochemical data, which indicated that MAP 2 was reduced but not absent from white matter (Vallee 1982; Vallee and Davis 1983).

MAP 1A was present in both gray matter and white matter, again consistent with our biochemical results (Bloom et al. 1984b). In gray matter, the protein was prominent in dendritic processes and more abundant than MAP 2 in cell bodies. Double labeling of tissue sections revealed that within gray matter, the distribution of MAP 1A and MAP 2 was virtually coincident. Thus, the two proteins coexist within the same dendrites.

In white matter, MAP 1A was found to be present both in axons and glial cells. In fact, oligodendrocytes were the most brightly staining structures in white matter. Further analysis of primary cultures of brain cells (Bloom et al. 1984b) and a wide variety of primary cultured cells and cell lines (Bloom et al. 1984a) indicated that MAP 1A was extensively distributed among cell types. MAP 1A was found associated with both interphase and mitotic microtubules. Thus, this protein was more uniformly distributed throughout the cytoplasm of neuronal cells, was more widely distributed among cell types, and was associated with a wider variety of microtubule-containing cellular structures than MAP 2.

Figure 5 shows a section of brain tissue stained with one of our more recently obtained monoclonal antibodies to MAP 1B. This antibody (and all of the other anti-MAP-1B antibodies that we have produced) reacts strongly with both white and gray matter. In fact, the relative abundance of MAP 1B in white matter appears to be even greater than that of MAP 1A, consistent with the apparently greater abundance of MAP 1B in microtubules isolated from white matter (Bloom et al. 1984b; G.S. Bloom et al., in prep.). Thus, the distribution of MAP 1B is distinct not only from that of MAP 2, but from that of MAP 1A as well. These results provide further evidence that MAP 1A and MAP 1B are distinct proteins. In addition, they identify yet another high-molecular-weight MAP component of axons. On the basis of the relative abundance of MAP 1A and MAP 1B in white-matter microtubule preparations (Figs. 2–4), we suspect that together these proteins make up the bulk of axonal MAPs.

As for MAP 1A, we have found that MAP 1B is quite widespread among cell types, both on interphase and mitotic microtubules (G.S. Bloom et al., in prep.). Thus, our results all indicate that the MAP-1 polypeptides are widespread, whereas the MAP-2 polypeptides are not. We are presently attempting to prepare antibodies to MAP 1C to determine whether this rule is true for this protein as well.

Distinct Nonmicrotubule Binding Sites for the MAPs in the Cell

The ultimate aim of our work is, of course, to understand the function of the MAPs in the cell. One approach to this problem has been to search for nontubulin binding sites for the MAPs in vitro, using biochemical and biophysical assays. This approach has resulted in the identification of a number of organelles to which the MAPs appear to bind but, in general, weakly or at low stoichiometry. Thus, MAPs, or MAP-containing microtubules, have been found to interact with muscle F-actin filaments (Griffith and Pollard 1978; Sattilaro et al. 1981), secretory granules (Sherline et al. 1977; Suprenant and Dentler 1982), coated vesicles (Sattilaro et al. 1980), and neurofilaments (Runge et al. 1981; Leterrier et al. 1982). Although each of these interactions is consistent with a known or postulated role for microtubules in the cell, it is not certain what relationship they bear

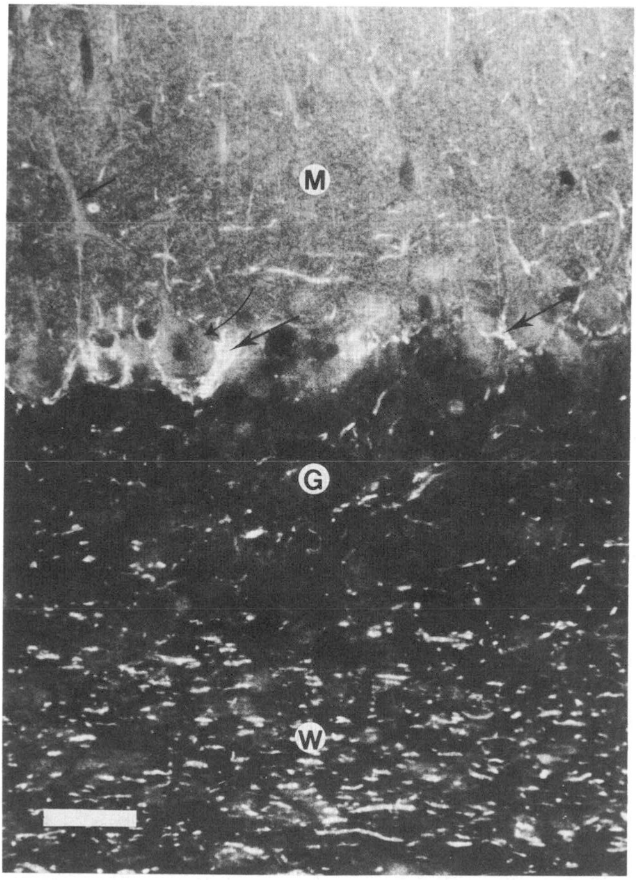

Figure 5
Preferential axonal localization of MAP 1B in neuronal tissue. A parasaggital section through a folium of adult rat cerebellum is shown stained by indirect immunofluorescence microscopy according to the method of Bloom et al. (1984b), using a different monoclonal antibody to MAP 1B than was used for Fig. 4. (M) Molecular layer; (G) granule-cell layer; (W) white-matter layer. In contrast to anti-MAP 2 (Miller et al. 1982; Bloom et al. 1984b; De Camilli et al. 1984) and anti-MAP 1A (Bloom et al. 1984b), immunoreactivity is marginal in the Purkinje cell dendrites (small straight arrow) and cell bodies (curved arrow) of the molecular layer. Axonal staining in the cerebellar white matter is even more pronounced than was observed for MAP 1A (Bloom et al. 1984b). Staining of the basket cell axons (large straight arrows) surrounding the Purkinje cell bodies is particularly striking. The other anti-MAP-1B antibodies that we have produced react intensely with both axons and dendrites. Bar, 40 μm.

to interactions actually occurring in the cell. It is quite conceivable that the appropriate combinations of organelle and microtubule preparation have yet to be assayed in the test tube or that the activity of the proteins involved in these interactions is lost during isolation. Thus, for example, it is possible that brain actin, but not muscle actin, interacts specifically with microtubules in the cell or even that the interaction of the actomyosin system with microtubules, if this occurs in the cell at all, is mediated by an association of myosin, rather than actin, with a component protein of microtubules.

It occurred to us that we might be able to use our anti-MAP antibodies to identify nonmicrotubule binding sites for the MAPs in cells under conditions involving minimal perturbation of the native interactions. One strategy to accomplish this would be to disassemble microtubules in cells and then determine the distribution of each of the MAPs by immunofluorescence microscopy (Bloom and Vallee 1983; G.S. Bloom et al., in prep.). In the absence of the microtubule binding site for a given MAP, we might find the protein in a distribution characteristic of some other organelle. Such an association might normally be obscured due to codistribution of the organelle with microtubules.

We used primary cultures of newborn or embryonic rat brain cells for this work. These preparations contained cells with high levels of MAP 2, MAP 1A, and MAP 1B. Of particular value for these studies were the presence of a class of flat, MAP-2-containing cells, which we believe represent undifferentiated neurons (Bloom and Vallee 1983; G.S. Bloom et al., in prep.). Our first experiments were with anti-MAP 2 alone (Bloom and Vallee 1983). Microtubules were disassembled in two different ways: Cells were pretreated with vinblastine or colchicine to disassemble microtubules in the living cell and then processed for immunofluorescence microscopy. Alternatively, the cells were permeabilized by detergent extraction. Microtubules were subsequently depolymerized by exposure to Ca^{++} or phosphate buffers, and immunofluorescence microscopy was then performed. In this way, the possibility was minimized that the MAP 2 might reassociate with an organelle to which it is not normally bound.

In the presence of microtubules, anti-MAP 2 gave a staining pattern virtually coincident with that obtained with anti-tubulin. In the absence of microtubules, MAP-2 staining was still observed. In cells that had been pretreated with vinblastine or colchicine, bright staining of cytoplasmic cables was observed. Such cables are characteristic of a number of types of intermediate filaments that are normally codistributed and, presumably, associated with microtubules, but collapse toward the nucleus in the absence of these structures. Staining with anti-vimentin antibody demonstrated that the cables indeed contained intermediate filaments. In cells extracted with detergent and then subjected to microtubule-depolymerizing conditions, vimentin filaments remain in their normal microtubulelike array in the cytoplasm. Under these conditions, anti-MAP 2 gave a staining pattern again coincident with that obtained with anti-vimentin. Thus, MAP 2 appeared to be associated with intermediate filaments when microtubules were absent from the cell.

In view of the normal coincident distribution of microtubules and vimentin filaments in the cell, a simple explanation of our results is that MAP 2 has binding sites both for microtubules and intermediate filaments. Perhaps, then, MAP 2 is the factor responsible for cross-linking the two fiber systems.

To test whether the association with intermediate filaments was specific to MAP 2, we labeled the primary rat brain cells with anti-MAP 1A (Bloom et al. 1984a; G.S. Bloom et al., in prep.) and anti-MAP 1B (G.S. Bloom et al., in prep.). Both antibodies showed a microtubule staining pattern in control cells. After pretreatment of the cells with vinblastine, staining became disperse. We also double-labeled cells with anti-MAP 1A and anti-MAP 2 to compare the distribution of the two MAPs in the same cells. Even in cells showing intermediate filament cables brightly stained with anti-MAP 2, the pattern of MAP-1A staining was disperse. Thus, the association with intermediate filaments appears to be specific to MAP 2. It may also be noted that these results represent yet another instance of the differential distribution of the MAP isoforms within the same cell.

The distribution of MAP 1A was itself interesting, both in control and in drug-treated cells. As noted above, anti-MAP 1A stained microtubules in control cells as determined by double labeling with anti-tubulin (Bloom et al. 1984a). However, staining was discontinuous and punctate. This condition was observed in a wide variety of cells, and under several staining conditions, and has not been observed for tubulin, MAP 2, or MAP 1B. Although the pattern of staining became disperse in the presence of microtubule-disrupting drugs, it appeared that the distribution of fluorescence was still punctate. These results suggested to us that MAP 1A might be associated with some discrete structures that were themselves aligned with microtubules. We do not as yet know whether this is the case. However, it seems reasonable to speculate that the spots could correspond to membranous organelles of some kind, in view of the large body of evidence indicating that such organelles can be localized along and transported by microtubules. We are at present attempting to pursue this problem further using immunoelectron microscopy.

One MAP, One Function

Our results so far suggest that each of the high-molecular-weight MAP isoforms may be uniquely distributed among cell types and even within the same cell. Our biochemical preparations of MAP 1 and MAP 2 show some properties in common, but many significant differences. This alone suggests that the MAPs may play different roles in the cell. Further indication that this may be true has come from our studies described in the preceding section, which have provided evidence for differences in the nature of the cytoplasmic organelles with which the various MAPs interact.

For these reasons we are coming to believe that each MAP isoform may prove to be responsible for a distinct cellular function. In other words, our data are pointing to a one MAP:one function model. An important implication of such a model is that a new level of care must be exercised in the design and interpretation of experiments aimed at reconstructing the interactions of microtubules with other organelles in vitro. Thus, it now seems desirable to attempt to identify interactions involving specific high-molecular-weight MAP isoforms.

Other Evidence Obtained from Monoclonal Antibodies

In raising monoclonal antibodies to the MAPs, we have frequently obtained antibodies that stain organelles other than microtubules. This has occurred despite the use of electrophoretically purified proteins obtained from purified microtubule preparations as immunogen (see, e.g., Bloom et al. 1984b). Two other reports have appeared describing monoclonal antibodies prepared against MAP 1 that showed anomalous cellular staining patterns (Sato et al. 1983; Thompson et al. 1983). In addition, reports have appeared describing polyclonal antibodies to

MAP 1 that stained a variety of nonmicrotubule structures (Hill et al. 1981; Sherline and Mascardo 1982).

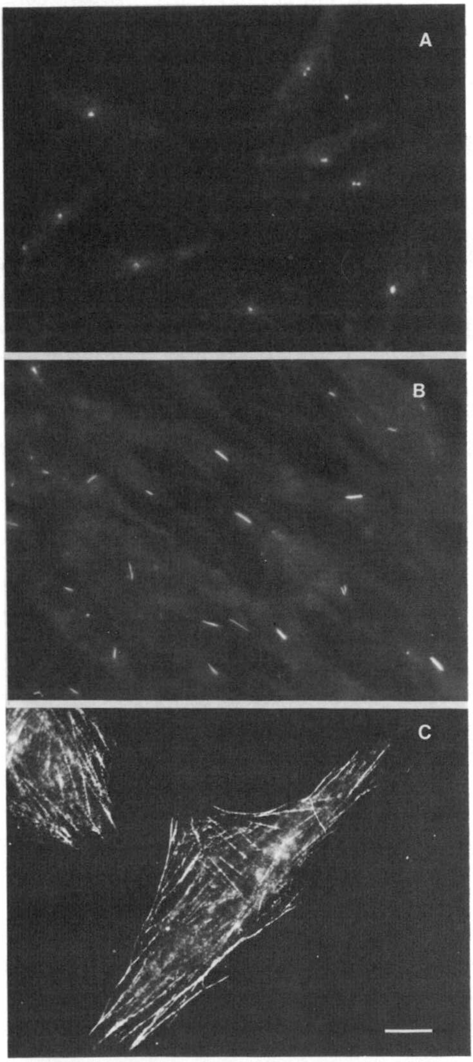

Figure 6
Reaction of anti-MAP monoclonal antibodies with centrosomes, primary cilia, and stress fibers. Cells were stained by indirect immunofluorescence microscopy with rhodamine-conjugated rabbit anti-mouse IgG used as the second antibody (according to the method of Bloom et al. [1984b]). (*A*) Reaction of monoclonal antibody with primary cultured brain cells (method of Bloom and Vallee 1983). A pair of juxtanuclear spots is seen, apparently representing the centrosomes or centrioles. (*B*) Same antibody reacted with C6 cells. Bright staining of linear elements that may represent primary cilia is seen. (*C*) Staining of CHO cells with another monoclonal antibody raised against the MAPs. This antibody stains stress fibers. Bar, 10 μm; same magnification for all three panels.

In the case of monoclonal antibodies, contamination with nonspecific immunoglobulins can be ruled out as an explanation for these effects, although other trivial explanations (epitopes shared between unrelated structures, impure antigen) cannot. However, it is interesting that certain staining patterns have appeared repeatedly in our laboratory, and between laboratories. One such staining pattern is shown in Figure 6A. This particular antibody stains a single juxtanuclear spot in most cells. Occasionally, the spot appears to split in two. These properties are characteristic of centrioles or centrosomes. In cells known to contain multiple centrosomes and centrioles, such as NIE 115 neuroblastoma cells (Spiegelman et al. 1979), the antibody stains multiple juxtanuclear spots. In yet other cell lines, bright staining of what appear to be primary cilia within the cytoplasm is seen (Fig. 6B). Since the antibody also stains cytoplasmic microtubules, albeit weakly and only in some cells, it is possible that it is recognizing an antigen common to multiple microtubule-containing structures.

Figure 6C shows a staining pattern obtained with another antibody raised in this case against electrophoretically purified MAP 1B. This antibody stains stress fibers. Microtubule staining has not been seen, although it is possible that the antibody recognizes a distinct class of microtubules that are codistributed (and associated with) stress fibers. A similar staining pattern has been observed by Thompson et al. (1983) using an independently isolated anti-MAP-1 monoclonal antibody.

Although it is too soon to evaluate these results completely, it is possible that they represent yet further evidence for the existence of MAPs involved in mediating the interaction of microtubules with other cellular organelles. This suggests that the number of MAP isoforms or the number of functional states of the MAPs may be even greater than we have as yet observed.

REFERENCES

Berlinrood, M., S.M. McGee-Russell, and R.D. Allen. 1972. Patterns of particle movement in nerve fibers in vitro—An analysis by photokymography and microscopy. *J. Cell Sci.* **11:** 875.

Bloom, G.S. and R.B. Vallee. 1983. Association of microtubule-associated protein 2 (MAP 2) with microtubules and intermediate filaments in cultured brain cells. *J. Cell Biol.* **96:** 1523.

Bloom, G.S., F.C. Luca, and R.B. Vallee. 1983. Identification of microtubule-associated proteins (MAPs) in purified sea urchin egg microtubules and in the mitotic spindle of dividing sea urchin eggs. *J. Cell Biol.* **97:** 202a.

———. 1984a. Widespread cellular distribution of MAP 1A in the mitotic spindle and on interphase microtubules. *J. Cell Biol.* **98:** 331.

Bloom, G.S., T.A. Schoenfeld, and R.B. Vallee. 1984b. Widespread distribution of the major polypeptide component of MAP 1 (microtubule associated protein 1) in the nervous system. *J. Cell Biol.* **98:** 320.

Bulinski, J.C. and G.G. Borisy. 1979. Self-assembly of microtubules in extracts of cultured HeLa cells and the identification of HeLa microtubule associated proteins. *Proc. Natl. Acad. Sci.* **76:** 293.

Burgoyne, R.D., and R. Cumming. 1984. Ontogeny of microtubule associated protein 2 in rat cerebellum: Differential expression of the doublet polypeptides. *Neuroscience* **11:** 157.

Caceres, A., L.I. Binder, M.R. Payne, P. Bender, L. Rebhun, and O. Steward. 1984. Differential subcellular localization of tubulin and the microtubule associated protein MAP 2 in brain tissue as revealed by immunocytochemistry with monoclonal hybridoma antibodies. *J. Neurosci.* **4:** 394.

Cleveland, D.W., S.-Y. Hwo, and M.W. Kirschner. 1977. Purification of tau, a microtubule associated protein that induces assembly of microtubules from purified tubulin. *J. Mol. Biol.* **116:** 207.

De Camilli, P., P. Miller, F. Navone, W.E. Theurkauf, and R.B. Vallee. 1984. Distribution of microtubule associated protein 2 (MAP 2) in the nervous system of the rat studied by immunofluorescence. *Neuroscience* **11:** 819.

Dentler, W.L., S. Granett, and J.L. Rosenbaum. 1975. Ultrastructural localization of the high molecular weight proteins associated with *in vitro* assembled brain microtubules. *J. Cell Biol.* **65:** 237.

Duerr, A., D. Pallas, and F. Solomon. 1981. Molecular analysis in cytoplasmic microtubules *in situ*: Identification of both widespread and specific proteins. *Cell* **24:** 203.

Griffith, L.M. and T.D. Pollard. 1978. Evidence for actin filament-microtubule interaction mediated by microtubule-associated proteins. *J. Cell Biol.* **78:** 958.

Hayden, J.H., R.D. Allen, and R.D. Goldman. 1983. Cytoplasmic transport in keratocytes: Direct visualization of particle translocation along microtubules. *Cell. Motil.* **3:** 1.

Herrmann, H., R. Pytela, J.M. Dalton, and G. Wiche. 1984. Structural homology of microtubule-associated proteins 1 and 2 demonstrated by peptide mapping and immunoreactivity. *J. Biol. Chem.* **259:** 612.

Herzog, W. and K. Weber. 1978. Fractionation of brain microtubule-associated proteins. *Eur. J. Biochem.* **92:** 1.

Hill, A.-M., R. Maunoury, and D. Pantaloni. 1981. Cellular distribution of the microtubule-associated proteins HMW (350K, 300K) by indirect immunofluorescence. *Biol. Cell.* **41:** 43.

Hirokawa, N. 1982. Cross-linker system between neurofilaments, microtubules, and membranous organelles in frog axons revealed by quick-freeze, deep-etching method. *J. Cell Biol.* **94:** 129.

Kim, H., L. Binder, and J.L. Rosenbaum. 1979. The periodic association of MAP 2 with brain microtubules *in vitro*. *J. Cell Biol.* **80:** 266.

Kuznetsov, S.A., V.I. Rodionov, V.I. Gelfand, and V.A. Rosenblat. 1981a. Purification of high M_r microtubule proteins MAP 1 and MAP 2. *FEBS Lett.* **135:** 237.

———. 1981b. Microtubule-associated protein MAP 1 promotes microtubule assembly *in vitro*. *FEBS Lett.* **135:** 241.

Laemmli, U.K. 1970. Cleavage of structural proteins during assembly of the head of bacteriophage T4. *Nature* **227:** 680.

Leterrier, J.F., R.K.H. Liem, and M.L. Shelanski, 1982. Interactions between neurofilaments and microtubule-associated proteins: A possible mechanism for intraorganellar bridging. *J. Cell Biol.* **95:** 982.

Matus, A., R. Bernhardt, and T. Hugh-Jones. 1981. High molecular weight microtubule-associated proteins are preferentially associated with dendritic microtubules in brain. *Proc. Natl. Acad. Sci.* **78:** 3010.

Miller, P., U. Walter, W.E. Theurkauf, R.B. Vallee, and P. De Camilli. 1982. Frozen tissue sections as an experimental system to reveal specific binding sites for the regulatory subunit of type II cAMP dependent protein kinase in neurons. *Proc. Natl. Acad. Sci.* **79:** 5562.

Murphy, D.B. and G.G. Borisy. 1975. Association of high-molecular-weight proteins with microtubules and their role in microtubule assembly *in vitro*. *Proc. Natl. Acad. Sci.* **72:** 2696.

Murphy, D.B., R.B. Vallee, and G.G. Borisy. 1977. Identity and polymerization-stimulatory activity of the non-tubulin proteins associated with microtubules. *Biochemistry* **16:** 2598.

Olmsted, J.B. and H.D. Lyon. 1981. A microtubule-associated protein specific to differentiated neuroblastoma cells. *J. Biol. Chem.* **256:** 3507.

Papasozomenos, S.C., L.I. Binder, P. Bender, and M.R. Payne. 1982. A monoclonal antibody to microtubule-associated protein 2 (MAP 2) localizes with neurofilaments in the beta,beta'-iminodipropionitrile (IDPN) model. *J. Cell Biol.* **95:** 341a.

Runge, M.S., T.M. Laue, D.A. Yphantis, M. Lifsics, A. Saito, M. Altin, K. Reinke, and R.C. Williams, Jr. 1981. ATP-induced formation of an associated complex between microtubules and neurofilaments. *Proc. Natl. Acad. Sci.* **78:** 1431.

Sato, C., K. Nishizawa, H. Nakamura, Y. Komagoe, K. Shimada, R. Ueda, and S. Suzuki. 1983. Monoclonal antibody against microtubule associated protein-1 produces immunofluorescent spots in the nucleus and centrosome of cultured mammalian cells. *Cell Struct. Funct.* **8:** 245.

Sattilaro, R.F., W.L. Dentler, and E.L. Lecluyse. 1981. Microtubule-associated proteins (MAPs) and the organization of actin filaments *in vitro*. *J. Cell Biol.* **90:** 467.

Sattilaro, R.F., E. Lecluyse, and W.L. Dentler. 1980. Associations between microtubules and coated vesicles *in vitro*. *J. Cell Biol.* **87:** 250a.

Sherline, P. and R. Mascardo. 1982. Epidermal growth factor induced centrosomal separation: Mechanism and relationship to mitogenesis. *J. Cell Biol.* **95:** 316.

Sherline, P., Y.C. Lee, and L.S. Jacobs. 1977. Binding of microtubules to pituitary secretory granules and secretory granule membranes. *J. Cell Biol.* **72:** 380.

Sloboda, R.D., S.A. Rudolph, J.L. Rosenbaum, and P. Greengard. 1975. Cyclic AMP-dependent endogenous phosphorylation of a microtubule-associated protein. *Proc. Natl. Acad. Sci.* **72:** 177.

Smith, D.S., U. Jarlfors, and B.F. Cameron. 1975. Morphological evidence for the participation of microtubules in axonal transport. *Ann. N.Y. Acad. Sci.* **253:** 472.

Spiegelman, B.M., M.A. Lopata, and M.W. Kirschner. 1979. Multiple sites for the initiation of microtubule assembly in mammalian cells. *Cell* **16:** 239.

Suprenant, K.A. and W.L. Dentler. 1982. Association between endocrine pancreatic secretory granules and *in vitro*-assembled microtubules is dependent upon microtubule associated proteins. *J. Cell Biol.* **93:** 164.

Theurkauf, W.E. and R.B. Vallee. 1982. Molecular characterization of the cAMP-dependent protein kinase bound to microtubule-associated protein 2. *J. Biol. Chem.* **257:** 3284.

———. 1983. Extensive cAMP dependent and cAMP independent phosphorylation of MAP 2 (microtubule-associated protein 2). *J. Biol. Chem.* **258:** 7883.

Thompson, W.C., D.J. Asai, C. Dresden, and D.L. Purich. 1983. A monoclonal antibody shows MAP 1 associated with stress fibers, but not microtubules. *J. Cell Biol.* **97:** 201a.

Vallee, R.B. 1980. Structure and phosphorylation of microtubule-associated protein 2 (MAP 2). *Proc. Natl. Acad. Sci.* **77:** 3206.

———. 1982. A taxol-dependent procedure of the isolation of microtubules and microtubule-associated proteins (MAPS). *J. Cell Biol.* **92:** 435.

Vallee, R.B. and G.S. Bloom. 1983. Isolation of sea urchin egg microtubules using taxol and identification of mitotic spindle MAPS with the use of monoclonal antibodies. *Proc. Natl. Acad. Sci.* **80:** 6259.

———. 1984. High molecular weight microtubule-associated proteins (MAPs). In *Modern cell biology* (ed. B. Satir), vol. 3, p. 21. A.R. Liss, New York.

Vallee, R.B. and G.G. Borisy. 1977. Removal of the projections from cytoplasmic microtubules *in vitro* by digestion with trypsin. *J. Biol. Chem.* **252:** 377.

Vallee, R.B. and S.E. Davis. 1983. Low molecular weight microtubule associated proteins are light chains of microtubule-associated protein 1 (MAP 1). *Proc. Natl. Acad. Sci.* **80:** 1342.

Vallee, R.B., M.J. DiBartolomeis, and W.E. Theurkauf. 1981. A protein kinase associated with the projection portion of MAP 2 (microtubule associated protein 2). *J. Cell Biol.* **90:** 568.

Weatherbee, J.A., R.B. Luftig, and R.R. Weihing. 1980. Purification and reconstitution of HeLa cell microtubules. *Biochemistry* **19:** 4116.

Weatherbee, J.A., P. Sherline, R.N. Mascardo, J.G. Izant, R.B. Luftig, and R.R. Weihing. 1982. Microtubule-associated proteins of HeLa cells: Heat stability of the 200,000 molecular weight HeLa MAPs and detection of the presence of MAP 2 in HeLa cell extracts and cycled microtubules. *J. Cell Biol.* **92:** 155.

Assembly and Morphogenesis of the Avian Erythrocyte Cytoskeleton

Elias Lazarides
Division of Biology
California Institute of Technology
Pasadena, California 91125

The genesis of biological ultrastructure is one of the most fundamental but also most poorly understood aspect of growth and development. How does supramolecular structure arise and how is the information supplied to direct its assembly? With the realization over the past decade of the cytoskeletal complexity of higher eukaryotic cells, it has become increasingly apparent that the multitude of cytoskeletal elements must be somehow responsible either directly for the genesis of form or indirectly for its maintenance. In simpler systems, such as the tobacco mosaic virus, classical studies have shown that for the most part morphogenesis of this virus proceeds by self-assembly (Butler and Klug 1971; Durham and King 1971; Ohno et al. 1971; Okada and Ohno 1972). Analyses of the assembly and morphogenesis of complex bacteriophages have corroborated the importance of self-assembly but have also pointed to the importance of temporal order in subunit assembly, which influences assembly both thermodynamically and kinetically (Wood 1973, 1979). We refer to self-assembly in this context to any assembly process requiring no more information than that contained in the constituent macromolecular subunits of the finished product (Wood 1973). The assembly of cytoskeletal elements and the morphogenesis of more complex structures clearly involve additional factors to self-assembly as more information and control are required to generate diversity in form and function. Three obvious additional requirements in higher eukaryotic cells are for temporal information, specifying the time at which constituents become available; spatial information, controlling the site at which assembly will occur; and preexisting structure, which may be required to direct the correct deposition of new material. This is certainly evident in the terminal differentiation program of a number of cell lineages as the cells proceed from a proliferative stage, to a gradual, often multistep, process of commitment, to a final postmitotic phase where they assemble specialized

Molecular Biology of the Cytoskeleton. © Cold Spring Harbor Laboratory. 0-87969-174-3/84 $1.00

cytoskeletal structures, modify their plasma and intracellular membranes, and assume their specialized functions. During this process, cells invariably modify their proliferative cytoskeletal phenotype by the synthesis of new cytoskeletal proteins, by the synthesis of variants or isoforms of the preexisting cytoskeletal polypeptides, and by either retaining their proliferative cytoskeletal complement in their postmitotic phase or gradually replacing it with new variants and new cytoskeletal polypeptides. All of these changes, which are highly cell-lineage-specific, give rise to the characteristic cytoskeletal suprastructure of a given differentiated cell. If then we are to understand the control of cytoskeletal morphogenesis, we need to define first the cytoskeletal domains and their constituent polypeptides, then the control of assembly of each domain, and subsequently the mechanism of joining of these domains to each other to bring about the ultimate supramolecular phenotype of the cell. Many of these general ideas are similar to those delineated earlier for complex bacteriophage morphogenesis. As in the case of different types of bacteriophages, where differences exist in the actual details of assembly and the polypeptides involved (Wood 1979; Wood and King 1979), so it is expected that a similar situation will prevail in the cytoskeletal morphogenesis of various cell lineages. However, ultimately the principles that emerge should be generalizable to all cell types, irrespective of their supramolecular complexity.

AVIAN ERYTHROCYTE AS A MODEL SYSTEM

The nucleated avian erythrocyte has provided an excellent model system in which to investigate biological assembly at the molecular level. Even though its cytoskeletal ultrastructure is rather simple, all three major cytoskeletal elements, actin filaments, intermediate filaments, and microtubules, are contained in this cell with a highly organized topological distribution (Granger and Lazarides 1982). This has enabled the development of electron microscopic and biochemical techniques that have allowed us initially to define the various cytoskeletal domains of this cell type and subsequently to study their assembly during erythroid development. These studies graphically illustrate the problems of spatial and temporal regulation of cytoskeletal morphogenesis, departures from simple self-assembly, and the various levels of regulation that might be operative in this type of morphogenesis in development. Furthermore, by comparing the control of assembly and morphogenesis of the same cytoskeletal domains in different cell lineages of the same species or in the same cell type but in different species (notably mammalian erythroid development where the cells become enucleated), we gain a unique insight in the way different lineages regulate cytoskeletal morphogenesis.

CYTOSKELETAL DOMAINS IN THE AVIAN ERYTHROCYTE

Membrane-Skeleton

Over the past several years, the mammalian erythrocyte has provided a model system for the elucidation of cytoskeletal structure and, in particular, the molecular details of the interaction of actin with the plasma membrane. The mammalian erythrocyte contains a dense anastomosing network of proteins composed of spectrin, actin, and several associated polypeptides (for review, see Branton et al. 1981; Bennett 1982). This network lines the inner surface of the erythrocyte's plasma membrane and serves as its cytoskeleton, contributing to the shape, integrity, and elasticity of the cell. The principal component of this membrane-skeleton is spectrin, a protein composed of two nonidentical polypeptides termed α-spectrin and β-spectrin. The formation of this skeleton involves the self-association of spectrin into $(\alpha,\beta)_2$ tetramers and $(\alpha,\beta)_{2n}$ oligomers (Liu et al. 1984) and the cross-linking of actin by the spectrin. The association of this skeleton with the membrane is mediated, at least in part, by the protein ankyrin, which binds both to the β subunit of spectrin and the cytoplasmically oriented amino-terminal segment of the transmembrane anion transporter (band 3). The resulting molar ratio of the complex between the anion transporter, ankyrin, β-spectrin, and α-spectrin is 1:1:2:2, even though the number of anion transporter polypeptides substantially exceeds those of ankyrin. Protein 4.1 is thought to strengthen this network by forming a stable ternary complex with spectrin and actin (for review, see Branton et al. 1981). In addition, protein 4.1 may be involved in the anchorage of the spectrin-actin skeleton to the lipid bilayer. Protein 4.1 remains associated with the membrane after removal of spectrin and actin, where it appears to associate at least with the cytoplasmic domain of glycophorin, a transmembrane sialoglycoprotein (Anderson and Lovrien 1984). Thus, protein 4.1 may complement the well-defined ankyrin-mediated linkage of spectrin to the bilayer (for review, see Bennett 1982). Mammalian erythrocytes also contain a protein structurally analogous to skeletal-muscle tropomyosin. Although its function in the erythrocyte is presently unknown, it is present in this cell in approximately the same ratio to actin as in skeletal muscle (seven actins to one tropomyosin), suggesting a structural role (Fowler and Bennett 1984).

Similarly to mammalian erythrocytes, avian erythrocytes contain a subcortical network of proteins composed predominantly of α- and β-spectrins (Granger et al. 1982). The avian erythrocyte membrane-skeleton also contains polypeptides analogous to actin, ankyrin, protein 4.1, and tropomyosin (Granger et al. 1982; Repasky et al. 1982; Georgatos and Lazarides 1983; Granger and Lazarides 1984), as well as the anion transporter (Granger et al. 1982; Jay 1983). Although the details of the interaction of these polypeptides with each other and with the membrane have not been studied in as great a detail as in mammalian erythrocytes, it is assumed that they serve an analogous function. In particular, the avian erythrocyte analog of ankyrin is a protein known as goblin. Goblin is characterized by hormone-dependent phosphorylation (Beam et al. 1979) by at least three endogenous protein kinases (Alper et al. 1980b). Phosphorylation of goblin, after hormone-induced increase of intracellular cAMP, is correlated quantitatively with a concomitant increase in Na^+-K^+ cotransport (Alper et al. 1980a). Goblin is serologically and by peptide mapping related to ankyrin, and it reassociates with the avian anion transporter and spectrin in vitro in approximately the same ratio as its mammalian erythroid counterparts (Nelson and Lazarides 1984). Other properties of these membrane-skeleton polypeptides characteristic of avian erythrocytes and pertinent to the discussion that follows are presented below.

Intermediate Filament Skeleton

Unlike mammalian erythrocytes, avian erythrocytes possess nuclei and a few mitochondria. In addition to the spectrin-based membrane-skeleton, these cells possess a circumferential band of microtubules known as the marginal band. Its mechanism of circumferential growth and potential attachment to the membrane will not be discussed further here (for discussion, see Granger and Lazarides 1982; Joseph-Silverstein and Cohen 1984). Chicken erythrocytes also contain a system of filaments that have been identified as intermediate filaments (Virtanen et al. 1979; Woodcock 1980; Granger et al. 1982) and that form a three-dimensional network in these cells, interlinking the nucleus to the membrane-skeleton on opposite sides of the cell (Granger and Lazarides 1982). The details of this transcytoplasmic interconnection have become evident with the development of a technique for visualizing these filaments in avian erythrocytes (Granger and Lazarides 1982; Granger et al. 1982). Cells adhering to a flat glass substrate were disrupted by sonication, and their exposed insoluble cytoskeletons were shadowed with platinum. Examination of the platinum replicas by transmission electron microscopy allowed the visualization of the three-dimensional distribution of the filaments inside the cell. An example of such a partially disrupted chicken erythrocyte is shown in Figure 1. Intermediate filaments are seen to attach side-on to the spectrin-based membrane-skeleton and extend and attach

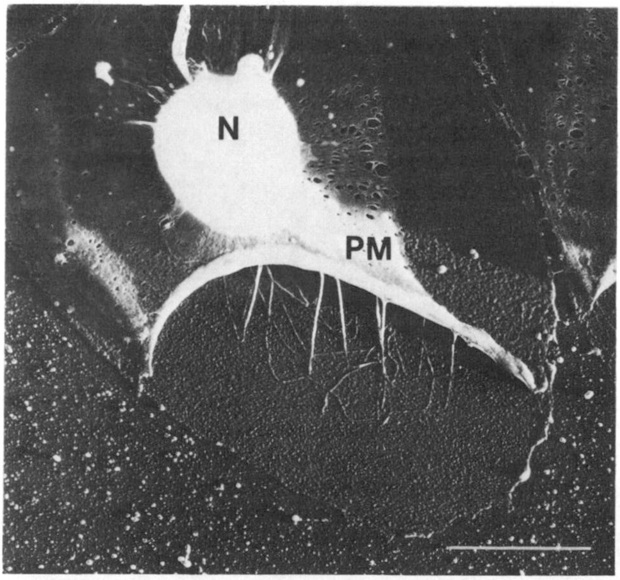

Figure 1
Visualization of the three-dimensional disposition of intermediate filaments in sonicated chicken erythrocytes. This partially disrupted erythrocyte shows filaments that span from the adhering to the overlying regions of the plasma membrane skeleton. This preparation was unidirectionally shadowed, not bleached, and printed with reverse contrast. (PM) Overlying plasma membrane; (N) position of the nucleus. Bar, 2 μm. (Reprinted, with permission, from Granger and Lazarides 1982.)

to the nucleus (not shown) (see Granger and Lazarides 1982; Lazarides and Granger 1983) and upper membrane-skeleton, forming a three-dimensional cytoplasmic network.

Biochemical and immunological analyses of total adult chicken erythrocyte (or membrane-associated) intermediate filaments have shown that they are composed of two polypeptides: vimentin (M_r = 52,000) and a high-molecular-weight protein (M_r = 230,000) known as synemin (Granger et al. 1982). Vimentin functions as the main core subunit of the filaments, whereas synemin is periodically associated along the vimentin filament backbone and may function to cross-link them. This became evident with the development of an immunoelectron microscopic technique to localize these antigens along intermediate filaments. Prolonged sonication of attached erythrocytes removes the nonadherent part of the cell, leaving attached to the coverslips the adhering part of the plasma membrane and the associated membrane-skeleton. These elliptical patches of membrane frequently contain mats of intermediate filaments, thus demonstrating the stable association of a subset of intermediate filaments with the membrane-skeleton. Immunoelectron microscopic studies on these membrane patches with anti-vimentin antibodies have revealed a uniform decoration of all of the intermediate filaments, showing that vimentin is the core polypeptide subunit of these filaments. An example of such images is depicted in Figure 2. Incubation of the erythrocyte patches with anti-synemin has revealed periodically spaced foci along the long axis of all of the filaments (Fig. 2C). Such micrographs have suggested that synemins of adjacent filaments can self-associate and thus mediate lateral cross-linking of the filaments (Granger and Lazarides 1982).

Anisotropic Association of Intermediate Filaments with the Membrane-Skeleton

One of the more important aspects in the morphogenesis of the avian erythrocyte cytoskeleton to become evident from the platinum replicas of these cells was the mechanism of attachment of intermediate filaments to the membrane-skeleton. This association is anisotropic in the sense that the vimentin filaments attach only to the center of the membrane-skeleton surrounded by a free margin at least 1 μm wide where attachment does not occur. Furthermore, the spatial relationships of the membrane-bound intermediate filaments and the marginal band to each other and to the membrane-skeleton have become evident, with sonications carried out in microtubule-stabilizing buffer. Such images have demonstrated that both intermediate filaments and the marginal band of microtubules interact with the plasma membrane indirectly via the spectrin-based membrane-skeleton and that they are spatially segregated from each other (Granger and Lazarides 1982). These observations have given rise to the important notion that the spectrin-based membrane-skeleton is inhomogeneous with respect to these two filament systems and that any interaction between these two systems is mediated by the spectrin-actin network (Granger and Lazarides 1982). Thus, at least for intermediate filaments, a mechanism must be operative during morphogenesis that dictates the joining of this filament system with the membrane-skeleton preferentially in the center of the spectrin-based membrane-skeleton and away from the marginal band. This issue is discussed in more detail below.

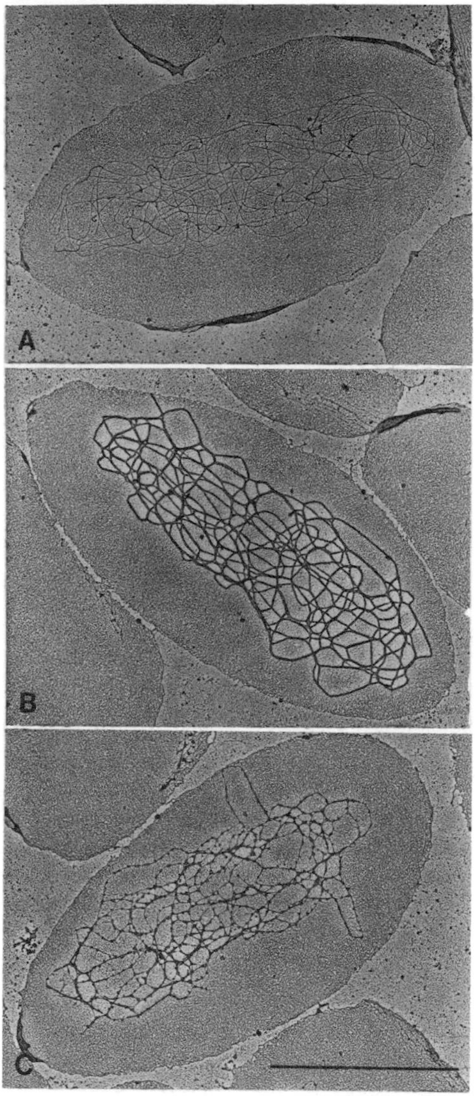

Figure 2
Antibody decoration of chicken erythrocyte intermediate filaments. When the filaments are preincubated with the synemin preimmune serum, the diameter of the filaments is indistinguishable from that of controls (*A*). Incubation with anti-vimentin antibodies (*B*) results in uniform decoration of the filaments with no obvious periodicity. Incubation of the filaments with anti-synemin antibodies (*C*) reveals a periodic distribution of the antigen. The average spacing of the anti-synemin foci is 180 μm. Bar, 5 μm. (Reprinted, with permission, from Granger and Lazarides 1982.)

Three Distinct, Interconnected Cytoskeletal Domains in Avian Erythrocytes

The above studies have established that avian erythrocytes contain three clearly identifiable cytoskeletal domains, whose protein compositions are distinct from one another: the spectrin-based membrane-skeleton, a transcytoplasmic system of intermediate filaments, and a marginal band of microtubules. These three domains are interlinked to one another through the submembranous-spectrin network at distinct sites on this network and to the nucleus through the intermediate filament network. A summary of the known protein constituents of the spectrin network and the intermediate filament network in adult erythrocytes is given in Table 1. A hypothetical diagram depicting the interrelationship of the spectrin network and the intermediate filament network in avian erythrocytes and a comparison with mammalian erythrocytes are shown in Figure 3 to facilitate the discussion that follows. At present, all detailed interactions within the spectrin network in avian erythrocytes are not known and are derived by analogy to those delineated in human erythrocytes. Similarly, the details of the interaction of intermediate filaments with the membrane-skeleton are not known.

PATHWAY OF ASSEMBLY OF THE ERYTHROCYTE CYTOSKELETON

Assembly and Topogenesis of the Membrane-Skeleton

In studying the mechanism of assembly of the erythroid membrane-skeleton, there are a few characteristics of the complex worth considering. The first is that it is composed of one or more transmembrane polypeptides (e.g., anion transporter and glycophorin) and a number of peripheral polypeptides. Assembly therefore requires a combining of both soluble and membrane-bound components, which are synthesized and transported to the sites of assembly by different pathways in the cell. To synthesize and transport to the membrane an anion transporter polypeptide, the cell requires 30–45 minutes (Braell and Lodish 1981), whereas to synthesize a polypeptide as large as α-spectrin, β-spectrin, or ankyrin, the cell requires less than 10 minutes (Blikstad and Lazarides 1983b). Assembly then of the erythroid membrane-skeleton requires temporal control, since the reacting polypeptides are processed at different rates, and spatial control, since they all are to assemble with each other, exclusively at the plasma membrane after their synthesis at distinct sites. Additionally, the polypeptides have to assemble at a precise ratio, often substoichiometric. Our studies on this subject have been summarized recently in detail (Lazarides and Moon 1984; Moon and Lazarides 1984; Moon et al. 1984) and only some of the more general conclusions are discussed here.

Posttranslational Control of Assembly

The earlier stages in the assembly of the membrane-skeleton, in particular the initiation of assembly, have not yet been studied, mostly because of the difficulty in obtaining large quantities of cells very early in the erythroid lineage. However, we assume, not necessarily correctly, that by studying the growth phase of this reaction, we also learn about the basic principles of the initiation of assembly in the early phases of erythroid development. Interpretation of the data is also complicated by the fact that erythroid cells from chicken embryos are heterogeneous in terms of differentiation and the type of lineage, and the number of mitotic cells of a given lineage varies from stage to stage of embryo development. The primitive series of erythroid cells dominates the circulation of the early embryo, becoming postmitotic by days 5–6 of development and gradually disappearing by

Table 1
Cytoskeletal Domains in the Avian Erythrocyte

Membrane-skeleton	*Intermediate filament skeleton*
α-Spectrin ($M_r = 240{,}000$)	vimentin ($M_r = 52{,}000$)
β-Spectrin ($M_r = 220{,}000$)	synemin ($M_r = 230{,}000$
Ankyrin (goblin; $M_r = 260{,}000$)	
Actin ($M_r = 42{,}000$)	
Protein 4.1 (family of polypeptides)	
Tropomyosin (family of polypeptides)	
Anion transporter	
Glycophorin[a]	

[a]Not yet identified in avian erythrocytes.

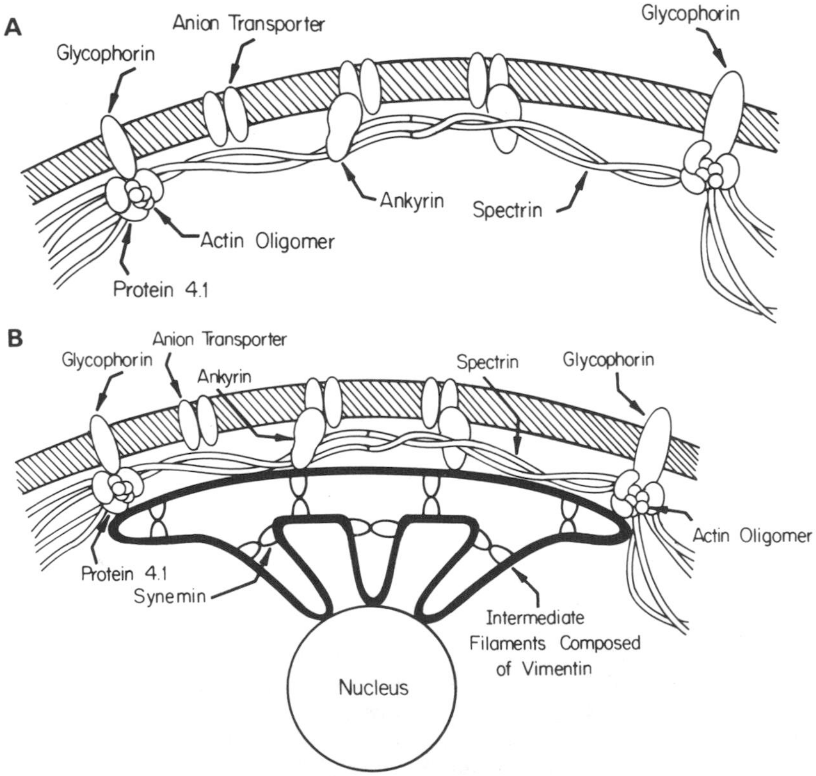

Figure 3
Highly schematic arrangement for the major polypeptides of the human red blood cell (*A*) and chicken red blood cell (*B*) erythrocyte cytoskeletons. The diagram of the human erythrocyte cytoskeleton is a modification from Branton et al. (1981). In the diagram of the chicken erythrocyte cytoskeleton, the position of the marginal band of microtubules is not shown.

day 16. The definitive series of erythroid cells appears in the circulation at about 5 days of incubation. They become postmitotic 2–3 days after the primitive series does and represent the sustained lineage that persists throughout the life of the organism (Bruns and Ingram 1973). Therefore, depending on the stage of development, the erythroid cells can be quite heterogeneous, and observations on a whole population of erythroid cells may not necessarily apply to all cells in the population. However, studies with either primitive cells or definitive cells at different stages of erythroid development have shown that all identifiable polypeptides of the membrane-skeleton, e.g., anion transporter, α-spectrin, β-spectrin, ankyrin, actin, protein 4.1, and presumably also tropomyosin, are synthesized simultaneously (Weise and Chan 1978; Blikstad et al. 1983; Moon and Lazarides 1983a, 1984; Granger and Lazarides 1984 and in prep.). Furthermore, within the resolution of techniques employed thus far (i.e., antigenic cross-reactivity and gel electrophoresis in denaturing media), the same variants of these polypeptides are expressed throughout development. A notable exception, which is discussed in more detail below, is protein 4.1, whose polypeptide pattern changes in postmitotic primitive and definitive erythroid cells (B.L. Granger and E. Lazarides, in prep.). In vivo and in vitro studies have indicated that the assembly of cytoplasmic components of the erythroid membrane-skeleton occurs posttranslationally. Assembly of each polypeptide onto the membrane proceeds rapidly after synthesis without an appreciable lag (Blikstad et al. 1983; Granger and Lazarides 1984; Moon and Lazarides 1984; see also Weise and Chan 1978), and inhibition of either initiation or elongation of protein synthesis abruptly stops their assembly (Weise and Chan 1978; Moon et al. 1984). Therefore, either synthesis in the cytoplasm and assembly onto the membrane of the polypeptides are coupled events or these two events are not necessarily coupled, and high-affinity binding sites on the membrane rapidly bind any available newly synthesized polypeptides. As discussed below, the data are consistent more with the latter hypothesis. Therefore, the erythroid cell does not appear to employ temporal control of protein synthesis to order the steps in the assembly of the components of the membrane-skeleton during development. This order then is not imposed at the level of gene expression; rather, it is imposed at the level of gene-product interaction by structural features of the intermediates.

Synthesis of Components in Excess of Amounts Assembled

Studies on the synthesis and assembly of α- and β-spectrins and ankyrin in chicken embryo erythroid cells have shown that each polypeptide is synthesized in excess of the amount assembled and not in the ratio observed at steady state (Blikstad et al. 1983; Moon and Lazarides 1983a, 1984). However, the molar ratio of newly assembled polypeptides is close to that at steady state in the membrane-skeleton. Therefore, the stoichiometry of the complex is determined posttranslationally through the binding of proteins to each other and to the membrane-skeleton, rather than by the molar ratio of the components synthesized (Lazarides and Moon 1984; Moon and Lazarides 1984). Thus, there is no evidence for feedback control of gene expression or of gene-product synthesis by the structural end products of the membrane-skeleton. However, the transcription patterns of the genes of each membrane-skeleton polypeptide impose built-in controls on the relative levels of each polypeptide available for assembly. Although not studied in detail, these transcription patterns appear to be regulated independently for each membrane-skeleton polypeptide. For example, the level of synthesis of the anion transporter appears to be substantially higher than those of the spectrins and ankyrin in primitive erythroid cells, but declines more precipitously than those of the spectrins and ankyrin in later stages of erythroid development (Weise and Chan 1978). Concurrently, the fraction of unassembled α- and β-spectrins and ankyrin increases (Blikstad et al. 1983). It therefore appears that the pathway of assembly of the membrane-skeleton has evolved to maximize rapid assembly of the constituent polypeptides rather than to economize reactants. A similar situation is observed also in the assembly of certain bacteriophages, where some of the substructures are produced in substantial excess over others and never end up on the infectious particles (Wood 1973).

Spatial Control of Assembly: Stabilization of Polypeptides Upon Assembly

One of the important regulatory features of assembly of the membrane-skeleton to emerge from these studies is that assembled polypeptides are rendered resistant to catabolism, whereas the excess unassembled polypeptides are catabolized rapidly in the cytosol (Blikstad et al. 1983; Moon and Lazarides 1984). Blocking assembly does not inhibit degradation of at least α- and β-spectrins and ankyrin and vice versa (C.M. Woods and E. Lazarides, in prep.), suggesting that assembly and degradation operate simultaneously, but independently. These observations imply the existence of a cytosolic mechanism that discriminates between assembled and unassembled α- and β-spectrins and ankyrin and catabolizes unassembled polypeptides. Since it is the peripheral membrane-skeleton polypeptides that are degraded if they remain unassembled, binding to their transmembrane receptor at the membrane (e.g., anion transporter or glycophorin) may be the key event that confers stability to them. This may also serve as a safeguard mechanism to ensure that the peripheral membrane-skeleton polypeptides assemble only at the correct cytoplasmic site (plasma membrane).

Ordering the Steps in Assembly

The conclusion that the stoichiometry of the membrane-skeleton components is obtained posttranslationally during the binding of components to each other and to the membrane-skeleton is consistent with the possibility that components in the assembly complex located on the membrane (e.g., anion transporter) limit the extent of assembly of the peripheral components, thus also successively ordering the steps in their assembly (Blikstad et al. 1983; Moon and Lazarides 1984). In vitro reconstitution studies have confirmed this prediction, showing that the amount of β-spectrin synthesized limits the amount of α-spectrin assembled (Moon and Lazarides 1983a). Similar arguments have also been put forward for the amount of ankyrin synthesized limiting the amount of β-spectrin assembled and the availability of transmembrane binding sites (e.g., anion transporter) limiting the extent of ankyrin assembled and hence the overall extent of assembly of the peripheral components of the membrane-skeleton (Lazarides and Moon 1984; Moon and Lazarides 1984). Studies with mouse mutants defective in the synthesis of various peripheral components of the membrane-skeleton have also confirmed some of these predictions and observations (Bodine et al. 1984). Similar to the situation in avian erythrocytes, in mouse erythrocytes, α-spectrin is synthesized in excess of β-spectrin, and the extent of synthesis and assembly of the latter limits the extent of assembly of the former; the excess unassembled polypeptides appear also to be rapidly degraded. Furthermore, mutants deficient in the synthesis of ankyrin still exhibit partial spectrin assembly, suggesting that other polypeptides in addition to ankyrin and the anion transporter, such as protein 4.1 and glycophorin, may mediate spectrin assembly and stabilization (Bodine et al. 1984). A similar situation is also observed in avian erythrocytes (Moon et al. 1984).

We have previously formulated these observations in the form of a hypothesis stating that assembly and stabilization of the membrane-skeleton in the erythrocyte is a receptor-mediated process (Lazarides and Moon 1984; Moon and Lazarides 1984). A summary of this hypothesis is shown in Figure 4. (For a detailed discussion of this hypothesis, see Moon and Lazarides [1984] and Lazarides and Moon [1984].) This hypothesis explains how assembly and localization of this membrane-skeleton multisubunit complex occur in the absence of coordinated transcriptional and translational regulation to synthesize the precise ratio of the subunits found at steady state. Two general principles for the assembly of multisubunit complexes have emerged from this hypothesis: (1) Stoichiometry of polypeptides in the assembled complex is predetermined by the properties of each of the subunits, and assembly occurs posttranslationally to the extent allowed by the availability of the limiting component, and (2) one or more subunits of the complex act as the assembly receptor(s) to stabilize the constituent polypeptides against catabolism, which also results in the accumulation of the complex at a specific cytoplasmic site.

Assembly of Intermediate Filaments

Lineage-specific Expression of Vimentin Filaments

Although many of the details of the assembly of the intermediate filament domain are different from those of the membrane-skeleton domain, many of the principles are the same. The regulation of the expression and assembly of the various

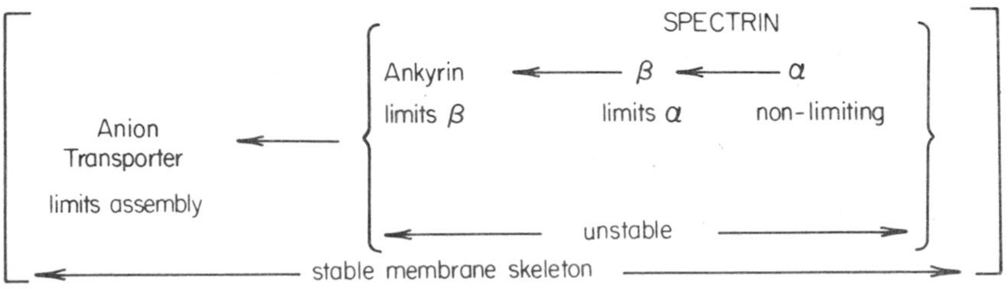

Figure 4
Receptor-mediated assembly and stabilization hypothesis of the erythrocyte membrane skeleton. (Reprinted, with permission, from Moon and Lazarides 1984.)

intermediate filament subunits in erythroid and nonerythroid cells has been discussed recently (Ngai et al. 1984a; Capetanaki et al., this volume) and is not discussed here except where it pertains to erythroid cells. Rather, we will concentrate on some of the changes that the filaments undergo during development and maturation of the organism and their implication for the morphogenesis of the erythroid cell.

Assembly of intermediate filaments proceeds independently from that of the spectrin-based membrane-skeleton. This conclusion is based on two main observations. The first is that whereas both primitive and definitive erythroid cells constitutively express the subunits of the membrane-skeleton, the primitive erythroid series expresses only basal levels of vimentin and very few identifiable vimentin-containing intermediate filaments. However, there is a tremendous induction in the accumulation of vimentin mRNA (Capetanaki et al. 1983), with a concomitant accumulation of vimentin filaments (Granger and Lazarides 1982) in the definitive erythroid series, reaching maximal levels late in chicken embryo development. Therefore, the accumulation of intermediate filaments, unlike that of the membrane-skeleton, appears to be lineage-specific (Capetanaki et al. 1983; Ngai et al. 1984b). This observation has emphasized that expression of intermediate filaments is an essential part of the terminal differentiation program of the definitive series of erythroid cells. On morphogenetic grounds, we have hypothesized that this accumulation of intermediate filaments may be linked to the need of the definitive erythroid cells to interlink the nucleus and the membrane-skeleton, thereby mechanically integrating membrane organelles with each other as the need for shape changes and deformability of the cells increase during circulation with the maturation of the organism (Granger and Lazarides 1982; Lazarides and Granger 1983; Capetanaki et al. 1983; Ngai et al. 1984a,b).

The second observation stems from studies on the expression of subunits of the membrane-skeleton and intermediate filaments in mammalian erythroid cells and, in particular, Friend murine erythroleukemia (MEL) cells, which have provided an excellent model system for studying the molecular events in mammalian erythropoiesis (Ngai et al. 1984a,b). When these cells are induced chemically to differentiate, one of the changes that ensues is an induction in the synthesis of the membrane-skeleton protein spectrin (Arndt-Jovin et al. 1976; Eisen et al. 1977) and the transmembrane anion transporter (Sabban et al. 1980). However, in striking contrast to the induction of these membrane-skeleton proteins, or the pattern of vimentin expression in chicken erythroid cells, MEL cells rapidly repress the expression of vimentin upon chemical induction to differentiate (Ngai et al. 1984a,b). Therefore, as in chicken erythroid cells, vimentin filament expression in this system is regulated by the abundance of vimentin mRNA and represents an early and rapid process in the developmental program of these cells. The dynamic positive and negative regulations of the vimentin gene in avian and mammalian erythropoiesis have suggested that the presence or absence of vimentin filaments represents an essential feature of each of these differentiation programs. Furthermore, their lineage- and species-specific modes of regulation may have evolved as part of the morphogenetic requirements of the cells. By analogy to avian erythroid cells, we have hypothesized that enucleation during mammalian erythroid development is facilitated by the loss of vimentin filaments, thus necessitating the evolution of opposite modes of regulation of their expression during avian and mammalian erythroid development (Ngai et al. 1984a,b).

Stable Filament Growth and Rapid Self-assembly

Growth of intermediate filaments in the definitive series of chicken erythroid cells occurs posttranslationally from a small saturable pool of soluble subunits. Polymerization occurs by self-assembly of vimentin polypeptides, and it is rapid, complete, and essentially irreversible (Blikstad and Lazarides 1983a). Therefore, during active growth of the cells, the filaments appear to grow continuously with little evidence of turnover. Furthermore, there is no evidence for feedback control of gene expression or of gene-product synthesis by the structural end product, namely, vimentin. Inhibiting the assembly of vimentin by incorporation of amino acid analogs in its backbone does not inhibit its rate of synthesis (Moon and Lazarides 1983b). Therefore, the levels of vimentin mRNA at any instance in development appear to determine directly the level of vimentin available for polymerization but also, indirectly, the level of vimentin in the form of filaments. This is probably also the case for synemin, since the ratio of newly synthesized synemin to vimentin is approximately the same as that at steady state, namely, 1:20–1:50 (Granger and Lazarides 1982; Moon and Lazarides 1983c; B.L. Granger and E. Lazarides, unpubl.), and most of the newly synthesized synemin assembles onto the growing filaments, similarly to vimentin (Moon and Lazarides 1983c). However, the kinetics of assembly of synemin are substantially slower than those of vimentin, which has given rise to the hypothesis that the rate of vimentin polymerization and filament growth limits the rate and extent of synemin binding and assembly (Moon and Lazarides 1983c). The mechanism by which synemin binds to vimentin is currently unknown, nor is it known how the periodicity of synemin binding along the backbone of the vimentin filaments is determined. The analysis of the kinetics of vimentin and synemin coassembly has been consistent with the hypothesis that polymerization of vimentin generates synemin-binding sites through some long-range suprastructure of the filament as a result of a higher-order change in the structure of vimentin upon polymerization (Moon and Lazarides 1983c). Perhaps this higher-order change in vimentin structure may also dictate the periodic binding site for synemin (for discussion, see Granger and Lazarides 1982; Moon and Lazarides 1983c).

Developmental Changes in the Structure of Intermediate Filaments

During the course of erythroid cell differentiation and maturation of the organism, intermediate filaments undergo at least three fundamental changes in their structure. These changes may represent deviations from simple self-assembly and may have direct implications in the regulation of morphogenesis of the cell and the function of the filaments in this process in particular. In all cells examined, including erythrocytes, a fraction of the vimentin and synemin molecules are phosphorylated (O'Connor et al. 1981; Granger et al. 1982; Sandoval et al. 1982; B.L. Granger and E. Lazarides, unpubl.). Only a small fraction of the population of the vimentin polypeptides in erythroid cells from young embryos are phosphorylated (Blikstad and Lazarides 1983a), but more than half of the vimentin polypeptides are stably phosphorylated in adult erythrocytes (Granger et al. 1982). Furthermore, during the assembly of intermediate filaments, phosphorylation of vimentin and synemin occurs well after their assembly on the filaments ation of vimentin and synemin occurs well after their assembly on the filaments (Blikstad and Lazarides 1983a; Moon and Lazarides 1983c). The function of vimentin phosphorylation is not currently understood but may be important in regulating the degree of cross-linking of the filaments by synemin (see below), in the stability of the filaments, or in regulating the site of interaction of synemin with vimentin. In a similar vein, some structural features of the vimentin filament backbone may dictate the fraction of vimentin polypeptides in the filament that become phosphorylated at any time in development. Irrespective of what the molecular basis of this selective phosphorylation might be, these observations indicate that the filaments may undergo a fundamental change in their structure during erythroid cell maturation and that chemical modifications such as phosphorylation play an important role in this process.

A second change in the structure of intermediate filaments during erythroid development and maturation is revealed also by a change in the spacing of synemin along the filaments. Although the synemin periodicity along the vimentin filament backbone from adult erythrocytes (as revealed by immunoelectron microscopy using the platinum replica technique) is approximately 180 nm, in filaments from embryonic erythrocytes, it is 25–30% greater than that of the adult filaments (Granger and Lazarides 1982). In addition to a change in the synemin spacing, the extent of synemin-mediated cross-linking of the filaments may also change. Barring experimental pitfalls and artifacts using immunoelectron microscopy on membrane patches isolated by sonication, the extent of synemin-mediated cross-linking observable in filaments from embryonic erythrocytes is substantially less than that observed in adult cells (Granger and Lazarides 1982). If this observation indeed reflects a maturation and aging-dependent increase in filament cross-linking mediated by synemin, then this process is also developmentally regulated. The dramatic increase in the number of filaments observed late in development, as well as the structural and chemical modifications that vimentin and synemin undergo during maturation, may directly influence this process.

A change in the structure of intermediate filaments during the development and maturation of the organism is also implied by a change in the subunit composition of the filaments themselves during this time. In addition to vimentin, chicken embryo erythroid cells also express small amounts of the major subunits of neurofilaments (NF70; M_r = 70,000), normally expressed only in neurons (for review, see Lazarides 1980). Compared to vimentin, these cells express substoichiometric amounts of NF70, and antibody decoration data have suggested that the two subunits randomly copolymerize. During maturation of the organism, the amount of NF70 declines, and it is nearly absent from the erythroid cells of adults, further suggesting that modulation of the ratio of these two subunits in a given erythrocyte is a function of their differential synthesis (Granger and Lazarides 1983). This decline in the incorporation of NF70 into filaments during development may also contribute to the reduction in the linear periodicity of the association of synemin with the core filament and a facilitation in the increase in the synemin-mediated cross-linking (Granger and Lazarides 1983). These studies have also revealed that individual erythroid cells do not modulate their vimentin:NF70 ratio during differentiation but that this ratio changes as the cell population changes during development and growth of the organism, culminating in the virtual absence of NF70 from most cells in the adult. Perhaps the changes in the linear periodicity of synemin, the degree of synemin-mediated cross-linking, and the increase in the steady-state levels of vimentin and synemin phosphorylation may also change as the cell population changes during development and growth of the organism concurrently with the change in the vimentin:NF70 ratio. It is apparent then that during the maturation of the organism, the intermediate filament domain undergoes a number of structural changes that deviate from simple principles of self-assembly and that undoubtedly contribute to the ultimate morphogenesis of the erythroid cell.

Anisotropic Joining of Intermediate Filaments with the Membrane-Skeleton

Two more issues in the biogenesis of the erythroid cytoskeleton are worth discussing. These are perhaps the least well understood but are very fundamental in erythroid morphogenesis. The first concerns the initiation of growth of intermediate filaments. Whether initiation of filament growth occurs directly from a nucleating site, such as the nuclear membrane, or occurs first in the cytoplasm and subsequently the elongating filaments become attached to the nuclear membrane, is unknown. Examination of embryonic erythroid cells using electron microscopy or immunofluorescence has suggested that very young cells have few, if any, intermediate filaments in contact with the membrane-skeleton and relatively few filaments lying on the nuclear membrane. As the cells mature, these nucleus-associated filaments appear to elongate and establish contacts with the plasma membrane by some sort of looping out process (Granger and Lazarides 1982). If this mechanism of filament growth is correct, it implies that the growing end of the filament remains in association with the nuclear membrane and that this is also the site of subunit addition. At present, these issues are far from being resolved experimentally but are important in understanding how filament growth is spatially and temporally regulated.

The second issue concerns the anisotropic joining of the filaments with the membrane-skeleton. As noted earlier (see Figs. 1 and 2), the joining of the two domains occurs on the membrane-skeleton 1–2 μm away from the marginal band of microtubules. This implies that during their biogenesis, either the membrane-skeleton or the intermediate filaments (or both) acquire some sort of localized chemical or structural inhomogeneity that acts as the regulatory signal to join anisotropically these two domains. What this regulatory signal might be is unknown. As discussed above, intermediate filaments undergo a number of structural and chemical changes during maturation, and any one of them alone or synergistically with the other changes might facilitate joining. In particular, the phosphorylation of only a subset of the vimentin and synemin polypeptides might contribute to a localized chemical inhomogeneity in the filaments, rendering only that segment of the filament competent to join with the membrane-skeleton.

On the other hand, the erythroid membrane-skeleton is generally thought of as being two-dimensionally isotropic from studies on the mammalian erythrocyte membrane-skeleton. One observation, however, is worth considering since it provides a hint that the membrane-skeleton may be structurally anisotropic. This stems from the pattern of expression of protein 4.1. In adult erythrocytes, this protein is composed of at least three to four major and two to three minor variants that are highly homologous to each other, but whose molecular weights and patterns of phosphorylation vary extensively (Granger and Lazarides 1984). A set of these variants is expressed and accumulates in proliferative primitive and definitive erythroid cells. A second set of variants begins to accumulate specifically in postmitotic cells in both the primitive and definitive series cells, and thereafter both classes of variants (with one exception) accumulate at steady state on the membrane, with the postmitotic variants eventually predominating (Granger and Lazarides 1984 and in prep.). The induction of accumulation of the new protein 4.1 variants in postmitotic definitive series cells coincides with that of vimentin in these cells.

At present, there is no evidence of a functional relationship between any of the protein 4.1 variants and intermediate filaments. However, it is apparent that after the final cytokinesis of an erythroid cell in the primitive or definitive series is complete, this cell will begin assembling both "premitotic" (constitutive) and "postmitotic" variants. Whether these variants assemble randomly with each other and also randomly on the membrane-skeleton or whether there is a site of preferred growth on the membrane where addition of newly synthesized protein 4.1 variants takes place (e.g., at the area previously occupied by the contractile ring) is unknown. The latter mechanism, however, could potentially provide areas of the membrane containing both pre- and postmitotic variants of protein 4.1, thereby establishing an asymmetry in the composition of the membrane-skeleton. Such a mechanism could provide the asymmetry in the membrane necessary for the anisotropic joining of the two domains.

GENERAL CONSIDERATIONS

Studies on the assembly and morphogenesis of the chicken erythrocyte cytoskeleton have allowed the formulation of certain principles that may be generally applicable to the morphogenesis of any differentiating cell if the gene products and the assembly domains they constitute can be defined. They have shown that assembly of two of the three major cytoskeletal domains, the membrane-skeleton and the intermediate filaments, is independent. The pathway of assembly is branched, with the branches being indeed functionally independent. Moreover, the levels of the finished product, either synthesized or assembled, do not appear to exert feedback controls on their own production or on the production of other components or domains. Self-assembly is the major driving force in the assembly of each domain, with the level of each reactant polypeptide available for assembly being regulated at the mRNA level in a lineage-specific manner independently for each polypeptide. All reactant polypeptides within a domain are synthesized simultaneously within a lineage and any excess unassembled polypeptides are rapidly degraded. The order of assembly of polypeptides within each domain and of domains with each other cannot be due to temporal regulation of gene-product synthesis, but is imposed on the level of gene-product inter-

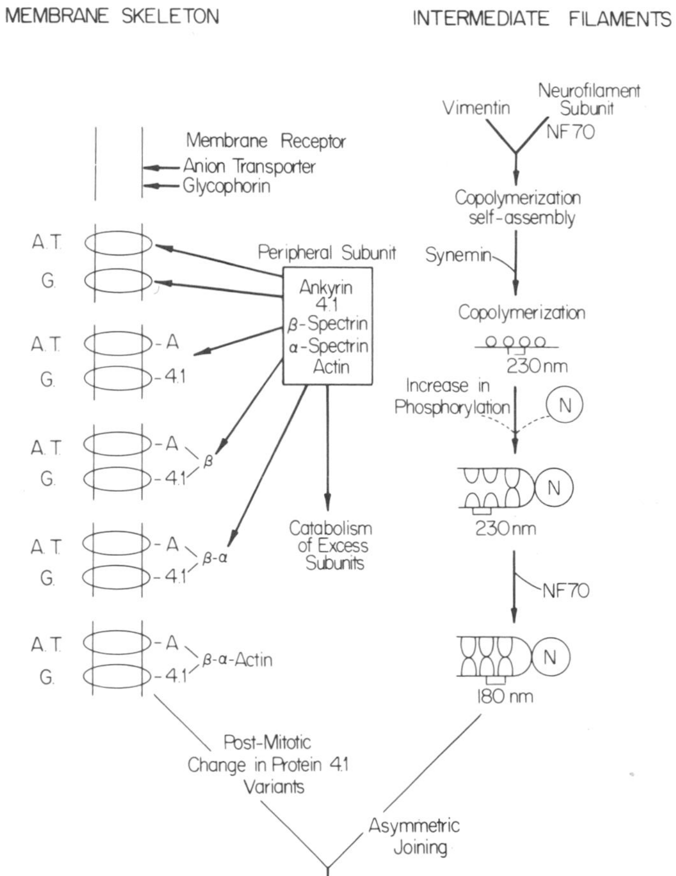

Figure 5
Partial sequence and order of steps in the assembly of the chicken erythrocyte cytoskeleton. The order of assembly steps denoted with a dashed line is uncertain. Assembly of glycophorin and protein 4.1 is by inference from mammalian erythrocytes.

action. As a consequence, the latter is also responsible for the spatial regulation of assembly so that domains assemble only from or at a particular cellular site. Therefore, assembly of polypeptides within a domain and of domains into a suprastructure are constrained to a unique sequential order. This is a principle in assembly previously established also in the morphogenesis of complex bacteriophages (Wood 1973, 1979). A summary of the pathway of assembly of the chicken erythrocyte cytoskeleton is presented in Figure 5.

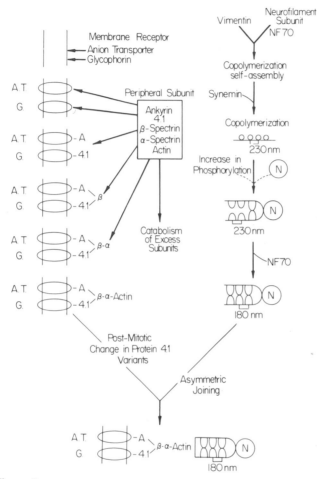

Another concept to emerge from the study of erythrocyte cytoskeletal morphogenesis is that the production of the final cytoskeletal phenotype is a multistep process that is generated progressively in a lineage-specific fashion. During this time, a domain can become "competent" to be linked to another domain, irrespective of the availability of the other domain in that lineage. This competence may be brought about by a number of mechanisms, including the synthesis of new variants of polypeptides present on the old template; by chemical modification of a subset of the polypeptides in the old template; or by removal of a polypeptide from the old template. For example, in the joining of the membrane-skeleton and the intermediate filament domains, a given erythroid lineage must have assembled its membrane-skeleton with an asymmetry of intermediate filament binding sites already built in and simultaneously express the intermediate filament domain before joining is possible. Yet, the two events are regulated independently; if any of these two events has not occurred in any given lineage, then the cells will be arrested at that particular stage in the morphogenetic pathway. This, then, may give rise to the cytoskeletal suprastructures of less differentiated cells. It also allows cells of different lineages to increase or decrease their cytoskeletal complexity by expressing or shutting off the expression of whole domains or of joining proteins, in accordance with their morphogenetic requirements.

ACKNOWLEDGMENTS

The work summarized here was supported by grants from the National Institutes of Health, the National Science Foundation, the Muscular Dystrophy Association of America, and a Research Career Development Award from the National Institutes of Health (E.L.).

REFERENCES

Alper, S.L., K.G. Beam, and P. Greengard. 1980a. Hormonal control of Na^+K^+ cotransport in turkey erythrocytes. Multiple site phosphorylation of goblin, a high molecular weight protein of the plasma membrane. *J. Biol. Chem.* **225:** 4864.

Alper, S.L., H.C. Palfrey, S.A. DeRiemer, and P. Greengard. 1980b. Hormonal control of protein phosphorylation in turkey erythrocytes. Phosphorylation by cAMP-dependent and Ca^{2+}-dependent protein kinases of distinct sites in globlin, a high molecular weight protein of the plasma membrane. *J. Biol. Chem.* **255:** 11029.

Anderson, R.A. and R.E. Lovrien. 1984. Glycophorin is linked by band 4.1 protein to the human erythrocyte membrane skeleton. *Nature* **307:** 655.

Arndt-Jovin, D.J., W. Ostertag, H. Eisen, K. Klimek, and T.M. Jovin. 1976. Studies of cellular differentiation by automated cell suspension. Two model systems: Friend virus-transformed cells and *Hydra attenuata. J. Histochem. Cytochem.* **24:** 332.

Beam K.G., S.L. Alper, G.E. Palade, and P. Greengard. 1979. Hormonally regulated phosphoprotein of turkey erythrocytes: Localization to plasma membrane. *J. Cell Biol.* **83:** 1.

Bennett, V. 1982. The molecular basis for membrane-cytoskeleton association in human erythrocytes. *J. Cell. Biochem.* **18:** 49.

Blikstad, I. and E. Lazarides. 1983a. Vimentin filaments are assembled from a soluble precursor in avian erythroid cells. *J. Cell Biol.* **96:** 1803.

———. 1983b. Synthesis of spectrin in avian erythroid cells: Association of nascent polypeptide chains with the cytoskeleton. *Proc. Natl. Acad. Sci.* **80:** 2637.

Blikstad, I., W.J. Nelson, R.T. Moon, and E. Lazarides. 1983. Synthesis and assembly of spectrin during avian erythropoiesis: Stoichiometric assembly but unequal synthesis of α- and β-spectrin. *Cell* **32:** 1081.

Bodine, D.M., C.S. Birkenmeier, and J.E. Barker. 1984. Spectrin deficient inherited homolytic anemias in the mouse: Characterization by spectrin synthesis and mRNA activity in reticulocytes. *Cell* **37:** 721.

Braell, W.A. and H.F. Lodish, 1981. Biosynthesis of the erythrocyte anion transport protein. *J. Biol. Chem.* **256:** 11337.

Branton, D., C.M. Cohen, and J. Tyler. 1981. Interaction of cytoskeletal proteins on the human erythrocyte membrane. *Cell* **24:** 24.

Bruns, G.A.P. and V.M. Ingram. 1973. The erythroid cells and haemoglobins of the chick embryo. *Philos. Trans. R. Soc. Lond. B* **266:** 225.

Butler, P.J.G. and A. Klug. 1971. Assembly of the particle of tobacco mosaic virus from RNA and dicks of protein. *Nat. New Biol.* **229:** 47.

Capetanaki, Y.G., J. Ngai, C.N. Flytzanis, and E. Lazarides. 1983. Tissue-specific expression of two mRNA species transcribed from a single vimentin gene. *Cell* **35:** 411.

Durham, A.C.H. and A. Klug. 1971. Polymerization of tobacco mosaic virus protein and its control. *Nat. New Biol.* **229:** 42.

Eisen, H., R. Bach, and R. Emery. 1977. Induction of spectrin in erythroleukemic cells transformed by Friend virus. *Proc. Natl. Acad. Sci.* **74:** 3898.

Fowler, V. and V. Bennett. 1984. Erythrocyte membrane tropomyosin. Purification and properties. *J. Biol. Chem.* **259:** 5978.

Georgatos, S.D. and E. Lazarides. 1983. Tropomyosin in mammalian and avian erythrocytes. *J. Cell Biol.* **97:** 279a.

Granger, B.L. and E. Lazarides. 1982. Structural associations of synemin and vimentin filaments in avian erythrocytes revealed by immunoelectron microscopy. *Cell* **30:** 263.

———. 1983. Expression of the major neurofilament subunit in chicken erythrocytes. *Science* **221:** 553.

———. 1984. Membrane skeletal protein 4.1 of avian erythrocytes is composed of multiple variants that exhibit tissue-specific expression. *Cell* **37:** 595.

Granger, B.L., E.A. Repasky, and E. Lazarides. 1982. Synemin and vimentin are components of intermediate filaments in avian erythrocytes. *J. Cell Biol.* **92:** 299.

Jay, D.G. 1983. Characterization of the chicken erythrocyte anion exchange protein. *J. Biol. Chem.* **258:** 9431.

Joseph-Silverstein, J. and W.D. Cohen. 1984. The cytoskeletal system of nucleated erythrocytes. III. Marginal band function in mature cells. *J. Cell Biol.* **98:** 2118.

Lazarides, E. 1980. Intermediate filaments as mechanical integrators of cellular space. *Nature* **283:** 249.

Lazarides, E. and B.L. Granger. 1983. Transcytoplasmic integration in avian erythrocytes and striated muscle: The role of intermediate filaments. In *Modern cell biology* (ed. J.R. McIntosh), vol. 2, p. 143. A.R. Liss, New York.

Lazarides, E. and R.T. Moon. 1984. Assembly and topogenesis of the spectrin-based membrane skeleton in erythroid development. *Cell* **37:** 354.

Liu, S.-C., P. Windisch, S. Kim, and J. Palek. 1984. Oligomeric states of spectrin in normal erythrocyte membranes: Biochemical and electron microscopic studies. *Cell* **37:** 587.

Moon, R.T. and E. Lazarides. 1983a. β-Spectrin limits α-spectrin assembly onto membranes following synthesis in a chicken erythroid cell lysate. *Nature* **305:** 62.

———. 1983b. Canavanine inhibits vimentin assembly but not its synthesis in chicken embryo erythroid cells. *J. Cell Biol.* **97:** 1309.

———. 1983c. Synthesis and post-translational assembly of intermediate filaments in avian erythroid cells: Vimentin assembly limits the rate of synemin assembly. *Proc. Natl. Acad. Sci.* **80:** 5495.

———. 1984. Biogenesis of the avian erythroid membrane skeleton: Receptor-mediated assembly and stabilization of ankyrin (goblin) and spectrin. *J. Cell Biol.* **98:** 1899.

Moon, R.T., I. Blikstad, and E. Lazarides. 1984. Regulation of assembly of the spectrin-based membrane skeleton in chicken embryo erythroid cells. In *Cell membranes: Methods and reviews* (ed. E.L. Elson et al.). Plenum Press, New York. (In press.)

Nelson, W.J. and E. Lazarides. 1984. Goblin (ankyrin) in striated muscle: Identification of the potential membrane receptor for erythroid spectrin in muscle cells. *Proc. Natl. Acad. Sci.* **81:** 3292.

Ngai, J., Y. Capetanaki, and E. Lazarides. 1984a. Expression of the genes coding for the intermediate filament proteins vimentin and desmin. In *Intermediate filaments* (ed. D. Fishman et al.). New York Academy of Science, New York. (In press.)

———. 1984b. Differentiation of murine erythroleukemia cells results in the rapid repression of vimentin gene expression. *J. Cell Biol.* **99:** 306.

O'Connor, C.M., D.L. Gard, and E. Lazarides. 1981. Phosphorylation of intermediate filament proteins by cAMP-dependent protein kinases. *Cell* **23:** 135.

Ohno, T., Y. Nozu, and Y. Okada. 1971. Polar reconstitution of tobacco mosaic virus (TMV). *Virology* **44:** 510.

Okada, Y. and T. Ohno. 1972. Assembly mechanism of tobacco mosaic virus particle from its ribonucleic acid and protein. *Mol. Gen. Genet.* **114:** 205.

Repasky, E.A., B.L. Granger, and E. Lazarides. 1982. Widespread occurrence of avian spectrin in non-erythroid cells. *Cell* **29:** 821.

Sabban, E.L., D.D. Sabatini, V.T. Marchesi, and M. Adesnik. 1980. Biosynthesis of erythrocyte membrane protein band 3 in DMSO-induced Friend erythroleukemia cells. *J. Cell. Physiol.* **104:** 261.

Sandoval, I.V., C.A.L.S. Colaco, and E. Lazarides. 1982. Purification of the intermediate filament-associated protein, synemin, from chicken smooth muscle. *J. Biol. Chem.* **25:** 2568.

Virtanen, I., M. Kurkinen, and V.-P. Lehto. 1979. Nucleus-anchoring cytoskeleton in chicken red blood cells. *Cell Biol. Int. Rep.* **3:** 157.

Weise, M.J., and L.L. Chan. 1978. Membrane protein synthesis in embryonic chick erythroid cells. *J. Biol. Chem.* **253:** 1892.

Wood, W.B. 1973. Genetic control of bacteriophage T4 morphogenesis. In *Genetic mechanisms of development* (ed. F.H. Ruddle), p. 29. Academic Press, New York.

———. 1979. Bacteriophage T4 assembly and the morphogenesis of subcellular structure. *Harvey Lect.* **73:** 203.

Wood, W.B. and J. King. 1979. Genetic control of complex bacteriophage assembly. In *Comprehensive virology* (ed. H. Fraenkel-Conrat and R.R. Wagner), vol. 13, p. 581. Plenum Press, New York.

Woodcock, C.L.F. 1980. Nucleus-associated intermediate filaments from chicken erythrocytes. *J. Cell Biol.* **85:** 881.

MUTANT ANALYSIS OF THE CYTOSKELETON

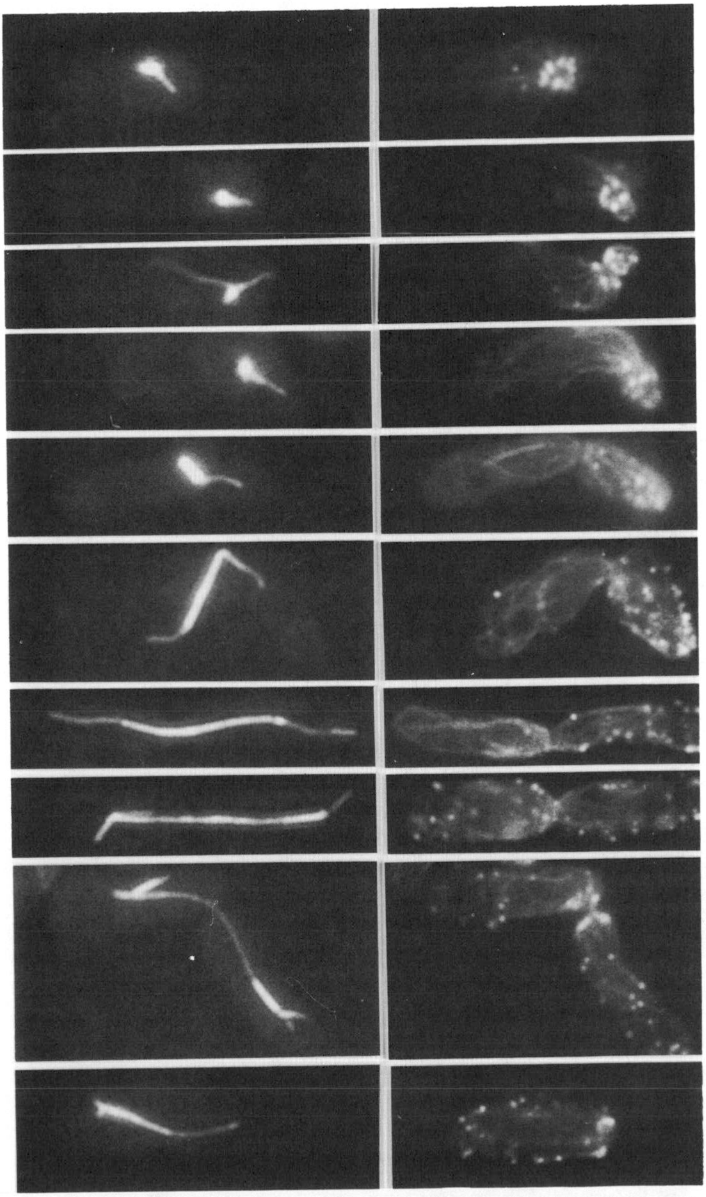

Recto: Double immunofluorescence of *S. uvarum* cells with anti-actin (*top*) and anti-tubulin (*bottom*). The cells have been placed in their cell-cycle order (top to bottom) (see J. Kilmartin, this volume).

Genetics of the Yeast Cytoskeleton

James H. Thomas, Peter Novick, and David Botstein
Department of Biology
Massachusetts Institute of Technology
Cambridge, Massachusetts 02139

The eukaryotic cell-division cycle consists of an ordered series of events, some of which entail major morphological changes. The cell's cytoskeleton is a central element in these changes, and it has been clear for many years that an understanding of mitosis, for example, will involve understanding the way in which tubulin and its associated proteins interact in the assembly, function, and disassembly of the mitotic spindle. The cell cycle and the cytoskeleton have traditionally been studied by cell biologists and biochemists, usually in higher eukaryotic cells, by observation, by application of drugs that interfere with cell-cycle functions or cytoskeletal assemblies, and by isolation and characterization of the major protein species (tubulin, microtubule-associated proteins, actin, actin-binding proteins, etc.).

The cell-division cycle has also been the subject of genetic analysis, principally in the budding yeast *Saccharomyces cerevisiae* and the fission yeast *Schizosaccharomyces pombe*. Through the pioneering work of Hartwell (1974), a class of conditional-lethal mutations of *S. cerevisiae* has been recognized that conditionally blocks the progress of the cell-division cycle at a particular point. Such *cdc* mutations define a large set of *CDC* genes, all of which affect progress of cell division; these genes can be ordered into dependent pathways of function by analysis of the phenotypes of double mutants.

Recently, it has become clear that yeasts have actin and tubulins that are strikingly similar in primary structure to their analogs in higher cells: There is 92% identity in the amino acid sequences of actin from *S. cerevisiae* and chicken brain and 71% identity between the β-tubulins from the same two species. Furthermore, as described by Kilmartin and Pringle et al. (both this volume), the visualization by immunofluorescence of microtubules and microfilaments in yeast shows a strong similarity in the organization of actin and tubulin in yeasts and other eukaryotic cells.

The conservation of primary structure and the apparent similarity in supramolecular structure between yeast and other eukaryotes, combined with the well-known advantages of yeast as a system for study of genetics by both classical and recombinant DNA methods (for review, see Botstein and Davis 1982), suggest that a true molecular genetics of the cell cycle and cytoskeleton can be achieved using yeast as a model system. It has recently been shown that yeast contains a single essential actin gene (Gallwitz and Sures 1980; Ng and Abelson 1980; Shortle et al. 1982a) and a single essential β-tubulin gene (Neff et al. 1983), making analysis of any mutations in these genes simple relative to organisms in which there are multiple copies of these genes.

The goal of such a molecular genetic approach is the association of genes specifying proteins involved in the cytoskeleton and cell cycle with their function. The connection between gene, product, and function is best made through the isolation of mutations: The mutant gene can be associated with the mutant protein by a number of criteria, often by direct determination of both the DNA and amino acid sequence, whereas the connection between the gene product and its function can be made by examination of the mutant phenotype in vivo or in a suitable in vitro system.

Even in yeast, however, one still faces technical problems in associating genes, products, and functions essential for growth, since the mutations must be conditional to permit growth of the organism. Straightforward biochemical analysis is also complicated by the fact that very little is known about the proteins that make up the cell structure. Thus, the geneticist can identify candidate genes by observing mutant properties that suggest failure in cell architecture or cell cycle. The biochemist and cell biologist, on the other hand, can find proteins (like actin and tubulin) that are abundant in structures implicated in basic cell functions. The challenge to the yeast molecular geneticist is to find ways to bring these two lines of endeavor together so that the geneticist's genes can be associated with the cell biologist's structures and functions.

In principle, there are two general ways in which to proceed. The classical genetic route (plan A) begins with mutations, which define a gene and whose properties indicate failure in a particular cellular function. Using mutations, one can, in yeast, readily isolate the gene as a DNA fragment, by using DNA transformation with gene libraries to complement the mutation. Given the isolated gene, one can analyze the gene as a physical entity and, using newly developed techniques (e.g., raising antibodies to fusion proteins [Shuman et al. 1980] or to synthetic peptides determined from DNA sequences [Sutcliffe et al. 1983a,b]), proceed to find the gene product. In favorable cases then, one can learn, through plan A, something about the way in which this protein contributes to the cellular function. By this route, one eventually obtains all the elements: the gene, the function, and the product.

The second approach (plan B) begins with a cellular process, let us say mitosis. One identifies a gene product (e.g., a protein like tubulin) whose involvement in this cellular process is known. Then, using recombinant DNA methods, one isolates the gene using information provided by the protein (as in the case of insulin; Villa-Komaroff et al. 1978). Once the gene is available as a DNA fragment, one can readily induce mutations in the gene by directly altering the DNA by chemical or biochemical means. In yeast, methods have been devised to replace the normal gene on the chromosome with the mutagenized DNA so that the effects of particular mutations can be studied. It should be clear that with plan B, one ends up with the same elements as with plan A, but in a different order: product, gene, and function.

It is vital to recognize that both routes of approach have in common the fusion of classical and modern genetics. It is not really a question of "new" and "old" but of finding a strategy using elements of both. Second, it should be noted that both plans A and B contain elements of modern recombinant DNA technology, but they are not all the same for both plans. Third, it should be remembered that both plans result in the achievement of the same goal: association of gene, product, and function. Finally, it is necessary to recognize that the most interesting functions in the cell are executed through the interaction of many genes and gene products so that the analysis of gene and protein interactions is a crucial element of any attempt to construct a true molecular picture of cell functions. The most successful approaches to such interactions in the cell are firmly based in classical genetics: the isolation and characterization of mutations in secondary genes that modify or "suppress" mutations in the primary genes of interest. These mutations in primary genes can be mutations obtained by either plan A or plan B and can be analyzed further by modifications of either plan.

Plan A and plan B also have important and complementary defects. The success of plan B is limited to proteins (and their genes) that have been identified by biochemical means. This means that they will strongly tend to be the abundant proteins and not necessarily the most interesting ones. Plan A overcomes this difficulty, since mutations do not respect the abundance of the product they affect. With plan A, however, to target specific functions is often difficult, and functions that mutate rarely to a conditional form may be difficult to identify. The β-tubulin gene is a good example of this problem: Despite the fact that β-tubulin mutants show a Cdc phenotype, none have been isolated among the hundreds of Cdc mutants.

The remainder of this paper describes recent progress in our laboratory in defining the molecular nature of the yeast cytoskeleton and its relationship to the cell-division cycle, using the complementary advantages of plans A and B.

Genetics of Yeast Actin

Background

The first evidence that yeast contains actin was the isolation of an actinlike protein from cell extracts based on its affinity for DNase (Koteleansky et al. 1979; Water et al. 1980). The tight binding of actin to calf thymus DNase in a 1:1 stoichiometry, resulting in its loss of enzymatic activity, had been established previously (Lazarides and Lindberg 1974). Water et al. (1980) used DNase coupled to Sepharose to purify yeast actin by affinity chromatography. Elution required denaturing conditions, although gentler elution conditions have now been developed (Zechel 1980). Native yeast actin has more recently been purified by conventional techniques (Greer and Schekman 1982a), using DNase inhibition as an assay. The purified protein shares many properties with rabbit muscle actin. Polymerization of yeast actin into 7-nm-thick filaments could be induced, and the filaments could be decorated with a proteolytic fragment of muscle myosin (HMM). Copolymerization of yeast and rabbit muscle actin has been demonstrated, although some differences were noted (Greer and Schekman 1982b).

Two laboratories have reported the presence of a single actin gene in yeast (Gallwitz and Seidel 1980; Ng and Abelson 1980), in contrast to higher eukaryotes, which typically contain up to 15 actin genes. Using the highly conserved coding sequence of the actin gene, these two groups cloned the yeast actin gene by screening a plasmid library of yeast genomic DNA for hybridization to a *Dictyostelium discoideum* actin gene. The complete nucleotide sequence of the cloned yeast gene has been determined (Gallwitz and Sures 1980; Ng and Abelson 1980). The 373 amino acid residues of the yeast actin protein are perfectly colinear with rabbit muscle actin and differ at only 44 residues. The yeast actin gene was found to contain a single intervening sequence of 309 bp, beginning after the fourth codon. The presence of this intron is not thought to be important for actin gene expression, since it can be precisely deleted with no effect on expression of the actin gene (Larson et al. 1983).

The biological function of actin in a simple nonmotile eukaryote such as yeast is unknown. However, its striking conservation during evolution suggests that yeast actin plays a role or roles similar to those of cytoplasmic actins in higher eukaryotes. Among the proposed functions are maintenance of a cytoskeleton, organization of cytoplasmic membrane proteins, and transport of material within cells.

Constructing Actin Mutants In Vitro

The construction and phenotypic characterization of actin mutants in vitro is an excellent example of plan B at work. The first actin mutant constructed was a null mutation generated by disruption of the chromosomal actin gene (Shortle et al. 1982a). A restriction fragment that was internal to the coding sequence of the actin gene was subcloned into a yeast integrating plasmid vector, YIp5. The resulting plasmid (called pBR111) carries the yeast selectable marker *URA3* and transforms *ura3*$^-$ yeast to Ura$^+$ only by integration into the yeast genome by homologous recombination between the yeast sequences found on the plasmid and a yeast chromosomal locus. Thus, when this plasmid transforms yeast, it can integrate *via* either its *URA3* homology or its actin gene homology. Integration at the *URA3* locus has no effect on the function of the endogenous actin gene. However, Figure 1 shows that integration into the actin gene results in disruption of the coding sequence, leaving a direct repeat, each copy of which contains an incomplete actin gene.

A disruptive integration into the actin gene might result in a lethal mutation in a haploid yeast. However, if the loss of the actin gene function due to disruption is recessive, then an integrant into only one chromosome of a diploid should be viable but cause a recessive lethal mutation (associated with the disrupted actin gene). Six independent Ura3$^+$ transformants of a homozygous *ura3*$^-$ diploid strain were analyzed. Gel-transfer hybridization experiments demonstrated that four of these transformants resulted from integration at the actin locus, and the

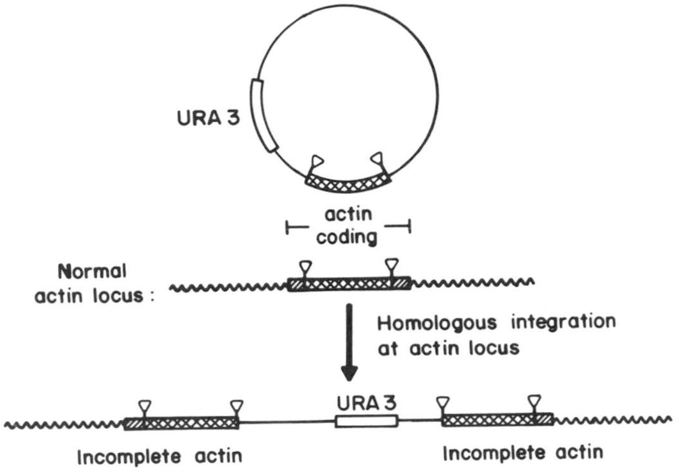

Figure 1
Disruption of the actin gene by integrative transformation. A fragment of the actin gene internal to the coding sequence subcloned on an integrating vector directs integration to the actin locus. This results in two incomplete actin sequences.

remaining two were integrated at the *URA3* locus (leaving the actin gene intact). Each of these diploids was sporulated, and the viability of the spores from the resulting tetrads was noted. All spores that had inherited the plasmid (marked by Ura3+) integrated at the actin locus were inviable, whereas those integrated at the *ura3* locus were viable. This indicated that the disruption of the actin gene was a recessive lethal event and that actin is required (at least) for growth of spores into a viable colony.

Constructing Conditional Mutants in the Actin Gene

Studying the function of actin by a genetic approach requires mutants conditional for actin function(s). Since we now knew that the actin gene was essential for viability, we possessed a phenotype by which to recognize conditional actin mutants (conditional-lethality). Our problem was to devise a simple method to recognize recessive conditional-lethal mutations in an actin gene introduced into yeast by transformation. Simple introduction of the mutant actin gene on a replicating plasmid was not suitable, since the endogenous actin gene would mask all but dominant plasmid actin mutants. To achieve immediate expression of the in-vitro-mutagenized actin gene, a variation of the method of gene disruption was used (D. Shortle et al., in prep.). Rather than a plasmid carrying a DNA fragment entirely internal to the actin-coding region, a plasmid containing a larger DNA fragment with only one end lying within the coding sequence was used. The result of integration of such a plasmid into the actin locus is shown in Figure 2. It can be seen that a duplication of the actin locus results, only one copy of which remains functional. If the plasmid contained a mutation in the actin region and the integrative recombinant event occurred on the appropriate side of the actin mutation, it would result in the immediate expression of the mutant gene in yeast. To ensure that the integration event occurred at the actin locus rather than at the

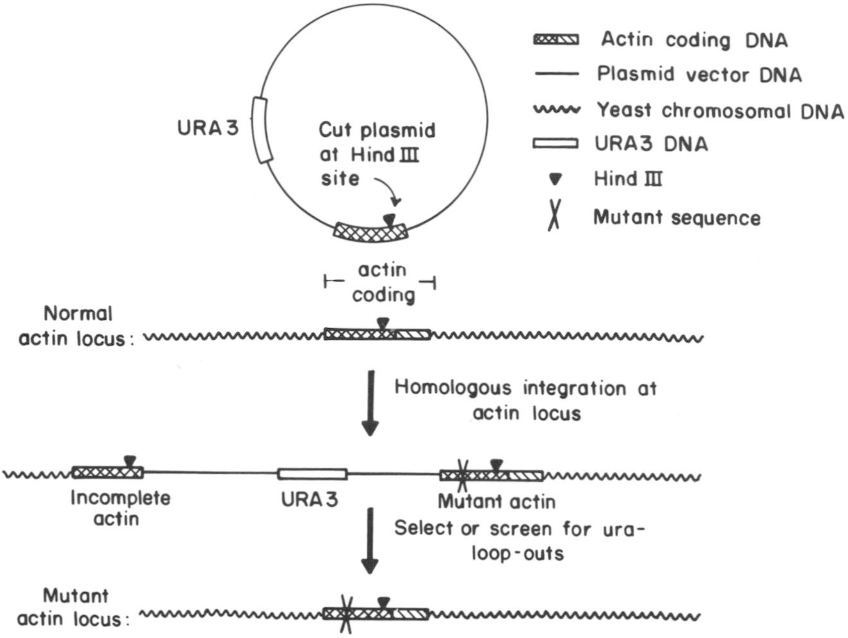

Figure 2

Construction of chromosomal actin mutants. A mutant actin sequence is transferred from a plasmid to the chromosomal actin locus by integration, followed by excision. The plasmid shown, carrying a portion of a mutant actin gene, is directed to integrate at the actin locus. This results in one copy deleted for the carboxyterminal coding sequence and a complete second copy that may contain a mutation. This structure allows expression of even recessive actin mutations, since a single functional actin gene is present in the transformant. After recognition of the mutation by its conditional-lethal phenotype, the plasmid sequences are excised by homologous recombination, leaving the actin mutation on an otherwise normal chromosome.

ura3 locus, the plasmid was cut at the unique *Hin*dIII restriction site located in the region of actin homology. The effect on yeast transformation of forming such a double-strand break is to stimulate greatly the frequency of transformation as well as to direct the integration exclusively to the chromosomal locus homologous to the region on the plasmid in which the double-strand break lies (Orr-Weaver et al. 1981).

The mutagenesis of the actin gene in vitro could, in principle, be achieved in a variety of ways. In fact, the mutagenesis was achieved by RecA-mediated D-loop generation and treatment of the resulting single-strand region with the single-strand-specific mutagen sodium bisulfite (Shortle et al. 1980). The details of this procedure are unimportant here and have in any case been largely supplanted by other methods that generate a wider variety of base changes (see, e.g., Shortle et al. 1982b). A large number of mutant plasmids were generated and pooled together for transformation into yeast. Ura3^{+} yeast transformants were screened for conditional-lethality and several temperature-sensitive strains were identified. As a first test of the identity of these mutants, tetrad analysis was used to test the linkage of the temperature-sensitive phenotype to the actin locus (marked by the plasmid *URA3* gene). Only 10% of the mutations mapped at the actin locus. The other mutations were presumably due to background mutations unrelated to actin.

Three mutations were found that were closely linked to the actin gene. The plasmid sequences were removed from these three mutant strains, leaving only the mutant actin gene behind at an otherwise normal chromosomal locus. This was achieved by screening for spontaneous Ura$^-$ segregants that result from looping out of the plasmid sequences from the chromosome by homologous recombination between the two copies of the actin locus. Such segregants arise at a frequency of about 10^{-4}, making screening laborious. A simple positive selection for Ura3$^-$ now exists (F. Lacroute, pers. comm.). The site of recombination in the loop-out event can lie on either side of the actin mutant site. Thus, some Ura$^-$ segregants are *ts*$^+$ and others are *ts*$^-$. This is another indication that the *ts*$^-$ lethal lies in the actin gene, since other mutations will not produce frequent *ts*$^+$ segregants during loop-out of the plasmid. Three actin temperature-sensitive mutants (*act1-1, act1-2,* and *act1-3*) were constructed using this protocol.

Phenotypes of Actin Mutants

In trying to understand the phenotype of these actin mutants, some features of the yeast cell-division cycle are important. *S. cerevisiae* divides by budding. Near the start of each cell cycle, a small round bud appears on the mother cell. This bud grows as the cell cycle progresses, remaining separated from the mother by a short neck region. As the bud enlarges, the nucleus migrates to the bud neck, elongates into the daughter cell, and divides. By this point, the bud is approaching the mother cell in size and cytokinesis and cell separation rapidly ensue, completing the cell cycle.

Several features characterize the neck separating the mother cell and the bud. A collar of chitin (a polymer of *N*-acetyl glucosamine) is formed in the cell wall surrounding the bud neck (Cabib 1975). This chitin ring can be visualized in the fluorescence microscope with the dye Calcofluor. Following cell division, the chitin ring remains associated with the mother cell and is stable for many generations, giving rise to a series of chitin rings (called bud scars) on "old" mother cells. The plasma membrane of the bud neck is completely lined on the cytoplasmic side by rings of a 10-nm filament visible by thin-section electron microscopy (Byers and Goetsch 1976). This internal collar disappears just prior to cytokinesis. The size of these filaments is inconsistent with that of either microfilaments (7 nm) or microtubules (30 nm). The rings are more likely to be equivalent to the intermediate filaments of higher eukaryotes.

The phenotype of the temperature-sensitive actin mutants has been studied primarily through the use of two fluorescence microscopic techniques for visualizing the intracellular distribution of actin in yeast. Adams and Pringle (1984) have used rhodamine-tagged phalloidin, a mushroom toxin known to interact with polymerized actin, whereas Kilmartin and Adams (1984) have used affinity-purified anti-actin antibody and a fluorescently tagged secondary antibody. In either case, the cells are first fixed with formaldehyde. Rhodamine-phalloidin requires no further permeabilization of the cells, but for antibody staining, removal of the cell wall is necessary to allow penetration of the antibody. These techniques produce similar results. Our observations using the anti-actin antibody, kindly supplied by J. Kilmartin (this volume), are discussed below.

Through much of the cell cycle, the actin-staining pattern is asymmetric, as shown in Figure 3. The mother cell shows cables of actin directed toward the bud neck, and the bud shows brightly staining dots just below the cell surface. Few dots are seen in the mother cell and few cables are seen in the bud. This actin asymmetry persists until shortly before cytokinesis, at which time randomly directed cables and scattered dots are seen in both mother and bud. At the time of bud emergence, in addition to the dots and cables, a ring of actin appears to form around the bud neck only on the mother side of the neck. This actin ring persists as a collar around the neck throughout the early portion of the cell cycle but disappears at later times. In contrast, the ring of 10-nm filaments seen in the electron microscope is found through the full length of the bud neck, supporting the conclusion that this ring is not composed of actin.

Several differences between the results presented above and the rhodamine-phalloidin staining pattern observed by Adams and Pringle (1984) should be noted. The asymmetry of the actin dots seen with phalloidin is maintained only through the early part of bud growth, and a collar around the neck is seen at the time of cytokinesis. Whether these differences reflect the use of a different sample preparation, different staining reagents, or differences in the yeast strains used is presently unclear.

Mutations in actin might affect its localization in the cell. The pattern of actin staining in *act1-2* and *act1-3* mutants has been examined using anti-actin antibody. At permissive temperature, *act1-3* cells show only a few faint cables, and the surface dots are found in both the mother cell and the bud. After 2 hours at the restrictive temperature, no cables are seen, only randomly distributed dots. A somewhat different phenotype is observed for *act1-2*. At permissive temperature, dots are seen in both mother and bud; however, many heavy bars of actin are also seen, often crossing the cell diametrically. After 2 hours at the restrictive temperature, the bars have dissipated, a meshwork of fine filaments appears to fill the cell, and randomly distributed dots lie near the cell surface. The aberrant actin-staining pattern of the mutants even at their permissive temperature is consistent with their poor growth at this temperature, and presumably reflects a partial actin defect.

The actin mutants affect other cell processes as well. At permissive temperature, the mutants appear considerably larger and more rounded than wild-type cells. At the restrictive temperature, this trend becomes more pronounced, leading to very large and nearly spherical cells. The vacuole swells, filling much of the cytoplasm. The percentage of budded cells drops to about 30% after 2 hours; a wild-type exponential culture contains 50–60% budded cells. After 4 hours, cell lysis becomes prevalent. Both actin mutants show a pronounced osmotic sensitivity; at temperatures approaching the restrictive temperature, addition of 0.5–1 M sorbitol, KCl, or ethylene glycol to the medium abolishes growth. This phenomenon is not specific to these alleles, but rather reflects a reduced level of functioning actin, since diploid strains with one wild-type copy of actin and one disrupted copy also display osmotic sensitivity.

The actin mutants also affect chitin localization. At permissive temperature,

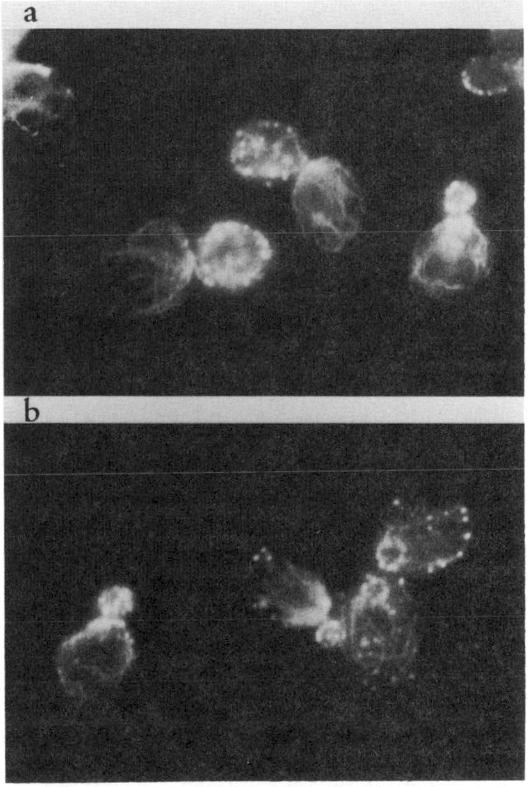

Figure 3
Immunofluorescent staining of actin in wild-type diploid yeast. Budded cells show cables in the mother cell and dots near the surface of the bud (*a*). A ring of actin dots forms just prior to bud emergence and persists as a collar around the neck of cells with a small bud (*b*).

chitin rings are seen, but their diameters are much larger than in wild type. After arrest for 2 hours at the restrictive temperature, the mutant cells are uniformly bright, suggesting delocalized chitin synthesis.

These observations suggest at least two possible roles for actin in yeast. Actin may play a role in bud emergence, as suggested by the actin collar formed at the base of the bud neck, the delocalized chitin staining, and the reduction in budded cells at restrictive temperature. Actin may also play a role in osmotic regulation of the cell, as suggested by the cell lysis, osmotic sensitivity, and swollen vacuoles of the actin mutants.

Suppressors of Actin Mutants

The elaborate cell-cycle-dependent pattern of actin localization must be controlled both temporally and spatially. Such control most likely involves proteins that regulate the site and extent of actin polymerization by means of direct or indirect physical interaction with actin. Candidates for such proteins have been identified in higher eukaryotes by in vitro polymerization assays (Hitchcock-DeGregori 1980). We hope to identify such proteins in yeast by genetic methods. The deleterious effects of a mutation in one protein may be compensated for by a mutation in a second protein that physically interacts with it (Jarvik and Botstein 1975; Morris et al. 1979). Among such extragenic suppressors, some will simultaneously confer a new phenotype that may be informative.

To perform this sort of analysis for actin, pseudorevertants of a temperature-sensitive (*ts*) actin mutant were screened for cold sensitivity (*cs*) for growth. Restricting our attention to this class of mutants allowed simple genetic manipulation and phenotypic analysis of the suppressor mutants, since they have a phenotype of their own that can be assessed in the absence of any actin defect.

An *act1-3* strain was incubated on plates at 37°C (the restrictive temperature) to allow the growth of revertants. Colonies that arose were tested for cold sensitivity for growth at 17°C. Tetrad analysis of these strains identified those that had a single locus responsible for both the *ts* suppressor and the *cs*-lethal phenotypes. Eighteen revertants (~50%) fell into this class; the others were the result of more than one mutation and were not pursued. These 18 mutants were arranged into complementation groups according to their *cs* phenotype. Three groups were defined, *sac1, sac2,* and *sac3* (*s*uppressor of *ac*tin). The largest group, *sac1*, contains 13 independently isolated members, *sac2* has 3 members, and *sac3* has 2 members. All of these mutants were recessive for both cold sensitivity and suppression of temperature sensitivity. Suppressors of this sort are typically dominant in yeast (see, e.g., Moir et al. 1982), suggesting that there is a special property of these actin suppressors that explains their recessivity. One model to explain this phenomenon is that *SAC*-gene products interact along the length of an actin polymer. If all copies of the *SAC* protein must be the suppressor type to function as a suppressor, then the mutant allele would show recessive suppression.

If the *SAC*-gene products physically interact with actin, two predictions can be made. First, *sac* mutants should exhibit allele specificity in suppression. In fact, one allele of each group has been shown to suppress *act1-3*, the mutation with which they were isolated, but not *act1-2*. The second prediction is that the *sac* mutants should show a phenotype at their restrictive temperature that is related to the phenotype of the original actin mutant in some way. Immunofluorescent staining with anti-actin antibody was performed on strains containing a mutant *sac* gene but wild-type actin. At the restrictive temperature, 14°C, *sac1* shows dots on the cell surface of both mother and bud and loss of actin cables. *sac2* also exhibits randomly localized dots as well as occasional heavy actin bars, whereas *sac3* contains randomly distributed dots and heavy bars extending the full length of budded cells. It is thus clear that mutations in the *SAC* genes can affect the localization of wild-type actin. The pattern of chitin deposition is also aberrant in the *sac* mutants; *sac2* and *sac3* show delocalized chitin synthesis at the restrictive temperature, whereas *sac1* exhibits bright patches of chitin at the tip of small buds or on the mother cell in addition to the normal collar around the neck. These results indirectly support the hypothesis that the *SAC*-gene products physically interact with actin. Future work will be directed toward identification and localization of the *SAC*-gene products.

Genetics of Yeast Tubulin

Background

Mitotic spindles consisting of microtubules were first observed in *Saccharomyces* in 1966 by electron microscopy (Robinow and Marak 1966). Extensive cytological observations of microtubules at all phases of the yeast life cycle have been made at the electron microscopic level (for a recent review, see Byers 1981), and more recently by the use of monoclonal antibodies against tubulin as an immunofluorescent probe for microtubules (Kilmartin et al. 1982; Adams and Pringle 1984; Kilmartin and Adams 1984). Yeast has a typical mitotic spindle that is organized by spindle-pole bodies (analogous to the higher eukaryotic centrosome) that are embedded in the nuclear envelope. All microtubules in yeast have one end closely associated with a spindle-pole body, radiating both into and out of the nucleus to form the spindle and the cytoplasmic microtubules, respectively. Duplication of the spindle-pole body is an important event in the yeast cell cycle that immediately follows "start" (commitment to a new cell cycle) and coincides with bud emergence (Byers and Goetsch 1975). Observation of the state of the spindle-pole bodies and associated microtubules in mutants that affect the yeast cell-division cycle indicates that these structures are closely regulated during the cell cycle (Byers and Goetsch 1974).

During karyogamy following conjugation in yeast, microtubules appear to play a key role. The spindle-pole body and associated cytoplasmic microtubules undergo a series of characteristic changes that appear to orient and mediate migration of the two nuclei, resulting in fusion of their nuclear envelopes and spindle-pole bodies (Byers and Goetsch 1974). Microtubules are also, of course, involved in meiosis, forming the meiotic spindle (Moens and Rapoport 1971).

Tubulin has been purified from yeast by self-assembly in vitro (Kilmartin 1981). The purified protein coassembles with mammalian tubulin and appears very similar in physical properties. The gene for β-tubulin has been cloned, sequenced, and demonstrated to be present in one copy in yeast (Neff et al. 1983), in contrast to the many copies of the gene found in higher eukaryotes. The amino acid sequence of yeast β-tubulin protein, inferred from the gene sequence, is identical to chicken β-tubulin at 71% of its residues. Using the gene-disruption technique described above in detail for actin, Neff et al. (1983) showed that this β-tubulin gene is essential for growth in yeast.

Mutations in the β-Tubulin Gene by Drug Resistance

At the time we set out to make mutants in the tubulin gene(s) of yeast, no clones of either α- or β-tubulin genes of yeast were available. We therefore chose to attempt to isolate such mutants by selecting resistance to the anti-mitotic drug benomyl (Davidse and Flach 1977). Some mutants resistant to benomyl in *Aspergillus nidulans* had been shown to lie in a β-tubulin gene (Sheirr-Neiss et al. 1978). Reasoning that this might also be true in yeast, we isolated a large number of spontaneous benomyl-resistant (Ben^R) mutants. These mutants were shown to lie in a single gene (J.H. Thomas et al., in prep.). Among the many Ben^R mutants, six also displayed cold sensitivity and three others displayed temperature sensitivity for growth. In crosses, these conditional-lethal phenotypes perfectly cosegregated with the Ben^R phenotype, and where two mutants shared a common recessive phenotype (Ben^R or conditional-lethality), they were shown to fail to complement.

The locus defined by these Ben^R mutants was identified as the gene encoding β-tubulin by use of a newly isolated clone of yeast β-tubulin (Neff et al. 1983). This clone was isolated by virtue of its DNA homology with a previously isolated chicken β-tubulin clone (Cleveland et al. 1980). It was positively identified as a gene for β-tubulin by sequence analysis. The connection between the clone of yeast β-tubulin and the Ben^R locus was made by two separate tests: a complementation test and a genetic linkage test. To test complementation, a recessive, cold-sensitive Ben^R mutant, *tub2-104*, was transformed with a centromere-containing vector (Clarke and Carbon 1980; Botstein and Davis 1982) carrying the entire wild-type β-tubulin gene. In every transformant, the wild-type (benomyl-sensitive, cold-resistant) phenotype was restored. Segregants of several transformants that had lost the plasmid were isolated, and the complementing activity always cosegregated with the plasmid. This demonstrated that the yeast insert in this plasmid can supply the functions lacking in the mutant. Another plasmid that contained only a part of the β-tubulin gene carried on a yeast-integrating vector was integrated into the chromosome at the β-tubulin locus and shown by tetrad analysis to be tightly linked to the Ben^R locus.

The gene-disruption method, described in detail for the actin gene (see Fig. 1), was also applied to the β-tubulin gene, with the same results: When the β-tubulin gene is disrupted by transformation, it creates a recessive lethal mutation that is linked to the gene disruption and the Ben^R locus. In this case, one further experiment was possible (since mutants in the β-tubulin gene existed) that demonstrated that the disrupted β-tubulin gene in the diploid was nonfunctional. The gene disruption was performed in a diploid heterozygous for a cold-sensitive Ben^R mutant that was recessive for both phenotypes (*tub2-104*). We found that when the β-tubulin gene disruption occurred in the wild-type homolog of this diploid, it resulted in the expression of both the cold-sensitivity and Ben^R phenotypes of the β-tubulin mutant gene on the other homolog. This confirmed that the DNA resulting in the gene disruption does indeed affect the same gene that is defined by the Ben^R mutations.

Phenotypes of β-Tubulin Mutants

Studies have suggested that microtubules are involved in a variety of cell processes in yeast, including chromosome segregation, bud emergence (Byers 1981), and nuclear fusion during conjugation (Byers and Goetsch 1975). Until recently, there has been no good experimental evidence addressing any of these points. The possession of β-tubulin mutants allowed us to ask directly what cellular defects are associated with a defect in microtubules.

If microtubules are required for progression through mitosis but at no other stage of the cell cycle, then β-tubulin mutants should show a cell-cycle-specific terminal arrest morphology (Hartwell 1974). We have observed the arrest of several different β-tubulin mutant alleles and have found a classic cell-division-cycle defect in mitosis. In the tightest mutant alleles, over 90% of the cells arrest as large-budded doublet cells. The nuclear DNA in these cells (as detected by fluorescence microscopy with a DNA-staining dye, DAPI) is undivided and usually remains unelongated and fails to migrate to the neck between the mother and bud. The mutant block is rapidly imposed, since cells arrest uniformly after only one cell-cycle time at the restrictive temperature. In general, the arrest closely resembles that observed with the anti-mitotic drugs methyl benzimidazole-2-yl-carbamate (Wood and Hartwell 1982) or nocodazole (J.H. Thomas et al., unpubl.). This cell-cycle arrest supports the hypothesis that microtubules are involved in chromosome segregation, thus resulting in a block in mitosis. At the same time, it argues against a microtubule requirement for bud emergence, since this cell-cycle event is apparently unaffected in the mutants.

Using indirect immunofluorescent staining with antibodies directed against yeast tubulin (Kilmartin et al. 1982; Kilmartin and Adams 1984), we determined that the microtubules in a *tub2-104* strain largely, but not entirely, disappear at the restrictive temperature. They are reduced to a dot or a faint short bar of staining, similar to the pattern of staining normally seen during part of the G_1 phase in wild type.

The mutant *tub2-104* also shows a karyogamy defect during conjugation. When two permissively grown *tub2-104* mutant strains were allowed to conjugate at their restrictive temperature, they underwent normal cytoplasmic fusion. However, if large-budded zygotes were recovered from such a conjugation and allowed to divide to form colonies at permissive temperature, over 90% of them segregated "cytoductants" (haploid cells of one parental genotype or the other), indicating that nuclear fusion had failed in these zygotes. When the same experiment was performed using a *tub2-104* mutant crossed to a *TUB2*$^+$ strain, the zygotes rarely formed cytoductants, indicating that the *tub2-104* karyogamy defect can be compensated for by the wild-type gene product contributed by the other parent. Thus, tubulin has the properties of a "bilateral" karogamy deficiency (both parent cells must be defective; see Fink and Conde 1977), in contrast to most other karogamy-defective mutants, which are unilateral (Fink and Conde 1977; Dutcher and Hartwell 1983).

When the zygotes produced by these matings were fixed and their nuclear DNA and microtubules were viewed by fluorescence microscopy, the nature of the karyogamy defect was clear. In the *tub2-104* cross by *TUB2*$^+$ at the restrictive temperature (as well as in wild-type crosses at either temperature or a *tub2-104* by *tub2-104* cross at permissive temperature), cytoplasmic microtubules are normally elaborated and the nuclei are closely associated or fused at very early times after cell fusion. In the *tub2-104* cross by *tub2-104* at the restrictive temperature, there is a general failure to elaborate cytoplasmic microtubules and the nuclei remain well separated.

Suppressors of a β-Tubulin Mutant

Assembly and function of tubulin, like actin, must be controlled spatially and temporally to result in the complex pattern of changes seen during the cell cycle. We have used pseudoreversion as a means of genetically identifying proteins that interact with tubulin. The cold-sensitive mutant *tub2-104* was reverted to *cs*$^+$ and screened for those revertants that had simultaneously acquired temperature sensitivity for growth (*ts*). Subsequent crosses established that the cold-sensitive suppressor and *ts*-lethal phenotypes were tightly linked and extragenic (except one, which was shown to be an intragenic *ts* revertant). Twenty-six such mutants were classified into 16 complementation groups, of which 6 have more than one member. When examined for cell-cycle phenotype, mutants from one complementation group (six members) showed a phenotype similar to that of β-tubulin temperature-sensitive mutants, even in a *TUB2*+ strain. This preliminary observation suggests that at least some of these mutants may define genes whose products are functionally related to tubulin.

Several of the suppressor mutants were tested for dominance of their suppressor phenotype. All were found to be recessive, supporting the argument that recessive suppressors may be common for proteins that form part of a highly polymerized structure (see above).

Map Position of the ACT1 *and* TUB2 *Genes*

The *ACT1* gene was mapped to the left arm of chromosome VI of yeast (Falco and Botstein 1983), using a new method involving the integration of a plasmid containing 2-μm sequences into the chromosome at the site of a cloned gene. When the same method was applied to the *TUB2* gene, identical results were obtained (J.H. Thomas et al., in prep.). Tetrad analysis of a cross between an actin locus marked by an integrated *URA3* gene and a *tub2* benomyl-resistant allele showed close linkage of the two genes (PD:NPD:TT, 22:0:0). Examination of the plasmid clones carrying each of these genes and gel-transfer hybridization experiments confirming the structure of the genomic locus demonstrated that *ACT1* and *TUB2* are immediately adjacent, with about 1 kb of DNA separating them. Analysis of transcription and DNA sequence studies of this region (Gallwitz 1982; Gallwitz et al. 1983; N.N. Neff and J.H. Thomas, unpubl.) indicate that the *ACT1* and *TUB2* genes are arranged in opposite orientation (Fig. 4) so that transcription is divergent. A small transcript of unknown function lies between the genes in the same orientation as *ACT1*.

Plan A Revisited

It is clear that plan B provides powerful tools for the genetic analysis of proteins already thought to be important in the function of the cytoskeleton. This leaves

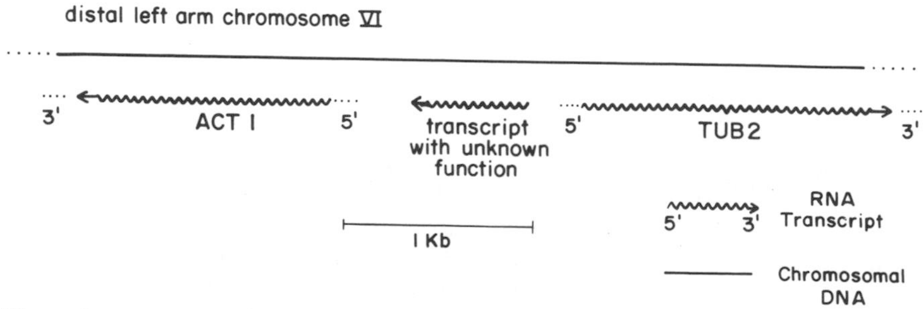

Figure 4
Arrangement of the actin and β-tubulin region of chromosome VI. The actin and β-tubulin messages are divergently transcribed and a third transcript of unknown function lies between them. The precise ends of the transcripts are not known.

the geneticist at the mercy of cell biologists for the identification of interesting problems. One solution to this difficulty is plan A: the isolation of mutations in random genes that display "interesting" properties, such as a cell-division-cycle-specific arrest. One of the crucial decisions to make when using this approach is what constitutes an interesting phenotype. The identification of Cdc mutants, for instance, has proved very fruitful (Pringle and Hartwell 1981); however, no Cdc mutant has yet been shown to be a defect in a structural component of the cytoskeleton. We now know that actin mutants would never have been found by this criterion, and although mutants in β-tubulin should have shown a Cdc phenotype, none were isolated. Recently, we have found a mutation in a gene whose protein product is unknown, but whose properties are intriguing. Here, we discuss this mutation, *ndc1-1*, as an example of the application of plan A to the identification of new genes related to cytoskeletal function.

Recognition of ndc1-1

The mutation *ndc1-1* was originally isolated by D. Moir as part of a search through random *cs*-lethal mutants for those that displayed a Cdc arrest (Moir et al. 1982). The *ndc1-1* mutant was recognized as having a Cdc-related arrest, although its terminal arrest morphology was not uniform. The mutant was not pursued by Moir et al., but recently we became interested in this mutant, since its nonpermissive arrest phenotype seemed related to the β-tubulin arrest.

We have characterized the terminal arrest phenotype of *ndc1-1* strains primarily by two methods: (1) Cells arrested in liquid culture can be fixed to allow observation of nuclear DNA, microtubules, or other structures by fluorescence microscopy, and (2) cells can be arrested on agar slabs, and the products of a single cell's lineage can be assessed. The first method does not permit assessment of the lineage of the cells being observed, and the second method does not permit the detection of nuclei and microtubules. When the two methods are applied together, they provide complementary information about the pattern of arrest.

When an asynchronous population of *ndc1-1* cells grown at the permissive temperature was shifted to the restrictive temperature on an agar slab and individual cells were followed by photomicroscopy, the result shown in Table 1 was seen. Initially, unbudded cells produced five to six cell bodies over several generation times. This suggests that either the arrest due to the *ndc1-1* defect is "leaky" (allowing cells to perform the mutant function slowly) or the cell cycle is not immediately arrested during the block. Also, the five or six cell bodies at the end of the arrest time might be part of one undivided cell or they might be due to several divided cells. Restrictive arrest in liquid medium distinguished these possibilities, as shown in Table 2. Very few multiply budded cells accumulated even at long arrest times, indicating that the cells are undergoing division prior to the production of a new bud. Furthermore, nuclear DNA division was completely blocked in the *ndc1-1* arrest, since no cells with divided nuclear DNA were seen. Together, these results indicate that although DNA division is defective in the *ndc1-1*-arrested cells, the cell cycle progresses for several generations. As would be expected for such an arrest, "aploid" cells (lacking detectable nuclear DNA staining) accumulated during the arrest, eventually becoming the predominant cell type. These cells were always unbudded and never succeeded in progressing

Table 1
Mean Number of Cell Bodies at Times after Shift of *ndc1-1* to Restrictive Temperature

Cell morphology at time of shift	*Generation times*		
	1.5	*3.3*	*4.4*
Unbudded	2.7	4.1	4.6
Small budded	4.2	6.2	7.5
Large budded	5.5	8.1	9.0

Cells were grown in liquid at permissive temperature and were lightly sonicated to disrupt clumps prior to shift. A sample of cells was spotted onto a prechilled slab of nutrient agar on a microscope slide and covered with a coverslip. A wire mesh screen embedded in the agar allowed the same field to be photographed at time 0 and several subsequent times. Wild-type cells, when subjected to this protocol, grow normally for several generations, with the same generation time as in log phase liquid growth.

through another cell cycle. Thus, we can infer that the daughter of an *ndc1-1*-defective cell division that retains the nuclear DNA must rebud and continue to divide several times over the course of the arrest, giving rise to the several cells seen during arrest on solid medium.

We tested whether DNA replication occurs in the *ndc1-1*-arrested cells by asking if the nucleated cells that arise from the first division have duplicated their genome, becoming 2N diploids. We arrested haploid *ndc1-1* cells at the nonpermissive temperature for about 1.5 generation times, at which point nearly all of the large-budded cells have the characteristic undivided nuclear DNA. These cells were plated onto agar medium at permissive temperature, and large-budded cells were micromanipulated out. These cells were allowed to divide, and

Table 2
Percentage of Various Cell Types at Times after Shift of Wild-type or *ndc1-1* Cells to Restrictive Temperature

		Cell types				
		unbudded			*large budded*	
Strain	*No. of generation times after shift*	*+ nuclear DNA*	*– nuclear DNA*	*small budded*	*1 region of nuclear DNA*	*2 regions of nuclear DNA*
Wild type	2.0	60	0	23	17	
ndc1-1	0	64	0	26	2	8
ndc1-1	0.7	58	0	22	10	10
ndc1-1	1.5	21.5	0.5	20	51	7
ndc1-1	2.6	10	22	7	59	2
ndc1-1	3.3	5	40	8	47	0

Cells were grown in liquid medium at permissive temperature and shifted in early log phase to the restrictive temperature for *ndc1-1* cells. At time points, cells were removed and prepared for observation by fixation and treatment as if for immunofluorescence (Kilmartin and Adams 1984). We found that this method gave DAPI staining superior to other techniques, which is important in the certain identification of aploid cells. A representative time point is shown for wild-type cells; these numbers do not change significantly at any time during the arrest and aploid cells are never seen. *ndc1-1* cells grown at permissive temperature are also indistinguishable from wild type.

the viability and genotype of the resulting daughters were assessed. Most cells divided shortly after shift to permissive temperature, giving rise to one viable and one inviable cell (this cell never budded). By crossing the colony that arose from the viable cell and dissecting tetrads, we were able to show that they were precise or nearly precise 2N diploids (12/13 cases). The two cells produced by the division presumably correspond to the aploid daughter (never budded) and the daughter that contains all of the nuclear DNA (diploidized). Thus, DNA replication occurred normally in these cells, but the chromosomes all segregated to one daughter.

Since microtubules are involved in chromosome segregation, we fixed and stained arrested *ndc1-1* cells with antibody directed against tubulin (Kilmartin et al. 1982), visualizing the microtubules by indirect immunofluorescence. To appreciate the aberrant patterns seen in these *ndc1-1* cells, it is necessary to be familiar with the normal cycle of nuclear division in yeast. Figure 5 shows a sketch

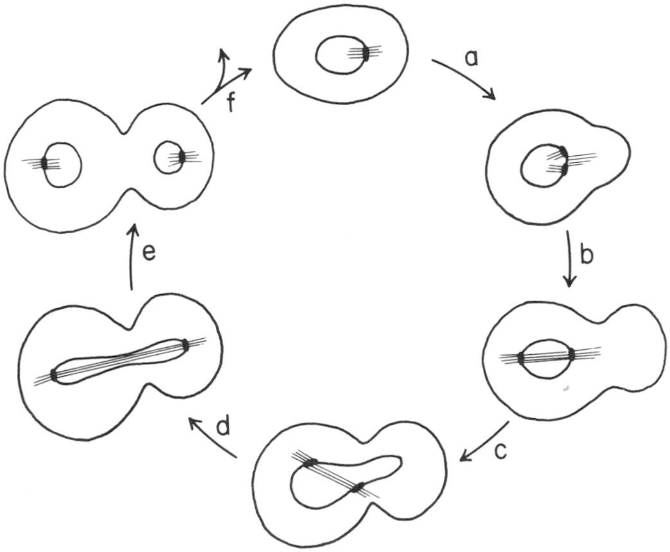

Figure 5
Schematic diagram of the yeast cell cycle, showing the major features of the nuclear division pathway, derived primarily from electron microscopic data (Byers 1981). Shown in each cell are the cell wall, the nuclear envelope, the spindle-pole body (embedded in the nuclear envelope), and the intra- and extranuclear microtubules radiating from the spindle-pole body. (*a*) One of the first steps in the cell cycle is the duplication of the spindle-pole body and the nearly simultaneous emergence of the bud. DNA replication initiates shortly thereafter. (*b*) As the bud grows, DNA replication is completed; the two spindle-pole bodies migrate to opposite poles of the nucleus, lining up roughly on the long axis of the cell; and a short intranuclear spindle forms between them. (*c*) The nucleus then migrates into the neck of the bud and begins elongation. (*d*) The spindle elongates rapidly until it extends nearly to the ends of the two cell bodies. The appropriate chromosomes remain closely associated with each spindle-pole body during this stage, resulting in their segregation. (*e*) The spindle breaks down from the middle toward the ends and the nuclear envelope divides. (*f*) Cytokinesis and cell separation ensue.

of the events pertinent to this discussion. All microtubules in yeast have one end associated with the spindle-pole body (or bodies), which remains embedded in the nuclear envelope throughout the cell cycle (in yeast, the nuclear envelope remains intact during mitosis). Some microtubules radiate into the nucleus to form the spindle, and others radiate out to form cytoplasmic microtubules. At the time of bud emergence, the single spindle-pole body duplicates and the two products migrate to opposite sides of the nearly spherical nuclear envelope, forming the spindle between them. At the same time, DNA replication occurs, followed by migration of the nucleus into the neck of the mother and bud. There, the spindle and the nucleus elongate through the neck of the bud, finally spanning the length of the two cell bodies. During this elongation, the nuclear DNA can be seen to separate into equal regions, which segregate to their respective poles. Finally, the elongated spindle breaks near the middle into half-spindles, and the nuclear envelope divides, followed shortly by cell division.

The appearance of typical *ndc1-1* cells after about 1.5 generations of arrest is shown in Figure 6. The microtubules remain intact and are roughly normal in number and general appearance. Nearly all of the large-budded cells have microtubule bundles radiating from a single point in each cell body. These structures appear similar to the half-spindles seen in normal cells just prior to cell division. In these cells, nuclear DNA is present in only one cell body. Judging from the appearance of the microtubules found in the aploid cell body, it appears that the spindle-pole body has divided and properly segregated, since the microtubules appear to originate from a single point and are morphologically normal in appearance. Also shown in Figure 6 is an aploid daughter cell, in which it appears that there is a spindle-pole body and its associated microtubules, but no nuclear DNA.

It seemed possible that cells defective for *ndc1-1* function proceed along an aberrant pathway not related to the normal cell-division cycle. To test this possibility, we imposed a double block with *ndc1-1* and another treatment that normally results in a block at a specific stage of the cell cycle. If the step blocked by this second agent were absent from the aberrant *ndc1-1* pathway, then it would not affect the arrest of the *ndc1-1*-defective cells. If, however, the aberrant *ndc1-1* pathway is also dependent on the second block, then that block should be epistatic to *ndc1-1*. The two secondary blocks we employed were α factor, a yeast pheromone that blocks cells in G_1, and hydroxyurea, which blocks DNA synthesis. Epistasis of these blocks was tested both by morphological criteria (whether the double-blocked cells followed the pattern of *ndc1-1* arrest or showed the α factor or hydroxyurea pattern) and by assessing the ploidy of cells that were allowed to recover from such a double block. Both α factor and hydroxyurea were epistatic to *ndc1-1* for both the morphology and diploidization phenotypes, producing only arrested cells that were typical of the drug block alone and preventing the diploidization of the *ndc1-1* cells. Thus, the aberrant pathway followed by cells during the *ndc1-1* block still passes through these two cell-cycle steps. This supports the argument that *ndc1-1*-arrested cells replicate their DNA.

The behavior of the *ndc1-1* cells has important implications for analysis of the yeast cell-division cycle. First, it suggests that not all defects in a cell-cycle function result in a uniform arrest morphology; there may be many other mutants that affect cell-cycle functions but have not been recognized for this reason. Second,

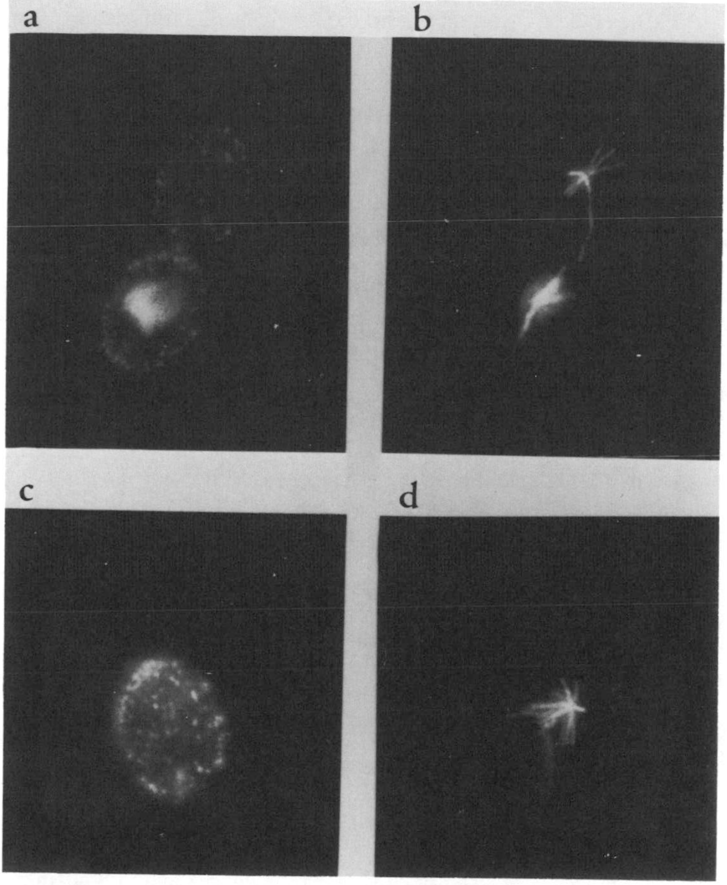

Figure 6
Pattern of DNA and microtubules found in *ndc1-1* mutants arrested at restrictive temperature for 1.5 generation times. *a* and *b* show the same large-budded cell stained with the DNA staining dye DAPI (*a*) and with monoclonal antibody against tubulin (*b*). The two stains can be viewed independently by the use of appropriate excitation and emission filters. The DAPI stain reveals the nuclear DNA (large bright area) clearly situated in only one cell body, whereas the microtubules are found in both cell bodies, apparently radiating from two properly segregated spindle-pole bodies. Small dots of DAPI stain distributed primarily on the periphery of the cell are mitochondrial DNA. (*c*,*d*) Analogous stains of an unbudded aploid cell from the same arrest. The DAPI-stained panel is overexposed to show the absence of nuclear DNA despite excellent staining of the mitochondrial DNA. The microtubules in this cell nonetheless appear normal, all appearing to radiate from a single site.

these results indicate that proper chromosome segregation is not required for progression through the cell cycle. In the parlance of Jarvik and Botstein (1975), chromosome segregation in the *ndc1-1* mutant cells is not on a dependent pathway with any other function required for the cell-division cycle.

In conclusion, we have presented examples of application of the combined methods of classical cell biology, genetics, and the newly developed recombinant DNA methods to *S. cerevisiae*. It seems clear that continued application of these methods, by plan A and plan B, holds real promise for the understanding of the cell cycle and cytoskeleton at the molecular level.

REFERENCES

Adams, A.E.M. and J.R. Pringle. 1984. Localization of actin and tubulin in wild-type and morphogenetic mutants of *Saccharomyces cerevisiae*. *J. Cell Biol.* **98:** 934.

Botstein, D. and R.W. Davis. 1982. Principles and practice of recombinant DNA research with yeast. In *The molecular biology of the yeast* Saccharomyces. *Metabolism and gene expression* (ed. J.N. Strathern et al.), p. 607. Cold Spring Harbor Laboratory, Cold Spring Harbor, New York.

Byers, B. 1981. Cytology of the yeast life cycle. In *The molecular biology of the yeast* Saccharomyces. *Life cycle and inheritance* (ed. J.N. Strathern et al.), p. 59. Cold Spring Harbor Laboratory, Cold Spring Harbor, New York.

Byers, B. and L. Goetsch. 1974. Duplication of spindle plaques and integration of the yeast cell cycle. *Cold Spring Harbor Symp. Quant. Biol.* **38:** 123.

———. 1975. The behavior of spindles and spindle plaques in the cell cycle and conjugation of *Saccharomyces cerevisiae. J. Bacteriol.* **124:** 511.

———. 1976. A highly ordered ring of membrane-associated filaments in budding yeast. *J. Cell Biol.* **69:** 717.

Cabib, E. 1975. Molecular aspects of yeast morphogenesis. *Annu. Rev. Microbiol.* **29:** 191.

Clarke, L. and J. Carbon. 1980. Isolation of a yeast centromere and construction of functional small circular chromosomes. *Nature* **287:** 504.

Cleveland, D., M. Lopata, R. MacDonald, N. Cowan, W. Rutter, and M. Kirschner. 1980. Number and evolutionary conservation of α- and β-tubulin and cytoplasmic α- and γ-actin genes using specific cloned cDNA probes. *Cell* **20:** 95.

Davidse, L. and W. Flach. 1977. Differential binding of methyl benzimidazol-2-yl carbamate to fungal tubulin as a mechanism of resistance to this antimitotic agent in strains of *Aspergillus nidulans. J. Cell Biol.* **72:** 174.

Dutcher, S.K. and L.H. Hartwell. 1983. Genes that act before conjugation to prepare the *Saccharomyces* nucleus for caryogamy. *Cell* **33:** 203.

Falco, S.C. and D. Botstein. 1983. A rapid chromosome-mapping method for cloned fragments of yeast DNA. *Genetics* **105:** 857.

Fink, G.R. and J. Conde. 1977. Studies on *KAR1*, a gene required for nuclear fusion in yeast. In *International cell biology* 1976–77 (ed. B.R. Brinkley and K.R. Porter), p. 414. Rockefeller University Press, New York.

Gallwitz, D. 1982. Construction of a yeast actin gene intron deletion mutant that is defective in splicing and leads to the accumulation of precursor RNA in transformed yeast cells. *Proc. Natl. Acad. Sci.* **79:** 3493.

Gallwitz, D. and R. Seidel. 1980. Molecular cloning of the actin gene from yeast *Saccharomyces cerevisiae. Nucleic Acids Res.* **8:** 1043.

Gallwitz, D. and I. Sures. 1980. Structure of a split yeast gene: Complete nucleotide sequence of the actin gene in *Saccharomyces cerevisiae. Proc. Natl. Acad. Sci.* **77:** 2546.

Gallwitz, D., C. Donath, and C. Sander. 1983. A yeast gene encoding a protein homologous to the human *c-has/bas* proto-oncogene product. *Nature* **306:** 704.

Greer, C. and R. Schekman. 1982a. Actin from *Saccharomyces cerevisiae. Mol. Cell. Biol.* **2:** 1270.

———. 1982b. Calcium control of *Saccharomyces cerevisiae* actin assembly. *Mol. Cell. Biol.* **2:** 1279.

Hartwell, L.H. 1974. *Saccharomyces cerevisiae* cell cycle. *Bacteriol. Rev.* **38:** 164.

Hitchcock-DeGregori, S.E. 1980. Actin assembly. *Nature* **288:** 437.

Jarvik, J. and D. Botstein. 1975. Conditional-lethal mutations that suppress genetic defects in morphogenesis by altering structural proteins. *Proc. Natl. Acad. Sci.* **72:** 2738.

Kilmartin, J. 1981. Purification of yeast tubulin by self-assembly *in vitro*. *Biochemistry* **20:** 3629.

Kilmartin, J. and A. Adams. 1984. Structural rearrangements of tubulin and actin during the cell cycle of the yeast *Saccharomyces*. *J. Cell Biol.* **98:** 922.

Kilmartin, J., B. Wright, and C. Milstein. 1982. Rat monoclonal antitubulin antibodies derived by using a new nonsecreting rat cell line. *J. Cell Biol.* **93:** 576.

Koteleansky, U., M. Gluckhova, M. Benjanian, A. Surguchov, and V. Smirnov. 1979. Isolation and characterization of actin-like protein from yeast *Saccharomyces cerevisiae*. *FEBS Lett.* **102:** 55.

Larson, G.P., K. Itakura, H. Ito, and J.J. Rossi. 1983. *Saccharomyces cerevisiae* actin-*Escherichia coli lacZ* gene fusions: Synthetic-oligonucleotide-mediated deletion of the 309 base pair intervening sequence in the actin gene. *Gene* **22:** 31.

Lazarides, E. and U. Lindberg. 1974. Actin is the naturally occurring inhibitor of deoxyribonuclease I. *Proc. Natl. Acad. Sci.* **71:** 4742.

Moens, P.B. and E. Rapoport. 1971. Spindles, spindle plaques, and meiosis in the yeast *Saccharomyces cerevisiae* (Hansen). *J. Cell Biol.* **50:** 344.

Moir, D., S. Stewart, B. Osmond, and D. Botstein. 1982. Cold-sensitive cell-division-cycle mutants of yeast: Isolation, properties, and pseudoreversion studies. *Genetics* **100:** 547.

Morris, N.R., M.H. Lai, and C.E. Oakley. 1979. Identification of a gene for α-tubulin in *Aspergillus nidulans*. *Cell* **16:** 437.

Neff, N.N., J.H. Thomas, P. Grisafi, and D. Botstein. 1983. Isolation of the β-tubulin gene from yeast and demonstration of its essential function *in vivo*. *Cell* **33:** 211.

Ng, R. and J. Abelson. 1980. Isolation and sequence of the gene for actin in *Saccharomyces cerevisiae*. *Proc. Natl. Acad. Sci.* **77:** 3912.

Orr-Weaver, T., J. Szostak, and R. Rothstein. 1981. Yeast transformation: A model system for the study of recombination. *Proc. Natl. Acad. Sci.* **78:** 6354.

Pringle, J.R. and L.H. Hartwell. 1981. The *Saccharomyces cerevisiae* cell cycle. In *The molecular biology of the yeast* Saccharomyces. *Life cycle and inheritance* (ed. J.N. Strathern et al.), p. 97. Cold Spring Harbor Laboratory, Cold Spring Harbor, New York.

Robinow, C.F. and J. Marak. 1966. A fiber apparatus in the nucleus of the yeast cell. *J. Cell Biol.* **29:** 129.

Sheirr-Neiss, G., M. Lai, and N. Morris. 1978. Identification of a gene for tubulin in *Aspergillus nidulans*. *Cell* **15:** 639.

Shortle, D., J. Haber, and D. Botstein. 1982a. Lethal disruption of the yeast actin gene by integrative DNA transformation. *Science* **217:** 371.

Shortle, D., P. Grisafi, S.J. Benkovic, and D. Botstein. 1982b. Gap misrepair mutagenesis: Efficient site-directed induction of transition, transversion, and frameshift mutations *in vitro*. *Proc. Natl. Acad. Sci.* **79:** 1588.

Shortle, D., D. Koshland, G. Weinstock, and D. Botstein. 1980. Segment-directed mutagenesis: Construction in vitro of point mutations limited to a small predetermined region of a circular DNA molecule. *Proc. Natl. Acad. Sci.* **77:** 5375.

Shuman, H.A., T.J. Silhavy, and J.R. Beckwith. 1980. Labelling of proteins with β-galactosidase by gene fusion. *J. Biol. Chem.* **255:** 168.

Sutcliffe, J.G., R.J. Miller, T.M. Shinnick, and F.E. Bloom. 1983a. Identifying the protein products of brain-specific genes with antibodies to chemically synthesized peptides. *Cell* **33:** 671.

Sutcliffe, J.G., T.M. Shinnick, N. Green, and R.A. Lerner. 1983b. Antibodies that react with predetermined sites on proteins. *Science* **219:** 660.

Villa-Komaroff, L.A., A. Efstratiadis, S. Broome, P. Lomedico, R. Tizard, S.P. Naker, W.L. Chick, and W. Gilbert. 1978. A bacterial clone synthesizing proinsulin. *Proc. Natl. Acad. Sci.* **75:** 3727.

Water, D., J. Pringle, and L. Kleinsmith. 1980. Identification of an actin-like protein and of its messenger ribonucleic acid in *Saccharomyces cerevisiae. J. Bacteriol.* **144:** 1143.

Wood, J.S. and L.H. Hartwell. 1982. A dependent pathway of gene functions leading to chromosome segregation in *S. cerevisiae. J. Cell Biol.* **94:** 718.

Zechel, K. 1980. Dissociation of the DNase I-actin complex by formamide. *Eur. J. Biochem.* **110:** 337.

Centromeric DNA Structure in Yeast Chromatin

Kerry Bloom, Enrique Amaya, and Elaine Yeh
Department of Biology
University of North Carolina
Chapel Hill, North Carolina 27514

The centromere is a specific region along eukaryotic chromosomes that interacts with the mitotic apparatus to direct chromosome movement during cell division. The yeast, *Saccharomyces cerevisiae*, has proved most tractable for the isolation of centromeric DNA. Fragments of DNA isolated by virtue of their ability to promote the proper segregation of replicating plasmid DNA molecules through mitosis and meiosis map to the centromeric region of yeast chromosomes (Clarke and Carbon 1980). The centromeric DNAs (*CEN*), isolated from five chromosomes in yeast (*CEN3*, Clarke and Carbon 1980; *CEN4*, Stinchcomb et al. 1982; *CEN5*, Maine et al. 1984; *CEN6*, Panzeri and Philippsen 1982; *CEN11*, Fitzgerald-Hayes et al. 1982b), are all *cis*-acting DNA elements that function as stabilizing elements for replicating molecules. The *CEN* sequences function in either orientation on plasmid molecules, and at least three centromeres, *CEN3*, *CEN11* (Clarke and Carbon 1983), and *CEN4* (A. Hill and K. Bloom, unpubl.), are interchangeable in yeast chromosome III. Yeast centromeres are therefore not necessarily chromosome-specific and may function as autonomous units stabilizing the plasmid or chromosome in which they reside.

To begin an analysis of the interacting components at the centromeric region, we have studied the chromatin structure of the centromeric DNA. The bulk of the DNA in yeast and other eukaryotic cells is wrapped around histone core particles to create a periodic array of 146-bp nucleosomal subunits and 20–50-bp linker sequences. The centromeric DNA, in contrast, is organized in a 220–250-bp core particle that is refractory to nucleolytic digestion (Bloom and Carbon 1982). This region of structural differentiation is associated with all of the centromeric sequences that have been examined to date, whether they are present within the

  0-87969-174-3/84 $1.00

yeast genome or on replicating plasmid molecules (Bloom and Carbon 1982; Bloom et al. 1984). The maintenance of this structural differentiation in chromatin containing a 289-bp *CEN3* fragment, in place of the normal 624-bp *CEN3* fragment, indicates that the DNA sequences within the truncated fragment provide the necessary information for the binding of interactive components. This region of the centromeric core includes the short regions of DNA (element I, 14 bp; element II, 82–89 bp; and element III, 11 bp) that are conserved in their nucleotide sequence and spatial arrangement in the centromeres from different chromosomes (Fitzgerald-Hayes et al. 1982a). We have examined the possibility that these elements are juxtaposed in the cell nucleus where they could provide a common binding site for components of the segregation apparatus.

Topological Conformation of Yeast Centromeric DNA

The winding of the DNA helix around the histone core in chromatin results in a highly organized structure that brings together regions of DNA that are separated in the linear molecule. Upon extraction of the chromosomal proteins, the turns of plasmid DNA around a protein core are manifested as superhelical turns in the covalently closed, circular DNA molecules. The number of these superhelical turns can be directly measured by counting the number of bands that are visualized when the plasmid molecules are electrophoretically separated in the presence of a DNA-intercalating agent (Shure et al. 1977). Measurement of the superhelicity of centromeric plasmids, in comparison to noncentromeric plasmids, will therefore provide an indication of how centromeric DNA sequences are folded in vivo.

We have constructed small, circular DNA plasmids that contain only the DNA sequences essential for their propagation as minichromosomes in yeast. Deletion of DNA sequences of bacterial origin reduces nonspecific protein binding to plasmids in vivo and therefore enhances their utility in studying specific protein-DNA interactions in transformed yeast cells. The plasmids shown in Figure 1 contain the wild-type 858-bp *CEN11* DNA fragment or deletion mutant *CEN11* fragments as indicated; a genetically selectable marker, *TRP1*; and sequences that provide autonomous replication for the plasmid in yeast, *ars1*. The ability of centromeric DNA to confer mitotic stability to these minichromosomes was assayed by following the plasmid-linked marker, *TRP1*, through several generations of nonselective growth. As shown in Figure 2, the plasmid, pYe(C11)1, with the 858-bp wild-type *CEN11* fragment was maintained in 70–80% of the population, whereas the noncentromeric plasmid, pYe(Y)2, was lost from more than 80% of the population. Deletion mutant plasmids with elements I and III intact, p*dl*(28)1, are as mitotically stable as the wild-type centromeric plasmids, whereas those deletion mutant plasmids lacking the key element I-III region, p*dl*(48)3 and p*dl*(6)3, are no longer mitotically stable and were lost very rapidly from the population. Deletion of the bulk of bacterial-derived sequences from centromeric plasmids in yeast therefore does not alter the ability of centromeric DNA to stabilize circular plasmids at low copy number through successive mitotic cell divisions.

We have measured the superhelicity of the small centromeric DNA plasmids and deletion mutant plasmids to determine whether the centromeric DNA helix is wound around the 220–250-bp nuclease-protected centromeric core in a manner similar to that of the bulk of chromatin DNA around a histone core. Total DNA

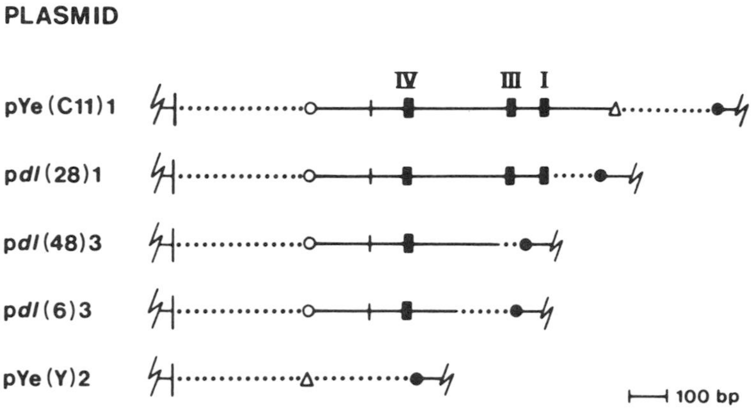

Figure 1
Comparative restriction maps of the small yeast chromosomal DNA plasmids. These plasmids are derived from pYe(*CEN11*)5, and the corresponding deletion mutant plasmids described by Bloom et al. (1983). pBR322 deletions were constructed by partial restriction enzyme cleavage of the parental plasmids with *Eco*RI and *Sal*I. The linear fragments containing *CEN11* DNA, or deletion mutant *CEN11* DNA, and *TRP1-ars1* sequences were blunt-ended with the Klenow fragment of *Escherichia coli* DNA polymerase. The fragments were circularized by ligation and used to transform yeast strain J17 to TRP$^+$. The plasmid pYe(C11)1 contains the 858-bp *CEN11 Sau*3A yeast DNA fragment (solid lines), 650 bp of bacterial-derived pBR322 DNA (dotted lines), and the 1453-bp *Eco*RI DNA fragment containing the yeast *TRP1-ars1* DNA sequences (extended solid lines). The deletion mutant plasmids contain the 1453-bp *TRP1-ars1* DNA sequences and the following *CEN11* and pBR322 DNAs. p*dl*(28)1: 644-bp *CEN11*, 535-bp pBR322; p*dl*(48)3: 530-bp *CEN11*, 441-bp pBR322; p*dl*(6)3: 396-bp *CEN11*, 540-bp pBR322. pYe(Y)2 contains the 1453-bp *Eco*RI *TRP1-ars1* yeast DNA sequences and 650 bp of pBR322. The regions of complete homology between *CEN3* and *CEN11*, elements I, III, and IV, are indicated by the solid boxes. Element II lies between I and III and contains 88 bp (>93% A + T). Restriction enzyme sites are *Bam*HI (△), *Sau*3A (○), *Sal*I/EcoRI junction (●), *Eco*RI (long vertical lines), and *Hin*fI (short vertical lines).

from yeast cells transformed with each of the small, yeast DNA plasmids described above was isolated, deproteinized, and separated electrophoretically on agarose gels in the presence of the DNA-intercalating agent, chloroquine phosphate. To visualize the plasmid molecules, the DNA was transferred to nitrocellulose as single strands and hybridized to radiolabeled *TRP1-ars1* sequences common to all of the plasmids. Each plasmid shown in Figure 3 was resolved into a set of bands, topoisomers, that differ only in the number of times one DNA strand crosses the other strand. The smallest plasmids, pYe(Y)2 and *TRP1-ars1* monomeric circles, were resolved into a set of four to five bands, whereas the larger plasmids, pYe(C11)1 and *TRP1-ars1* dimeric circles, were resolved into a set of eight to nine bands. The same number of topoisomers in the population of centromeric and noncentromeric plasmids of similar size indicates that the primary structural organization of these DNA sequences around a protein core must be very similar in the yeast cell nucleus. All of the noncentromeric yeast plasmids studied to date, including the endogenous 2-μm plasmid (Livingston and Hahne 1979), *TRP1-ars1* circles (Zakian and Scott 1982), and bacterial-derived plasmids

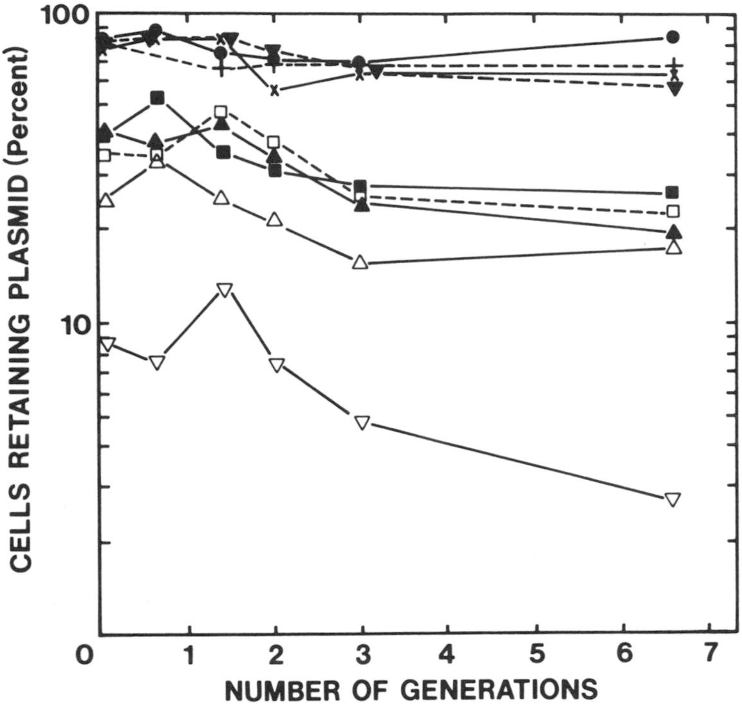

Figure 2
Mitotic stabilities of pYe(C11)1, deletion-mutant plasmids, and the *TRP1-ars1* circles in yeast. The mitotic stabilities of the plasmids pYe(C11)1 (●), pYe(*CEN11*)5 (▼), p*dl*(28)1 (×), p*dl*(48)3 (▲), p*dl*(6)3 (■), pYe(Y)2 (△), YRp7′ (▽), *TRP1-ars1* monomeric circle (+), and *TRP1-ars1* dimeric circle (□) were measured in yeast strain J17 as described previously (Bloom et al. 1983). pYe(C11)1 and p*dl*(28)1 were maintained at low copy number through mitosis (see Fig. 3D,E, plasmid and chromosomal bands) as were their parental plasmids pYe(*CEN11*)5 and *dl*31-28 (Bloom et al. 1983). The *TRP1-ars1* monomeric circle was also stably maintained through mitosis in the absence of selective pressure. However, the stability of the *TRP1-ars1* monomeric circle is probably not a function of centromeric activity, since the *TRP1-ars1* circle is maintained at an elevated copy number (approximately 100 copies/cell) (see Fig. 3A,J, plasmid and chromosomal bands) (Zakian and Scott 1982).

(Bloom and Carbon 1982), form nucleosomal complexes indistinguishable from that of the bulk of yeast chromatin. The centromeric DNA may therefore wind around a protein core the same number of times these other DNA sequences wind around a histone core in chromatin.

Protein-DNA Interaction at the Centromeric Core

The winding of centromeric DNA in chromatin around a histone or histonelike core could easily result in the alignment of the key elements I and III. In fact, from the dimensions of the nucleosome core, one can estimate that short DNA elements (10–20 bp) juxtaposed on the face of a nucleosome are spaced at approximately 80-bp intervals (Weintraub 1980). The distance between elements I

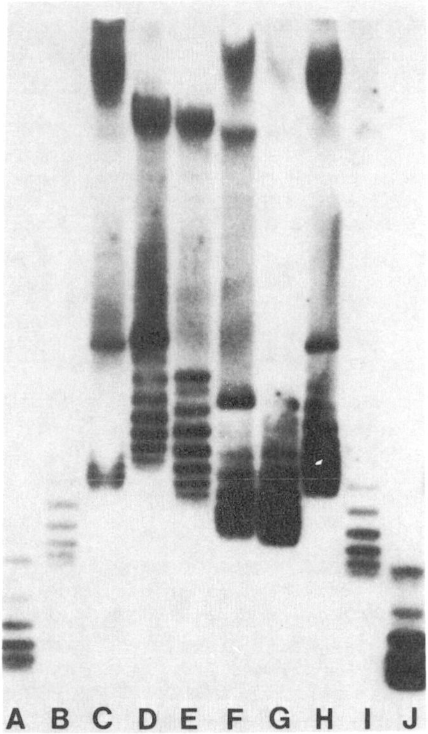

Figure 3
Comparison of the intracellular superhelicity of pYe(C11)1, deletion-mutant plasmids, and the *TRP1-ars1* circles in yeast. Total DNA was prepared from yeast strain J17 transformed with *TRP1-ars1* monomeric circles (*A,J*), *TRP1-ars1* dimeric circles (*C,H*), pYe(C11)1 (*D*), p*dl*(28)1 (*E*), p*dl*(48)3 (*F*), p*dl*(6)3 (*G*), and pYe(Y)2 (*B,I*). The DNA was electrophoresed in the presence of 1.5 μg/ml of chloroquine phosphate in a 1.4% agarose gel at 100 volts for 6 hr. DNA was transferred as single strands to nitrocellulose and hybridized to radiolabeled *TRP1-ars1* DNA. *TRP1-ars1* sequences are common to all of the plasmids (lower half of the autoradiograph), as well as chromosomal DNA sequences (most visible in the centromeric plasmid lanes *D* and *E*, upper half of the autoradiograph). Multimers of some of the noncentromeric plasmids can be seen in the uppermost region of the gel in lanes *C* and *H*.

and III ranges from 82 to 89 bp in the centromeres of four chromosomes that have been sequenced. This spatial conservation between the key elements indicates that their proximity in chromatin may play an essential role in centromeric function. We have therefore measured the affinity of DNA-binding proteins at the centromeric core to determine whether any of these components are dissociated from centromeric DNA at the same ionic strengths used to remove histones from bulk chromatin.

Chromosomal proteins can be removed from DNA by treatment of chromatin with NaCl (Paulson and Laemmli 1977; Bloom and Anderson 1978), and the structure of the protein-depleted chromatin complex can be determined (Bloom and Carbon 1982). We examined the dissociation pattern of the centromeric core in both the wild-type and substitution-mutant strains shown in Figure 4. Chromatin from isolated yeast nuclei was washed extensively with standard digestion buffer (SPCM) or with buffer containing 0.4, 0.75, or 1.25 M NaCl. Following the salt washes, chromatin was reequilibrated with the standard digestion buffer and partially cleaved with DNase I. The DNA was isolated from these samples, deproteinized, and restricted with *Hin*dIII. The DNA fragments were separated electrophoretically, blotted to nitrocellulose, and probed with the 900-bp *Hin*dIII-*Bam*HI fragment extending from the restriction site toward the centromere on chromosome III (Fig. 4). Only those restriction fragments that extend toward the centromere from the *Hin*dIII site are visualized following hybridization and autoradiography. The lengths of the hybridizing sequences therefore provide a direct

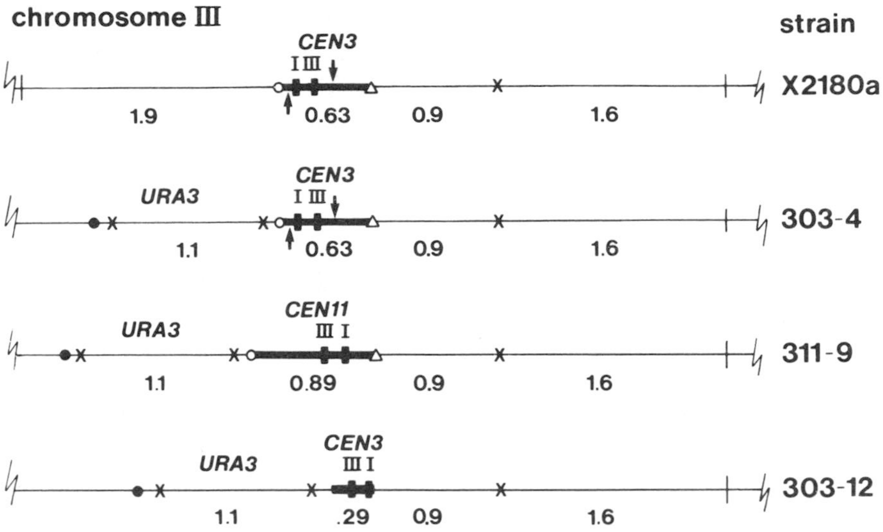

Figure 4
Comparative restriction maps of the centromeric region in chromosome III in wild-type and substitution-mutant yeast strains. The location of the 624-bp *CEN3* fragment (darkened line) along the wild-type yeast chromosome III is shown from strain X2180a. A series of substitution mutants have been constructed using site-directed substitution vectors as described by Clark and Carbon (1983), to replace the host sequences at the centromeric region of yeast chromosome III. The physical map in selected constructions include the 624-bp *CEN3* fragment in the proper orientation (303-4), the 858-bp *CEN11* fragment in the opposite orientation (with respect to the element I-III region in the wild-type chromosome III) (311-9), and a truncated 289-bp *CEN3* fragment in the opposite orientation in chromosome III (303-12). The *CEN* fragments are indicated by the darkened lines. Restriction sites are *Eco*RI (|), *Bam*HI (△), *Hin*dIII (×), and selected *Alu*I (down arrow), *Rsa*I (up arrow), and *Sau*3A (○) sites. Numbers denote kilobase pairs. The substitution chromosomes contain the selectable genetic marker, *URA3*, that allowed genetic identification of the colonies in which the substitution event initially occurred (Clarke and Carbon 1983).

map of the points of nucleolytic cleavage within the chromatin or DNA fiber relative to the restriction site (Wu 1980).

The dissociation pattern of nucleoproteins from the centromeric chromatin of the wild-type strain, X2180a, is shown in Figure 5. The pattern visualized after exhaustive washing in standard digestion buffer (Fig. 5, X2180a, no NaCl) revealed two prominent cutting sites flanking elements I-III, with the characteristic 220–250-bp spacing, indistinguishable from the pattern previously obtained with the conventional nuclei preparations (see Fig. 6B in Bloom and Carbon 1982). The pattern visualized after treatment of the chromatin complex with 0.4 M NaCl was comparable to that of the "no-salt" lanes. Thus, the protein or RNA components that confer this unique structure to the centromeric chromatin remain bound following dissociation of loosely bound chromosomal proteins. More tightly bound chromosomal proteins, including the core histone proteins, are not dissociated until higher salt concentrations (1–2 M NaCl) are employed (see Paulson and

Laemmli 1977; Bloom and Anderson 1978). Upon dissociation of these more tightly bound chromosomal proteins, the protected region of chromatin becomes accessible to nucleolytic digestion (Fig. 5, X2180a, 0.75 M and 1.25 M NaCl lanes), and the specific cleavage pattern in chromatin begins to resemble the cleavage pattern of naked DNA.

Similar results were obtained using chromatin from the substitution-mutant strains. These strains include the 624-bp *CEN3* fragment replaced back into the host chromosome III (303-4), a truncated 289-bp *CEN3* fragment replacing the 624-bp *CEN3* fragment in the opposite orientation (303-12), and an 858-bp *CEN11* fragment replacing the 624-bp *CEN3* fragment in the opposite orientation in chromosome III (311-9) (Fig. 4). These *CEN* fragments are all completely capable of stabilizing the entire yeast chromosome III through the processes of mitosis and meiosis (Clarke and Carbon 1983; Bloom et al. 1984; Carbon and Clarke 1984). The protected centromeric core structure encompassing elements I-III in each strain was intact following exhaustive washing in SPCM (Fig. 5, 303-12, 303-4, and 311-9, no NaCl lanes) and was very similar to the chromatin patterns previously obtained (Fig. 4) (Bloom et al. 1984). Following treatment with 0.4 M NaCl, again dissociating loosely bound chromosomal proteins, centromeric chromatin in these altered strains was unaffected. Washes with higher salt concentrations disrupted the protected centromeric core (Fig. 5, 303-12, 303-4, and 311-9, 0.75 M and 1.25 M NaCl lanes). The unique structure of the centromere is therefore dependent on the association of tightly bound chromatin components in the cell nucleus. Since the affinity of these chromatin components to *CEN* regions is independent of their orientation or host chromosome, the same components may mediate attachment of the centromeres from different chromosomes to the spindle apparatus.

DISCUSSION

The centromeric region of higher eukaryotes can be visualized at the morphological level as a primary constriction along the chromatin fiber. In yeast, although the chromosomes are not readily visible at the light microscope level, the centromeric region can also be structurally distinguished along the chromatin fiber. The centromeric region is organized in a 220–250-bp core particle quite distinct from the 160-bp nucleosomal core of bulk chromatin. The centromeric DNA may be folded around a histone or histonelike particle such that the key sequence elements, I and III, are brought into proximity in chromatin where they serve as the chromatin template for components of the mitotic apparatus. The size of the protected centromeric core may therefore reflect the size of one of these components bound to the centromere in chromatin. Once putative components of the segregation apparatus are identified, their molecular interactions can be analyzed biochemically and genetically in this simple eukaryotic system.

ACKNOWLEDGMENT

This work was supported by research grant GM-32238 from the National Institutes of Health (to K.S.B.).

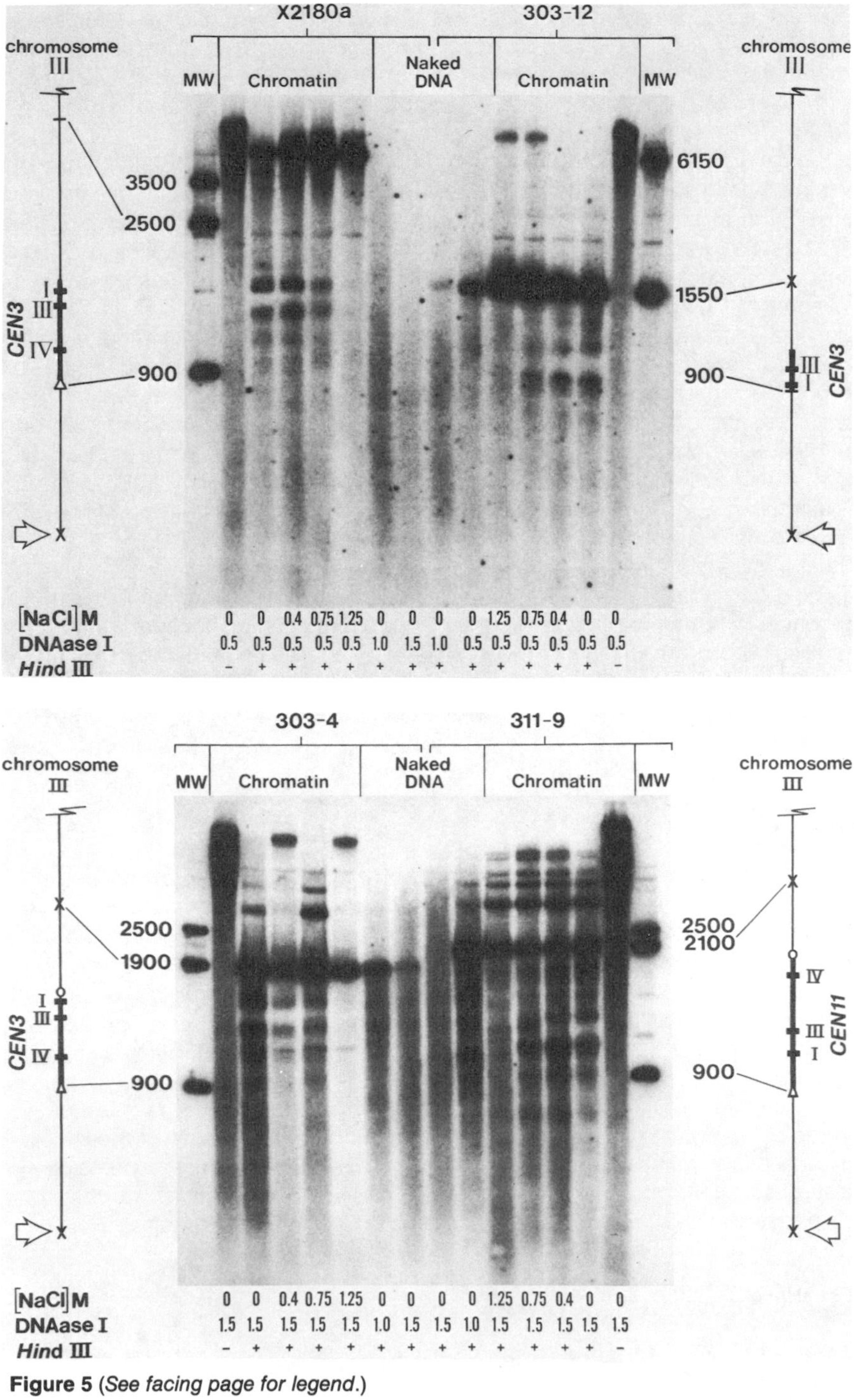

Figure 5 (*See facing page for legend.*)

REFERENCES

Bloom, K.S. and J.N. Anderson. 1978. Fractionation and characterization of chromosomal proteins by the hydroxyapatite dissociation method. *J. Biol. Chem.* **253:** 4446.

Bloom, K.S. and J. Carbon. 1982. Yeast DNA is in a unique and highly ordered structure in chromosomes and small circular minichromosomes. *Cell* **29:** 305.

Bloom, K.S., M. Fitzgerald-Hayes, and J. Carbon. 1983. Structural analysis and sequence organization of yeast centromeres. *Cold Spring Harbor Symp. Quant. Biol.* **47:** 1175.

Bloom, K.S., E. Amaya, J. Carbon, L. Clarke, A. Hill, and E. Yeh. 1984. Chromatin conformation of yeast centromeres. *J. Cell Biol.* (in press).

Carbon, J. and L. Clarke. 1984. Structural and functional analysis of a yeast centromere (*CEN3*). *J. Cell Sci.* (in press).

Clarke, L. and J. Carbon. 1980. Isolation of a yeast centromere and construction of functional small circular chromosomes. *Nature* **287:** 504.

———. 1983. Genomic substitutions of centromeres in *Saccharomyces cerevisiae. Nature* **305:** 23.

Fitzgerald-Hayes, M., L. Clarke, and J. Carbon. 1982a. Nucleotide sequence comparisons and functional analysis of yeast centromere DNAs. *Cell* **29:** 235.

Fitzgerald-Hayes, M., J.-M. Buhler, T. Cooper, and J. Carbon. 1982b. Isolation and subcloning analysis of functional centromere DNA (*CEN11*) from yeast chromosome XI. *Mol. Cell. Biol.* **2:** 82.

Livingston, D.M. and S. Hahne. 1979. Isolation of a condensed, intracellular form of the 2 μm DNA plasmid of *Saccharomyces cerevisiae. Proc. Natl. Acad. Sci.* **116:** 249.

Maine, G.T., R.T. Surosky, and B.-K. Tye. 1984. Isolation and characterization of the centromere from chromosome V (*CEN5*) of *Saccharomyces cerevisiae. Mol. Cell. Biol.* **4:** 86.

Panzeri, L. and P. Philippsen. 1982. Centromeric DNA from chromosome VI in *Saccharomyces cerevisiae* strains. *EMBO J.* **1:** 1605.

Paulson, J.R. and U.K. Laemmli. 1977. The structure of histone-depleted metaphase chromosomes. *Cell* **12:** 817.

Stinchcomb, D.T., C. Mann, and R.W. Davis. 1982. Centromeric DNA from *Saccharomyces cerevisiae. J. Mol. Biol.* **158:** 157.

Figure 5

Mapping nuclease-sensitive sites on centromeric chromatin from protein-depleted yeast chromosomes. Nuclei were prepared from wild-type strain X2180a, or the genomic substitution strains 303-12, 303-4, and 311-9 as described by Bloom et al. (1984). Prior to nuclease digestion, samples were resuspended in standard digestion buffer (SPCM) or in SPCM plus 0.4 M NaCl, 0.75 M NaCl, or 1.25 M NaCl. Each sample was immediately sedimented at 10,000*g* for 10 min and washed twice in the same solution of SPCM plus salt. The resulting salt-washed pellets were resuspended in SPCM and sedimented at 10,000*g* for 10 min. The final nuclear pellets, equilibrated with SPCM, were digested with DNase I (5 μg/ml) for the times (min) indicated. Naked DNA samples were deproteinized before nucleolytic digestion and reveal any sequence specificity inherent in DNase I. Purified DNA samples were electrophoresed on 1.4% agarose gels after digestion in the presence (+) or absence (−) of *Hin*dIII. The gel was blotted to nitrocellulose and hybridized to the 900-bp *Hin*dIII-*Bam*HI fragment shown in Fig. 4. Molecular-weight markers (MW) indicate yeast DNA fragments containing regions complementary to the probe, cut with *Hin*dIII-*Bam*HI, *Bam*HI-*Eco*RI, and *Hin*dIII-*Eco*RI for X2180a; *Hin*dIII and *Eco*RI for 303-12; and *Hin*dIII, *Hin*dIII-*Bam*HI, and *Bam*HI-*Eco*RI for 303-4 and 311-9. The portion of the chromosome containing *CEN3* or *CEN11* is shown at the side of the appropriate autoradiograph. The *CEN* fragments are indicated by the darkened line. Restriction sites are *Eco*RI (−), *Bam*HI (△), *Hin*dIII (×), and *Sau*3A (○).

Shure, M., D.E. Pulleyblank, and J. Vinograd. 1977. The problems of eukaryotic and prokaryotic DNA packaging and *in vivo* conformation posed by superhelical density heterogeneity. *Nucleic Acids Res.* **4:** 1183.

Weintraub, H. 1980. Recognition of specific DNA sequences in eukaryotic chromosomes. *Nucleic Acids Res.* **8:** 4745.

Wu, C. 1980. The 5′ ends of *Drosophila* heat shock genes in chromatin are hypersensitive to DNase I. *Nature* **286:** 854.

Zakian, V.A. and J.F. Scott. 1982. Construction, replication, and chromatin structure of *TRP1 R1* circle, a multiple-copy synthetic plasmid derived from *Saccharomyces cerevisiae* chromosomal DNA. *Mol. Cell. Biol.* **2:** 221.

Microtubules and Actin Filaments during the Yeast Cell Cycle

John V. Kilmartin
MRC Laboratory of Molecular Biology
Cambridge CB2 2QH, England

Yeast is a simple eukaryote with very well developed genetics. In spite of its small size, it has some potential for the investigation of cytoskeletal proteins because the analysis of mutant phenotype is a useful approach to the understanding of protein function. Of the large number of proteins that comprise the mammalian cytoskeleton, only tubulin and actin have thus far been found in yeast. Both proteins appear to be closely similar in functional properties to their counterparts in higher organisms (Kilmartin 1981; Greer and Scheckman 1982). It seems likely that many other yeast cytoskeletal proteins remain to be identified and characterized.

The distribution of microtubules during the yeast cell cycle has been extensively studied by electron microscopy (Byers 1981), and, as expected, these studies show a role for tubulin in mitosis and possibly in bud formation. Yeast has a mitotic spindle similar in structure to those of higher eukaryotes. There are cytoplasmic and astral microtubules as well. In yeast, the nuclear envelope remains intact throughout mitosis, and functions of the centrosomes are served by spindle-pole bodies (SPBs) that are embedded in the nuclear envelope. Cytoplasmic microtubules, which grow from the cytoplasmic face of the SPB, are present during both mitosis and interphase. Yeast (*Saccharomyces cerevisiae*) has one β-tubulin gene whose disruption is lethal (Neff et al. 1983). In another yeast species, *Schizosaccharomyces pombe*, mutation in the α-tubulin gene results in the absence of a spindle and in abnormally positioned SPBs (Toda et al. 1983).

In contrast to tubulin, little is known from electron microscopy of the distribution of actin filaments in yeast, possibly due to fixation problems or to the high density of ribosomes in the yeast cytoplasm. Thus, to study the distribution of actin during the yeast cell cycle, it seemed worthwhile to develop an alternative method, immunofluorescence, which has proved very successful in investigating the cytoskeleton of higher cells. Immunofluorescence is a simple and rapid method and is particularly suited to yeast research, where many cells containing different mutations and at many stages of the cell cycle need to be investigated.

Immunofluorescence has two main requirements: a fixation protocol to preserve cell structure while still allowing penetration by antibodies, and the preparation of antibodies monospecific for the particular yeast antigen of interest. Unfortunately, most antisera to mammalian cytoskeletal proteins do not appear to cross-react with their counterparts in yeast, so it is usually necessary to prepare antisera against the particular yeast protein. A fixation protocol has been developed (Kilmartin and Adams 1984) that is suitable for immunofluorescence in yeast. Using this protocol, microtubules were found to have a distribution similar to that previously found by electron microscopy (Byers 1981). This suggests that other antigens may also be well preserved. However, for any fixation protocol there will always be a small minority of antigens that are not well preserved. Such fixation artifacts can be avoided partly by employing several independent fixation methods. Ultimately, the best test of immunofluorescence results is the observation of live cells microinjected with fluorescently labeled antigens.

Microtubules and Actin Filaments during the Yeast Cell Cycle

Yeast cells (*Saccharomyces uvarum*) double-labeled with both anti-tubulin and anti-actin are shown in Figure 1. The actin-staining pattern is somewhat unusual: As expected, fibers are seen, but the overall pattern is dominated by the presence of dots. The dots almost certainly contain actin because an independent probe for actin filaments, rhodamine phalloidin, gave very similar results (Adams and Pringle 1984; Kilmartin and Adams 1984). The cells shown in Figure 1 were from an asynchronous population; the budded cells (Fig. 1b–i) were placed in a probable temporal sequence by a combination of mitotic spindle length and the ratio of mother to bud cell lengths. During the yeast cell cycle, the mother cells remain relatively constant in size, and most of the cell-wall growth takes place in the bud.

Cells with small buds (Fig. 1b–d) have cytoplasmic microtubules pointing into the bud, as found previously by electron microscopy (Byers 1981); these microtubules grow from the cytoplasmic face of the SPB. Occasionally (Fig. 1c), other cytoplasmic microtubules also pass down into the mother cell. Actin dots are present as a ring around the neck of the bud (Fig. 1b,c), and actin fibers appear to pass from the dots into the mother cell; but because of the relatively large depth of focus in light microscopy, such connectivity cannot be definitively established. As the bud continues to grow, the actin dots disappear from the neck region and transfer to the bud (Fig. 1c,d). With further bud growth, the dots seem to concentrate in the tip region of the bud (Fig. 1e–g). During active bud growth, few dots are present in the mother cell, although in some strains of *S. cerevisiae*,

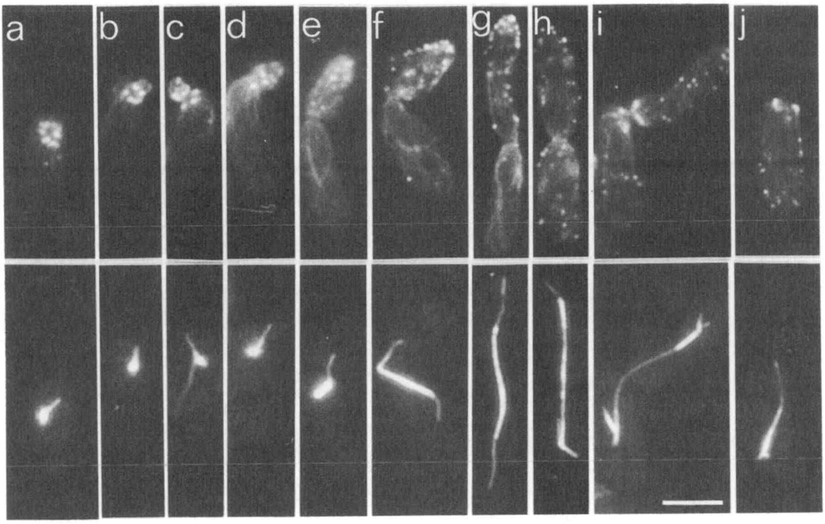

Figure 1
Double immunofluorescence of *S. uvarum* cells with anti-actin (*top*) and anti-tubulin (*bottom*). The cells have been placed in their cell-cycle order. Bar, 5 μm.

the asymmetry of staining between mother and bud is not so marked as that shown here (Adams and Pringle 1984). At some point during bud growth, the spindle forms (Fig. 1e) and elongates (Fig. 1f,g), but the actin-staining pattern appears to remain unaltered during spindle growth. After spindle elongation, the distribution of actin dots changes to a more equal one between mother and bud (Fig. 1h). As the spindle begins to break down (Fig. 1i), the actin dots become concentrated in the septum region between mother and bud.

The apparently unbudded cells in this asynchronous population could be classified into about five categories on the basis of their combination of tubulin and actin staining (Kilmartin and Adams 1984). Two of these (Fig. 1a,k) resembled the staining patterns for cells with small buds (Fig. 1b) and cells with disintegrating spindles (Fig. 1i). Thus, these unbudded cells could be classified as cells where budding is imminent (Fig. 1a) or cells that had just completed cytokinesis (Fig. 1j). The conclusion that the cells in Figure 1a were about to bud and that the ring of actin dots represents the budding site was confirmed by double labeling of fixed yeast cells with intact walls using rhodamine phalloidin and Calcofluor (Fig. 2). Rhodamine phalloidin stains actin filaments, and Calcofluor is a polysaccharide-specific dye that stains rings of chitin in the yeast cell wall (Hayashibe and Katohda 1973). The chitin rings form prior to budding and stain weakly with Calcofluor. The staining intensity in the neck region between mother and bud increases as the bud grows, until after separation when an intensely staining bud scar is left. In Figure 2a, the ring of actin dots around the neck of a small bud coincides with the weakly staining chitin ring. In an apparently unbudded cell in Figure 2b, the actin dot ring again coincides with the weakly staining chitin ring, indicating the site of bud formation. This cell has already produced at least five daughters, as indicated by the five intensely staining bud scars.

Possible Functions of Actin and Tubulin in Yeast

The actin-containing dots detected by anti-actin and rhodamine phalloidin change their distribution in a characteristic way during the yeast cell cycle (Fig. 1). They are present as a ring at the site of budding (Figs. 1a and 2), migrate to the tip of the bud during bud growth (Fig. 1e–g), and are concentrated in the neck region during septum formation (Fig. 1i). These regions of the yeast cell are all places where new cell-wall deposition is occurring, suggesting a role for the actin dots in this process (Adams and Pringle 1984; Kilmartin and Adams 1984). In addition, if the actin fibers are actually connected to the actin dots, then another possible function for the actin dots would be to ensure that, even in cells with side buds, the actin fibers pass from mother to bud, rather than along the length of the mother cell. The function of the actin fibers might be to transfer material from mother to bud by cytoplasmic streaming during bud growth.

One obvious role for microtubules in yeast is in the process of chromosome separation, but the immunofluorescence results suggest an additional possible function. During bud growth and prior to full elongation of the spindle in the vast majority of cells, a bundle of cytoplasmic microtubules passes into the bud from the cytoplasmic face of the SPB, which is positioned on the mother side of the neck region (Fig. 1b–e). This bundle of cytoplasmic microtubules may be involved in transport of material into the bud as previously suggested (Byers 1981). In addition, it would ensure that spindle elongation took place between mother and bud, rather than within the mother cell. The long cytoplasmic microtubules

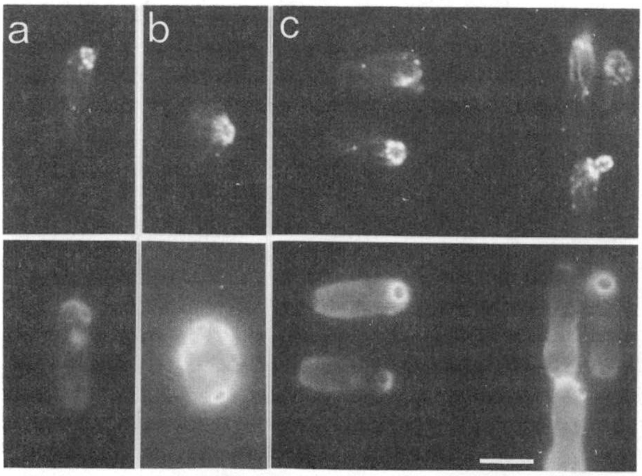

Figure 2
Double labeling of fixed *S. uvarum* cells with intact walls with rhodamine phalloidin (*top*) to show distribution of actin and with Calcofluor (*bottom*) to show distribution of chitin rings. Note that the rings of actin dots coincide with the weakly staining chitin rings where bud formation is starting. Bar, 5 μm.

at either end of the spindle (Fig. 1g) would also ensure a symmetric positioning of the spindle within the cell.

Control of the Distribution of Microtubules during the Yeast Cell Cycle: The Spindle Pole Body

As described above, tubulin and actin change their cellular localizations in a dynamic and predictable way during the various phases of the yeast cell cycle. At present, little is known about the molecules responsible for these changes in localization. In the case of microtubules, there is a specific organelle, the SPB, that must play a role in their distribution because it appears to be the sole initiator of microtubule assembly in yeast. The behavior of the SPB during the yeast cell cycle defines some interesting questions concerning the molecular biology of the SPB: (1) What are the molecules in the SPB to which the microtubules are attached and how do they function? (2) What is the mechanism of the replication of the SPB at the start of mitosis to produce two, and only two, copies? Answers to these questions are likely to require a combination of biochemical and genetic experiments. It is unlikely that SPBs can ever be isolated in sufficient quantity for detailed biochemical experiments, so one approach toward answering these questions would be to enrich for SPBs as far as possible, prepare antisera to these fractions, and then screen a λgt11 yeast genomic DNA expression bank (see reference to M. Snyder in Young and Davis 1983) to isolate SPB genes. The biochemistry of SPB proteins could then be studied after their preparation in large quantities by expression of the genes in *Escherichia coli* or yeast.

The initial stages in SPB enrichment involve preparation of yeast nuclei and a crude nuclear-envelope fraction, since SPBs are embedded in the nuclear envelope. The SPBs were assayed by quantitative dark-field light microscopy. SPBs were mixed with brain tubulin and warmed on a slide to initiate microtubule assembly from the SPBs. The resulting number of "asters," or foci of microtubule assembly, were counted in a measured area of the slide to give the active SPB concentration (Kilmartin and Fogg 1982). The SPBs were detached from the envelope by sonication, and further enrichment was attempted by sucrose gradient centrifugation. This approach was not very successful, so an affinity purification step was employed. Pig brain microtubules were grown off the SPBs (Fig. 3), and the SPBs with microtubules attached were spun through a sucrose step gradient to remove tubulin monomer. A monoclonal anti-tubulin antibody YL1/2 (Kilmartin et al. 1982) coupled to colloidal gold particles (5 nm dia) was then attached to the microtubules, and the complex was sedimented through another sucrose step gradient. The binding of colloidal gold to the microtubules resulted in a noticeable density shift (Fig. 4), separating the SPBs from contaminants. When the fraction containing SPBs was collected from the gradient, pelleted, fixed, and embedded in plastic, a section cut through the very thin pellet from top to bottom (hence, representative of all the components in this fraction) showed a substantial enrichment for SPBs (Fig. 5). The components recognizable in the pellet are SPBs (arrowheads), sheets of nuclear envelope still attached to the SPBs, and microtubules with antibody and gold particles attached. The yield of active SPBs was about 40% (40 liters of yeast harvested at 1×10^7 cells/ml gave 1.5×10^{11}

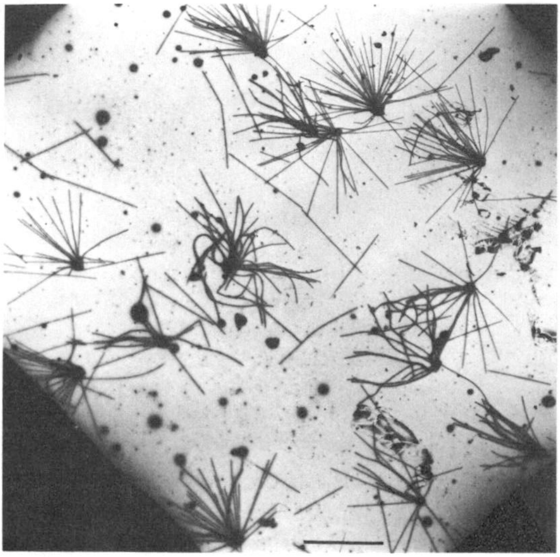

Figure 3
A grid square (400-mesh grid) of negatively stained SPBs after initiation of brain microtubule assembly. Bar, 5 μm.

SPBs after enrichment). The purification factor was estimated from the number of SPBs and from the amount of protein in that SPB fraction after subtracting the tubulin and YL1/2 antibody protein content (measured from SDS-gel scans). This indicated a purification of 1000–5000-fold. Currently, we are attempting to raise antisera to the SPB components.

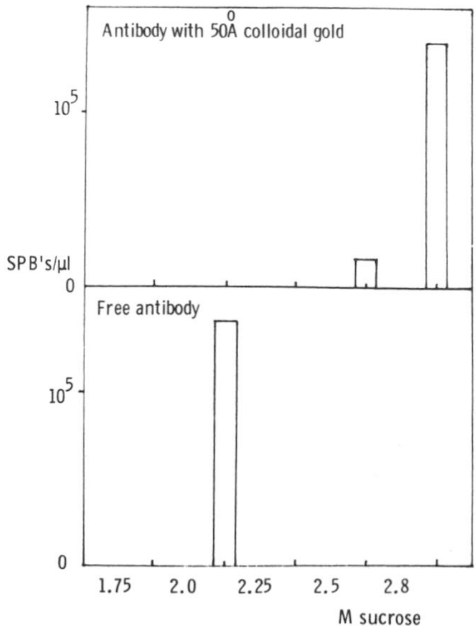

Figure 4
Demonstration of the density shift on a sucrose step gradient of SPBs after attachment of microtubules and anti-tubulin coupled with colloidal gold (*top*) compared with control SPBs attached to microtubules and unconjugated anti-tubulin (*bottom*).

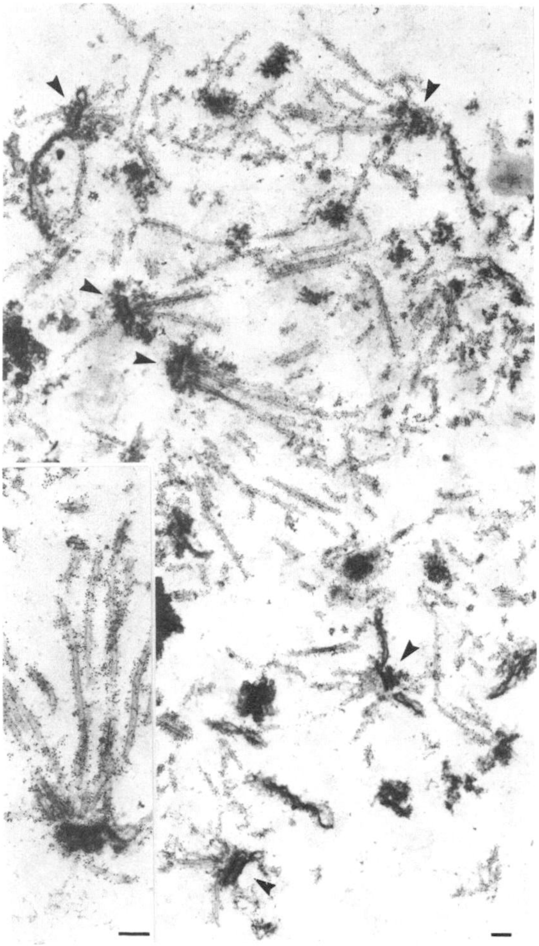

Figure 5
Thin section passing from the top to the bottom of the pellet containing the final fraction of the enriched SPBs (arrowheads). (*Inset*) One SPB at higher magnification. Bar, 0.1 μm.

REFERENCES

Adams, A.E.M. and J.R. Pringle. 1984. Relationship of actin and tubulin distribution to bud growth in wild-type and morphogenetic-mutant *Saccharomyces cerevisiae. J. Cell Biol.* **98:** 934.

Byers, B. 1981. Cytology of the yeast cell cycle. In *The molecular biology of the yeast* Saccharomyces (ed. J.N. Strathern et al.), vol. 1, p. 59. Cold Spring Harbor Laboratory, Cold Spring Harbor, New York.

Greer, C. and R. Scheckman. 1982. Actin from *Saccharomyces cerevisiae. Mol. Cell Biol.* **2:** 1270.

Hayashibe, M. and S. Katohda. 1973. Initiation of budding and chitin ring. *J. Gen. Appl. Microbiol.* **19:** 23.

Kilmartin, J.V. 1981. Purification of yeast tubulin by self-assembly in vitro. *Biochemistry* **20:** 3629.

Kilmartin, J.V. and A.E.M. Adams. 1984. Structural rearrangements of tubulin and actin during the cell cycle of the yeast *Saccharomyces. J. Cell Biol.* **98:** 922.

Kilmartin, J.V. and J. Fogg. 1982. Partial purification of yeast spindle pole bodies. In *Microtubules in microorganisms* (ed. P. Cappuccinelli and N.R. Morris), p. 157. Marcel Dekker, New York.

Kilmartin, J.V., B. Wright, and C. Milstein. 1982. Rat monoclonal antitubulin antibodies derived by using a new nonsecreting rat cell line. *J. Cell Biol.* **93:** 576.

Neff, N.F., J.H. Thomas, P. Grisafi, and D. Botstein. 1983. Isolation of the β-tubulin gene from yeast and demonstration of its essential function in vivo. *Cell* **33:** 211.

Toda, T., K. Umesono, A. Hirata, and M. Yanagida. 1983. Cold-sensitive nuclear division arrest mutants of the fission yeast *Schizosaccharomyces pombe. J. Mol. Biol.* **168:** 251.

Young, R.A. and R.W. Davis. 1983. Yeast RNA polymerase II genes: Isolation with antibody probes. *Science* **222:** 778.

Cellular Morphogenesis in the Yeast Cell Cycle

John R. Pringle, Kevin Coleman, Alison Adams, Sue Lillie, Brian Haarer, Charles Jacobs, Jane Robinson,* and Christopher Evans

Department of Cellular and Molecular Biology
Division of Biological Sciences
The University of Michigan, Ann Arbor
*Michigan 48109; *Department of Genetics*
Trinity College, Dublin, Ireland

During the past few years, our studies of the yeast cell cycle (Pringle and Hartwell 1981) have focused on its morphogenetic processes. These processes determine the shape of the cells and their three-dimensional organization; they generate membrane and cytoplasmic anisotropies and involve a conspicuous polarization of secretion and growth. The specific processes of interest (Fig. 1A,B) (see also Sloat et al. 1981; Adams and Pringle 1984; and references cited therein) are (1) the selection of a nonrandom site of the cell surface at which budding will occur; (2) the formation of a ring of chitin (the "bud scar") in the largely nonchitinous cell wall at that site; (3) the localization of new cell-wall growth to the region bounded by the chitin ring, resulting in the appearance and selective growth of a bud; (4) the localization of secretion of other materials to the surface of the bud; (5) the localization of new cell-wall growth to the tip of the bud during much of the period of bud growth; (6) the balancing of tip growth against periods of uniform (isotropic) growth of the bud cell wall, resulting in the normal ellipsoidal shape of the daughter cell; (7) the migration of the nucleus from a position within the mother cell into the neck connecting mother and bud; and (8) cytokinesis and the formation of the septal cell wall (Roberts et al. 1983). Although relatively simple and, in their details, unique to yeast, these processes appear to involve the same general principles as morphogenetic processes in other types of eukaryotic cells and seem likely to employ similar mechanisms. Thus, in our studies, we hope to exploit the great experimental advantages of yeast to yield generally useful information about the mechanisms of eukaryotic cellular morphogenesis as well as about the mechanisms of eukaryotic cellular reproduction.

Studies of such morphogenetic processes might take either of two rather dif-

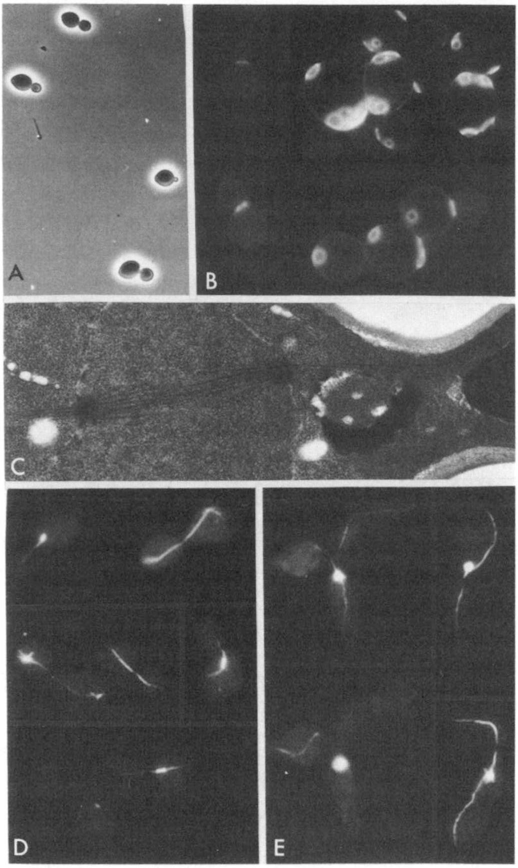

Figure 1

(*A*) Phase-contrast micrograph of wild-type *S. cerevisiae* cells, illustrating the ellipsoidal shape of the cells, the selective growth of the bud relative to the mother cell (four different cells are shown, but four successive views of the same cell would look much the same), and the nonrandomness of bud position (note that all buds have formed near a pole of the cell). (*B*) Fluorescence micrographs of wild-type cells stained with Calcofluor, a dye that binds specifically to chitin in the yeast cell wall (Sloat et al. 1981). Note that each developing bud has a ring of chitin at its base. These chitin rings remain on the mother cell after division; their distribution provides a further illustration of the nonrandomness of bud position. The bipolar pattern shown is typical of cells expressing both *MAT***a** and *MAT*α information at the mating-type locus; cells expressing only *MAT***a** or *MAT*α information typically show a unipolar pattern of budding. (*C*) Electron micrograph of a budding cell (the bud is on the right), illustrating spindle-pole bodies (microtubule-organizing centers in the nuclear envelope), intranuclear or spindle microtubules (recall that the nuclear envelope does not break down during mitosis in yeast), and cytoplasmic microtubules (note that one set of these runs from the nucleus out into the bud). (*D,E*) Fluorescence micrographs illustrating the distribution of microtubules in wild-type and *cdc4* mutant cells, as revealed by indirect immunofluorescence using a monoclonal anti-tubulin antibody (Kilmartin and Adams 1984). (*D*) Proliferating wild-type cells; note that cytoplasmic microtubules extend into the bud in each of the cells in which a complete spindle exists within the mother cell. (*E*) Temperature-sensitive *cdc4* cells stained after several hours at restrictive temperature. Under these conditions, the cells form multiple, abnormally elongated buds that fail to divide, whereas the nuclear cycle is arrested prior to DNA synthesis and with the spindle-pole body duplicated but not separated. Three cells are shown; that on the left is shown at two levels of focus.

ferent approaches. First, we could make the very reasonable hypothesis that known cytoskeletal elements such as microtubules and actin microfilaments are involved, and thus proceed to study such elements directly by morphological, biochemical, and genetic methods. Second, we could undertake a genetic approach that involves no a priori assumption that the processes of interest are controlled by the known cytoskeletal elements. In this paper, we summarize briefly our progress with these two approaches.

RESULTS AND DISCUSSION

Tubulin and Actin Distribution in Relation to Bud Growth

Yeast contains microtubules that are similar in morphology (Fig. 1C) (for review, see Byers 1981; King and Hyams 1982) and in the structure of their constituent tubulins (Kilmartin 1981; Kilmartin et al. 1982; King and Hyams 1982; Neff et al. 1983) to microtubules from other eukaryotic cells. Intranuclear (spindle) and cytoplasmic microtubules emanate from the opposite faces of the spindle-pole bodies in the nuclear envelope. Several lines of evidence from the serial-section electron microscopic studies of Byers and Goetsch (1974, 1975; see also Byers 1981) suggested that the cytoplasmic microtubules were involved in the selection of the budding site and/or in directing secretory vesicles to that site and into the growing bud. First, in both wild-type cells and various mutants (notably, *cdc4*; Fig. 1E), bud emergence and the early stages of bud growth occurred in cells with duplicated but unseparated spindle-pole bodies. In such cells, the double spindle-pole bodies were always oriented toward the budding site, and the cytoplasmic microtubules ran from the spindle-pole body into the bud, often seeming (at least superficially) to be associated with the secretory vesicles that were accumulated there. Second, centrifugation of newly formed zygotes altered both the typical position of the nuclei (and hence the spindle-pole bodies) within the zygotes and the typical location of the zygotes' first buds, but the orientation of the spindle-pole bodies and cytoplasmic microtubules toward the budding sites was retained (Byers 1981). Third, involvement of these microtubules in the directed movement of secretory vesicles was also suggested by their orientation toward the site of localized alterations of the cell wall in cells undergoing zygote formation (Byers and Goetsch 1975).

The development of effective immunofluorescence procedures for yeast (Kilmartin and Adams 1984; Adams and Pringle 1984) allowed extension of these correlations in two ways. First, it became clear that cytoplasmic microtubules continued to run from the nuclear envelope into the growing bud as the spindle-pole bodies separated, the spindle formed, and the nucleus migrated into the neck (Fig. 1D; cf. Fig. 1C). Second, it was observed that *cdc4* mutant cells (1) often had active cell-wall growth occurring at two or more sites simultaneously (see Fig. 2F) and (2) typically had multiple bundles of cytoplasmic microtubules emanating from the one double spindle-pole body in each cell and running out to the multiple bud tips (Fig. 1E). These observations appeared to strengthen the case for involvement of the cytoplasmic microtubules in the selective growth of the bud and the polarization of secretion.

At the same time, observations on the intracellular localization of actin in relation to the sites of cell-wall growth suggested that the actomyosin system of the cells might be involved in these processes. Although it was clear that yeast contained substantial amounts of an actin that resembled other eukaryotic actins (Water et al. 1980; Ng and Abelson 1980; Gallwitz 1982; Greer and Schekman 1982) and that the single actin gene per haploid cell was essential (Shortle et al. 1982), electron microscopy had provided no good clues as to the localization and function of actin in the cells. This problem was alleviated with the application of immunofluorescence and staining of actin with fluorochrome-labeled phallotoxins (Adams and Pringle 1984; Kilmartin and Adams 1984). Actin was seen to be disposed in cytoplasmic fibers and in numerous "dots" that appeared to be cortical in location (Fig. 2A,B,G). Although the exact nature of the actin dots is not known (sites of anchorage of actin fibers to the membrane? actin-coated vesicles?), their distribution and that of the actin fibers strongly suggested an involvement in localized growth of the cell wall. Specifically, during the cell cycle of wild-type cells, the dots clustered at sites of active cell-wall deposition, whereas the fibers tended to run along the long axis of the cell between mother and bud (Fig. 2A). Both the clustering of actin dots and the longitudinality of the actin fibers appeared accentuated in morphogenetic mutants that showed an exaggerated tip growth of abnormally elongated buds (Fig. 2B,G). Especially suggestive was the correlation between active cell-wall growth and clustering of actin dots at particular bud tips on *cdc4* cells (Fig. 2F,G).

Effects on Morphogenesis of Microtubule-disrupting Drugs

Several groups (Quinlan et al. 1980; Wood and Hartwell 1982) reported that treatment of yeast with the microtubule inhibitor methyl benzimidazole-2-yl-carbamate (MBC) arrested cell division, apparently by blocking nuclear division. The arrested cells terminated development as mother cells with large buds, suggesting that bud emergence and selective growth of the bud could occur in the absence of functional cytoplasmic microtubules. However, it was not clear that the cytoplasmic microtubules were really effectively disrupted by the drug. Indeed, examination of MBC-treated cells using anti-tubulin immunofluorescence revealed that extended, tubulin-containing structures persisted in such cells even after several hours of exposure to the drug (C. Jacobs, unpubl.). Although the structures visualized looked abnormal and might be nonfunctional, this was difficult to test, and the validity of inferences from such experiments about the functions of cytoplasmic microtubules was uncertain.

Fortunately, another inhibitor, nocodazole (Hoebeke et al. 1976), produced less equivocal results (C. Jacobs et al., in prep.). Examination of nocodazole-treated cells by anti-tubulin immunofluorescence revealed that microtubules began to be lost within a few minutes of exposure to the drug. By 30 minutes of exposure, most cells showed no localized staining, although a few cells showed small residual spots of fluorescence (Fig. 3A); by 60 minutes, even the spots were gone from virtually all cells. Electron microscopy confirmed this rapid loss of both spindle and cytoplasmic microtubules. Corresponding to this rapid loss of microtubules was a rapid arrest of cell division: Exposure of an asynchronous, exponentially growing population to nocodazole arrested division within about 40 minutes (residual increase in cell number was about 25%). The arrested cells were uni-

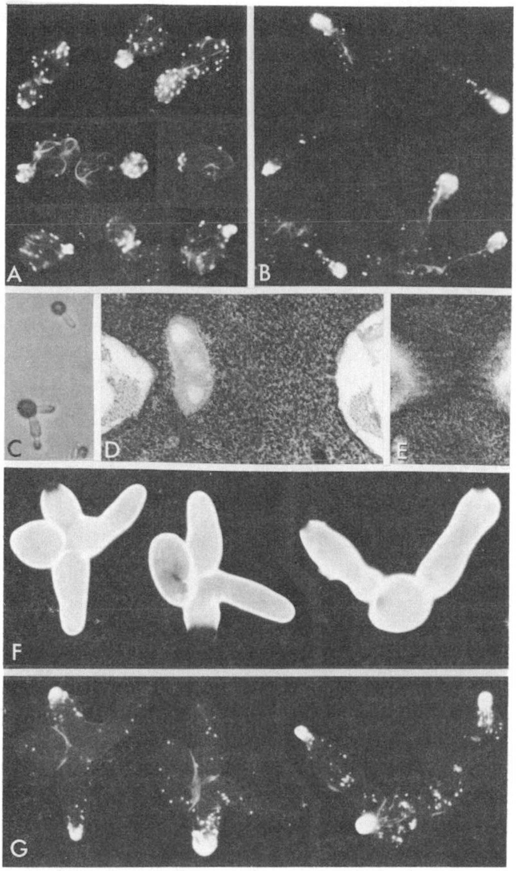

Figure 2
(*A,B*) Fluorescence micrographs illustrating the distribution of actin in wild-type and mutant cells, as revealed by staining with rhodamine-conjugated phalloidin (Adams and Pringle 1984). (*A*) Wild-type cells. Note the clustering of actin "dots" over buds (particularly small buds), at a site of presumed incipient budding (Kilmartin and Adams 1984) on an unbudded cell, and in the neck region of a cell that may be beginning septum formation. (*B*) Temperature-sensitive *cdc3* (upper cell) and *cdc12* (lower cells) mutants after several hours growth at restrictive temperature. Under these conditions, these mutants (also *cdc10* and *cdc11* mutants) form multiple, abnormally elongated buds (*C*) that fail to divide while the nuclear cycle continues to yield multinucleate cells. These mutants also fail to form a normal chitin ring (A. Adams and J. Pringle, unpubl.); they also are defective (Byers and Goetsch 1976b) in forming the ring of 10-nm filaments (biochemical nature unknown) that normally lies immediately subjacent to the plasma membrane in the neck region (*D,E*) (Byers and Goetsch 1976a). (*C*) Phase-contrast micrograph of *cdc10* cells budding at restrictive temperature. (*D,E*) Electron micrographs showing transverse (*D*) and glancing (*E*) sections of the ring of 10-nm filaments. (*F*) Fluorescence micrographs of *cdc4* mutant cells whose surface mannan has been stained with fluorescein-conjugated concanavalin A to reveal (as dark patches) the sites of active cell-wall deposition (Tkacz and Lampen 1972; Adams and Pringle 1984). (*G*) Same cells as in *F*, stained with rhodamine-conjugated phalloidin to reveal the distribution of actin.

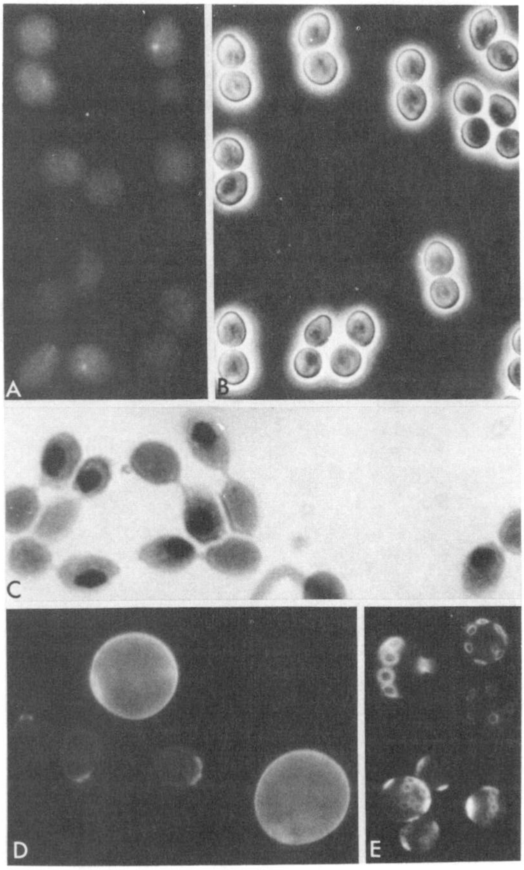

Figure 3
(*A*) Anti-tubulin immunofluorescence of wild-type cells that had been exposed for 30 min to 15 μg/ml of nocodazole in growth medium (cf. Fig. 1D). (*B*) Phase-contrast micrograph of cells from an exponentially growing population to which nocodazole had been added (to 15 μg/ml) 3 hr previously. (*C*) Bright-field micrograph of cells from the same population as in *B*, stained with Giemsa to reveal the positions of the nuclear DNA. In a similar experiment, cells were double-stained with DAPI (to reveal nuclear DNA) and Calcofluor (to reveal bud scars and thus identify the mother cells); this experiment demonstrated that the nuclei indeed were in the mother cells and not in the buds. (*D*) Fluorescence micrograph of Calcofluor-stained cells of a temperature-sensitive *cdc42* mutant strain. Cells from a culture incubated 5 hr at restrictive temperature (the two large, generally bright cells) were mixed after fixation with cells from the parent culture grown at permissive temperature. (*E*) Fluorescence micrographs of Calcofluor-stained, *MAT***a** and *MAT***a***/MATα cdc24* cells that had been grown at permissive temperature, illustrating the loss of the normal unipolar or bipolar budding pattern (cf. Fig. 1B).

formly large-budded (Fig. 3B). A similar arrested population was obtained when a population enriched by centrifugation for small unbudded cells was inoculated into fresh medium containing nocodazole. Thus, it seems clear that both bud emergence and selective bud enlargement can occur in the absence of cytoplasmic microtubules. Moreover, polarized deposition of new cell wall at the tips of buds on *cdc4* cells (similar to that in Fig. 2F) could be observed in nocodazole-treated cells that appeared devoid of microtubules by immunofluorescence. Finally, examination of small unbudded cells producing their first buds in the presence of nocodazole indicated that formation both of the chitin ring and of the ring of 10-nm filaments that normally is present in the neck region (Fig. 2D,E) was not impaired by the absence of cytoplasmic microtubules. No evidence was obtained as to whether the microtubules might be involved in determining bud position. Probably significantly, nocodazole was not observed to affect the intracellular distribution of actin in these experiments.

In contrast to their apparent noninvolvement in the processes of polarized growth and secretion, the cytoplasmic microtubules may be involved in nuclear migration. Thus, the nuclei in nocodazole-arrested cells appeared to be randomly positioned within the mother cells (Fig. 3C). This impression of randomness was strengthened by determining the "nuclear-migration index" (defined as the shortest distance from any part of the nucleus to the mother-bud neck divided by the length of the mother cell) for many individual nocodazole-arrested and control (exponentially growing) cells. In the control population, 93% of the uninucleate, budded cells had nuclear-migration indices ≤ 0.25, whereas in the arrested populations, only 52% of the cells had indices ≤ 0.25 and 27% had indices ≥ 0.4. These differences were accentuated by repeating the experiment with a *cdc2* strain. When this temperature-sensitive mutant is incubated at restrictive temperature, the cells arrest with large buds and nuclei that have typically migrated to, but not into, the necks (Culotti and Hartwell 1971); in our experiment, 90% of the cells had nuclear-migration indices ≤ 0.1. In contrast, when the cells were exposed to nocodazole during their incubation at restrictive temperature, only 25% had indices ≤ 0.1. Interestingly, when nocodazole was added to *cdc2* cells that had already arrested at restrictive temperature, the nuclei appeared to remain at the necks: 88% of the cells still had nuclear-migration indices ≤ 0.1 after 2 hours of exposure to nocodazole at restrictive temperature. It should be noted that interpretation of all these experiments is subject to the caveats (1) that the position of the nuclear envelope may not be accurately reflected in all cases by the position of the DNA (as revealed by the Giemsa and DAPI stains; see Thomas et al., this volume) and (2) that disruption of a process by nocodazole does not necessarily result from the action of the drug on microtubules. However, the tentative conclusion that nuclear migration may be dependent on the cytoplasmic microtubules appears consistent with the role of microtubules in nuclear movement in the filamentous fungus *Aspergillus* (Oakley and Morris 1980).

Genetic Approaches to Morphogenesis in Yeast

Studies of morphological correlations and of the effects of inhibitors, as described above, can be revealing but have many limitations. These limitations are particularly acute with regard to the establishment of causality relations and the identification of the (presumably) multiple components of the machinery responsible for the morphogenetic processes of interest. For these reasons, it is crucial also to investigate these processes using the powerful formal and molecular genetic methods that can be applied to yeast. Two quite different approaches are possible. First, one could seek mutations in genes coding for known components of the cytoskeleton, such as actin and the tubulins. The phenotypes of these mutants should reveal the functions of the affected components, whereas the isolation and analysis of "pseudorevertants" of these mutants (carrying extragenic suppressors of the original mutations) should reveal interactions among the known cytoskeletal components and identify new components that interact with those previously known (Botstein and Maurer 1982). This vital approach is illustrated by the elegant studies of D. Botstein and his collaborators (see Thomas et al., this volume). Their conclusions about the functions of actin and of microtubules in yeast agree well with those reached from morphological and inhibitor studies (see above), and they have already identified numerous new genes whose products may interact with actin and tubulin.

The alternative genetic approach, which we are engaged in, makes no a priori assumption that the major proteins and conspicuous structures that we presently know something about are the only, or even the most important, components of the machinery responsible for the morphogenetic processes of interest. We begin by seeking mutants using aberrant morphogenesis as our criterion for isolation; in effect, this represents an attempt to allow the organism to identify for us the genes encoding significant components of the morphogenetic machinery. We then use molecular-genetic techniques to attempt identification of the products of these genes. We hope that information about the structures of the gene products and about their intracellular localizations (the latter obtained immunocytochemically using antibodies raised against the gene products themselves or against appropriate fusion proteins) will provide clues to guide the early stages of biochemical investigation of gene-product function. Simultaneously, we analyze extragenic suppressors of the original morphogenetic mutations in the hope (1) of identifying interactions among the genes (or their products) already known from the analysis of morphogenetic mutants and (2) of identifying new genes that participate in the processes of interest and interact with the known genes but that, for one reason or another, have been refractory to identification by the analysis of morphogenetic mutants (for consideration of some possible reasons for such refractoriness, see Kaback et al. 1984). Given that the original mutations are presumably mostly missense, the types of interactions likely to allow suppressor mutations should be mostly at the level of regulation or of protein-protein interaction. For example, a mutation in a regulatory gene that increases the amount of some other gene product should sometimes suppress a mutation that reduces the efficiency of function of that gene product. Alternatively, if gene products A and B interact physically in the cell, a mutation in A should sometimes be able to compensate for the structure-destabilizing effects of a mutation in B (Botstein and Maurer 1982).

Our attraction to this second genetic approach, despite its nontrivial difficulties, springs fundamentally from the conviction that living cells and their constituent processes are very complex. This conviction is based on numerous recent studies at both the protein and nucleic acid levels (for review, see Kaback et al. 1984), of which perhaps the most relevant are the elegant genetic and biochemical studies indicating that the flagellar axoneme of *Chlamydomonas* contains ≥280 polypeptides (Luck et al. 1982; R. Segal and D. Luck, pers. comm.). The conclusion that living systems are complex has two corollaries. First, the components of the cell that we know something about (which are, by and large, the biochemically and/or morphologically conspicuous components) may not be the only ones that play interesting and important biological roles. It is sobering to realize that even in the lowly and simple yeast, we know or can guess at the functions of only perhaps 30–40% of the more than 5000 genes (Hereford and Rosbash 1977; Kaback et al. 1984). A particular possibility with some precedent that appears relevant to our interests is that the polarization of growth and secretion in yeast may depend on intracellular ion currents (Stump et al. 1980), rather than (or in addition to) the actions of cytoskeletal components. However, an even more intriguing possibility is that mechanisms are involved in morphogenesis that are as yet undreamt of in the philosophies of biochemists and cell biologists. Such novel mechanisms would probably be missed by a genetic approach that works "outward" from mutations affecting known cytoskeletal components, but should be identifiable by working "inward" from morphogenetic mutants. Second, even if all of the cell's morphogenetic machinery is related to the presently known cytoskeletal elements, some individual components may interact only very indirectly with the major components such as actin and tubulin. Such peripheral components may be easier to identify by starting with morphogenetic mutants than by working outward through suppressors, and suppressors of suppressors (etc.), starting with actin and tubulin mutations.

Morphogenetic Mutants of Yeast

We have focused to date on three classes of morphogenetic mutants. The outstanding characteristics of the prototype for one class, the *cdc4* mutant, are presented briefly in Figures 1E and 2F,G. Mutations in two other known genes (*CDC34* and an as yet unnamed gene defined by the JPT175 and JPTA1473 temperature-sensitive (*ts*) mutations recently isolated in our laboratory) produce phenotypes similar to that of *cdc4* mutants. The outstanding characteristics of the second class of mutants (*cdc3, cdc10, cdc11,* and *cdc12*) are presented briefly in Figure 2B,C. The prototype for the third class of mutants is the intensively studied *cdc24* mutant (Sloat et al. 1981); the *cdc42* and *cdc43* mutants are similar (A. Adams and J. Pringle, in prep.). At restrictive temperature, these mutants continue growing but fail to form buds and exhibit an apparently complete loss of ability to localize secretion; thus, huge round cells showing a nonpolarized deposition of new cell-wall material are formed (Fig. 3D). The *CDC24*-gene product, at least, also appears to be involved in the selection of budding sites, as two different *cdc24* mutants exhibit a loss of the normal pattern of budding during growth *at permissive temperature*, where other aspects of budding and cell-wall deposition appear normal (Fig 3E). Presumably, the mutant gene product is not working quite normally even at the "permissive" temperature.

The original *cdc4, cdc34, cdc3, cdc10, cdc11, cdc12,* and *cdc24* mutants were isolated by L. Hartwell and his collaborators by morphological screening of a collection of approximately 2000 *ts*-lethal mutants (Hartwell et al. 1973; Pringle and Hartwell 1981). More recently, we have screened about 5000 additional *ts*-lethal mutants in an attempt to identify more genes whose products are involved in the morphogenetic processes of the cell cycle. Numerous additional isolates defective in the known genes were obtained, but only three new genes (*CDC42,*

CDC43, and that represented by JPT175) were identified. This might mean that all the genes of interest are known, but it appears much more likely (see Kaback et al. 1984) to reflect the limitations of an approach based solely on *ts*-lethal mutations. In particular, it now seems likely that many genes do not readily make *ts* mutations and that mutations of many genes are not lethal, either because the function encoded is not absolutely essential or because it is duplicated elsewhere in the genome. Accordingly, we have begun to search for additional morphogenetic mutants using cold-sensitive (*cs*)-lethal and suppressible nonsense mutations. Early results with the *cs* mutants are encouraging; the mutants producing multiple, abnormally elongated buds fall into six complementation groups. As expected (Moir et al. 1982), one of these is *cdc11*, but at least some of the others appear to be genes not previously identified using *ts* mutants.

Pseudorevertants of Morphogenetic Mutants

Another approach to the identification of additional genes involved in morphogenesis is the analysis of extragenic suppressors of the available *ts* and *cs* morphogenetic mutations. A special attraction of this approach is that it also reveals interactions among the previously known and newly identified genes or their products (see above). To identify the suppressor-carrying strains, each apparent revertant to be analyzed is crossed to a wild-type strain and tetrads are dissected. The appearance of segregants with the phenotype of the parental mutant strain (*ts* or *cs* lethality; the appropriate aberrant cell morphology; Fig. 4) demonstrates that the reversional event was neither a true (original-site) reversion nor an intragenic second-site alteration. The segregation ratios then indicate whether the suppression was due to a single linked or unlinked suppressor mutation, to a more complex set of mutations producing suppression, or to an increase in copy number of the mutant gene resulting from aneuploidy (see footnotes e and g in Table 1). Additional genetic analysis (see c and d in Table 1) allows the suppressors to be sorted into genes and these genes to be tested for possible identity with those known previously as the sites of morphogenetic mutations.

Some ongoing pseudorevertant analyses (Botstein and Maurer 1982; Thomas et al., this volume) demand that each suppressor to be analyzed confers a new temperature-conditional lethality (e.g., that a suppressor of a *ts* mutation be itself a *cs*-lethal mutation) in addition to its suppression. This strategy has two important advantages. First, it facilitates complementation tests among the suppressor mutations and allows such tests between the suppressor mutations and the available *ts* or *cs* morphogenetic mutations. Second, the presence of a new mutant phenotype (e.g., arrest with morphologically aberrant cells) makes it much easier to be confident that the suppression being analyzed is due to some interesting form of gene-gene interaction and not to something trivial (e.g., stabilization of the originally metastable product of the mutant gene by a second mutation that alters slightly the intracellular physiological milieu). However, in our own studies, we have sacrificed these advantages in order to avoid overlooking potential suppressor genes that do not readily give rise to temperature-conditional lethal mutations. This seems important as recent studies have suggested that such genes may be the great majority of those in the genome (see above) (Kaback et al. 1984). We hope to maximize our chances of detecting suppressors that are in-

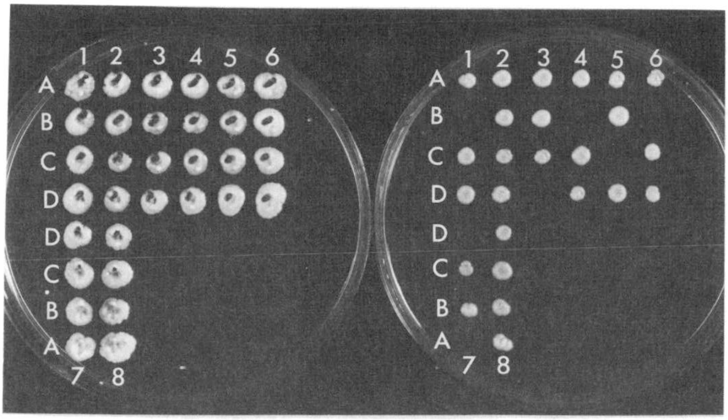

Figure 4
Segregants obtained when an apparent revertant of a *cdc10* strain was crossed by wild type. Numbers refer to the tetrads, and letters *A* through *D* refer to the four spores from a given tetrad. The permissive-temperature (23°C) master plate is on the left, the restrictive-temperature (36°C) replica is on the right. Dead cells scraped from the nongrowing colonies on the 36°C plate had the aberrant cell morphology characteristic of *cdc10* mutants. The segregation pattern obtained indicates that the original revertant carried a single extragenic suppressor that was not linked to *CDC10* itself. Note that all four segregants are growing about equally well in the tetrads showing four Ts^+:0 Ts^- segregations. Thus, the suppressed *cdc10* strains grow about as well as $CDC10^+$ strains for this suppressor.

teresting by demanding that suppression be efficient (i.e., that the suppressed mutant strains grow approximately as well as wild type; cf. Fig. 4). We are further reassured when the suppressors recovered fall into a modest number of genes, as has so far always been the case (see Table 1). However, whether any given gene identified as a suppressor (and not identical to an already known gene) is really involved directly in the processes of interest will remain uncertain until more is learned about that gene (e.g., by observing the phenotype obtained after mutagenesis in vitro, followed by gene replacement in vivo; see Botstein and Davis 1982; Botstein and Maurer 1982).

The results to date of our pseudorevertant analyses are summarized in Table 1. It is clear that in some cases (notably, *cdc10* and *cdc3*), simple extragenic suppressors are found easily and their analysis leads promptly to comprehensible and useful results. In other cases (e.g., *cdc12*), the analyses are laborious, at least with the alleles used so far as parents. The fact that the suppressors have fallen into modest numbers of genes (where so far analyzed) both suggests that these genes are worth studying further (see above) and makes such further study (by cloning and in vitro mutagenesis; see Botstein and Davis 1982; Botstein and Maurer 1982) appear to be a manageable goal.

A result that was not expected (Moir et al. 1982), but that now clearly parallels the results with suppressors of temperature-conditional actin and tubulin mutations (Thomas et al., this volume), is that most of the suppressors analyzed to date have been recessive (i.e., diploids homozygous for the *cdc* allele and heterozygous for the suppressor fail to grow at the restrictive temperature). This is both an interesting clue as to the mechanisms of the suppression and a great experimental convenience in two different ways. First, it allows complementation

Table 1
Pseudorevertant Analyses of Morphogenetic Mutants

Gene	No. of alleles studied[a]	No. of revertants analyzed	No. of simple extragenic suppressors[b]	No. of suppressor genes[c]
CDC3	2	14	6	2 (5[d],1)
CDC10	6[e]	40	26	≥1 (20[f])
CDC11	2	37	7	?
CDC12	2	50	1	1
CDC24	1	27	6	1 (6)
CDC42	1	31	15	4 (5,4,4,2)
X[g]	2	12	0	—

All of these studies are still in progress; presented here is a tentative summary of results to date. Most revertants analyzed to date have been spontaneous, but we plan eventually to analyze also mutagen-induced revertants of most or all mutants, on the grounds that this might reveal a different spectrum of suppressor mutations.

[a]Because the suppressors of interest are likely to be allele-specific (Botstein and Maurer 1982), the chances of identifying all genes that interact with a particular gene are increased by analyzing suppressors obtained with several different mutant alleles of that gene.

[b]Number of cases in which a single, extragenic suppressor mutation appears to account for the original "reversion." The remaining revertants involve intragenic events, a pair of mutations required for suppression, etc. (See also *e* and *g*.)

[c]Based on linkage analyses plus, in some cases, the complementation patterns among recessive suppressors. (Since none of the suppressors analyzed to date have conferred a new *ts*- or *cs*-lethal phenotype [see text], complementation tests based on such phenotypes have not yet been possible.) Numbers in parentheses are the numbers of independent isolates per suppressor gene.

[d]Linkage analysis shows this gene to be tightly linked to *CDC10*. As a large class of *cdc10* suppressors also maps at or near *CDC3*, it seems almost certain that mutations in these genes can suppress each other. Further evidence for interaction of these two genes comes from the observation that all *cdc3 cdc10* double mutants tested to date have been inviable at temperatures that are permissive for the single mutants.

[e]One of the *cdc10* strains tested yielded spontaneous revertants at frequencies at least 100× greater than those obtained with other strains. In the several cases analyzed, the "reversion" was apparently due to the strains having become disomic for chromosome III (the site of *CDC10*). Evidently, doubling the copy number of the mutant gene sufficed to allow growth at the formerly restrictive temperature. The other entries in this row refer to results with the other five *cdc10* strains examined.

[f]Linkage analysis shows this gene to be tightly linked to *CDC3*. See also note *d*.

[g]This as yet unnamed gene is defined by the JPT175 and JPTA1473 *ts* mutants (see text). JPTA1473 yielded spontaneous revertants at a normal frequency, but all 12 analyzed so far have proved to be due to intragenic events. JPT175 yielded high-frequency revertants that (from preliminary evidence) are probably disomic for the chromosome carrying the mutant gene (cf. note *e*).

analyses of the suppressors. (Diploids homozygous for the *cdc* mutation and heterozygous for each of two independently isolated suppressors are tested for growth at restrictive temperature.) Such analyses supplement linkage analyses in assigning the suppressors to genes and have also provided a further unexpected and interesting clue as to the mechanisms of suppression. In most cases, recessive suppressors that appear from linkage analysis to be in the same gene fail to complement each other (i.e., the diploids cited above grow at restrictive temperature) and complement other unlinked recessive suppressors (i.e., the diploids fail to grow), as expected. However, in some cases, all the suppressors in one gene have failed to complement all the suppressors in another, unlinked gene. Presumably, this reflects some interaction between the recessive suppressors, the nature of which can be elucidated as we learn more about the system. The second convenience offered by the recessiveness of the suppressors is that it will facilitate the cloning of the suppressor genes; i.e., libraries of wild-type DNA in suitable plasmid vectors can be screened for sequences that can convert a suppressed mutant strain back to its original temperature sensitivity.

Attempts to Identify Gene Products

We hope eventually to understand morphogenesis in yeast at the molecular level. Realization of this hope will depend on our ability to identify and elucidate the functions of the products of the genes defined by analysis of morphogenetic mutants and their suppressors. Fortunately, many of the techniques needed to move systematically from mutant to gene product have become available during the past few years. We have been applying these techniques to several of the genes of interest; our strategy and progress to date are summarized in Table 2.

The major strategic decision in this scheme is the method chosen to proceed from the identified, cloned gene to identification of the gene product. As identification of a particular spot on a gel as the product of a particular gene is not in itself very useful, our efforts are directed at obtaining an immunogen that can be used to raise antibodies specific for the gene product of interest. These antibodies will then be used to explore the synthesis, intracellular localization, and function of the gene product. The several methods available for obtaining a suitable immunogen include overproduction of the normal polypeptide in yeast or *Escherichia coli*, production of any of several types of fusion proteins containing antigenic regions of the protein of interest, and synthesis of oligopeptide haptens based on nucleotide sequence information. We expect that flexibility in choosing methods will be necessary to adapt to such exigencies of particular cases as the presence of inconvenient introns, instability of normal or fusion polypeptides, or difficulties in raising antisera of sufficiently high titer. At present, our method of first choice employs the "open-reading-frame vectors" developed by Weinstock et al. (1983) (see legend to Fig. 5). In successful cases such as *CDC12* (Fig. 5), a fusion protein is obtained that contains a long segment of the protein of interest, is readily purified based on its high molecular weight and/or β-galactosidase activity, and is contaminated, if at all, by *E. coli* proteins that are unlikely to elicit antibodies that will cause confusion by cross-reacting with yeast proteins. In other cases, application of this method proceeds less smoothly. Numerous apparently suitable fusions of *CDC24* fragments into pORF1 have been examined, but none have yet been found to produce a detectable fusion protein, for reasons that remain obscure.

A major question is how informative the antibodies we obtain will be. If a gene product of interest is not sufficiently abundant or tightly localized to give an unequivocal result by immunofluorescence microscopy, more problematic approaches involving cell fractionation and/or immunoelectron microscopy will have to be employed. Furthermore, success in localizing a gene product does not guarantee major insights into its biochemical function. However, it appears likely

Table 2
Progress of Attempts to Obtain Antibodies Specific for the Products of Genes Involved in Morphogenesis

	Gene					
Step	CDC3	CDC10	CDC11	CDC12	CDC24	X[a]
Unstable yeast transformant[b]	+	+	+	+	+	+
Yeast→*E. coli*→yeast[b]	+	+	+	+	+	+
Fragment size (kb)[c]	4.4	11.0	4.5	1.2	2.7	–
Transcript localization[c]	+	–	+	+	+	–
Gene identification[d]	M	–	I/M	I/M	I/D	–
Gene-fusion construction[e]	–	–	–	+	+	–
Fusion-protein isolation[e]	–	–	–	+	–	–
Antibody production	–	–	–	–	–	–

+ implies that a particular step has been accomplished for that gene; – implies that this step has not yet been accomplished.

[a]This is the as yet unnamed gene defined by the JPT175 mutant (see text).

[b]Libraries of wild-type yeast DNA in "shuttle vectors" (plasmids capable of replicating, and being selected for, both in yeast and in *E. coli*) are screened for sequences capable of complementing *ts* mutations in the genes of interest. Instability of the transformed phenotype during growth under nonselective conditions confirms that the transforming information is indeed plasmid-borne. Plasmid DNA is then recovered from the yeast transformant into *E. coli*; efficient retransformation of yeast confirms that the desired plasmid has been obtained.

[c]Plasmid purified from *E. coli* is used for restriction mapping and localization of transcribed regions by Northern blotting and/or analysis of R loops. The transcribed region responsible for the complementing activity is identified by subcloning and/or gene-disruption experiments (Botstein and Davis 1982). These steps are crucial because of the high density of transcribed regions in yeast (typically one per 2.5 kb of genomic DNA). Shown for each gene is the size of the smallest fragment obtained so far that retains complementing activity and whether or not the relevant transcribed region has been localized.

[d]Evidence that the plasmid that complements a particular *cdc* mutation actually contains that *CDC* gene is obtained by showing that the plasmid integrates (by homologous recombination) at the proper map position (I), by gene-disruption experiments (D), or by marker-rescue experiments (M) (Botstein and Davis 1982). This step is crucial because sometimes genes other than that harboring the original mutation can complement that mutation when plasmid-borne (particularly when high-copy-number plasmids are used). For example, a plasmid-borne *CDC12* gene can complement *cdc11* mutations as well as *cdc12* mutations (S. Lillie et al., in prep.).

[e]Of the many possible methods for proceeding from a cloned, identified gene to a suitable immunogen, we have been working so far primarily with the open-reading-frame vectors of Weinstock et al. (1983; see text and Fig. 5).

that at least many of the gene products we are studying will be found to localize to the nucleus (Hall et al. 1984), the cell membrane, or one of the known cytoskeletal systems. Any of these results would be suggestive as to possible functions and help to guide the next stage of experimental investigation. It should also be noted that nucleotide sequence information about the genes of interest may reveal homologies with known proteins that would provide further guidance for subsequent experimentation (see, e.g., Lorincz and Reed 1984).

Finally, it should be noted that the molecular analyses also can provide information about interactions among the genes of interest (e.g., the interaction between *CDC12* and *CDC11* revealed by the *cdc11*-complementing activity of *CDC12*-carrying plasmids; Table 2, footnote d). Such information should also contribute to our eventual understanding of how the genes we are studying function and cooperate to help produce appropriate morphogenetic behavior during the cell cycle.

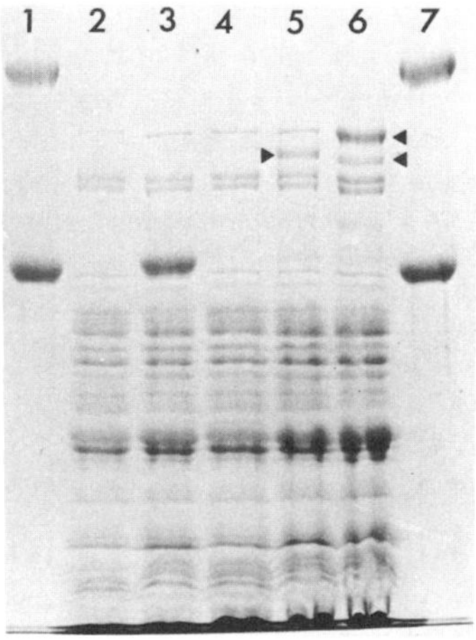

Figure 5
Proteins produced by *E. coli* strains with and without plasmids encoding *CDC12-lacZ* fusion proteins. Total cellular proteins were separated by SDS-polyacrylamide gel electrophoresis and stained with Coomassie blue. (*1,7*) Molecular-weight standards: myosin (205,000) and β-galactosidase (116,000). (*2,4*) Proteins from strain TK1046 harboring (*4*) or not harboring (*2*) plasmid pORF1 (Weinstock et al. 1983). This plasmid encodes an out-of-frame fusion between the 5′ terminus of the *ompF* gene (which encodes a major outer-membrane protein) and all but the 5′ terminus of the *lacZ* gene (which encodes β-galactosidase). A deletion or insertion that puts the *lacZ*-gene fragment into frame with the *ompF*-gene fragment leads to synthesis of a fusion protein possessing β-galactosidase activity. (*3*) Proteins from strain TK1046 harboring a plasmid in which *ompF* and *lacZ* have been put into frame by making a small deletion in the polylinker that connects them in pORF1. A fusion protein with a molecular weight slightly greater than that of normal β-galactosidase is produced, as expected. (*5,6*) Proteins from strain TK1046 harboring plasmids encoding *ompF-CDC12-lacZ* "tribrid" fusion proteins (indicated by the arrowheads). The lower-molecular-weight polypeptide in lane *6* is probably a degradation product of the higher-molecular-weight polypeptide.

SUMMARY AND CONCLUSIONS

It seems virtually certain that microtubules, the actomyosin system, and the 10-nm filaments of the neck region are involved in the morphogenetic processes of the cell cycle. Thus, direct study of these cytoskeletal elements by morphological, biochemical, and genetic methods should contribute to an understanding of these processes. However, it seems likely that these known cytoskeletal elements function as parts of complex systems, not all components of which may be easily revealed by direct study of the known components. In addition, it is possible that the morphogenetic processes also depend on more novel mechanisms. Thus, it seems crucial also to study the processes of interest using an approach that involves no assumptions as to what the important mechanisms are. Fortunately, the techniques necessary to make such an approach practical have been developing rapidly during the past few years. Although development of a detailed molecular understanding of morphogenesis in yeast will doubtless require much hard work and ingenuity, it at least appears to be a realistic goal.

ACKNOWLEDGMENTS

Other past and present members of our laboratory have also made significant contributions to the developing story described here. Barbara Sloat conducted the detailed characterization of *cdc24* mutants. Rich Longnecker, Almuth Tschunko, and Anne Stapleton have played major roles in the screening for *ts* and *cs cdc* mutants. Rob Preston, Paul Shiels (Trinity College, Dublin), Laird Bloom, and Paul Oeller have been involved in the pseudorevertant analyses.

Mike Kelley and Rich Longnecker assisted in the early characterization of plasmids complementing *cdc* mutants. We thank John Kilmartin, Breck Byers, T. Wieland, and many other colleagues for their gifts of materials and advice, and David Botstein and his group for sharing results with us prior to publication. Our work has been supported by grants from the U.S. Public Health Service (GM-31006) and the National Science Foundation (PCM-78-25607), a Biomedical Research Support Grant to The University of Michigan, a USPHS postdoctoral fellowship to C.J., USPHS predoctoral training grant support to K.C. and B.H., fellowships to A.A. from The Rackham Graduate School and the Cancer Research Committee at The University of Michigan, and travel assistance to J.R. from the American-Irish Foundation.

REFERENCES

Adams, A.E.M. and J.R. Pringle. 1984. Relationship of actin and tubulin distribution to bud growth in wild-type and morphogenetic-mutant *Saccharomyces cerevisiae. J. Cell Biol.* **98:** 934.

Botstein, D. and R.W. Davis. 1982. Principles and practice of recombinant DNA research with yeast. In *The molecular biology of the yeast* Saccharomyces: *Metabolism and gene expression* (ed. J.N. Strathern et al.), p. 607. Cold Spring Harbor Laboratory, Cold Spring Harbor, New York.

Botstein, D. and R. Maurer. 1982. Genetic approaches to the analysis of microbial development. *Annu. Rev. Genet.* **16:** 61.

Byers, B. 1981. Cytology of the yeast life cycle. In *The molecular biology of the yeast* Saccharomyces: *Life cycle and inheritance* (ed. J.N. Strathern et al.), p. 59. Cold Spring Harbor Laboratory, Cold Spring Harbor, New York.

Byers, B. and L. Goetsch. 1974. Duplication of spindle plaques and integration of the yeast cell cycle. *Cold Spring Harbor Symp. Quant. Biol.* **38:** 123.

———. 1975. Behavior of spindles and spindle plaques in the cell cycle and conjugation of *Saccharomyces cerevisiae. J. Bacteriol.* **124:** 511.

———. 1976a. A highly ordered ring of membrane-associated filaments in budding yeast. *J. Cell Biol.* **69:** 717.

———. 1976b. Loss of the filamentous ring in cytokinesis-defective mutants of budding yeast. *J. Cell Biol.* **70:** 35a.

Culotti, J. and L.H. Hartwell. 1971. Genetic control of the cell division cycle in yeast. III. Seven genes controlling nuclear division. *Exp. Cell Res.* **67:** 389.

Gallwitz, D. 1982. Construction of a yeast actin gene intron deletion mutant that is defective in splicing and leads to the accumulation of precursor RNA in transformed yeast cells. *Proc. Natl. Acad. Sci.* **79:** 3493.

Greer, C. and R. Schekman. 1982. Actin from *Saccharomyces cerevisiae. Mol. Cell. Biol.* **2:** 1270.

Hall, M.N., L. Hereford, and I. Herskowitz. 1984. Targeting of *E. coli* β-galactosidase to the nucleus in yeast. *Cell* **36:** 1057.

Hartwell, L.H., R.K. Mortimer, J. Culotti, and M. Culotti. 1973. Genetic control of the cell division cycle in yeast. V. Genetic analysis of *cdc* mutants. *Genetics* **74:** 267.

Hereford, L.M. and M. Rosbash. 1977. Number and distribution of polyadenylated RNA sequences in yeast. *Cell* **10:** 453.

Hoebeke, J., G. van Nijen, and M. De Brabander. 1976. Interaction of nocodazole (R17934), a new antitumoral drug, with rat brain tubulin. *Biochem. Biophys. Res. Commun.* **69:** 319.

Kaback, D.B., P.W. Oeller, H.Y. Steensma, J. Hirschman, D. Ruezinsky, K.G. Coleman, and J.R. Pringle. 1984. Temperature-sensitive lethal mutations on yeast chromosome I appear to define only a small number of genes. *Genetics* **108:** 67.

Kilmartin, J.V. 1981. Purification of yeast tubulin by self-assembly in vitro. *Biochemistry* **20:** 3629.

Kilmartin, J.V. and A.E.M. Adams. 1984. Structural rearrangements of tubulin and actin during the cell cycle of the yeast *Saccharomyces. J. Cell Biol.* **98:** 922.

Kilmartin, J.V., B. Wright, and C. Milstein. 1982. Rat monoclonal antitubulin antibodies derived by using a new nonsecreting rat cell line. *J. Cell Biol.* **93:** 576.

King, S.M. and J.S. Hyams. 1982. The mitotic spindle of *Saccharomyces cerevisiae*: Assembly, structure and function. *Micron* **13:** 93.

Lorincz, A.T. and S.I. Reed. 1984. Primary structure homology between the product of yeast cell division control gene *CDC28* and vertebrate oncogenes. *Nature* **307:** 183.

Luck, D.J.L., B. Huang, and G. Piperno. 1982. Genetic and biochemical analysis of the eukaryotic flagellum. *Symp. Soc. Exp. Biol.* **35:** 399.

Moir, D., S.E. Stewart, B.C. Osmond, and D. Botstein. 1982. Cold-sensitive cell-division-cycle mutants of yeast: Isolation, properties and pseudoreversion studies. *Genetics* **100:** 547.

Neff, N.F., J.H. Thomas, P. Grisafi, and D. Botstein. 1983. Isolation of the β-tubulin gene from yeast and demonstration of its essential function *in vivo. Cell* **33:** 211.

Ng, R. and J. Abelson. 1980. Isolation and sequence of the gene for actin in *Saccharomyces cerevisiae. Proc. Natl. Acad. Sci.* **77:** 3912.

Oakley, B.R. and N.R. Morris. 1980. Nuclear movement is β-tubulin-dependent in *Aspergillus nidulans. Cell* **19:** 255.

Pringle, J.R. and L.H. Hartwell. 1981. The *Saccharomyces cerevisiae* cell cycle. In *The molecular biology of the yeast* Saccharomyces: *Life cycle and inheritance* (ed. J.N. Strathern et al.), p. 97. Cold Spring Harbor Laboratory, Cold Spring Harbor, New York.

Quinlan, R.A., C.I. Pogson, and K. Gull. 1980. The influence of the microtubule inhibitor, methyl benzimidazol-2-yl-carbamate (MBC) on nuclear division and the cell cycle in *Saccharomyces cerevisiae. J. Cell Sci.* **46:** 341.

Roberts, R.L., B. Bowers, M.L. Slater, and E. Cabib. 1983. Chitin synthesis and localization in cell division cycle mutants of *Saccharomyces cerevisiae. Mol. Cell. Biol.* **3:** 922.

Shortle, D., J.E. Haber, and D. Botstein. 1982. Lethal disruption of the yeast actin gene by integrative DNA transformation. *Science* **217:** 371.

Sloat, B.F., A. Adams, and J.R. Pringle. 1981. Roles of the *CDC24* gene product in cellular morphogenesis during the *Saccharomyces cerevisiae* cell cycle. *J. Cell Biol.* **89:** 395.

Stump, R.F., K.R. Robinson, R.L. Harold, and F.M. Harold. 1980. Endogenous electrical currents in the water mold *Blastocladiella emersonii* during growth and sporulation. *Proc. Natl. Acad. Sci.* **77:** 6673.

Tkacz, J.S. and J.O. Lampen. 1972. Wall replication in *Saccharomyces* species: Use of fluorescein-conjugated concanavalin A to reveal the site of mannan insertion. *J. Gen. Microbiol.* **72:** 243.

Water, R.D., J.R. Pringle, and L.J. Kleinsmith. 1980. Identification of an actin-like protein and of its messenger ribonucleic acid in *Saccharomyces cerevisiae. J. Bacteriol.* **144:** 1143.

Weinstock, G.M., C. ap Rhys, M.L. Berman, B. Hampar, D. Jackson, T.J. Silhavy, J. Weisemann, and M. Zweig. 1983. Open reading frame expression vectors: A general method for antigen production in *Escherichia coli* using protein fusions to β-galactosidase. *Proc. Natl. Acad. Sci.* **80:** 4432.

Wood, J.S. and L.H. Hartwell. 1982. A dependent pathway of gene functions leading to chromosome segregation in *Saccharomyces cerevisiae. J. Cell Biol.* **94:** 718.

Tubulins of *Aspergillus nidulans*: Genetics, Biochemistry, and Function

N. Ronald Morris, James A. Weatherbee, John Gambino, and Lawrence G. Bergen*

Department of Pharmacology
University of Medicine and Dentistry
of New Jersey–Rutgers Medical School
Piscataway, New Jersey 08854

It has become apparent that tubulin is more complex than originally believed. Whereas 10 years ago the degree to which tubulin sequences were conserved was a new and important finding (Luduena and Woodward 1975), the recent description of multiple structurally different α- and β-tubulin polypeptides in a variety of organisms has focused attention on tubulin heterogeneity. The finding and characterization of multiple tubulins and of mutations affecting structural genes for tubulins in *Aspergillus* (Sheir-Neiss et al. 1978; Morris et al. 1979), *Drosophila* (Kemphues et al. 1979), *Physarum* (Burland et al. 1984), *Saccharomyces* (Neff et al. 1983), *Schizosaccharomyces* (Umesono et al. 1983), and mammalian cells (Cabral et al. 1980; Cabral et al. 1981; Keates et al. 1981) have demonstrated that, frequently, multiple genes code for different α- and β-tubulin polypeptides. Cloned tubulin genes have also provided evidence for multiple transcriptionally active copies of α- and β-tubulins in many organisms (see Cleveland 1983). In other cases, however, heterogeneity has been shown to represent posttranslational modification of tubulin proteins (Brown et al. 1982; Brunke et al. 1982; Lefebvre et al. 1980; L'Hernault and Rosenbaum 1983).

Mictotubules serve a variety of functions, which mostly concern cell motility, cell shape, and organellar movement. In most organisms, a number of different types of microtubules can be distinguished. In *Aspergillus nidulans*, there are two types of spindle microtubules (polar and kinetochore microtubules) and two

*Present address: Department of Bacteriology, University of Wisconsin-Madison, Madison, Wisconsin 53706.

types of cytoplasmic microtubules (interphase network and mitotic astral microtubules). In other organisms, various other types of microtubules (ciliary, flagellar, subcortical, etc.) are seen. The parallel between the complexity of structurally different tubulins and functionally different microtubules leads to the following questions: What is the relationship between tubulins and microtubules? Do functionally different microtubules contain different tubulins, as originally suggested by the multitubulin hypothesis (Fulton and Simpson 1976)? If so, are the tubulins in different classes of microtubules totally different or are some tubulins common to more than one type of microtubule? Does the tubulin composition of a microtubule specify the binding of its particular microtubule-associated proteins? Does tubulin composition affect microtubule polymerization or stability, and how, exactly, is tubulin composition related to microtubule function?

Our laboratory has undertaken a molecular genetic analysis of tubulin as related to microtubule structure and function in the fungus *A. nidulans*. The attributes that make *Aspergillus* particularly useful for this purpose are its powerful genetic system and the fact that microtubule behavior during interphase and mitosis is similar to the behavior of microtubules in cells of higher organisms. Other useful attributes of *Aspergillus* are its ability to grow in simple defined media and to reach a biomass greater than 10 g/liter, making it a suitable organism for biochemical investigation.

Aspergillus has four genes for tubulin, two for α-tubulin and two for β-tubulin. The *tubA* gene codes for two α-tubulin polypeptides, α_1- and α_3-tubulin (Morris et al. 1979). A second, as yet unidentified, α-tubulin gene codes for a third polypeptide, α_2-tubulin. Similarly, the *benA* gene codes for two β-tubulin polypeptides, β_1- and β_2-tubulin (Sheir-Neiss et al. 1978), and in this paper, we demonstrate the existence of a third β-tubulin peptide, β_3, which is a product of another, as yet unidentified, gene for β-tubulin.

benA- and *tubA*-gene products are involved in both mitotic spindle function and nuclear movement in *Aspergillus*. We have shown that spindle microtubules fail to disassemble in strains carrying the temperature-sensitive mutation, *benA33*, which causes a block in mitosis at restrictive temperature (Oakley and Morris 1981). Since the *benA33* mutation is in the structural gene that codes for the β_1/β_2-tubulin polypeptides, this constitutes direct evidence that spindle microtubule stability is increased by the *benA* mutation and that at least one of the two *benA*-gene products is a spindle microtubule component. Additional evidence that the *benA33* mutation causes hyperstabilization of microtubules is that strains carrying this mutation are multiply resistant to unrelated antimicrotubule drugs, suggesting that the increased microtubule stability makes it more difficult for these drugs to cause microtubule depolymerization (Oakley and Morris 1981). It should be noted that this mutationally caused hyperstabilization of microtubules defines a new mechanism of drug resistance, which also occurs in mammalian cells (Cabral et al. 1980).

Another effect of the *benA33* mutation is that it causes an anaphase block: Chromosomes do not move to the spindle poles in *benA33* strains blocked at restrictive temperature (Oakley and Morris 1981). Because the mitotic block is relieved by antimicrotubule drugs that are known to destabilize microtubules, we believe that depolymerization of spindle microtubules is required for chromosome movement in *Aspergillus*. Nuclear migration is also inhibited by the *benA33* mutation at restrictive temperature. Since we have shown that *benA* β-tubulins are involved in nuclear movement in *Aspergillus* (see Oakley and Morris 1980), the failure of nuclei to move under restrictive conditions suggests that the cytoplasmic microtubules might also be hyperstable in *benA33* mutants and that this hyperstability might be responsible for the inhibition of nuclear migration.

Two α-tubulin mutations, *tubA1* and *tubA4*, suppress the temperature sensitivity of *benA33*. We have hypothesized that these mutations suppress *benA33* by causing microtubules to be less stable than normal. In support of this hypothesis is the observation that strains containing the *tubA* mutations are hypersensitive to structurally unrelated microtubule inhibitors (Oakley and Morris 1981), which suggests that the lower intrinsic stability of *tubA* microtubules makes them more easily depolymerizable by these drugs. However, no direct evidence has been presented to demonstrate that these mutations actually destabilize microtubules. In this paper, we show by direct visualization, using anti-tubulin fluorescence immunocytochemistry, that the *benA33* mutation causes both spindle and cytoplasmic microtubules to be hyperstable and that the *tubA1* and *tubA4* mutations destabilize spindle and cytoplasmic microtubules. These data demonstrate that both *benA*- and *tubA*-gene products must be components of both spindle and cytoplasmic microtubules.

It is technically difficult to visualize microtubules by anti-tubulin immunofluorescence in intact hyphal germlings of *A. nidulans* because of the cell wall. However, it has been possible to visualize both spindle and cytoplasmic microtubules in protoplasts (Gambino et al. 1984). Conditions have been developed that generate protoplasts that grow and undergo nuclear division at the same logarithmic rate as walled germlings. This indicates that the protoplasts are essentially physiologically normal. The purpose of this paper is twofold: to present definitive evidence for multiple tubulins and multiple tubulin genes in *Aspergillus* and to review the behavior of spindle and cytoplasmic microtubules during the cell cycle in wild-type and mutant protoplasts.

RESULTS

Aspergillus Has Two β-Tubulin Genes

The tubulins of *A. nidulans* were initially characterized by copolymerization with brain tubulin, by two-dimensional gel electrophoresis, and by peptide mapping (Sheir-Neiss et al. 1978). We have now extended this characterization by purifying the *Aspergillus* tubulins and using specific monoclonal antibodies to identify the α- and β-tubulins. One advantage of the initial purification is that it allows us to load larger amounts of tubulin for two-dimensional gel electrophoresis and therefore to detect minor tubulin species more easily. We have also used mutationally induced electrophoretic alterations in β-tubulin to find a minor tubulin species, β_3, that is ordinarily occluded by the wild-type β_1/β_2-tubulin polypeptides.

The concentration of tubulin in *Aspergillus* is too low to allow purification by the conventional method of microtubule polymerization/depolymerization. Therefore, *Aspergillus* tubulin was purified on DEAE-cellulose followed by further fractionation by ammonium sulfate precipitation. Each step of the purification procedure was optimized by testing dot blots on nitrocellulose with anti-tubulin antibodies to make certain that there were no major losses at any step. The initial extraction was tested for completeness in the same way. On two-dimensional gel electrophoretograms of purified wild-type tubulin, three protein spots, previously designated α_1, α_2, and α_3, could be seen in the α-tubulin region and two spots, β_1 and β_2, could be seen in the β-tubulin region (Fig. 1A).

All of the tubulins visible on two-dimensional gels of wild-type tubulin preparations reacted on Western blots with affinity-purified polyclonal antibody raised

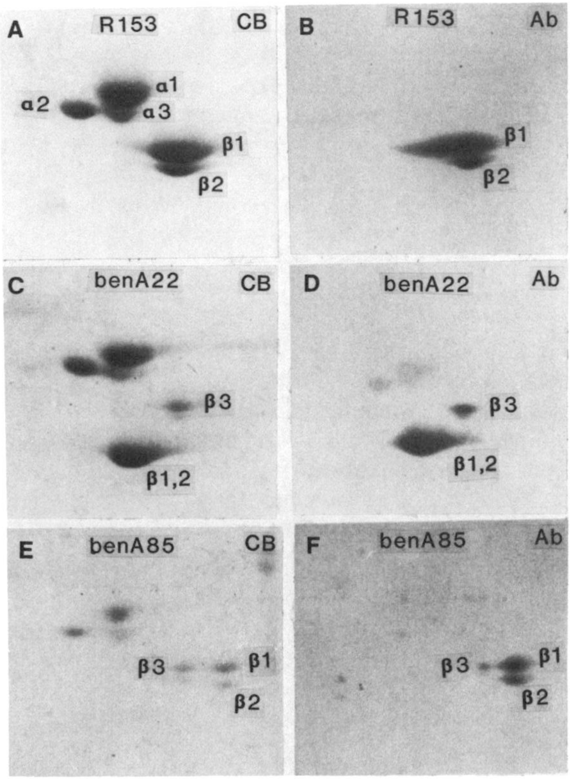

Figure 1
Demonstration of the presence of β_3 on two-dimensional gels. (*A*) Coomassie-blue-stained (CB) two-dimensional gel of a tubulin-enriched fraction from R153 (containing wild-type tubulins). (*B*) Western blot of a gel similar to *A* reacted with a polyclonal, affinity-purified anti-tubulin antibody (Ab). (*C*) Coomassie-blue-stained (CB) gel of a tubulin-enriched fraction from BEN 20 (*benA22*). (*D*) Western blot of a gel similar to *C* reacted with the anti-tubulin antibody (Ab). (*E*) Coomassie-blue-stained (CB) gel of a tubulin-enriched fraction from BEN 19R7 (*benA85*). (*F*) Western blot of a gel similar to *E* reacted with the anti-tubulin antibody (Ab). It can be seen that the major β-tubulins (β_1 and β_2) of BEN 20 show a 2+ shift and an apparent decrease in molecular weight and are no longer distinguishable as a doublet. In BEN 19R7, these tubulins show only a 2− shift. For both mutants, it can be seen that the shifts of the β_1- and β_2-tubulins uncover a spot that has the proper molecular weight and isoelectric point for a β-tubulin and reacts with the anti-tubulin antibody (labeled β_3).

against porcine brain tubulin (generously provided by D. Cleveland, Johns Hopkins Medical School). Specific monoclonal antibodies against α-tubulin (generously provided by J. Kilmartin, Medical Research Council, England) or β-tubulin (generously provided by S. Blose, Cold Spring Harbor Laboratory) were used to characterize the tubulins more closely. The proteins labeled in Figure 1 as α-tubulins all stained with anti-α-tubulin antibody but not with anti-β-tubulin antibody, and, conversely, the proteins labeled as β-tubulins stained with anti-β-tubulin antibody, but not with anti-α-tubulin antibody.

Wild-type strains of *Aspergillus* appear to contain only two β-tubulin spots, β_1 and β_2. However, in some *benA* mutants with electrophoretically altered β-tubulins (in which the β_1/β_2-tubulins are shifted from their normal positions), a minor protein spot can be seen (Sheir-Neiss et al. 1978). The availability of a monoclonal antibody against β-tubulin allowed us to determine whether this protein was also a β-tubulin. Tubulin was purified from two different *benA* mutants, *benA22*, which exhibits a 2+ shift of the β_1/β_2-tubulins, and *benA85*, which exhibits a 2− shift of the β_1/β_2-tubulins. In both cases, when these purified tubulins were subjected to two-dimensional gel electrophoresis, a spot was again seen underlying the original position of the β_1/β_2-tubulins (Fig. 1A,C,E). Thus, this protein behaved like tubulin through the purification procedure. On Western blots, the spot stained with polyclonal anti-tubulin antibody (Fig. 1B,D,F), but because of its very low abundance it did not stain with monoclonal antibody against either α-tubulin or β-tubulin. A faint reaction with anti-β-tubulin antibody could be seen on one-dimensional gels of purified *benA22* tubulins (data not shown).

To demonstrate more definitively a reaction of this spot (β_3-tubulin) with anti-β-tubulin monoclonal antibody, it was necessary to purify it away from β_1/β_2-tubulins, which reacted very strongly with the antibody. This was accomplished by cutting and pooling the β_3 bands (identified by briefly staining with Coomassie blue) from several lanes of a one-dimensional SDS gel, electrophoresing the material on a second SDS gel, and staining transfers from this gel to nitrocellulose with the β-tubulin-specific monoclonal antibody. Under these conditions, it was possible to demonstrate clearly that β_3 reacts with the β-tubulin-specific antibody (Fig. 2). Additional evidence that β_3 is a β-tubulin was provided by one-dimensional peptide mapping, which showed that the β_3 peptides are similar to total β-tubulin peptides (β_1, β_2, and β_3) and different from α-tubulin peptides (data not shown).

Aspergillus Has Two α-Tubulin Genes

We have also used tubulin purification and monoclonal antibodies to verify earlier results from this laboratory which suggested that there was more than one gene for α-tubulin in *Aspergillus* (Morris et al. 1979). Purified tubulin from a wild-type strain and from another strain, BEN 9R7, which contains an electrophoretically silent mutation in the *benA* gene and a second mutation in the *tubA* gene, was subjected to two-dimensional gel electrophoresis followed by Western blotting of the gel with α-tubulin-specific monoclonal antibody (Fig. 3). The α_1- and α_3-tubulins showed a coordinate 1+ shift in this mutant, indicating that these proteins are both products of the *tubA* gene. The α_2-tubulin retained the same position that it occupies in the wild type; therefore, it must be the product of another, as yet unidentified, gene for α-tubulin. The fact that all three α-tubulins purify as

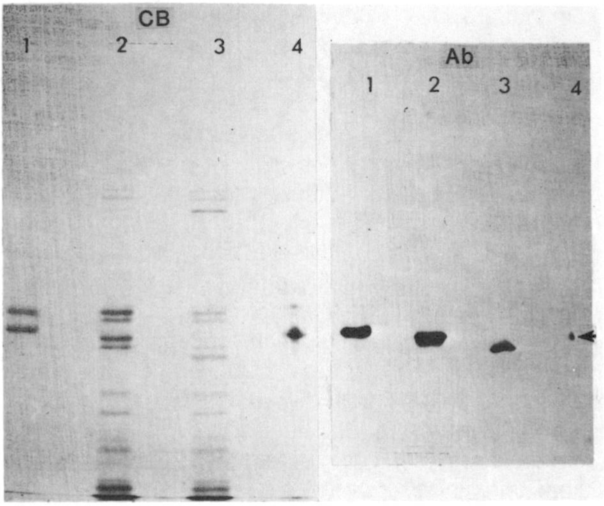

Figure 2
Demonstration of the reaction of a β-tubulin-specific monoclonal antibody (DM1B) with the β_3-tubulin of BEN 20. To demonstrate a reaction of the β_3-tubulin of BEN 20 with the β-tubulin-specific monoclonal antibody, it was necessary to separate this species from the major β-tubulins (see text). This was accomplished by cutting β_3 bands from several gels, reequilibrating the bands with sample buffer, and running on fresh gels. (*Left*) Coomassie-blue-stained gel (CB). (Lane *1*) Porcine brain tubulin (1 μg); (lane *2*) tubulin-enriched preparation from R153 (10 μg); (lane *4*) β_3-tubulin from BEN 20, cut from three gels and rerun. (*Right*) Western blot of a similar gel reacted with the monoclonal antibody. Under these conditions, it is possible to demonstrate clearly that the antibody does react with β_3-tubulin (arrow).

tubulins and stain with monoclonal antibody against α-tubulin confirms the previously tentative identification of these proteins.

benA- and *tubA*-gene Products Are Required for Both Mitotic Spindle and Cytoplasmic Microtubules

Previous experiments had indicated that the temperature-sensitive *benA33* β-tubulin mutation causes spindle microtubules to be hyperstable at restrictive temperature and, in addition, suggested that cytoplasmic microtubules might be similarly affected by this mutation. Other experiments suggested that the *tubA1* and *tubA4* α-tubulin mutations destabilize spindle and cytoplasmic microtubules. Staining of microtubules in fixed *Aspergillus* protoplasts with anti-tubulin antibodies has now provided direct evidence in support of these postulated effects of the *benA* and *tubA* mutations on microtubule stability.

Aspergillus protoplasts were prepared by enzymatically removing the cell wall with Novozyme 234 in nutrient medium osmotically buffered with magnesium sulfate (Fig. 4A). These protoplasts appear to be physiologically normal, as evidenced by the fact that nuclear division proceeds at the same rate in these protoplasts as in the walled hyphal germlings from which they are derived (Gambino

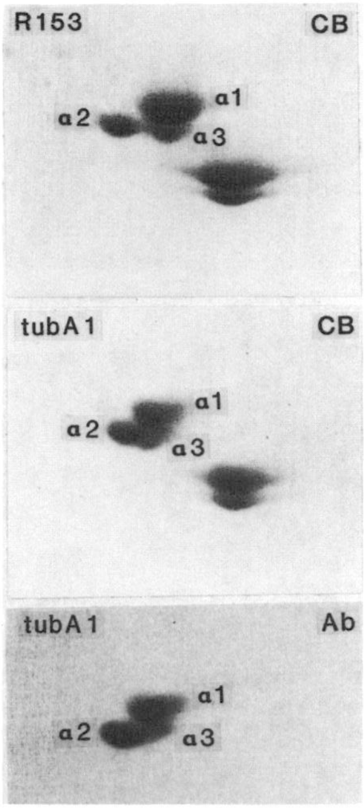

Figure 3
Evidence for multiple α-tubulin genes in *Aspergillus*. (*Top*) Coomassie-blue-stained (CB) gel of the tubulins of R153; (*middle*) Coomassie-blue-stained (CB) gel of the tubulins of BEN 9R7 (*tubA1*); (*bottom*) Western blot of a similar preparation of BEN 9R7 tubulins reacted with an α-tubulin-specific monoclonal antibody (Ab) (YOL1/34). As shown, the α-tubulins designated α_1 and α_3 show a coordinated 1+ shift in this mutant, indicating that these two species are products of the *tubA* gene and that α_2-tubulin is the product of another α-tubulin gene.

et al. 1984). Nuclear division, which is synchronous in germlings, is also synchronous in the protoplasts.

Protoplast microtubules stain well with anti-tubulin antibodies. In the wild-type protoplasts, a complex cytoplasmic microtubule network can be seen during interphase (Fig. 4B). At mitosis, when the mitotic spindle appears, the cytoplasmic microtubule network disappears, except for a few astral microtubules that emanate from the spindle poles (Fig. 4C). Thus, the observed behavior of the microtubules at mitosis is similar to that seen in animal cells, despite the fact that *Aspergillus* has a closed mitosis (i.e., the nuclear membrane does not break down during mitosis). In contrast to the wild type, the cytoplasmic microtubules of *benA33* mutants persist during mitosis, providing direct evidence that both the spindle microtubules and the cytoplasmic microtubules are hyperstabilized by this mutation (Fig. 5A). In *tubA1* and *tubA4* mutants, however, neither cytoplasmic nor spindle microtubules are seen when protoplasts are stained with anti-tubulin antibody (Fig. 5B). There are several possible explanations for our inability to observe microtubules in these *tubA* mutants. One possibility is that there are no microtubules in these mutants. Another is that there are microtubules, but they are less stable than wild-type microtubules and are depolymerized during the fixation process. A third possibility is that both the *tubA1* and *tubA4* mutations alter the epitope recognized by the anti-α-tubulin monoclonal antibody such that the antibody no longer reacts with the mutant α-tubulin.

To control for the possibility that these mutations altered a specific epitope, we examined protoplasts from doubly mutant strains carrying both *tubA1* or *tubA4* and the *benA33* mutation, which is known to suppress the hypersensitivity of these mutations to benomyl and therefore might be expected to suppress *tubA*-

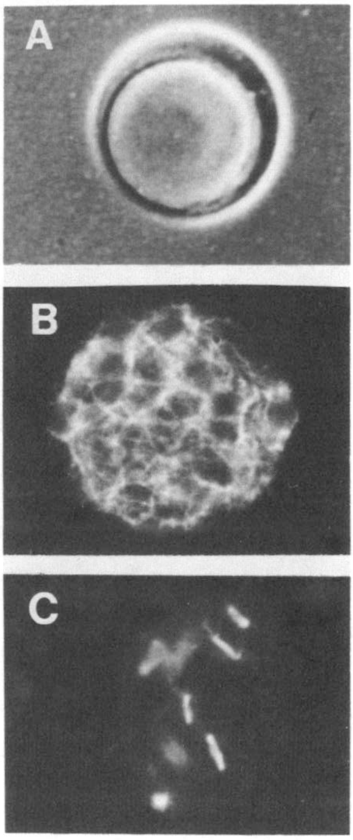

Figure 4
Visualization of microtubules in wild-type protoplasts. Protoplasts were allowed to grow for several cell cycles before fixation and immunofluorescent staining of tubulin. *(A)* Phase-contrast micrograph of a protoplast; (*B*) anti-tubulin staining pattern of an interphase cell; (*C*) anti-tubulin staining pattern of a mitotic cell. Note that there are several mitotic spindles in the mitotic cell and that there are very few astral cytoplasmic microtubules. Magnification, 6000×.

induced microtubule instability. In the two *benA, tubA* double mutants that we examined, staining of cytoplasmic microtubules was again apparent; therefore, the α-tubulin epitope is not altered by the *tubA* mutations (Fig. 5C,D).

We next asked whether there was any nuclear division occurring in *tubA1* and *tubA4* protoplasts. This was done by determining the rate of increase in nuclear number over time and also by determining the mitotic index of DAPI-stained protoplasts. These experiments showed clearly that nuclear division was occurring in these strains at a rate not very different from the wild type. Since a mitotic spindle is a sine qua non for mitosis, the best explanation for our failure to observe spindle microtubules in *tubA1* and *tubA4* protoplasts is that spindles are unstable to fixation in these mutant strains. It would seem likely that the apparent absence of cytoplasmic microtubules in these protoplasts has the same explanation. These data provide direct evidence that the *tubA* mutations destabilize microtubules as previously hypothesized.

DISCUSSION

The data presented in this paper demonstrate that *Aspergillus* has four tubulin genes, *benA, tubA*, and two other unnamed genes, which among them code for six tubulin polypeptides. Mutations in *tubA* and *benA* cause electrophoretic alterations in two α- and two β-tubulin polypeptides, respectively, but do not affect a third α- or β-tubulin polypeptide. Thus, the third α- and β-tubulin polypeptides are the products of two other tubulin genes, in which we have not yet found any

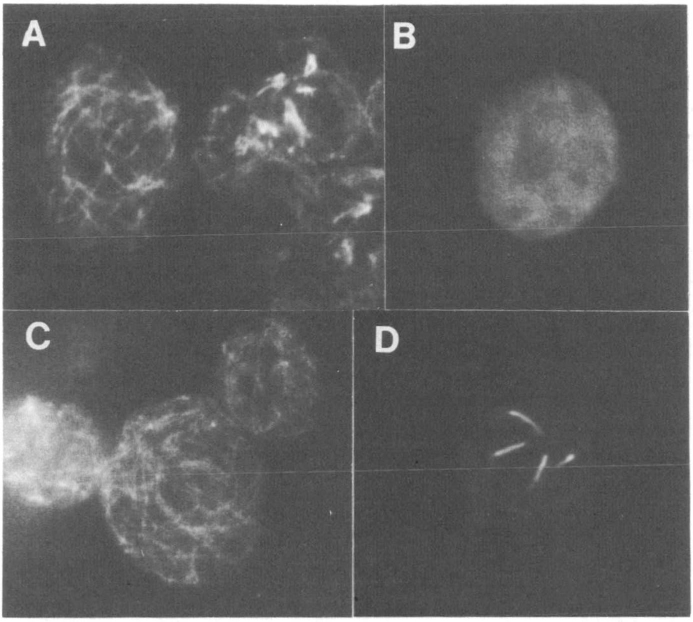

Figure 5
Microtubule patterns of tubulin mutants. (*A–D*) Immunofluorescent staining patterns of microtubules in protoplasts prepared from strains containing mutant tubulins. (*A*) *benA33*, containing hyperstable microtubules as the result of a mutation in the gene coding for its β_1- and β_2-tubulins, shows complex cytoplasmic microtubule patterns in both interphase and mitotic cells. (*B*) Microtubules of *tubA4*, which are hypostable as the result of a mutation in the gene coding for its α_1- and α_3-tubulins, are not preserved during fixation. (*C*) Interphase protoplasts of the double mutant containing both *benA33* and *tubA4*. (*D*) Mitotic protoplasts of this double mutant. Both interphase and mitotic microtubule patterns are wild type in appearance.

mutations. We have shown that mutations in *benA* and *tubA* can affect microtubule stability, and we have used these effects on microtubule stability to assign functions to the gene products.

The data demonstrate that the *benA*- and *tubA*-gene products are involved in both spindle and cytoplasmic microtubule function. However, since each of these genes codes for two polypeptides, we still do not know which polypeptide or polypeptides are found in which microtubules. It is possible that β_1 is found only in spindle microtubules and β_2 only in cytoplasmic microtubules (or vice versa), or even that both β_1 and β_2 are found in spindle and cytoplasmic microtubules. The same is true for α_1 and α_3.

To clarify further the function of the *benA*- and *tubA*-gene products, it will be necessary to prepare spindles and cytoskeletons from protoplasts to analyze the composition of spindle and cytoplasmic microtubules. The availability of the temperature-sensitive mutant *bimE7* (Morris 1976a), which at restrictive temperature becomes blocked in mitosis with a stable spindle and no cytoplasmic microtubules (Morris 1976b) and which exhibits the same block in protoplasts (Gambino et al. 1984), should allow the preparation and analysis of spindle microtubules, whereas cytoplasmic microtubules can be prepared from protoplasts of wild-type interphase cells. By using a protocol similar to that used by Solomon et al. (1979) to analyze the microtubules of mammalian cells, it should be possible not only to determine the tubulin composition of particular classes of microtubules, but also to identify idiotypic microtubule-associated proteins.

What is the function of the other two tubulin gene products, the α_2- and β_3-tubulins? There are no known mutations in structural genes for these proteins to help us determine function; however, there are some mutations, which are potential candidates for these structural genes, that have not yet been evaluated. Since *benA* and *tubA* code for known tubulins, it is possible that other genes that affect benomyl resistance code for the unknown tubulins. For example, in *Physarum polycephalum*, mutations in several different tubulin genes cause resistance to methyl benzimidazole-2-yl-carbamate, the active break-down product of benomyl (Burland et al. 1984; see also this volume). In *Aspergillus*, there are three known genes for benomyl resistance, *benA, benB,* and *benC*. It is possible that *benB* or *benC* may code for α_2 or β_3, and we are in the process of evaluating this possibility.

Another possibility that we have considered is that these proteins may be involved in some special aspect of the *Aspergillus* life cycle, e.g., asexual spore formation. Preliminary evidence suggests that this guess may be correct. Certain *Aspergillus* strains selected for resistance to benomyl were able to grow in the presence of benomyl but were unable to produce conidia (asexual spores) in the presence of the drug. From one of these strains, we selected conidiation-resistant mutants that were able to conidiate in the presence of benomyl. The mutants were generated in a *benA22* background, in which the β_1- and β_2-tubulins are electrophoretically shifted 2+ to reveal the β_3-tubulin polypeptides, so that electrophoretic abnormalities in this protein could be detected. Tubulin was purified from these mutants and subjected to two-dimensional gel electrophoresis. Of 21 such mutants examined in this way, all had lost the β_3-tubulin polypeptide and acquired a new protein spot.

Remarkably, in all 21 cases, the new protein spot was in the same location on the gel. We have not yet demonstrated that the new spot is tubulin; but, if it is, the fact that all 21 conidiation-resistant mutants exhibit the same electrophoretic shift would be highly unusual. Mutations in a structural gene for β_3-tubulin would be expected to cause amino acid substitutions such that the new spots would exhibit a variety of shifts of the magnitude of one or two plus or minus charges (see, e.g., Sheir-Neiss et al. 1978). The fact that all of the β_3-tubulin shifts are identical suggests that all of the mutations to conidiation resistance affect the same process, presumably posttranslational modification of β_3 tubulin.

These data, although incomplete, suggest strongly that β_3-tubulin functions during spore formation. It is not difficult to see how the loss of the normal, wild-type form of β_3-tubulin might cause conidiation to be resistant to benomyl. It is necessary to postulate either that β_3 and one or both of the *benA* tubulins are normally involved in sporulation; that benomyl sensitivity is dominant and is normally conferred by wild-type β_3 and that if β_3-tubulin is defective, the resistance of the *benA22* mutation is expressed; or that sporulation is normally mediated only by β_3-tubulin and that the *benA* tubulin can, if β_3-tubulin is defective, assume the function of the β_3-tubulin, thereby conferring benomyl resistance upon sporulation. Since the conidial resistance mutations were selected in a strain carrying the *benA22* mutation, a plausible explanation for the development of conidial resistance is that the loss of the normal benomyl-sensitive β_3-tubulin allowed the *benA22* β-tubulins to take over and to confer benomyl resistance on conidiation.

The results presented here, as well as in many of the other papers in this volume, demonstrate the usefulness of the molecular genetic approach for the analysis of cytoskeletal function. The special utility of genetics per se is that it allows the investigator to test the postulated functions of specific cytoskeletal proteins in the living cell. *A. nidulans*, because of the fact that its spindle and cytoplasmic microtubules behave during the mitotic cycle essentially like those of higher eukaryotes, because of its powerful genetics, and because its biology is fascinating, is extremely well suited to this type of analysis. Particularly encouraging for the future development of the system is that fact that *Aspergillus* has recently been added to the list of transformable organisms (Balance et al. 1983). The addition of recombinant DNA technology to our combined genetic, biochemical, and morphological approach should greatly facilitate the elucidation of microtubule function in this organism.

ACKNOWLEDGMENT

This work was supported by National Institutes of Health grants GM-29228-03 and PCM-8202690 to N.R.M. and GM-08709 to L.G.B.

REFERENCES

Balance, D.J., F.P. Buxton, and G. Turner. 1983. Transformation of *Aspergillus nidulans* by the orotidine-5′-phosphate decarboxylase gene of *Neurospora crassa*. *Biochem. Biophys. Res. Commun.* **112:** 284.

Brown, B.A., R.A. Nixon, and C.A. Marotta. 1982. Post-translational processing of α-tubulin during axoplasmic transport in CNS axons. *J. Cell Biol.* **94:** 159.

Brunke, K.J., P.S. Collis, and D.W. Weeks. 1982. Post-translational modification of tubulin dependent on organelle assembly. *Nature* **297:** 516.

Burland, T.G., T. Schedl, K. Gull, and W.F. Dove. 1984. Genetic analysis of resistance to benzimidazoles in *Physarum*: Differential expression of β-tubulin genes. *Genetics* (in press).

Cabral, F., I. Abraham, and M.M. Gottesman. 1981. Isolation of a taxol-resistant Chinese hamster ovary cell mutant that has an alteration in β-tubulin. *Proc. Natl. Acad. Sci.* **78:** 4388.

Cabral, F., M.E. Sobel, and M.M. Gottesman. 1980. CHO mutants resistant to colchicine, colcemid or griseofulvin have an altered β-tubulin. *Cell* **20:** 29.

Cleveland, D.W. 1983. The tubulins: From DNA to RNA and back again. *Cell* **34:** 330.

Fulton, C. and P.A. Simpson. 1976. Selective synthesis and utilization of flagellar tubulin. The multitubulin hypothesis. *Cold Spring Harbor Conf. Cell Proliferation* **3:** 987.

Gambino, J., L.G. Bergen, and N.R. Morris. 1984. The effects of mitotic and tubulin mutations on microtubule architecture in actively growing protoplasts of *Aspergillus nidulans*. *J. Cell Biol.* (in press).

Keates, R.A.B., F. Sarangi, and V. Ling. 1981. Structural and functional alterations in microtubule protein from Chinese hamster ovary cell mutants. *Proc. Natl. Acad. Sci.* **78:** 5638.

Kemphues, K., R.A. Raff, T.C. Kaufman, and E.C. Raff. 1979. Mutation in a structural gene for a β-tubulin specific to testis in *Drosophila melanogaster. Proc. Natl. Acad. Sci.* **76:** 3991.

Lefebvre, P.A., C.D. Silflow, E.D. Wieben, and J.L. Rosenbaum. 1980. Increased levels of mRNAs for tubulin and other flagellar proteins after amputations or shortening of *Chlamydomonas* flagella. *Cell* **20:** 469.

L'Hernault, S.W. and J.R. Rosenbaum. 1983. *Chlamydomonas* α-tubulin is posttranslationally modified in the flagella during flagellar assembly. *J. Cell Biol.* **97:** 258.

Luduena, R.F. and D.O. Woodward. 1975. α- and β-tubulin: Separation and partial sequence analysis. In *The biology of cytoplasmic microtubules* (ed. D. Soifer), p. 272. NY Academy of Sciences, New York.

Morris, N.R. 1976a. Mitotic mutants of *Aspergillus nidulans. Genet. Res.* **26:** 237.

———. 1976b. A temperature-sensitive mutant of *Aspergillus nidulans* reversibly blocked in nuclear division. *Exp. Cell Res.* **98:** 204.

Morris, N.R., M.H. Lai, and C.E. Oakley. 1979. Identification of a gene for α-tubulin in *Aspergillus nidulans. Cell* **16:** 437.

Oakley, B.R. and N.R. Morris. 1980. Nuclear movement is β-tubulin dependent in *Aspergillus nidulans. Cell* **19:** 255.

———. 1981. A β-tubulin mutation in *Aspergillus nidulans* that blocks microtubule function without blocking assembly. *Cell* **24:** 837.

Neff, N.F., J.H. Thomas, P. Grisafi, and D. Botstein. 1983. Isolation of the β-tubulin gene from yeast and demonstration of its essential function *in vivo. Cell* **33:** 211.

Sheir-Neiss, G., M.H. Lai, and N.R. Morris. 1978. Identification of a gene for β-tubulin in *Aspergillus nidulans. Cell* **15:** 639.

Solomon, F., M. Magendantz, and A. Salzman. 1979. Identification with cellular microtubules of one of the co-assembling microtubule-associated proteins. *Cell* **18:** 431.

Umesono, K., T. Toda, S. Hayashi, and M. Yanagida. 1983. Two cell division cycle genes *NDA2* and *NDA3* of the fission yeast *Schizosaccharomyces pombe* control microtubular organization and sensitivity to anti-mitotic benzimidazole compounds. *J. Mol. Biol.* **168:** 271.

Cell Types, Microtubular Organelles, and the Tubulin Gene Families of *Physarum*

Anne Roobol, Eileen C.A. Paul, Christopher R. Birkett, Kay E. Foster, and Keith Gull
Biological Laboratory, University of Kent
Canterbury, Kent, CT2 7NJ, England

Timothy G. Burland, William F. Dove, Larry Green, Lawrence Johnson, and Tim Schedl
McArdle Laboratory, University of Wisconsin
Madison, Wisconsin 53706

Early reports dealing with the biochemistry of microtubules often emphasized the conservation of tubulins isolated from different sources. However, more recent evidence obtained at the level of both the tubulin polypeptide and the tubulin gene has begun to document tubulin diversity. The intriguing suggestion of these studies is that multiple tubulins can be expressed between the different cell types found in the life cycle of some eukaryotic microbes and within the different tissues of metazoan organisms. Furthermore, it appears that multiple tubulins might exist within an individual cell, raising the possibility that different microtubules or microtubule-based organelles might be composed of distinct tubulins. A central issue in the molecular biology of the cytoskeleton is to know the nature of these structural specializations in tubulin and the significance of these for microtubule organization and function. We have chosen to address the problem of organization and expression of a tubulin multigene family, using the slime mold *Physarum polycephalum*. This myxomycete presents attractive opportunities as an experimental system in which distinct, characteristic arrangements of microtubule types are exhibited by three distinct major cell types: the myxameba, the flagellate, and the plasmodium (see Fig. 1).

The myxameba is a microscopic form, approximately 10 μm in diameter, with a single haploid nucleus. Nuclear division is accompanied by cell division; thus, myxamebae are cell populations in which selectional genetics can be performed. Under moist conditions, the myxamebae can transform directly into motile flagellates; however, these are nonproliferative cells and so must return to the myxamebal form to continue growth and division. The plasmodium is a macroscopic syncytium, growing on nutrient surfaces to diameters in excess of 10 cm, at

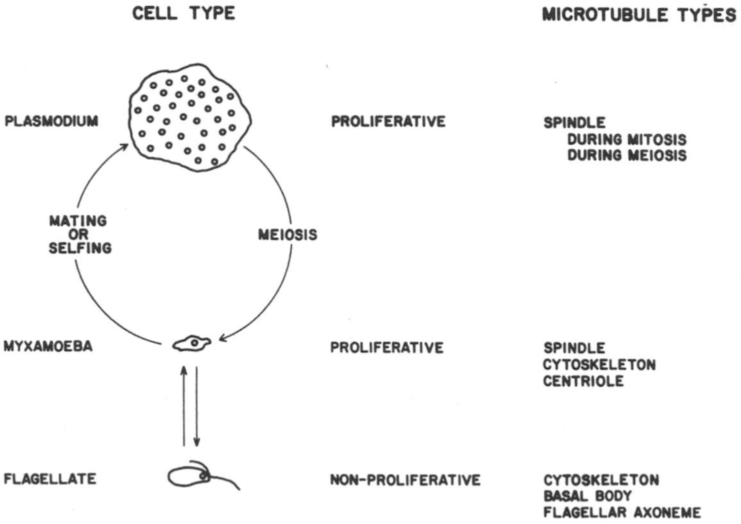

Figure 1
Major cell types and microtubular organelles in the *Physarum* life cycle. The syncytial plasmodium grows by nuclear division without cell division; the cellular myxameba grows in mass and number by coupling nuclear division and cell division. The plasmodial nucleus is diploid, except in selfing strains when it is haploid. The myxamebal nucleus is ordinarily haploid. Under moist conditions, the myxameba transforms into a nonproliferative flagellate.

which size it contains in excess of 10^9 nuclei. Plasmodia possess the remarkable property of synchronous cell cycle emphasized by the fact that all nuclei can pass through mitotic metaphase within 5 minutes of one another in a 10-hour cell cycle (see Schedl et al., this volume).

The transition between plasmodium and myxameba is a meiotic sporulation induced by starvation and illumination of the plasmodium. The transition between myxameba and plasmodium is usually zygotic, achieved by the mating between myxamebae that differ in genotype at the highly polymorphic mating-type (*mt*) locus. However, of great utility in genetic analysis is the possibility for direct transition from myxameba to plasmodium without mating. This occurs at low frequency in selfing strains, generating plasmodia with haploid nuclei.

In this paper, we explore the relationships among cell types, microtubular organelles, and the tubulin gene families of *Physarum*. We then indicate the issues that remain to be resolved between redundancy and specialization of the members of these gene families.

DISCUSSION

Microtubular Organelles of *Physarum*

Physarum myxamebae possess a cytoplasmic microtubule-organizing center (MTOC)/centriole complex that is closely associated with the nucleus. The anterior centriole is linked to the main MTOC by a distinct bridge structure, and the

MTOC is then linked to the nucleus by a group of unusually stable microtubules (Havercroft et al. 1981; Roobol et al. 1982). The three-dimensional architecture of the highly ordered cytoplasmic microtubule arrays nucleated by this MTOC has been analyzed in detail by Wright et al. (1980). This cytoplasmic network of microtubules breaks down at mitosis when chromosome segregation is mediated via an open spindle, with cell division following soon after telophase. The reversible transformation from myxameba to flagellate is accompanied by specific changes in cell shape and the arrangement of microtubules (Wright et al. 1979). The nucleus migrates to one end of the cell, and the two centrioles act as basal bodies for the growth of the two flagella. In the plasmodium, the nuclear envelope remains intact during the major part of mitosis; there are no centrioles, and microtubules are absent from the cytoplasm (Havercroft and Gull 1983). Thus, the three major *Physarum* cell types—myxameba, flagellate, and plasmodium—all contain very different arrangements of microtubules. The myxameba forms cytoplasmic, centriolar, and mitotic spindle microtubules; the flagellate contains cytoplasmic, basal body, and flagellar axoneme microtubules; and the plasmodium contains only spindle microtubules during mitosis and meiosis (Fig. 1).

Tubulin Isotypes of *Physarum*

We discuss here whether a correlation exists between the expression of different tubulin isotypes and the existence of particular cell types or microtubule organelles. The tubulins of *Physarum* myxamebae and plasmodia can be self-assembled into microtubules in vitro (Roobol et al. 1980, 1983). This affords a primary experimental definition of the *Physarum* tubulins. Further characterizations have included two-dimensional gel electrophoresis, peptide mapping by partial proteolysis, reaction with well-defined anti-tubulin monoclonal antibodies, and the in vitro translation of specific mRNAs selected by hybridization to cloned tubulin DNA sequences. Myxamebae possess one α-tubulin electrophoretic species (α_1) and one β-tubulin (β_1), whereas the plasmodia possess four electrophoretically separable tubulin isotypes (α_1, β_1, α_2, and β_2). These differences in display of tubulin isotypes seen in Figure 2 for purified proteins are also found when RNA,

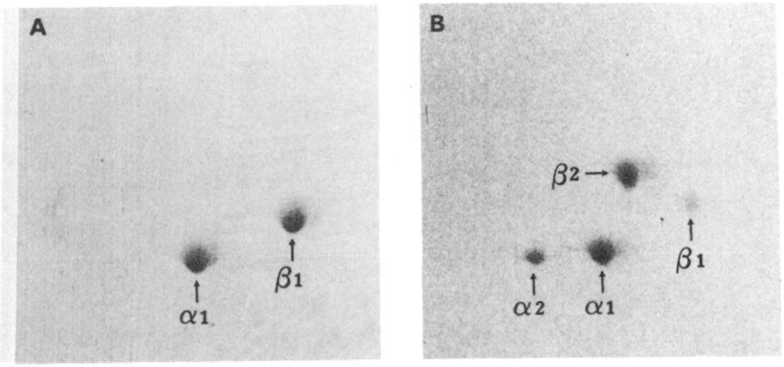

Figure 2
Self-assembled tubulins of the myxameba and the plasmodium. Two-dimensional gels of the tubulin components of microtubules purified by successive cycles of assembly and disassembly, from myxamebae (*A*) and plasmodia (*B*). Separable electrophoretic spots are given individual isoform designations.

extracted from each cell type, is translated in vitro (Fig. 3) (see Burland et al. 1983). Thus, the differences in tubulin isotypes between myxameba and plasmodium preexist at the mRNA level. Do these differences represent the use of different genes? Do individual electrophoretic species from a particular cell type contain the products of more than one gene? Does electrophoretic identity of isotypes expressed by different cell types imply the use of the same structural gene in both cell types? We discuss these issues below.

The flagellate displays another α-tubulin isotype, α_3. In Figure 4, myxamebae were radiolabeled and then allowed to transform into flagellates, and their proteins were analyzed by two-dimensional gel electrophoresis of lysates. The formation of α_3 is strictly correlated with the flagellate. In contrast to the differences in isotype discussed above, however, α_3 cannot be found when RNA from cultures of flagellates is translated in vitro. Thus, α_3 seems to form by a posttranslational modification. This conclusion is reinforced by pulse-chase experiments showing that incorporated radiolabel is efficiently transferred into α_3 even when the pool of labeled amino acids is exhausted (Green and Dove 1984). A very similar process has been shown to operate in the alga *Chlamydomonas* and the trypanosome *Crithidia fasciculata* (L'Hernault and Rosenbaum 1983; McKeithan

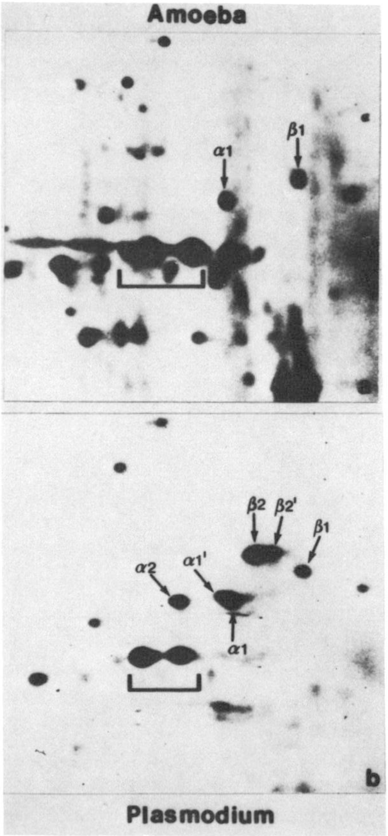

Figure 3
Tubulins synthesized by translation in vitro, directed by myxamebal and plasmodial RNA. Two-dimensional gel electrophoresis of polypeptides synthesized in vitro in a wheat-germ extract from total cellular RNA. (*a*) Myxamebal RNA; (*b*) plasmodial RNA. The horizontal bracket indicates the position of actin in the translation products. (Reprinted, with permission, from Burland et al. 1983.)

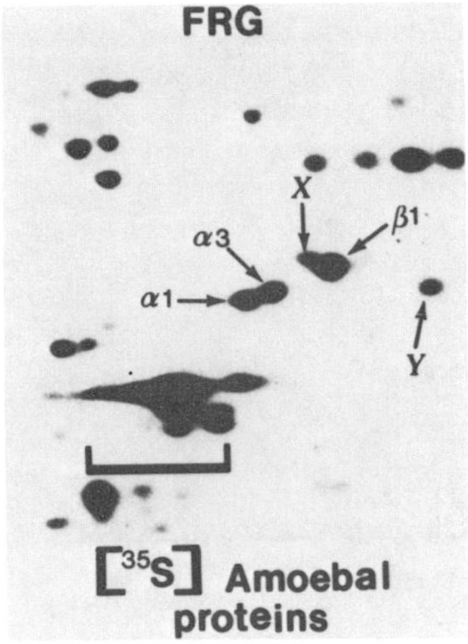

Figure 4
Tubulins labeled in vivo in cultures of myxamebae transforming into flagellates. Myxamebae grown on [^{35}S]methionine-labeled *E. coli* were harvested under conditions promoting flagellate formation. Polypeptides from cell lysates were resolved by two-dimensional gel electrophoresis and then detected by fluorography, as described by Burland et al. (1983). As in Fig. 3, the actin region is denoted by the bracket.

et al. 1983; Russell et al. 1983; Russell and Gull 1984). The electrophoretic mobility of α_3-tubulin relative to its precursor α-tubulin (α_1) is similar to that found in both *Chlamydomonas* and *Crithidia*, where the posttranslationally modified tubulin appears to be specific to the flagellum axonemal microtubules. Three major points come from this investigation of tubulin isotypes. First, specific constellations of tubulin electrophoretic species are associated with each of the three *Physarum* cell types. Second, the novel isotype (α_3) associated with the rapid and reversible transition to the flagellate is produced by a posttranslational modification of existing α-tubulin polypeptides. However, the differences in isotypes of the myxameba and plasmodium preexist at the mRNA level and are thus possibly orchestrated by the expression of different genes. Third, there is no simple relationship between the complexity of tubulin isotype expression and the range of microtubular organelles elaborated by a cell type: The more complex set of isotypes is found in the plasmodium, which exhibits only spindle microtubules.

Utility of the Tubulin Isotypes

Since a variety of tubulin isotypes have been shown to be present in *Physarum*, it is of some concern to know that they are all capable of participating in microtubule formation and that the cell does in fact make use of the complete constellation of available isotypes. All the isotypes detected in cell lysates of myxamebae and plasmodia are also found in preparations of microtubules purified from these cell types by cycles of in vitro self-assembly (Roobol et al. 1980, 1983). Furthermore, the ratio of specific tubulin isotypes in the purified microtubule preparations reflects those ratios that exist in the cell lysate. We have also asked whether the four plasmodial tubulin isotypes are able to enter the nucleus and participate in the mitotic spindle. Nuclei can be isolated from plasmodia at metaphase of mitosis using a protocol that preserves the intranuclear mitotic spindle microtubules (Roobol et al. 1984); all four plasmodial tubulin isotypes are present in these preparations. Similarly isolated interphase nuclei do not contain microtubules and show a complete absence of all four tubulin isotypes. Thus, by these criteria it appears that all the tubulin isotypes that are expressed in a particular cell type can be utilized in microtubules.

Tubulin Genes

Restriction fragments detected by Southern blotting of DNA isolated from different strains of *Physarum* show length polymorphisms, thus allowing tubulin restriction fragments to be used for a genetic analysis of α- and β-tubulin DNA sequences in the *Physarum* genome (cf. Schedl and Dove 1982; Schedl et al. 1984b).

DNA sequences for α-tubulin have been enumerated by Southern blotting of *Physarum* DNA, utilizing a cDNA probe synthesized upon RNA isolated from late G_2 plasmodia (see Schedl et al. 1984a and this volume). DNA from haploid myxamebae show seven or eight bands in an *Eco*RV digest. In Figure 5, the myxamebal gametes 275 and CLd originate from two different natural isolates. Their *Eco*RV hybridization spectra for α-tubulin DNA sequences are completely distinct. The diploid plasmodium (CLd × 275) shows the sum of these two spectra. Haploid myxamebal meiotic segregants from this plasmodium then display which DNA fragments come from alternative alleles of a locus (the law of mutual exclusion) and which come from the same allele of a complex locus (the law of cosegregation). Such mapping methods have established that there are four α-tubulin loci: *altA*, *altB*, *altC*, and *altD*. *altB* contains more than one α-tubulin sequence. Similar analysis using β-tubulin-specific DNA sequences as probes indicates that *Physarum* has at least three β-tubulin loci, *betA*, *betB*, and *betC* (Schedl et al. 1984b). No linkage can be detected between any of the α-tubulin loci and any of the β-tubulin loci.

Thus, the number of DNA sequence loci for tubulin is in excess of the number of tubulin isotypes resolved on two-dimensional gels. Is the excess accounted for by pseudogenes? Or, are we underestimating the number of distinct tubulin gene products? In the remainder of this paper, we indicate two lines of investigation that favor the latter explanation.

Plasmodial α-Tubulin Complexity Revealed by Monoclonal Antibodies

We have initiated a search for monoclonal antibodies that recognize specific tubulin isotypes. We injected purified *Physarum* myxamebal tubulin into mice, performed fusions, and screened the resulting hybridomas using a micro-ELISA technique. Two hybridoma cell lines were established and characterized, KMX-1 and KMP-1. The KMX-1 antibody is an IgG that recognizes the *Physarum* myxamebal β-tubulin. It appears to have wide cross-reactivity, recognizing the β-tubulin of a wide range of organisms including mammalian cells. When used to probe Western blots of two-dimensional gels of myxamebal and plasmodial proteins, KMX-1 detects the β_1-tubulin of myxamebae and both the β_1- and β_2-tubulin species in plasmodia. Thus, KMX-1 appears to recognize all of the β-tubulin species of *Physarum*.

KMP-1 is an IgM antibody that recognizes the α-tubulin of myxamebae. However, it has the useful property of being unable to recognize the α-tubulin of many other organisms including mammalian cells. Western blots of two-dimensional gels of *Physarum* cell extracts and purified tubulins show that KMP-1 also has a limited specificity within even the *Physarum* α-tubulin family. Although KMP-1 recognizes the α_1-tubulin species in *Physarum* myxamebae, it does not recognize the plasmodium-specific α_2-tubulin species, and surprisingly, it does not detect the complete plasmodial α_1-tubulin species. Only the two outer portions of this two-dimensional gel spot are recognized. That this specificity is not an artifact can be seen by counterstaining these same blots with Amido Black and so show-

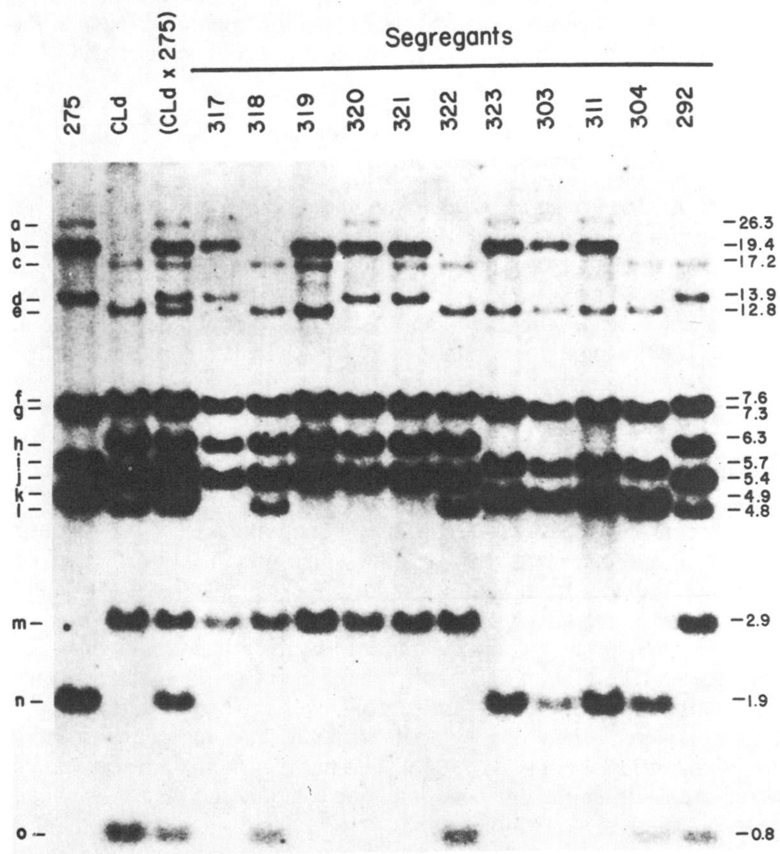

Figure 5

Southern-blot mapping of the α-tubulin DNA sequences of *Physarum.* Myxamebal gametes CLd and MA275 arise from the independent natural isolates Wis 1 and Wis 2, respectively. They were intercrossed to produce the heterozygous diploid plasmodium (CLd × 275). That plasmodium was sporulated to generate haploid meiotic progeny 317, 318, ..., 292. DNA from plasmodia (4 μg/lane) and myxamebal gametes and segregants (2 μg/lane) was digested with the restriction enzyme *Eco*RV, electrophoresed on a 0.7% agarose gel, blotted, and then hybridized with a nick-translated α-tubulin cDNA probe.

ing the presence of unreacted protein in the α_2-tubulin position and in the middle of the α_1 two-dimensional gel spot (C. Birkett, unpubl.). Also, another monoclonal antibody specific for α-tubulins, YOL1/34 (kindly supplied by J. Kilmartin, this volume), is able to detect the plasmodial α_2 spot and all of the α_1 spot on Western blots of the plasmodial proteins. Evidence from Western blots of single-dimension isoelectric focusing (IEF) gels (giving an isoelectric resolution superior to two-dimensional gels) indicates that there may be four separate α_1-tubulin isotypes that comprise the plasmodial two-dimensional gel spot. KMP-1 does not recognize one of the middle pair of plasmodial α_1-tubulins on Western blots of these IEF gels. The KMP-1 antibody has revealed a further complexity of α-tubulin isotypes in *Physarum*. The plasmodial α_1 species is complex and contains up to four different electrophoretic subspecies that can be resolved using monoclonal antibodies. One of the plasmodial α_1 subspecies appears similar to the plasmodial α_2-tubulin isotype in that it lacks the recognition epitope of KMP-1.

Mutations within Tubulin Structural Genes Selected by Resistance to Benzimidazoles

The assembly of purified *Physarum* tubulin into microtubules in vitro is sensitive to the benzimidazole group of drugs but not to colchicine (Quinlan et al. 1981). Therefore, benzimidazoles are the preferred drugs for investigating microtubule function in *Physarum*. Resistance to benzimidazoles has been extremely valuable in defining tubulin loci in other organisms (Sheir-Neiss et al. 1978; Neff et al. 1983; Toda et al. 1984). Mutant myxamebae of *Physarum* resistant to MBC (methyl benzimidazole-2-yl-carbamate) have been isolated, and genetic analysis of these mutants has shown that mutation to MBC resistance can occur at any one of at least four unlinked loci, *benA*, *benB*, *benC*, and *benD* (Burland et al. 1984). MBC resistance tests showed that *benB* and *benD* are expressed in both myxamebae and plasmodia, whereas *benA* and *benC* appear to be myxameba-specific.

Do any of the benzimidazole-resistance loci contain tubulin DNA sequences? The MBC-resistant mutants have been isolated in the canonical Wis 1 haploid strain CLd. These mutants have been crossed to the canonical Wis 2 gamete MA275 to form a heterozygous diploid plasmodium analogous to that discussed in Figure 5. We have then investigated whether MBC resistance mutations segregate together with CLd tubulin DNA restriction fragments in meiotic progeny. We have found that mutations in *benA* cosegregate with the *betA* restriction fragment of CLd; mutations in *benC* cosegregate with *altC*, and mutations in *benD* cosegregate with *betB* (Schedl et al. 1984b). The number of meiotic segregants analyzed in each case is limited, but cosegregation is highly suggestive evidence for allelism (see Schedl et al. 1984b).

A superficial interpretation of these assignments of *ben* mutations to tubulin structural genes, together with the patterns of phenotypic expression of the *ben* mutations, would suggest that the *betA* and *altC* loci are expressed only in the myxameba, whereas the *betB* locus is expressed in both the myxameba and the plasmodium. We must be cautious, however, in concluding from drug resistance tests that a gene is silent in the plasmodium, because multiple tubulin polypeptides are coexpressed and MBC-sensitive polypeptides may create a sensitive phenotype even if resistant polypeptides are expressed. We define this issue more precisely below.

More certain evidence assigning a benzimidazole resistance mutation to a tubulin structural gene comes from alterations in electrophoretic mobility of tubulin polypeptides. The mutant BEN210 is particularly informative in that it produces a β-tubulin with novel electrophoretic mobility as well as a β-tubulin with normal β_1 mobility (see Fig. 6) (Burland et al. 1984). The mutation *ben-210* lies in the *benD* locus and cosegregates with a *betB* β-tubulin DNA fragment (Schedl et al. 1984b). The novel β-tubulin isoform of BEN210 myxamebae is also found in plasmodia derived by selfing of this mutant, confirming that the *benD* (*betB*) locus is expressed in both the myxameba and the plasmodium. We infer that the β-tubulin with normal β_1 mobility in the BEN210 mutant is encoded by the *betA* locus, because BEN210 plasmodia lack a tubulin isotype in the β_1 position, and *benA* resistance mutations, mapping to *betA* (above), do not confer resistance on selfed plasmodia. By further extension, we argue that the plasmodium-specific β-tubulin, β_2, is encoded by a third locus, possibly *betC*. Thus, rigorous assignment of genes to individual tubulin isotypes can be achieved when mutants display an electrophoretically altered tubulin isotype. We can already see that a single isotype can contain the products of more than one gene.

Multiple Tubulin Gene Products Coexpressed in a Single Cell Type

We have seen that multiple α-tubulin and β-tubulin isoforms are expressed in the plasmodium, despite its simple microtubule utilization. Now we also see that the myxameba expresses at least two β-tubulin genes. It is unlikely that the products of these two genes have identical structures, as found in *Chlamydomonas* (see Burland et al. 1984; C. Silflow, pers. comm.). Do coexpressed isoforms freely mix in microtubules or do they sort into distinct structures?

We have found that diploid myxamebae heterozygous for resistance mutations at the *betB (benD)* β-tubulin locus are fully sensitive to benzimidazoles; i.e., *benD* mutations are fully recessive (Burland et al. 1984; Schedl et al. 1984b). But hap-

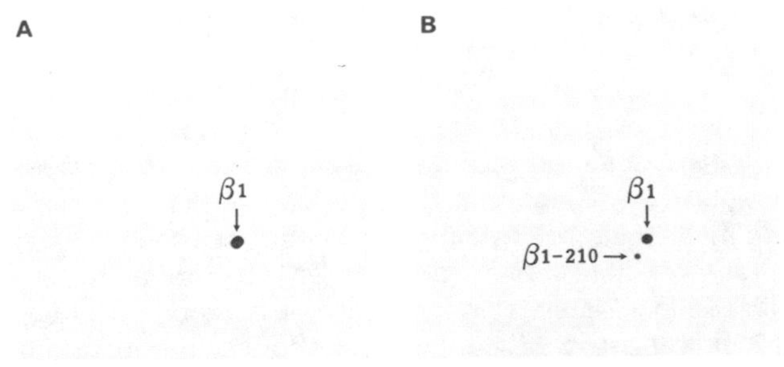

Figure 6
β-Tubulins expressed by wild-type and by BEN210 myxamebae. Lysates from wild-type and from BEN210 myxamoebal cultures were resolved on two-dimensional gels and then transferred to nitrocellulose filters. These transfers were then immunostained with a monoclonal antibody specific for β-tubulin. Immune complexes were detected by an ELISA reaction utilizing horseradish peroxidase (C. Birkett, unpubl.).

loid myxamebae carrying a resistance mutation at either locus are resistant, and haploid plasmodia carrying resistance in the *betB* locus, encoding a β-tubulin species that is minor in the plasmodium, are also resistant. Thus, drug resistance at one β-tubulin locus is epistatic to drug sensitivity at other wild-type loci. One possible interpretation of these observations is that coexpressed isoforms sort into distinct microtubules but that these microtubules are redundant for essential functions (see Schedl et al. 1984b). A second possible interpretation is that resistance is expressed when a critical level of resistant tubulin polypeptides is exceeded. The ways in which this issue need further investigation have been discussed by Schedl (1984) and will be indicated below.

CONCLUSIONS AND FURTHER PROSPECTS

We have concentrated on three cell types in the life cycle of *Physarum*. The plasmodium displays only mitotic spindle microtubules, yet it expresses at least two distinct β-tubulin isotypes and up to five α-tubulin isotypes. The myxameba displays cytoplasmic, centriolar, and spindle microtubules, yet it expresses a single isotype each of α-tubulin and β-tubulin. The mutant BEN210 shows that the β_1 isotype of the myxameba contains the products of two β-tubulin genes, one of which is also expressed in the plasmodium. The differences in isotype expression can be found when RNA is translated in vitro and seem to arise from differential expression of members of the tubulin gene families.

The flagellate, to which the myxameba can transform rapidly and reversibly, displays an additional tubulin isoform, α_3. In this case, posttranslational modification seems to underlie production of α_3.

Enumeration of DNA loci in the *Physarum* genome indicates that there are four α-tubulin loci, one with multiple sequences, and at least three β-tubulin loci. Tubulin isotypes can be further subdivided into distinct species, either by high-resolution electrophoresis combined with high-specificity antibodies or by mutations that shift the electrophoretic mobility of individual gene products. Progress in this direction indicates that there may be up to five active α-tubulin genes and three active β-tubulin genes.

These results dictate that the plasmodium and the myxameba each coexpress more than one β-tubulin molecule. A crucial issue now is to learn how these coexpressed molecules perform their functions. Do they freely mix as fully redundant gene products? Or do they sort into distinct microtubules, perhaps with subtly distinct functions? Two lines of investigation are useful in pursuing this issue. One is to obtain mutants affected in individual members of each gene family, and then to study what, if any, functional alterations ensue. The structural differences seen between the tubulin isoforms may be important for function, or they may represent neutral drift within a multigene family. The other active line of investigation is to obtain antibodies specific for individual gene products and then to study cytologically how these products are deployed within the plasmodium, the myxameba, and the flagellate.

ACKNOWLEDGMENTS

Work at the Biological Laboratory in Canterbury has been funded by grants from the Science and Engineering Research Council, the Agricultural and Food Research Council, the Wellcome Trust, and the Royal Society. That at the University of Wisconsin at Madison has been supported by core grant CA-07175, Program-Project grant CA-23076 in Tumor Biology, predoctoral training grant 5-T32-CA-09135, and postdoctoral training grant 2-T32-CA-09230, all from the National Cancer Institute.

REFERENCES

Burland, T.G., T. Schedl, K. Gull, and W.F. Dove. 1984. Genetic analysis of resistance to benzimidazoles in *Physarum*: Differential expression of β-tubulin genes. *Genetics* **108**: 123.

Burland, T.G., K. Gull, T. Schedl, R.S. Boston, and W.F. Dove. 1983. Cell type-dependent expression of tubulins in *Physarum. J. Cell Biol.* **97:** 1852.

Green, L. and W.F. Dove. 1984. Tubulin proteins and RNA during the myxamoeba-flagellate transformation of *Physarum. Mol. Cell. Biol.* **4**: 1706.

Havercroft, J.C. and K. Gull. 1983. Demonstration of different patterns of microtubule organization in *Physarum polycephalum* myxamoebae and plasmodia using immunofluorescence microscopy. *Eur. J. Cell Biol.* **32:** 67.

Havercroft, J.C., R. Quinlan, and K. Gull. 1981. Characterization of microtubule organizing centre from *Physarum polycephalum* myxamoebae. *J. Ultrastruct. Res.* **74:** 313.

L'Hernault, S.W. and J.L. Rosenbaum. 1983. *Chlamydomonas* α-tubulin is posttranslationally modified in the flagella during flagellar assembly. *J. Cell Biol.* **97:** 258.

McKeithan, T.W., P.A. Lefebvre, C.A. Silflow, and J.L. Rosenbaum. 1983. Multiple forms of tubulin in *Polytomella* and *Chlamydomonas*: Evidence for a precursor of flagellar α-tubulin. *J. Cell Biol.* **96:** 1056.

Neff, N.F., J.H. Thomas, P. Grisafi, and D. Botstein. 1983. Isolation of the β-tubulin gene from yeast and demonstration of its essential function in vivo. *Cell* **33:** 211.

Quinlan, R.A., A. Roobol, C.I. Pogson, and K. Gull. 1981. A correlation between *in vivo* and *in vitro* effects of the microtubule inhibitors colchicine, parbendazole and nocodazole on myxamoebae of *Physarum polycephalum. J. Gen. Microbiol.* **122:** 1.

Roobol, A., J.C. Havercroft, and K. Gull. 1982. Microtubule nucleation by the isolated microtubule organizing centre of *Physarum polycephalum* myxamoebae. *J. Cell Sci.* **55:** 365.

Roobol, A., E.C.A. Paul, and K. Gull. 1984. Isolation of nuclei containing mitotic spindles from the plasmodium of the slime mould *Physarum polycephalum. Eur. J. Cell Biol.* **34:** 52.

Roobol, A., C.I. Pogson, and K. Gull. 1980. *In vitro* assembly of microtubule proteins from myxamoebae of *Physarum polycephalum. Exp. Cell Res.* **130:** 203.

Roobol, A., M. Wilcox, E.C.A. Paul, and K. Gull. 1983. Identification of tubulin isoforms in the plasmodium of *Physarum polycephalum* by *in vitro* microtubule assembly. *Eur. J. Cell Biol.* **33:** 24.

Russell, D. and K. Gull. 1984. Flagellar regeneration in the trypanosome *Crithidia fasciculata* involves the post-translational modification of cytoplasmic alpha tubulin. *Mol. Cell. Biol.* **4:** 1182.

Russell, D.G., D. Miller, and K. Gull. 1983. Tubulin heterogeneity in the trypanosome, *Crithidia fasciculata. Mol. Cell. Biol.* **4:** 779.

Schedl, T. 1984. "Genetic organization and the expression of actin and α- and β-tubulin gene families in *Physarum polycephalum*." Ph.D. thesis, University of Wisconsin, Madison, Wisconsin.

Schedl, T. and W.F. Dove. 1982. Mendelian analysis of the organization of actin sequences in *Physarum polycephalum. J. Mol. Biol.* **160:** 41.

Schedl, T., T.G. Burland, K. Gull, and W.F. Dove. 1984a. Cell cycle regulation of tubulin RNA level, tubulin protein synthesis and assembly of microtubules in *Physarum. J. Cell Biol.* **99:** 155.

Schedl, T., J. Owens, W.F. Dove, and T.G. Burland. 1984b. Genetics of the tubulin gene families of *Physarum. Genetics* **108**: 143.

Sheir-Neiss, G., M.H. Lai, and N.R. Morris. 1978. Identification of a gene for β-tubulin in *Aspergillus nidulans. Cell* **15:** 639.

Toda, T., Y. Adachi, Y. Hiraoka, and M. Yanagida. 1984. Identification of the pleiotropic cell cycle gene *NAD 2* as one of two different α-tubulin genes in *Schizosaccharomyces pombe. Cell* **37:** 223.

Wright, M., L. Mir, and A. Moisand. 1980. The structure of the proflagellar apparatus of the amoebae of *Physarum polycephalum*; relationship to the flagellar apparatus. *Protoplasma* **103:** 69.

Wright, M., A. Moisand, and L. Mir. 1979. The structure of the flagellar apparatus of the swarm cells of *Physarum polycephalum. Protoplasma* **100:** 231.

Tubulin Expression and the Cell Cycle of the *Physarum* Plasmodium

Tim Schedl, Timothy G. Burland, and William F. Dove
McArdle Laboratory, University of Wisconsin Madison, Wisconsin 53706

Anne Roobol, Eileen C.A. Paul, Kay E. Foster, and Keith Gull
Biological Laboratory, University of Kent Canterbury, Kent CT2 7NJ, England

The high mitotic synchrony and macroscopic size of the *Physarum* plasmodium present a favorable opportunity to study molecular events in an unperturbed eukaryotic cell cycle. This proliferative form displays several biological features that we must appreciate at the outset:

1. As a syncytium, the plasmodium employs closed nuclear mitoses. The spindle is nucleated within the nucleoplasm just prior to prophase. The nuclear membrane seems to remain intact throughout mitosis, with the possible exception of localized disruption near the poles in telophase (Havercroft and Gull 1983).
2. No microtubules can be found in the plasmodial cytoplasm (Havercroft and Gull 1983). Thus, tubulin is used only in the spindle and is a minor component, representing on the order of 10^{-3} of total plasmodial protein (Laffler et al. 1981).
3. The plasmodium is distinguished by its natural system of somatic fusion (Poulter and Dee 1968) that permits rapid formation of heterokaryons between genetically compatible partners that differ in mitotic schedule and/or in size. Analysis of such fusions has indicated that nuclear division is controlled by a cytoplasmic condition that can be averaged between partners (Rusch et al. 1966; Tyson 1982).
4. There is no G_1 phase in the cell cycle of the plasmodium, just as in the rapid cleavage cycles of early embryos (Blumenthal et al. 1974; Harland and Laskey 1980). In a total cycle of 10–12 hours, S phase occupies 2–3 hours (Kubbies and Pierron 1983), G_2 8–9 hours, and mitosis 30 minutes.

  0-87969-174-3/84 $1.00

5. Key events that coordinate the plasmodial cell cycle seem to lie immediately before mitosis rather than after. Mitotic averaging in heterokaryons (see above) operates until a point of nuclear commitment arises in late G_2 phase (Loidl and Sachsenmaier 1982; Tyson 1982). Furthermore, mutants arrested before metaphase show that reentry into DNA replication is not dependent on nuclear division (Laffler et al. 1979; Burland and Dee 1980).

DISCUSSION

We have investigated whether any of the macromolecules that participate in cell-cycle events are preferentially synthesized at the stage when they are used. The tubulins of the mitotic spindle have drawn our initial attention. Several experimental advances have been crucial in this work: (1) development of conditions for microtubule assembly in vitro from both myxamebal and plasmodial tubulins, permitting unambiguous identification of these molecules (Roobol et al. 1980, 1983); (2) development of conditions for growing plasmodia up to 30 cm in diameter (the "preparative plasmodium"), permitting one to isolate multiple samples for biochemical and cytological analysis over the cell cycle of a single plasmodium (Schedl et al. 1984); and (3) isolation conditions for plasmodial nuclei that retain the organization of microtubules (Roobol et al. 1984).

Cytological Analysis of Mitosis

Cytological analysis of the plasmodial cell cycle has been carried out both by indirect immunofluorescence, utilizing polyclonal and monoclonal anti-tubulin antibodies (Havercroft and Gull 1983; E.C.A. Paul and K.E. Foster, unpubl.), and by electron microscopy of thin sections (Schedl et al. 1984). One major conclusion of this work is that no organized tubulin polymer can be detected in the plasmodium outside of the mitotic interval beginning 30–60 minutes before metaphase and ending 30 minutes after metaphase in the 10–12-hour cell cycle. A second conclusion is that microtubules start to be organized within the nucleoplasm 30–60 minutes before metaphase. A bipolar spindle then matures at metaphase. In Figure 1, immunofluorescence of a smear from a metaphase plasmodium illustrates the high synchrony and intranuclear nature of the mitotic spindles. Finally, the intranuclear spindle is disassembled after division, with short microtubules detectable at 15 minutes after metaphase and none at 30 minutes. In summary, the plasmodium displays microtubules only within the nucleus and only within the mitotic period, 30–60 minutes before metaphase to 15–30 minutes after.

The earliest time at which microtubules can be detected corresponds well to the time discussed above at which plasmodial nuclei seem to be committed to proceed through mitosis when introduced by fusion into a lagging partner.

Premitotic Labeling of Spindle Tubulins

Over the past several years, we have investigated whether the tubulins for the mitotic spindles of the plasmodial nuclei are selectively synthesized during mi-

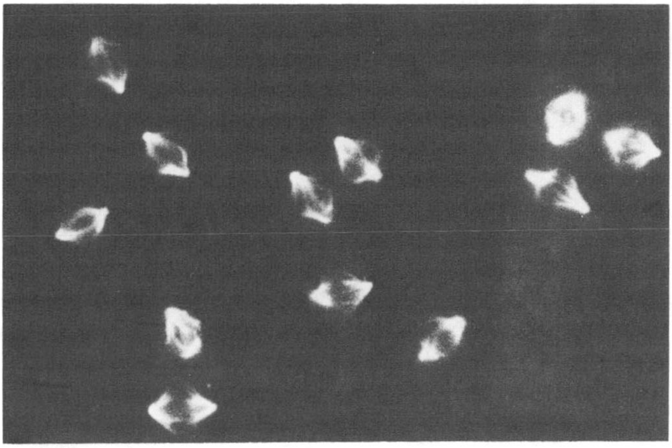

Figure 1
Spindles of synchronous metaphase nuclei. Nuclei isolated from a metaphase plasmodium were stained for indirect immunofluorescence. The primary antibody was the monoclonal anti-β-tubulin DM1B (Blose et al. 1984).

tosis. These experiments were initiated by T.G. Laffler and M. Polanshek (University of Wisconsin, Madison) and, independently, by G. Turnock (University of Leicester, United Kingdom). An account of this series of studies is given by Schedl (1984) and by Schedl et al. (1984). In this paper, we excerpt the points salient to our understanding of tubulin expression in the plasmodial cell cycle.

When plasmodia are pulse-labeled with [^{35}S]methionine for different 15-minute intervals in the 10-hour cell cycle and the labeled polypeptides are then resolved on two-dimensional gels and detected by fluorography, the plasmodial tubulins are unusual in showing periodic labeling. They are labeled effectively during late G_2 phase of the cell cycle but not after mitosis (Fig. 2, a vs. b). Pulse-chase and double-label control experiments have shown that the selective premitotic labeling cannot be explained by selective recovery at mitosis. Labeled tubulin polypeptides are metabolically stable and can be detected at all times in the cell cycle by silver staining the electrophoretograms from plasmodial lysates (Fig. 2, c vs. d).

The rise and fall in tubulin synthesis seen in vivo can also be demonstrated by translation in vitro. When total plasmodial RNA is extracted at different times in the cell cycle and then translated in vitro, tubulins are produced by premitotic, not postmitotic, RNA.

Multiple tubulin isotypes are expressed in the plasmodium and are found in isolated metaphase plasmodial nuclei (Burland et al. 1983; Roobol et al. 1984 and this volume). Is the selective premitotic synthesis of tubulins coordinate over this set of isotypes? The synthesis of all isoforms rises and then decreases around metaphase; however, the plasmodium-specific β_2 isotype seems to be turned off earlier than the other tubulins, as judged from labeling in vivo or from translation in vitro (Schedl 1984; Schedl et al. 1984). The newly synthesized tubulins are able to participate in the spindle; molecules radiolabeled in G_2 phase are found in spindles isolated from the next metaphase (E.C.A. Paul and A. Roobol, unpubl.).

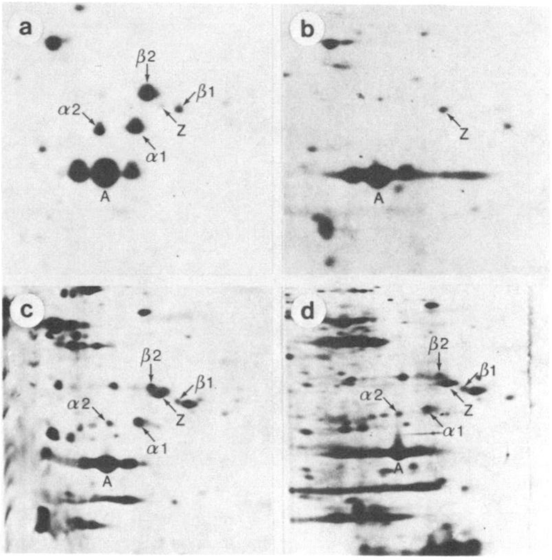

Figure 2
Pulse-labeled and total tubulin polypeptides before and after plasmodial mitosis. Plasmodia were labeled for 15 min with [^{35}S]methionine, beginning either 60 min before (−60) or 60 min after (+60) metaphase. Proteins were resolved on two-dimensional gels as described by Burland et al. (1983). Pulse-labeled polypeptides were detected by fluorography, and total polypeptides by silver staining of parallel gels (Wray et al. 1981). (*a*) Fluorograph of −60 sample; (*b*) fluorograph of +60 sample; (*c*) silver stain of −60 sample; (*d*) silver stain of +60 sample. A indicates the position of actin. Spot Z is not a tubulin (Burland et al. 1983).

Kinetics of the Rise and Fall of α-Tubulin RNA Level

Quantitative studies of the kinetics of tubulin gene expression over the cell cycle have become feasible with the isolation of a *Physarum* cDNA clone for α-tubulin, templated by premitotic plasmodial RNA (Schedl et al. 1984). It has been possible from a single preparative plasmodium to extract sufficient RNA for quantitation at a series of time points in the cell cycle and also to assess cytologically the state of organization of the mitotic spindle microtubules. Levels of α-tubulin RNA were measured by dot-blot hybridizations in which a radiolabeled α-tubulin cDNA probe was hybridized to total plasmodial RNA (Schedl et al. 1984). The changes in α-tubulin RNA levels fit very closely to exponential kinetics, with a doubling time of 80 minutes for the rise and a half-life of 20 minutes for the fall. The amplitude of modulation is at least 40-fold (Fig. 3).

The very high synchrony of the plasmodium, both in S phase (Kubbies and Pierron 1983) and in mitosis (Schedl et al. 1984), makes it unlikely that asynchrony contributes significantly to the exponential character of this rise and fall. It seems more likely that the apparently exponential behavior is an intrinsic feature of the processes whereby levels of tubulin RNA are increased and then decreased.

These kinetics of modulation in tubulin RNA level can be explained by any of a broad range of models. Changes in rate of synthesis or in rate of turnover can

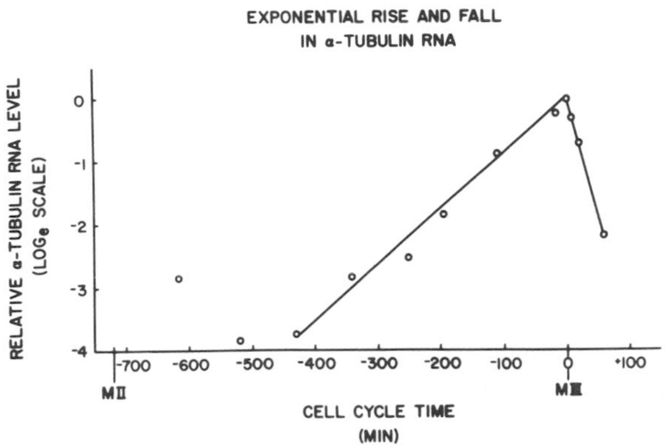

Figure 3
Relative α-tubulin RNA level over the plasmodial cell cycle. Total RNA samples were extracted from segments of one preparative plasmodium at different times in the cell cycle. Progress and synchrony of the cycle were assessed by phase and electron microscopy of the same plasmodium. The level of RNAs homologous to probes for α-tubulin, actin, and *Physarum* cDNA clones 16 and 42 were measured by quantitative dot-blot hybridization. The relative level of α-tubulin RNA was calculated by normalizing to the level of actin, clone 16, or clone 42; each normalizing sequence gave the same result (cf. Schedl et al. 1984).

be invoked. Here, we assess one plausible model for tubulin RNA accumulation over the cell cycle. In this model, the transcription of tubulin genes would be switched on to a constant rate in G_2 (zero-order synthesis) and then switched off at metaphase. Tubulin RNA would have a constant half-life, so that after metaphase, levels would fall by first-order kinetics. In such a model, RNA would accumulate asymptotically toward a plateau prior to mitosis and then fall exponentially after mitosis. The data shown in Figure 3 are quantitatively at odds with such a model, because the accumulation curve is apparently exponential. Thus, we seriously consider that the G_2 phase synthesis of tubulin RNA involves an accelerative component, e.g., an autocatalytic one. Below, we suggest one possible mechanism out of many with this feature.

The fall in α-tubulin RNA after metaphase is very rapid. The upper limit of the postmitotic half-life is 20 minutes, corresponding to only 3% of the cell-cycle time. Here, we consider the possibility of a facilitated decay. Established examples of regulation by facilitated decay include yeast histone RNAs (Hereford et al. 1981; Osley and Hereford 1982) and adenoviral early RNAs (Babich and Nevins 1981). Again, we return below to one example of such a process.

Cell-cycle Modulation of Tubulin RNA and Autoregulation

Two decades ago, it was suggested that gene activity might fluctuate over the cell cycle in response to a negative feedback control exerted by metabolites created by cell-cycle activity (see Mitchison 1969). This formalism, oscillatory repression, might be adapted to explain the rise and fall in tubulin gene expression over the plasmodial cell cycle. Indeed, many animal cell lines seem to display a negative feedback response, wherein tubulin RNA levels fall when the levels of tubulin protomer are increased, either by treatment with the microtubule-depolymerizing agent Colcemid or by microinjection of tubulin (Ben-Ze'ev et al. 1979; Cleveland et al. 1981, 1983). For the *Physarum* plasmodium, the time course of the fall in α-tubulin RNA is coincident with the depolymerization of microtubules, and so is concordant with negative feedback by protomer. In contrast, the time course of the rise in α-tubulin RNA is not directly consistent with release from negative autoregulation by tubulin protomer. As summarized above, tubulin polymerization is initiated only in the interval between 30 and 60 minutes before metaphase. But α-tubulin RNA levels rise perceptibly as early as 6 hours before metaphase, long before polymerization into microtubules would be extensive enough to deplete the total pool of protomer. Thus, negative autoregulation by tubulin protomer over tubulin gene expression is a possible explanation for the postmitotic fall in plasmodial α-tubulin RNA. But the initiation of the premitotic rise does not seem to be caused by a release from negative autoregulation by tubulin protomer.

Temporal Restriction of Macromolecular Synthesis

What is the significance of the restriction of tubulin synthesis to the G_2 phase of the plasmodial cell cycle? This restriction is particularly puzzling because the tubulin polypeptides seem to have a long half-life (see Fig. 2). Thus, it is not obvious how restriction in time of synthesis can lead to restriction in time of action, as occurs with the homothallism endonuclease of *Saccharomyces cerevisiae*, which is thought to be unstable (Nasmyth 1983).

Perhaps there is a cell-cycle restriction in the localization of the plasmodial tubulins. For the *Physarum* plasmodium, an appropriate model states that tubulins are imported into the nucleus during G_2 phase and then are exported after mitosis. Indeed, tubulin molecules radiolabeled in G_2 phase can be found in nuclei isolated from metaphase plasmodia but not those isolated from interphase (A. Roobol, E.C.A. Paul, and K.E. Foster, unpubl.). Since the plasmodial mitoses are closed, some mechanism of import is necessary (cf. De Robertis 1983). One might expect that a shuttling of proteins within the cell would be accompanied by reversible or irreversible modifications. There is no evidence, however, for a molecular modification of tubulin coincident with a cycle of nuclear import and export. It is crucial to recognize, also, that the buffers used for nuclear isolation contain detergent and may well not permit retention of protomeric tubulin.

An import/export model has a particularly simple intersection with the analysis above of the kinetic data in Figure 3. If there is an autocatalytic component to the synthesis in G_2, it may be served by imported nuclear tubulins, acting to increase nuclear tubulin RNA levels. On the other side of metaphase, facilitated decay of cytoplasmic RNA, if it exists, might be directed by tubulin molecules released from the nucleus after mitosis. This model, even if correct in outline, clearly needs further provisions to be complete. How is synthesis initiated in G_2? How would the facilitated decay be blocked within the nucleus and during the G_2 increase?

Numerous other models can be devised that would account for the exponential properties of the data in Figure 3. The one advanced here has the virtue of direct experimental test.

CONCLUSIONS AND PROSPECTS FOR DEEPER STUDY

Analysis of the synthesis of tubulins for the mitotic spindles of the plasmodium has shown that the pronounced modulation in synthesis reflects changes in levels of tubulin RNA. Tubulins synthesized in G_2 are utilized to construct the upcoming mitotic spindle. The rise and the fall each follow apparently exponential kinetics. The rise is initiated hours before the pool of protomer is depleted by microtubule formation, whereas the fall is contemporaneous with spindle disassembly after metaphase. The rapidity of the fall in tubulin RNA suggests a process of facilitated decay. One model to explain the exponential character of the rise in tubulin RNA states that tubulin molecules imported into the nucleus during synthesis potentiate further tubulin RNA synthesis. The cell-cycle regulation of tubulin gene expression in the plasmodium is, of course, not the only well-studied example of cyclic modulation of gene expression. However, there is a useful set of experimental challenges and opportunities available in *Physarum* by which to deepen our understanding. A number of issues need to be investigated:

1. What are the relative contributions of changes in rate of synthesis and rate of decay of tubulin RNA to the kinetics of α-tubulin RNA level given in Figure 3?
2. What are the patterns of entry and exit of tubulins from the nucleus? Can tubulin molecules used in the spindle in one cell cycle be re-used in subsequent cycles?
3. Are there any functional differences between premitotic and postmitotic tubulin polypeptides, involving structural differences too subtle to detect by our current methods?
4. Are there any true autoregulatory components to the modulation of tubulin gene expression in the plasmodial cell cycle, rigorously demonstrable by mutations lying within a tubulin structural gene (cf. Smith and Magasanik 1971)?
5. Is there a cell-cycle modulation of plasmodial spindle components other than tubulin?
6. What features of the tubulin modulation seen in the *Physarum* plasmodium are unique to the special biology of this form and what features are general to other eukaryotes?

Several laboratories are addressing these issues. On the molecular side, it is clearly essential to capitalize on the high synchrony of the plasmodium to obtain an unambiguous pulse-chase analysis of the rates of tubulin RNA synthesis and decay. It is clearly also important to master the injection of cytoplasmic fractions into plasmodia. Injections of volumes up to 1 μl have been reported (Adams et al. 1981) but are not yet routine due to the wounding and gelling reactions of the plasmodium. On the genetic side, the capability for selectional genetics in the uninucleate, haploid myxamebal phase of *Physarum* needs to be utilized to obtain mutants affected in the tubulin structural genes and their regulators.

ACKNOWLEDGMENTS

We thank Larry Johnson for his steadfast attention to the treatment of our kinetic data. Work at the Biological Laboratory in Canterbury is supported by grants from the Science and Engineering Research Council, the Agricultural and Food

Research Council, the Wellcome Trust, and the Royal Society. That at the University of Wisconsin at Madison is supported by core grant CA-07175, Program-Project grant CA-23076 in Tumor Biology, and predoctoral training grant 5-T32-CA-09135, all from the National Cancer Institute.

REFERENCES

Adams, D.S., D. Noonan, and W.R. Jeffrey. 1981. Cytoplasmic polyadenylate processing events accompanying the transfer of mRNA from the free mRNP particles to the polysomes in *Physarum. Proc. Natl. Acad. Sci.* **78:** 83.

Babich, A. and J.R. Nevins. 1981. The stability of early adenovirus mRNAs is controlled by the viral 72 kd DNA-binding protein. *Cell* **26:** 371.

Ben-Ze'ev, A., S.R. Farmer, and S. Penman. 1979. Mechanism of regulating tubulin synthesis in cultured mammalian cells. *Cell* **17:** 319.

Blose, S.H., D.I. Meltzer, and J.R. Feramisco. 1984. 10 nm filaments are induced to collapse in living cells microinjected with monoclonal and polyclonal antibodies against tubulin. *J. Cell Biol.* **98:** 847.

Blumenthal, A.B., H.J. Kriegstein, and D.S. Hogness. 1974. The units of DNA replication in *Drosophila melanogaster* chromosomes. *Cold Spring Harbor Symp. Quant. Biol.* **38:** 205.

Burland, T.G. and J. Dee. 1980. Isolation of cell cycle mutants of *Physarum polycephalum. Mol. Gen. Genet.* **179:** 43.

Burland, T.G., K. Gull, T. Schedl, R.S. Boston, and W.F. Dove. 1983. Cell type-dependent expression of tubulins in *Physarum*. *J. Cell Biol.* **97:** 1852.

Cleveland, D.W., M.F. Pittenger, and J.R. Feramisco. 1983. Elevation of tubulin levels by microinjection suppresses new tubulin synthesis. *Nature* **305:** 738.

Cleveland, D.W., M.A. Lopata, P. Sherline, and M.W. Kirschner. 1981. Unpolymerized tubulin modulates the level of tubulin mRNAs. *Cell* **25:** 537.

De Robertis, E.M. 1983. Nucleocytoplasmic segregation of proteins and RNAs. *Cell* **32:** 1021.

Harland, R.M. and R.A. Laskey. 1980. Regulated replication of DNA microinjected into eggs of *Xenopus laevis. Cell* **21:** 761.

Havercroft, J.C. and K. Gull. 1983. Demonstration of different patterns of microtubule organization in *Physarum polycephalum* myxamoebae and plasmodia using immunofluorescence microscopy. *Eur. J. Cell Biol.* **32:** 67.

Hereford, L.M., M.A. Osley, J.R. Ludwig, and C.S. McLaughlin. 1981. Cell-cycle regulation of yeast histone mRNA. *Cell* **24:** 367.

Kubbies, M. and G. Pierron. 1983. Mitotic cell cycle control in *Physarum*. *Exp. Cell Res.* **149:** 57.

Laffler, T.G., M.T. Chang, and W.F. Dove. 1981. Periodic synthesis of microtubule proteins in the cell cycle of *Physarum. Proc. Natl. Acad. Sci.* **78:** 5000.

Laffler, T.G., A. Wilkins, S. Selvig, N. Warren, A. Kleinschmidt, and W.F. Dove. 1979. Temperature-sensitive mutants of *Physarum polycephalum*: Viability, growth, and nuclear replication. *J. Bacteriol.* **138:** 499.

Loidl, P. and W. Sachsenmaier. 1982. Control of mitotic synchrony in *Physarum polycephalum*. Phase shifting by fusion of heterophasic plasmodia contradicts a limit cycle oscillator model. *Eur. J. Cell Biol.* **28:** 175.

Mitchison, J.M. 1969. Enzyme synthesis in synchronous cultures. *Science* **165:** 657.

Nasmyth, K. 1983. Molecular analysis of cell lineage. *Nature* **302:** 670.

Osley, M.A. and L.M. Hereford. 1982. Identification of a sequence responsible for periodic synthesis of yeast histone H2A mRNA. *Proc. Natl. Acad. Sci.* **79:** 7689.

Poulter, R.T.M. and J. Dee. 1968. Segregation of factors controlling fusion between plasmodia of the true slime mould *Physarum polycephalum. Genet. Res.* (*Cambridge*) **12:** 71.

Roobol, A., E.C.A. Paul, and K. Gull. 1984. Isolation of nuclei containing mitotic spindles from the plasmodium of the slime mould *Physarum poycephalum. Eur. J. Cell Biol.* **34:** 52.

Roobol, A., C.I. Pogson, and K. Gull. 1980. *In vitro* assembly of microtubule proteins from myxamoebae of *Physarum polycephalum. Exp. Cell Res.* **130:** 203.

Roobol, A., M. Wilcox, E.C.A. Paul, and K. Gull. 1983. Identification of tubulin isoforms in the plasmodium of *Physarum polycephalum* by *in vitro* microtubule assembly. *Eur. J. Cell Biol.* **33:** 24.

Rusch, H.P., W. Sachsenmaier, K. Behrens, and V. Gruber. 1966. Synchronization of mitosis by fusion of plasmodia of *Physarum polycephalum. J. Cell Biol.* **31:** 204.

Schedl, T. 1984. "Genetic organization and expression of actin and alpha- and beta-tubulin gene families in *Physarum polycephalum*." Ph.D. thesis, University of Wisconsin, Madison.

Schedl, T., T.G. Burland, K. Gull, and W.F. Dove. 1984. Cell cycle regulation of tubulin RNA level, tubulin protein synthesis and assembly of microtubules in *Physarum. J. Cell Biol.* **99:** 155.

Smith, G.R. and B. Magasanik. 1971. Nature and self-regulated synthesis of the repressor of the *hut* operons in *Salmonella typhimurium. Proc. Natl. Acad. Sci.* **68:** 1493.

Tyson, J.J. 1982. Periodic phenomena in *Physarum.* In *Cell biology of* Physarum *and* Didymium (ed. H.C. Aldrich and J.W. Daniel), vol. I, p. 61. Academic Press, New York.

Wray, W., T. Boulikas, V.P. Wray, and R. Hancock. 1981. Silver staining of proteins in polyacrylamide gels. *Anal. Biochem.* **18:** 197.

Approaches to Flagellar Assembly and Size Control Using Stumpy- and Short-flagella Mutants of *Chlamydomonas reinhardtii*

Dorothy A. Baldwin, Michael R. Kuchka, Bonnie Chojnacki, and Jonathan W. Jarvik
Department of Biological Sciences
Carnegie-Mellon University
Pittsburgh, Pennsylvania 15213

In this paper, we summarize some investigations into *Chlamydomonas reinhardtii* flagellar morphogenesis, using mutants in which the extent of flagellar growth is abnormal. Some of the mutants have extremely short nonfunctional flagella (discussed below under Stumpy Mutants) and others have short but fully functional flagella (discussed under Genetic Analysis of Flagellar Size Control).

STUMPY MUTANTS: POTENTIAL SOURCES OF UNASSEMBLED FLAGELLAR PROTEINS

All Axoneme Proteins Spend Some Time in the Flagellar Matrix

It has been known for some time that the bulk of axoneme assembly occurs at the distal ends of growing flagella (Rosenbaum et al. 1969). Since most of the axoneme's mass is in its doublet microtubules, it follows that the doublets assemble at the tip, and since most of the lateral components of the axoneme are linked to the doublets, these probably also assemble at the tip. Furthermore, dikaryon complementation experiments have demonstrated that radial spokes, spoke heads, inner dynein arms, or outer dynein arms can assemble in situ onto preformed axonemes missing these lateral structures (Huang 1984; Luck 1984). For this to happen, the structural proteins must of necessity travel through the flagellar matrix to the sites of their incorporation into the axoneme. Therefore, it can be stated with confidence that at different times during its lifetime, each axoneme protein resides in two distinct flagellar compartments—the flagellar matrix and the axoneme. If we are to understand flagellar morphogenesis, then, it is important to examine the biochemical and supramolecular states of the axonemal proteins in the matrix compartment of the flagellum.

Stumpy-flagella Mutants

It occurred to us that some mutants that fail to assemble flagella might still attempt to transfer flagellar proteins from the cytoplasm to the flagellar matrix compartment. As a result, such mutants might have grossly defective flagella containing unassembled flagellar proteins. In fact, a mutant with the expected "stumpy" phenotype was isolated by McVittie some years ago and its ultrastructure was described (McVittie 1972). As a prelude to screening our own mutant collection, we obtained this mutant (*HA1*) and examined it closely. By light microscopy, the flagellar stumps were barely discernible, whereas by thin-section transmission electron microscopy, they were dramatically visible. Therefore, we used electron microscopy to examine a number of mutants that had been previously characterized by light microscopy as either stumpy or flagellaless. Twenty mutants were examined (four of these were temperature-sensitive mutants grown at nonpermissive temperature) and were found to fall into four broad classes: (1) those with no flagellum to speak of; (2) those with very short flagella; (3) those with short bulbous flagella containing electron-dense material; and (4) those with flagella trapped within the cell wall. We believe that the electron-dense material is actively transported into the bulbous flagella, because when wild-type cells were deflagellated and allowed to attempt regeneration in the presence of colchicine, the abortive flagella that resulted had a class 1, not a class 3, ultrastructure. We call members of classes 2 and 3 stumpy mutants. Electron micrographs of representative stumpy mutants are shown in Figure 1.

Stumpy Flagella Contain Bona Fide Flagellar Proteins

To analyze the protein composition of the stumpy flagella, we wanted to detach them from the cell body. In preliminary experiments, representative stumpy mutants were grown to relatively high cell density, conditions under which a large fraction of the cells exist in "palmelloid" clusters enclosed by the mother wall. This was done so that detached flagella would remain near their former sites of attachment. The cells were treated with the deflagellating agent dibucaine and observed by thin-section electron microscopy. It was found that the stumpy flagella were efficiently released from the cells, although the flagellar membranes did not remain intact (Fig. 2). Nonpalmelloid cultures of eight different stumpy mutants were then grown in large volumes, concentrated, and treated with dibucaine. Cell bodies were removed by centrifugation at low speed, and the proteins in the supernatants were subjected to SDS-gel electrophoresis, transferred to nitrocellulose, and probed with monospecific or monoclonal antibodies directed against a variety of axoneme and nonaxoneme flagellar proteins. As negative controls in these experiments, samples derived from fully flagellaless mutant cells (*bald-2*) (Goodenough and St. Clair 1975) were included. The results, summarized in Table 1, indicate clearly that stumpy mutant flagella contain bona fide flagellar proteins. We have also probed blots of stumpy flagella with complex polyspecific antibodies made against total axoneme protein. The results show

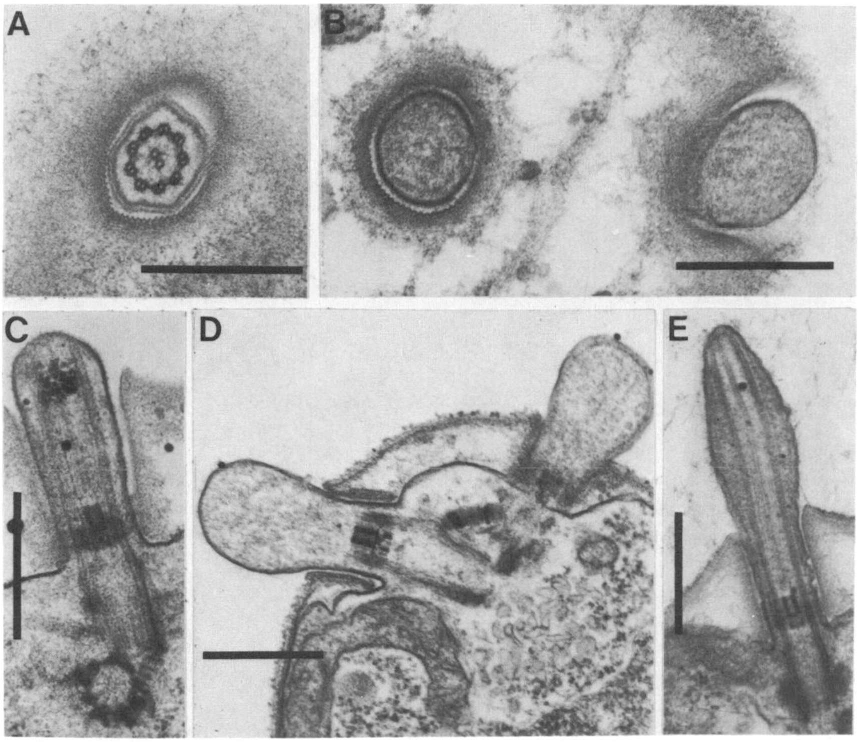

Figure 1
Stumpy-flagella ultrastructure. (*A*) Wild-type: transverse section just above the transition region. The axoneme is surrounded concentrically by the flagellar membrane, the flagellar surface coat, and the grooved tunnel through the cell wall. (*B*) Class-3 mutant *Fla⁻283:* transverse section as in *A*. (*C*) Class-2 mutant *Fla⁻162*. (*D,E*) Class-3 mutants *Fla⁻350* and *Fla⁻1719*. Bars, 0.5 μm.

that many axoneme protein species, in addition to those in Table 1, are present in the stumpy flagella.

Why Not Examine the Wild-type Flagellar Matrix?

At first thought, it might seem that the wild-type flagellar matrix would be at least as good a source of unassembled axonemal proteins as any stumpy mutant. However, there are some conceptual and practical liabilities to examining the wild-type matrix. First, as ^{35}S-pulse-labeling results indicate, the wild-type axoneme is in dynamic equilibrium with respect to its proteins. This means that the matrix contains both *un*assembled proteins and *dis*assembled versions of the same protein species. Sorting these out would be difficult. Second, the steady-state concentrations of axoneme proteins in the matrix are very low. Additional problems would arise were we to examine the matrix of regenerating wild-type flagella. Regenerating axonemes appear to be structurally labile as compared with fully mature axonemes (S. L'Hernault, pers. comm.); again, this means that

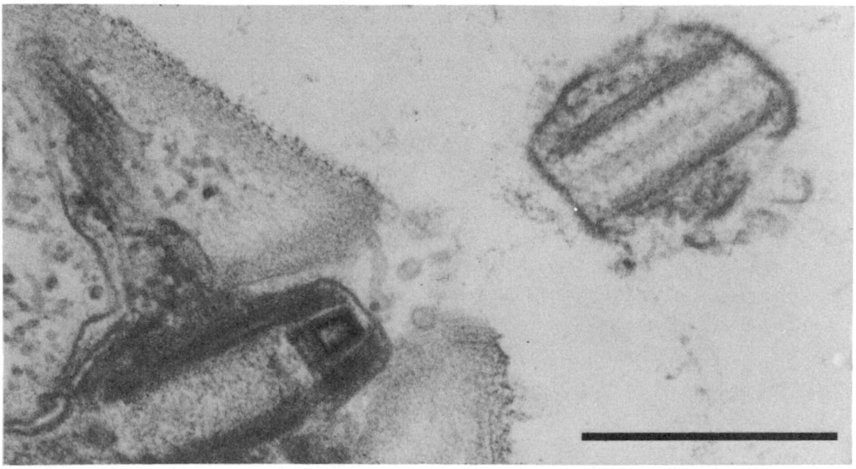

Figure 2
Stumpy flagellum released from class-3 mutant *Fla⁻1719*. Cells were fixed 5 sec after dibucaine treatment. Bar, 0.5 μm.

matrix fractions would include unassembled and disassembled proteins in undetermined ratios. Stumpy flagella, on the other hand, probably contain virgin flagellar proteins that have never had the chance to assemble into or onto an axoneme.

Prospects: Stumpy Mutants as Sources of Unassembled or Preassembled Flagellar Protein

Having demonstrated that stumpy flagella contain a variety of flagellar proteins, we are now in a position to look at the proteins' supramolecular states. For example, we would like to know whether radial spoke proteins assemble into spokes in the absence of doublet microtubules or whether, on the contrary, spoke assembly requires initiation sites on the doublets. By fractionating and analyzing the dibucaine supernatants described here, it ought to be possible to answer such questions. We also want to find conditions for releasing stumpy flagella from cells without rupturing the flagellar membrane. It then would be possible to purify the flagella so as to analyze their contents with high resolution. In addition, the contents might prove amenable to in vitro assembly under appropriate conditions.

GENETIC ANALYSIS OF FLAGELLAR SIZE CONTROL

Size control is a fundamental but mysterious property of all cells and organelles. In wild-type *C. reinhardtii*, flagellar length averages about 12 μm, with a standard deviation of only about 1 μm. How is flagellar length controlled? We cannot answer this question in molecular terms at present, but we have obtained evidence against some otherwise attractive hypotheses. We have also isolated and characterized a number of mutants with size-control defects. We believe it probable

Table 1
Flagellar Antigens Detected in Stumpy Flagella Using Monospecific or Monoclonal Antibodies

	Fla⁺	Fla⁻	*Class 2*		*Class 3*					
	wild type	bald-2	*238*	*241*	*162*	*236*	*267*	*283*	*350*	*1719*
α-Tubulin[a]	+ +	–	+ +	+ +	+	+ +	+ +	+ +	+	+
350K flagellar[b] membrane protein	+ +	+ / –	+ +	+ +	+ +	+ +	+ +	+ +	+ +	+ +
190K flagellar[b] membrane protein	+ +	–	+	+ +	+ +	+ +	+ +	+ +	+ +	+
18S dynein HMW3[c]	+	–	–	–	–	+	+	–	+	–

The wild-type signal from dynein HMW3 was quite weak, and so we do not rule out the possibility that all stumpy mutants contain some of the protein.
[a]Antibody was obtained from B. Huang (Baylor University).
[b]Antibody was obtained from R. Bloodgood (University of Virginia).
[c]Antibody was obtained from D. Mitchell (Yale University).

that analysis of these mutants will reveal the identity of key molecular components of the size-control system.

What the Mechanism of Size Control Probably Is Not

We have taken advantage of certain mutants to test two general size-control models. The results, summarized below, argue against both.

Does the Size of the Intracellular Flagellar Protein Precursor Pool Govern the Extent of Flagellar Assembly?

One rather simple hypothesis for flagellar size control holds that if the concentration of some flagellar protein (e.g., tubulin) in the cellular pool is above some critical value there is flagellar growth and that otherwise there is not. Thus, flagellar growth proceeds until the concentration of the relevant flagellar component in the pool has dropped to the critical concentration. We tested this notion using a mutant (*vfl-2*) that has normal cell size but a variable number of flagella per cell (Kuchka and Jarvik 1982). The flagella of *vfl-2* are of normal length irrespective of the number of flagella per cell, and so, by the above hypothesis, pool concentration should be equal in all cells. We determined the pool concentration operationally, by measuring the extent of flagellar regeneration in the absence of new protein synthesis, and found that the pool concentration was not equal in all cells but instead was proportional to flagellar number (e.g., the pool in cells with three flagella was three times the size of that in cells with one flagellum). The most straightforward interpretation of this result is that flagellar length is not regulated by limiting amounts of unassembled flagellar protein in the cellular pool.

Do Fundamental Kinetic Properties of the Flagellar Assembly Process Govern the Extent of Flagellar Assembly?

A number of models can be formulated in which flagellar size is intrinsically self-limiting, with flagellar growth rate slowing progressively as the flagellum lengthens because longer flagella inherently grow more slowly than short flagella (Levy 1974; Child 1978). The observation that the flagellar growth zone is at the distal tip makes such models particularly attractive, since it is easy to imagine that flagellar subunits have increasing difficulty in getting to their assembly sites as the flagellum gets longer and longer. We tested this category of size-control model using *shf-1* short-flagella mutants. More will be said about these mutants later in this paper; for now, all one need know is that these mutations are recessive in quadriflagellate *mutant/wild-type* dikaryons, so that after mating between mutant and wild type to form quadriflagellate cells, the two short flagella grow out to normal length. The kinetics of outgrowth of the two short flagella in *shf-1*/wild-type dikaryons was measured and compared to the kinetics of wild-type flagellar regeneration. We observed that the two short flagella grew out with kinetics that closely resembled the growth of wild-type flagella from zero to half length. If the above hypothesis for size control were correct, however, then the two half-length flagella should have elongated to full length with kinetics that resembled the outgrowth of wild-type flagella from half to full length. We concluded that flagellar length is probably *not* regulated by intrinsic kinetic properties of the flagellar assembly process itself (Jarvik et al. 1984).

Finally, we should mention a result of Lefebvre et al. (1978) that is strongly suggestive of the above conclusions. Wild-type cells were induced to resorb their flagella by adding a calcium chelator to the medium; then, when the flagella had resorbed most of the way, the cells were deflagellated, calcium was restored, and the kinetics of flagellar outgrowth were determined. It was found that the cells regrew their flagella very rapidly, but to wild-type length and no further. Since the flagellar precursor pools in these cells were augmented substantially, both by the resorbed proteins and by new proteins synthesized in response to deflagellation, the fact that cells regenerated to normal length implies that pool size does not regulate flagellar length. Furthermore, the cells' very rapid regeneration demonstrates that there is no inherent inverse relation between flagellar length and flagellar growth rate.

Is There a Special Flagellar Length-equalization Mechanism?

When *Chlamydomonas* loses its two flagella, both regenerate in synchrony. When it loses only one flagellum, the remaining one shortens as a new one grows; then, when the two are of about equal length, they typically grow out together until they reach full length (Rosenbaum et al. 1969). This result might reflect the existence of two distinct size-control mechanisms: One to ensure that the final flagellar length is correct and the other to ensure that both flagella are of equal length.

We have obtained results which suggest that a mechanism of the latter type exists and that length equalization is largely flagellar pair-autonomous. We define this last term operationally: Length equalization is autonomous if, in a cell with two pairs of flagella, the flagellar length of one pair is not affected by the length adjustment going on in the other pair. In one set of experiments, we took advantage of the fact that in dense cultures of some wild-type strains, as many as 5% of the cells are quadriflagellate bipolar twin cells that appear to have failed to complete cytokinesis. Careful observation of these cells by differential interference contrast microscopy revealed no evidence of a septum separating the two halves in most of them. Cultures containing twin cells were briefly exposed to flagella-amputating shear forces in a Waring Blendor, and the cells were immobilized in agar on microscope slides and observed over the next hour. In biflagellates that lost a single flagellum, one flagellum typically shortened as the other regenerated, just as originally reported by Rosenbaum et al. (1969). In quadriflagellates that lost a single flagellum, the lost flagellum's partner typically shortened, but the other flagellar pair showed very little response. It thus appears that the unit for rapid length adjustment is the *flagellar pair*, which, it must be remembered, includes two basal bodies linked by striated fibers as well as the flagella themselves. Two additional observations support this conclusion. First, similar experiments using quadriflagellates produced by the mating of gametes give similar results; when one of the four flagella was amputated, one of the three remaining flagella usually shortened dramatically while the other two did not. Second, in the striated-fiber-defective mutant *vfl-3* (Wright et al. 1983), in which flagella are not physically linked together in pairs, regeneration of one flagellum is not accompanied by resorption of others (R. Wright, pers. comm.). These observations indicate that physical proximity is not sufficient for pair-autonomous length adjustment, and they encourage us to conjecture that direct physical communication between the basal bodies is required. Indeed, such direct communication does exist in *C. reinhardtii*: Favorable sections show not only distal and proximal striated-fiber connections between the two basal bodies, but also a conjoining of the two basal bodies at their cell proximal ends (see, e.g., Fig. 2c in Wright et al. 1983).

Size Control Is an Ongoing Cellular Concern

The various regeneration results referred to above indicate clearly that the cell's response to the loss of one or more flagella is homeostatic in nature, resulting in a return to flagellar length equality and to normal flagellar length. But it should be emphasized that size control is probably a full-time cellular concern, not one only employed in response to a gross assault such as deflagellation. We know that in the *Chlamydomonas* cell body, there exists a substantial pool of unassembled flagellar protein (Rosenbaum et al. 1969) and that this pool is in dynamic equilibrium with the assembled flagellum. We take the fact that flagellar length is maintained while flagellar components continuously exchange as evidence for continuous flagellar length control by the cell. Further evidence on this point is discussed below.

Genetic Analysis

Short-flagella Mutants

Through the analysis of flagellar length mutants, we hope to identify genes and gene products involved in the flagellar size-control process. A major virtue of this approach is that we can proceed without any prejudice about the molecular basis of size control. Instead, we can let mutants defective for size-control function lead us to relevant molecules and/or structures.

By screening mutagenized cells for slow-swimming mutants, we have isolated

Table 2
Short-flagella Mutants

	Mean flagellar length	*Standard deviation*
Wild type	12.0	1.1
shf-1-236	6.1	1.1
shf-1-253	6.6	1.1
shf-1-277	7.1	1.0
shf-2-158	6.4	1.4
shf-2-1249	6.4	1.1
shf-3-1851	5.7	1.6

Mean flagellar length was calculated from measurements of 100 or more flagella. All units are micrometers.

a number of strains with short but motile flagella. We call these mutants *shf* for *sh*ort *f*lagella. Six mutants were chosen for initial detailed analysis. As indicated in Table 2, the mutants regulate flagellar length with a precision comparable to that shown by wild type. They all have functionally normal flagella in that they are capable of phototaxis, backward swimming when appropriately stimulated, and flagellar agglutination in mating, and so we think it likely that they are specifically defective in the size-control process. All of the mutants have substantial cellular pools of flagellar protein, ruling out the trivial possibility that their flagella are short because they synthesize insufficient quantities of flagellar protein.

Dominance tests were performed by mating *shf* and wild-type gametes and observing flagellar length in the resulting quadriflagellate dikaryons. In all cases, the two short flagella elongated to wild-type length, indicating that expression of the *shf* mutations is recessive. Complementation tests among the *shf* mutants were also performed by observing *shf/shf* dikaryons. By definition, allelic combinations show no flagellar elongation, whereas nonallelic combinations show outgrowth of all four flagella. Results placed the mutations in three distinct complementation groups designated *shf-1* (three alleles), *shf-2* (two alleles), and *shf-3* (one allele).

Tetrad analysis demonstrated that in each strain, the mutant phenotype is due to a determinant that segregates 2:2 in *shf/shf*$^{+}$ tetrads. Mutant × mutant intercrosses placed the mutations at three distinct loci, as expected. Mapping crosses to marked strains demonstrated that *shf-1* is very near the centromere of chromosome VI and *shf-2* is approximately 16 map units from *msr-1* on chromosome I. *shf-3* has not yet been mapped.

Reversion Analysis

For several distinct reasons, we have isolated phenotypic revertants of *shf* mutants. First, some revertants might carry extragenic suppressors; analysis of the suppessors might allow us to identify genes whose products functionally interact with the products of the known *shf* genes (Jarvik and Botstein 1975). Second, some revertants might be intragenic pseudorevertants; these carry new mutations in the identified *shf* gene—mutations that might have useful new molecular phenotypes (e.g., altered electrophoretic properties for the *shf*-gene product [Luck et al. 1977]) or new functional phenotypes (e.g., temperature-sensitive size control).

The phenotypes of several *shf* mutants have allowed us to obtain revertants by applying strong selections for motile phototactic cells. Specifically, (1) *shf-1-253* is temperature-sensitive for flagellar assembly (cells immotile at 34°C) and (2) *shf-1-253, shf-2-1249* and *shf-1-253, shf-3-1851* double mutants are flagellaless (cells immotile). By selecting motile revertants, we have obtained 44 independent *shf-1-253* revertants, 24 independent *shf-1-253, shf-2-1249* revertants, and 17 independent *shf-1-253, shf-3-1851* revertants. Analysis of these revertants is just beginning. So far, we have found the following. Of four *shf-1-253* revertants analyzed, two appear to be either true revertants or intragenic pseudorevertants and two carry unlinked extragenic suppressors. One of the suppressors is dominant to its wild-type allele in quadriflagellate dikaryons and one is recessive. (The assessment of dominance is made by constructing *shf/shf, sup* dikaryons. If the two short flagella lengthen, we conclude that the suppressor is dominant; if the two long flagella shorten, we conclude that it is recessive.) Interestingly, the recessive suppressor does not suppress the *shf-1-277* mutation, whereas the dominant suppressor suppresses *shf-1-277* as well as *shf-1-253*. The allele-specific effect of the recessive suppressor makes it an excellent candidate to work by altering a protein that functionally interacts with the *shf-1*-gene product. Initial analysis of the motile revertants of *shf-1-253, shf-2-1249* has demonstrated that some of them have either reverted at the *shf-2* locus or carry specific suppressors of *shf-2*. Further analysis of these suppressors is in progress.

Protein Composition of Short Flagella

Electrophoretic analysis of the proteins in the axonemes of *shf* mutants has revealed a number of distinct abnormalities. Autoradiographs of two-dimensional gels of ^{35}S-labeled *shf-1* axonemes are missing three protein spots. In one full-length revertant analyzed to date, that carrying *shf-1-253* and the allele-specific suppressor described above, two of the three proteins are still missing, but the third, an acidic protein with a molecular weight of approximately 110,000, is restored. This last protein is thus a candidate for having a direct role in size control. Before pursuing any of these leads biochemically, however, we wish first to make certain, through tetrad analysis, that the biochemical differences truly cosegregate with the size-control defects. The appropriate experiments are in progress.

Long-flagella Mutants

We have done some work with long-flagella mutants, although the analysis is not so complete as for the short mutants. Two *lf* mutations were isolated by McVittie (1972) and were shown to map to two loci, *lf-1* and *lf-2*; we have isolated five others and have received a sixth from P. Lefebvre (University of Minnesota). All *lf* mutations show recessive expression in dikaryons with the exception of those at *lf-1*. These are codominant to wild type, in that the quadriflagellate persists with two long and two normal flagella (McVittie 1972). (If the cell is deflagellated, however, it regenerates four normal flagella.) We have examined complementation behavior in almost all possible *shf/lf* dikaryons. Each behaved as expected from the dominance properties of the mutations present; for example, in a recessive *shf*/codominant *lf* dikaryon, the short flagella elongated to wild-type length while the long flagella did not shorten.

Not all *shf,lf* double mutants have yet been constructed, but we have already identified one that is flagellaless: *lf-3-89, shf-1-236* at 13°C. (*lf-3-89* is cold-sensitive, having long flagella at 13°C but not at 25°C.) This means that it should be possible to select revertants of *lf-3-89* in the *lf-3-89, shf-1-236* genetic background.

ACKNOWLEDGMENT

This work was supported by grant PCM-8216337 from the National Science Foundation and by RCDA award K04-AM00710 to J.W.J. from the National Institutes of Health.

REFERENCES

Child, F.M. 1978. The elongation of cilia and flagella: A model involving antagonistic growth zones. *ICN-UCLA Symp. Mol. Cell. Biol.* **12:** 351.

Goodenough, U.W. and H.S. St. Clair. 1975. *bald-2*: A mutation affecting the formation of doublet and triplet sets of microtubules in *Chlamydomonas reinhardtii*. *J. Cell Biol.* **66:** 480.

Huang, B. 1984. Genetic analysis of flagellar structure and motility. *J. Protozool.* **31:** 25.

Jarvik, J. and D. Botstein. 1975. Conditional-lethal mutations that suppress genetic defects in morphogenesis by altering structural proteins. *Proc. Natl. Acad. Sci.* **72:** 2738.

Jarvik, J., M. Kuchka, F. Reinhart, and S. Adler. 1984. Altered flagellar size control in *shf-1* short-flagella mutants of *Chlamydomonas reinhardtii*. *J. Protozool.* **31:** 199.

Kuchka, M.R. and J.W. Jarvik. 1982. Analysis of flagellar size control using a mutant of *Chlamydomonas reinhardtii* with a variable number of flagella. *J. Cell Biol.* **92:** 170.

Lefebvre, P.A., S.A. Nordstrom, J.E. Moulder, and J.L. Rosenbaum. 1978. Flagellar elongation and shortening in *Chlamydomonas*. V. Effects of flagellar detachment, regeneration and resorption on the induction of flagellar protein synthesis. *J. Cell Biol.* **78:** 8.

Levy, E.M. 1974. Flagellar elongation: An example of controlled growth. *J. Theor. Biol.* **43:** 133.

Luck, D.J.L. 1984. Genetic and biochemical dissection of the eucaryotic flagellum. *J. Cell Biol.* **98:** 789.

Luck, D.J.L., G. Piperno, Z. Ramanis, and B. Huang. 1977. Flagellar mutants of *Chlamydomonas*: Studies of radial spoke-defective strains by dikaryon and revertant analysis. *Proc. Natl. Acad. Sci.* **74:** 3456.

McVittie, A.C. 1972. Flagellum mutants of *Chlamydomonas reinhardtii*. *J. Gen. Microbiol.* **71:** 525.

Rosenbaum, J.L., J. Moulder, and D. Ringo. 1969. Flagellar elongation and shortening in *Chlamydomonas*. *J. Cell Biol.* **41:** 600.

Wright, R.L., B. Chojnacki, and J. Jarvik. 1983. Abnormal basal body number, location and orientation in a striated fiber-defective mutant of *Chlamydomonas reinhardtii*. *J. Cell Biol.* **96:** 1697.

Tubulin and Actin: Yin-Yang Gene Expression during *Naegleria* Differentiation

Elaine Y. Lai, Stephen P. Remillard, and Chandler Fulton
Department of Biology, Brandeis University
Waltham, Massachusetts 02254

yin-yang, *in Chinese thought, the two complementary forces, or principles, that make up all aspects and phenomena of life.*
Encyclopaedia Britannica,1980

To understand the regulation of gene expression during cell differentiation, one must first choose a cell differentiation. The choice of a paradigm in turn determines which genes are suitable for analysis. Our choice led us irresistibly to the genes for tubulin and actin.

Our Paradigm: The Quick-change Act of *Naegleria*

We sought a differentiation in which clonal populations of cells could be induced to undergo, rapidly and as individuals, a single, major phenotypic change. Our search ended with a remarkable unicellular eukaryote, *Naegleria gruberi*, which can alternate between two forms, amebae and flagellates (Fulton 1977a). *Naegleria* reproduces as an ameba about 15 μm in diameter. It can reproduce very rapidly for a eukaryote, with doubling times as short as 1.7 hours. The amebae are as easy to grow and work with as *Escherichia coli*. They are haploid and have a small genome, features that are convenient for the study of gene structure and function. The unique feature of these amebae is their ability, when transferred from a growth environment to a nutrient-free aqueous environment, to undergo a 1:1 conversion to streamlined swimming flagellates. The flagellates are temporary, and after a time that can vary from seconds to days, depending on the environmental conditions, they revert to amebae again.

The differentiation of amebae to flagellates can be made to occur rapidly, synchronously, and temporally reproducibly (Fig. 1A). Several procedures can be used to initiate and manipulate differentiation (Fulton 1977a). Various morphological changes of the differentiation can be quantitated by counting the percentage of particular quantal phenotypes among fixed cells. Under our standard differentiation conditions, half of the cells form flagella 60 ± 1 minutes after the environmental shift.

The cells change from amebae with an actin-based motility system to flagellates with a tubulin-based system. Thus, they alternate between two fundamental motility forms (Fulton 1977b). The actin-based motility system remains latent in the flagellates, but can be reactivated almost instantly by suitable stimuli (Fulton 1977b). In addition to the microtubules of the two flagella and basal bodies, the flagellates have a cage of microtubules just beneath the cell surface (Fulton 1977b; Walsh 1984). The tubulin-based system is more impressive than the actin-based system in the sense that flagellates swim about 100 times as fast as amebae walk.

This description of a yin-yang change between two motility forms is consistent with the current understanding of cell shape and motility as it applies to this differentiation, but it is also simplistic. Although no cytoplasmic microtubules have been observed in *Naegleria* amebae, the amebae use tubulin for mitosis. The capacity for ameboid motility remains, albeit latent, in flagellates.

Tubulin and Its Synthesis

When we sought to bring our analysis of the differentiation to a molecular level, a natural component to examine was the tubulin that makes up the microtubules of the flagella. We characterized outer-doublet tubulin and found that in all tested respects, this tubulin is like the tubulins of other eukaryotes (Kowit and Fulton 1974a). To our surprise, we found that the tubulin subunits found in the flagella are synthesized during differentiation. This was shown rigorously using isotope-dilution experiments, which indicate that at least 60%, and probably nearly 100%, of the α and β subunits of both outer-doublet and central-pair tubulin is synthesized during differentiation (Kowit and Fulton 1974b; Fulton and Simpson 1976; P.A. Simpson et al., in prep.).

The extent of synthesis was also examined using antibodies. Polyclonal rabbit antibodies to outer-doublet tubulin recognize the α and β subunits of both outer-doublet and central-pair tubulin (Kowit and Fulton 1974a,b; Lai et al. 1979b; C. Fulton et al., unpubl.). These antibodies have not recognized any other tubulin tested, including mitotic tubulin in extracts of *Naegleria* amebae or tubulins from several other organisms. This unusual specificity allowed us to measure the amount of flagellar tubulin antigen in cells by radioimmunoassay (Fig. 1A). Amebae have, at most, 3% as much flagellar tubulin antigen as flagellates. The amount of antigen begins to increase by about 20 minutes of differentiation and reaches a maximum about the time the flagellates have formed full-length flagella (Kowit and Fulton 1974b).

A next step was to move backward toward gene expression by asking whether flagellar tubulin mRNA increased in abundance during differentiation. RNA extracted from cells at successive times during differentiation was translated in the

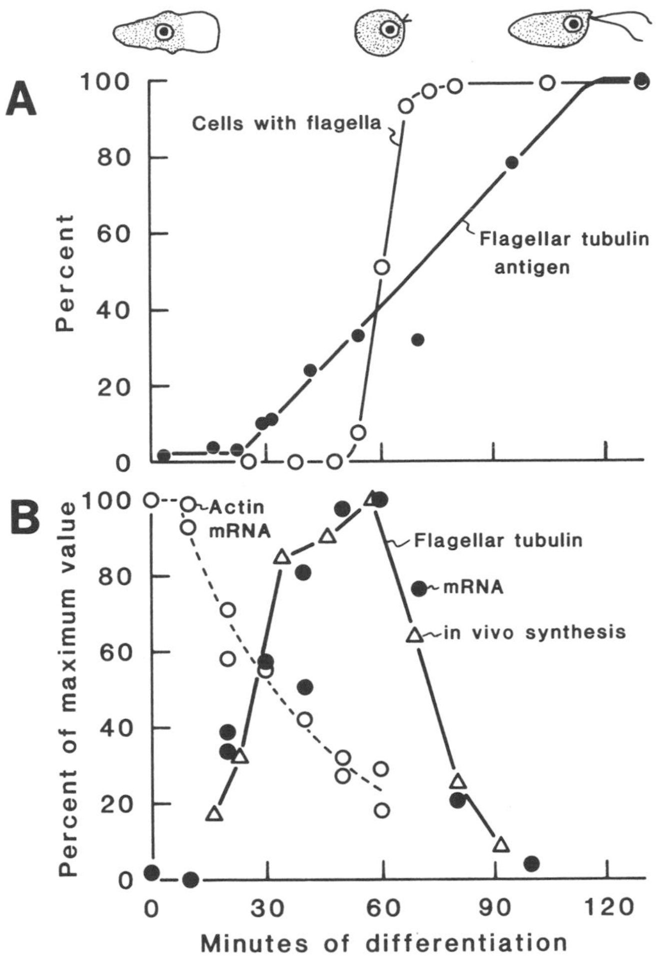

Figure 1
Aspects of *Naegleria* differentiation that motivate our research. Primary among these is the rapid and temporally reproducible differentiation (*A*), shown here by the time course of appearance of cells with flagella (○). Another aspect is the temporally programmed appearance of flagellar tubulin, here assayed as flagellar tubulin antigen by radioimmunoassay of cell extracts (●). The amount of antigen reaches a maximum as the flagella reach full length. (*B*) There are marked rises and falls in the abundance of translatable mRNAs for three families of defined proteins that are relevant to the differentiation process: actin (○) and the α and β subunits of flagellar tubulin (●). The in vivo rate of flagellar tubulin synthesis, as measured by pulse-chase experiments (△), parallels the rise and fall of translatable flagellar tubulin mRNA. (Redrawn from Kowit and Fulton 1974b; Lai et al. 1979b; Fulton 1983; Sussman et al. 1984a.)

wheat-germ cell-free system. Translation products with the electrophoretic mobility of the α and β subunits of tubulin were shown to be flagellar tubulin (Lai et al. 1979b). These products also were recognized by the antibodies to flagellar tubulin, which permitted us to quantitatively immunoprecipitate flagellar tubulin from translation products and thus to quantitate the amount of translatable flagellar tubulin mRNA present at various times during differentiation. As shown in Figure 1B, translatable flagellar tubulin mRNA is not detected in amebae. It rises to maximum abundance at about 60 minutes and then declines. The rate of in vivo flagellar tubulin synthesis, measured by pulse-chase experiments, follows precisely the same curve (Fig. 1B). The simplest interpretation is that flagellar tubulin synthesis during differentiation is regulated by the programmed transcription, and subsequent degradation, of flagellar tubulin mRNA. This interpretation is supported by experiments using actinomycin D, which selectively inhibits both RNA synthesis and differentiation in *Naegleria* (Fulton and Walsh 1980). Actinomycin D, added during differentiation, arrests the accumulation of translatable flagellar tubulin mRNA (Lai et al. 1979a).

The α and β subunits of flagellar tubulin are encoded on separate mRNAs. Translatable mRNAs for both subunits increase and decrease in abundance during differentiation together and in direct proportion to one another (Lai et al. 1979b). This suggests the possibility that both the transcription and stability of the mRNAs for these two subunits are controlled coordinately. Concurrent expression of α- and β-tubulin genes has been described in *Chlamydomonas* (Silflow and Rosenbaum 1981; Brunke et al. 1982).

Actin and Its Synthesis

Naegleria actin is a typical actin (Sussman et al. 1984b). For example, it forms F-actin filaments that decorate with subfragment 1 of rabbit myosin, and it also activates heterologous myosin Mg^{++}-ATPase. This actin is, however, unusual in certain respects. The actins of vertebrates, *Drosophila*, and several other metazoa have several forms that differ in isoelectric point; in vertebrates and *Drosophila*, these isoforms are the products of multiple distinct actin genes (for review, see Firtel 1981). In contrast, all lower eukaryotes examined to date have a single species of actin that accounts for at least 95% of the actin. *Naegleria* actin shows at least three isoforms. All actins characterized previously have a single residue of N^{τ}-methylhistidine as residue 73 of the protein. Although the amino acid composition of *Naegleria* actin is very similar to those of other characterized actins, we found no N^{τ}-methylhistidine. *Naegleria* actin lacks this posttranslationally added residue or at least has less than one residue per five actin molecules. Reconstruction experiments indicated that N^{τ}-methylhistidine is not lost during the purification of *Naegleria* actin; we conclude that it is not present in the actin at the time amebae are lysed. The absence of N^{τ}-methylhistidine indicates that at least some of the measured properties of actin, such as polymerization to microfilaments or interaction with myosin, do not require the presence of this posttranslationally modified residue. Antibodies to *Naegleria* actin are specific to actin of *Naegleria*; they do not recognize even the actins of other ameboid organisms such as *Acanthamoeba, Dictyostelium,* and *Physarum* (C. Fulton et al., in prep.).

Isotope-dilution experiments indicate that only about 1% of the actin in the cells is synthesized de novo during differentiation (P.A. Simpson et al., in prep.), so the synthesis of this major cell protein is selectively restricted beginning early in differentiation. The actin present at the beginning of differentiation is conserved during the 100 minutes of differentiation.

Quantitative cell-free translation, followed by immunoprecipitation of the product using the antibodies to *Naegleria* actin, was employed to measure the abundance of translatable actin mRNA at successive times during differentiation (Sussman et al. 1984a). In the experiment shown in Figure 1B, after a lag of about 7 minutes, translatable actin mRNA decreased in abundance exponentially with a half-life of 24–25 minutes. The simplest interpretation is that actin gene expression is turned off, and actin mRNA is selectively destroyed, as early events of differentiation. However, such experiments do not indicate whether the actin mRNA is degraded or somehow rendered untranslatable.

Hence, during the time that the cells are changing from an actin-based to a tubulin-based motility, actin, the major protein of both amebae and flagellates, is neither synthesized nor degraded, whereas tubulin for the flagella is synthesized de novo. The abundance of translatable mRNAs for tubulin and actin rise and fall in parallel with these changes.

cDNA Clones and Conservation of Tubulin and Actin Genes

DNA clones provide the means to continue dissection of the structure and expression of these genes in relation to differentiation. Since we are primarily interested in expressed genes, we prepared cDNA clones in a vector developed by Okayama and Berg (1982). We prepared libraries of cDNA clones to both 0-minute RNA, when translatable actin mRNA is most abundant, and to 60-minute RNA, when the flagellar tubulin mRNAs are most abundant. We used the chicken cDNA clones for tubulin and β-actin of Cleveland et al. (1980) as probes in our search for tubulin and actin clones. To make a lengthy story brief, the former proved very difficult and the latter easy. Our primary α-tubulin clone, pNαT1, contains a 1.6-kb insert and selects mRNA that translates α-tubulin, as characterized by one-dimensional and two-dimensional gel electrophoresis and by immunoprecipitation with our antibodies to flagellar tubulin. Partial DNA sequencing verified that the clone contains the full coding region of an α-tubulin. Using pNαT1, we found additional α-tubulin clones in the 60-minute cDNA library at a frequency of 0.5%.

One full-length actin clone, pNA1, which appears to encode the major expressed actin, has been completely sequenced. There are 376 deduced amino acids, with one more at the aminoterminal end than most comparable actins. When compared with *Acanthamoeba* actin (Vandekerckhove et al. 1984), *Naegleria* actin shows over 50 substitutions, whereas when *Acanthamoeba* actin is compared with mammalian β-actin, it shows only 15 substitutions. Thus, *Naegleria* actin is among the most substituted actins that have been sequenced.

Our first results in studying these α-tubulin and actin cDNA clones led to an interesting point: These two proteins, both present in the first eukaryote and both among the most conserved proteins over billennia of evolution, and thought to be comparably conserved, have genes that appear to differ in their degree of conservation. The frequent statement that tubulin genes are conserved misled us, and may have misled others. In the case of *Naegleria*, there is minimal homology between pNαT1 and the chicken α-tubulin clone T1 of Cleveland et al.

(1980). In fact, under conditions of minimal stringency, the insert of T1 recognizes certain clones that contain *Naegleria* inserts that are not α-tubulin as well as it recognizes pNαT1. The results of our sequencing of pNαT1 so far show that the longest stretch of homology between the *Naegleria* insert and the chicken T1 is 11 bases. Similar results were reported by Neff et al. (1983) in their comparison of the yeast β-tubulin clone with the chicken β-tubulin clone. Yet in some cases, reasonable homology has been conserved. For example, the chicken tubulin clones hybridize strongly to *Chlamydomonas* tubulin genes (Ares and Howell 1982; Brunke et al. 1982), even though the alga and bird separated early in evolution.

In contrast, the homology of actin DNAs appears to be well preserved over more than 10^9 years of evolution, and probes from distantly related organisms have regularly served to isolate new actin clones, as the chicken probe did for the *Naegleria* actin genes. We know of no exception to this generalization.

Simple explanations, such as codon preference, do not suffice. Codon usage in *Naegleria* is strongly and similarly biased in the actin and α-tubulin genes, much in the fashion that has been described for yeast genes (Bennetzen and Hall 1982), yet one retains homology and the other does not. It is possible that tubulin subunits, as proteins, will prove to be less conserved than actin (cf. Toda et al. 1984); it is also possible that the actin DNA sequence is conserved for some reason other than coding requirements.

Multiple Genes for Tubulin and Actin

Genomic blots of *Naegleria* DNA probed with the insert of either the *Naegleria* actin cDNA clone or, at reduced stringency, the chicken β-actin cDNA clone reveal a similar array of restriction fragments. The number of discrete bands leads us to conclude that *Naegleria* has about a dozen actinlike sequences, which fits into the pattern observed in other organisms (Firtel 1981). Partial sequencing of a second actin cDNA clone indicates that at least two actin genes are expressed in amebae, but so far the differences found are either silent substitutions or in the 3′-untranslated region.

In contrast, genomic blots probed with the insert of *Naegleria* α-tubulin clone pNαT1 indicate the presence of about eight very similar, but not identical, α-tubulinlike sequences in the genome. The similarity of these sequences suggested the possibility that they might be arrayed in a closely spaced tandem repeat, as has been observed in *Leishmania* and trypanosomes (Landfear et al. 1983; Seebeck et al. 1983; Thomashow et al. 1983), but this does not appear to be the case. If the *Naegleria* α-tubulin genes are arrayed in a tandem repeat, they must be separated by either long or heterogeneous spacers. Overall, these results suggest an α-tubulin gene family that has a different architecture than those described so far in other eukaryotes (Cleveland 1983).

Several of the α-tubulin cDNA clones have been partially characterized, and we find that their restriction maps and sequences show many similarities. A detailed study of two clones has revealed, so far, two silent nucleotide differences in the coding region (about 100 codons compared) and a major difference in the 3′-untranslated region. These results indicate that at least two α-tubulin genes are expressed during differentiation, but the similarities are more striking than the differences.

Gene Expression during Differentiation

The cDNA clones make it possible to measure directly RNA sequence abundance during differentiation. The results of quantitative RNA dot hybridization are shown in Figure 2. It is evident that the changes in abundance during differentiation of the physical transcripts for α-tubulin and actin (Fig. 2) are directly proportional to the previously measured abundance of translatable mRNAs for these polypeptides (Fig. 1B). In this experiment, actin mRNA decayed exponentially, after a lag of 3 minutes, with a half-life of 20 minutes. The results indicate that the decrease in translatable actin mRNA is due to a rapid destruction of the actin mRNA sequences, rather than to some masking of the sequences.

The results for α-tubulin show that the synthesis of flagellar tubulin is controlled by the abundance of its mRNA. In turn, this abundance must be controlled by the relative rates of transcription and of degradation. The observed rates are remarkable. α-Tubulin mRNA abundance increases at least 80-fold in less than 50 minutes, and later declines with a half-life of about 8 minutes, faster even than the earlier disappearance of actin mRNA.

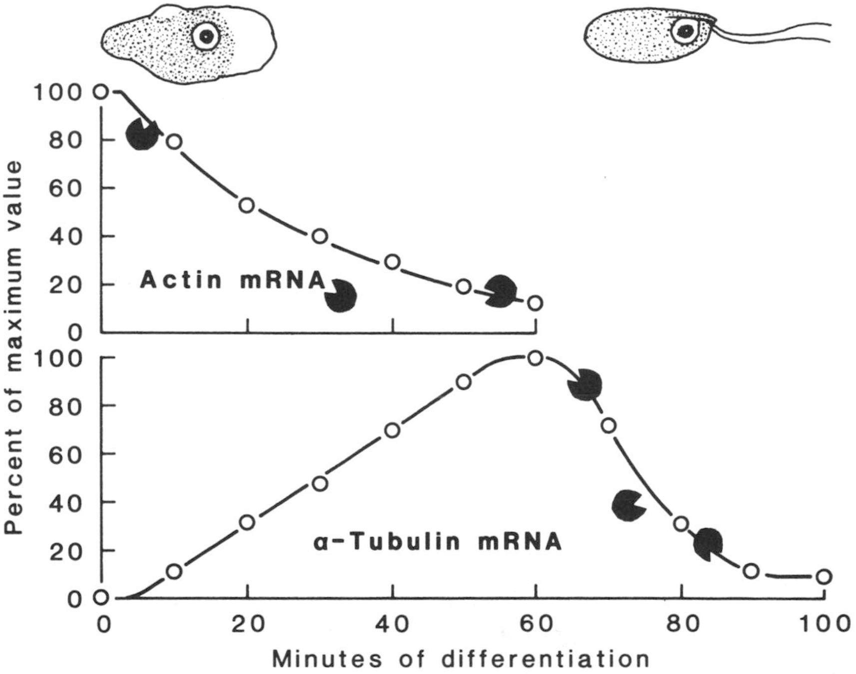

Figure 2

Abundance of actin and α-tubulin mRNA sequences during differentiation, as measured by quantitative RNA dot hybridization (Kafatos et al. 1979). Aliquots of 2 μg of total RNA isolated at the indicated times were dotted in triplicate onto nitrocellulose filters. The filters were probed using the insert of either actin cDNA clone pNA1 or α-tubulin cDNA clone pNαT1 labeled with ^{32}P by nick translation. The dots, localized by autoradiography, were then cut from the filter to measure the radioactivity. Results are shown as the percentage of the maximum amount of hybridization observed with each probe.

The regulation of the striking rise and fall of these mRNAs during differentiation provides our main challenge for future research. In addition, we are intrigued by the difference in timing of the destruction of the two mRNAs. What controls the timing and activity of the Pac men (Fig. 2) and other intracellular gremlins that are responsible for the rises and falls of these mRNAs?

Multiple Genes, Multiple Products?

There are multiple isoforms of *Naegleria* actin, and similar isoforms are produced when *Naegleria* actin mRNA is translated in the wheat-germ system (Sussman et al. 1984a,b). It is possible that these isoforms indicate the presence of three to four species of actin in this unicellular eukaryote. There are multiple actin genes in *Naegleria*—as in other amebae that appear to produce a single form of actin (Firtel 1981; Nellen and Gallwitz 1982)—and more than one of these genes are expressed in *Naegleria*. Further work will be necessary to decide whether actins of different primary structures are produced.

The multitubulin hypothesis, as originally proposed (Fulton and Simpson 1976), was based on the following evidence, here brought up to date. Amebae, at least at the time of mitosis, have more than enough tubulin to build flagella (Kowit and Fulton 1974b; Fulton and Dingle 1971 and unpubl.). Yet most, if not all, of the tubulin that makes up the α and β subunits of both outer-doublet and central-pair microtubules is synthesized during differentiation (Kowit and Fulton 1974b; Fulton and Simpson 1976; P.A. Simpson et al., in prep.). Flagellar tubulin is also antigenically distinct from the tubulin of amebae (Kowit and Fulton 1974a), even when it is synthesized in a cell-free system (Lai et al. 1979b). These results suggested the synthesis and utilization of a distinct tubulin for flagella, and we propose that cells may use different tubulins, made by different genes, for different functions. There is now abundant evidence that many organisms, indeed all eukaryotes examined so far except *Saccharomyces,* have and express multiple tubulin genes (for review, see Cleveland 1983). There is also evidence for tissue-specific expression of particular tubulin genes (see, e.g., Kalfayan and Wensink 1982; Raff et al. 1982; Burland et al. 1983), but there still is no definitive evidence whether or not the products of different tubulin genes are used for distinct microtubular organelles.

We find a low abundance of α-tubulin mRNA in amebae (Fig. 2, time 0). One possible explanation for this is that the tubulin mRNA for mitosis is made and degraded rapidly during a short period in the cell cycle, as it is during differentiation. Changes in tubulin mRNA abundance during the cell cycle have been observed in *Chlamydomonas* (Ares and Howell 1982), and tubulin synthesis peaks just before mitosis in *Physarum* (Laffler et al. 1981). Another possibility is that tubulin mRNA is present in the amebae but is not recognized well by the flagellar α-tubulin DNA probe, i.e., there might be two distinct α-tubulin gene families or subfamilies.

As a first test of these possibilities, we examined the abundance of α-tubulin mRNA during the preparations for and execution of a highly synchronous mitosis in *Naegleria* amebae (Fulton and Guerrini 1969). To our surprise, the abundance of α-tubulin mRNA, measured using pNαT1, remained at a similar low level throughout the mitotic cycle. We are continuing to explore this provocative result.

SUMMARY

Naegleria gruberi undergoes a remarkable differentiation from amebae with an actin-based motility system to flagellates with a tubulin-based system. During this 100-minute differentiation, synthesis of actin is selectively restricted while flagellar tubulin is synthesized as a programmed event of differentiation. The abundance of actin mRNA decreases rapidly during the first hour of differentiation, while the abundance of α-tubulin mRNA increases about 80-fold. Then α-tubulin mRNA disappears rapidly, with a half-life of less than 10 minutes. Thus, this organism displays a yin-yang utilization of two fundamental motility systems—actin-based and tubulin-based—correlated with the synthesis of these structural proteins and the abundance and stability of their mRNAs. The rapid changes in mRNA abundance offer two aspects of the regulation of gene expression—transcription and mRNA stability—in stark relief for analysis. Both α-tubulin and actin are encoded by multigene families. In each case it is clear that more than one gene is expressed, but the cause of the multiple isoforms of actin and the issue of whether one or two gene families encode the mitotic and flagellar α-tubulins remain challenges for the future.

ACKNOWLEDGMENT

This research was supported by grants from the National Science Foundation and the National Institutes of Health.

REFERENCES

Ares, M. and S.H. Howell. 1982. Cell cycle stage-specific accumulation of mRNAs encoding tubulin and other polypeptides in *Chlamydomonas. Proc. Natl. Acad. Sci.* **79:** 5577.

Bennetzen, J.L. and B.D. Hall. 1982. Codon selection in yeast. *J. Biol. Chem.* **257:** 3026.

Brunke, K.J., E.E. Young, B.U. Buchbinder, and D.P. Weeks. 1982. Coordinate regulation of the four tubulin genes of *Chlamydomonas reinhardi. Nucleic Acids Res.* **10:** 1295.

Burland, T.G., K. Gull, T. Schedl, R.S. Boston, and W.F. Dove. 1983. Cell type-dependent expression of tubulins in *Physarum. J. Cell Biol.* **97:** 1852.

Cleveland, D.W. 1983. The tubulins: From DNA to RNA to protein and back again. *Cell* **34:** 330.

Cleveland, D.W., M.A. Lopata, R.J. MacDonald, N.J. Cowan, W.J. Rutter, and M.W. Kirschner. 1980. Number and evolutionary conservation of α- and β-tubulin and cytoplasmic β- and γ-actin genes using specific cloned cDNA probes. *Cell* **20:** 95.

Firtel, R.A. 1981. Multigene families encoding actin and tubulin. *Cell* **24:** 6.

Fulton, C. 1977a. Cell differentiation in *Naegleria gruberi. Ann. Rev. Microbiol.* **31:** 597.

———. 1977b. Intracellular regulation of cell shape and motility in *Naegleria.* First insights and a working hypothesis. *J. Supramol. Struct.* **6:** 13.

———. 1983. Macromolecular synthesis during the quick-change act of *Naegleria. J. Protozool.* **30:** 192.

Fulton, C. and A.D. Dingle. 1971. Basal bodies, but not centrioles, in *Naegleria. J. Cell Biol.* **51:** 826.

Fulton, C. and A.M. Guerrini. 1969. Mitotic synchrony in *Naegleria* amebae. *Exp. Cell Res.* **56:** 194.

Fulton, C. and P.A. Simpson. 1976. Selective synthesis and utilization of flagellar tubulin.

The multi-tubulin hypothesis. In *Cell motility* (ed. R. Goldman et al.), vol. 3, p. 987. Cold Spring Harbor Laboratory, Cold Spring Harbor, New York.

Fulton, C. and C. Walsh. 1980. Cell differentiation and flagellar elongation in *Naegleria gruberi*: Dependence on transcription and translation. *J. Cell Biol.* **85:** 346.

Kafatos, F.C., C.W. Jones, and A. Efstratiadis. 1979. Determination of nucleic acid sequence homologies and relative concentrations by dot hybridization procedure. *Nucleic Acids Res.* **7:** 1541.

Kalfayan, L. and P.C. Wensink. 1982. Developmental regulation of *Drosophila* α-tubulin genes. *Cell* **29:** 91.

Kowit, J.D. and C. Fulton. 1974a. Purification and properties of flagellar outer doublet tubulin from *Naegleria gruberi* and a radioimmune assay for tubulin. *J. Biol. Chem.* **249:** 3638.

———. 1974b. Programmed synthesis of tubulin for the flagella that develop during cell differentiation in *Naegleria gruberi. Proc. Natl. Acad. Sci.* **71:** 2877.

Laffler, T.G., M.T. Chang, and W.F. Dove. 1981. Periodic synthesis of microtubular proteins in the cell cycle of *Physarum. Proc. Natl. Acad. Sci.* **78:** 5000.

Lai, E.Y., D.J. Sussman, and C. Fulton. 1979a. The rise and fall of translatable mRNAs for tubulin and actin during cell differentiation in *Naegleria*: Dependence on transcription. *J. Cell Biol.* **83:** 416a.

Lai, E.Y., C. Walsh, D. Wardell, and C. Fulton. 1979b. Programmed appearance of translatable flagellar tubulin mRNA during cell differentiation in *Naegleria. Cell* **17:** 867.

Landfear, S.M., D. McMahon-Pratt, and D.F. Wirth. 1983. Tandem arrangement of tubulin genes in the protozoan parasite *Leishmania enriettii. Mol. Cell. Biol.* **3:** 1070.

Neff, N.F., J.H. Thomas, P. Grisafi, and D. Botstein. 1983. Isolation of the β-tubulin gene from yeast and demonstration of its essential function in vivo. *Cell* **33:** 211.

Nellen, W. and D. Gallwitz. 1982. Actin genes and actin messenger RNA in *Acanthamoeba castellanii*. Nucleotide sequence of the split actin gene I. *J. Mol. Biol.* **159:** 1.

Okayama, H. and. P. Berg. 1982. High-efficiency cloning of full-length cDNA. *Mol. Cell. Biol.* **2:** 161.

Raff, E.C., M.T. Fuller, T.C. Kaufman, K.J. Kemphues, J.E. Rudolph, and R.A. Raff. 1982. Regulation of tubulin gene expression during embryogenesis in *Drosophila melanogaster. Cell* **28:** 33.

Seebeck, T., P.A. Whittaker, M.A. Imboden, N. Hardman, and R. Braun. 1983. Tubulin genes of *Trypanosoma brucei*: A tightly clustered family of alternating genes. *Proc. Natl. Acad. Sci.* **80:** 4634.

Silflow, C.D. and J.L. Rosenbaum. 1981. Multiple α- and β-tubulin genes in *Chlamydomonas* and regulation of tubulin mRNA levels after deflagellation. *Cell* **24:** 81.

Sussman, D.J., E.Y. Lai, and C. Fulton. 1984a. Rapid disappearance of translatable actin mRNA during cell differentiation in *Naegleria. J. Biol. Chem.* **259:** 7355.

Sussman, D.J., J.R. Sellers, P. Flicker, E.Y. Lai, L.E. Cannon, A.G. Szent-Györgyi, and C. Fulton. 1984b. Actin of *Naegleria gruberi:* Absence of N^{τ}-methylhistidine. *J. Biol. Chem.* **259:** 7349.

Thomashow, L.S., M. Milhausen, W.J. Rutter, and N. Agabian. 1983. Tubulin genes are tandemly linked and clustered in the genome of *Trypanosoma brucei. Cell* **32:** 35.

Toda, T., Y. Adachi, Y. Hiraoka, and M. Yanagida. 1984. Identification of the pleiotropic cell division cell cycle gene *NDA*2 as one of two different α-tubulin genes in *Schizosaccharomyces pombe. Cell* **37:** 233.

Vandekerckhove, J., A.A. Lai, and E.D. Korn. 1984. Amino acid sequence of *Acanthamoeba* actin. *J. Mol. Biol.* **172:** 141.

Walsh, C. 1984. Synthesis and assembly of the cytoskeleton of *Naegleria gruberi* flagellates. *J. Cell Biol.* **98:** 449.

Genetic Techniques for Analysis of Nematode Muscle

Amy Bejsovec, David Eide, and Philip Anderson
Department of Genetics
University of Wisconsin
Madison, Wisconsin 53706

Current knowledge concerning the protein components of muscle is based largely on biochemical analysis of myofibrillar preparations. Such in vitro studies are limited because direct evidence for the in vivo function of isolated proteins is difficult to obtain. In vitro techniques, furthermore, are restricted often to the study of abundant proteins. Very little is known about minor sarcomeric components or how the sarcomere is assembled overall. Certainly, the assembly and function of a structure as complex as the sarcomere require many more than the dozen or so proteins commonly studied.

Genetic techniques provide an alternative approach to the study of muscle. Mutations that cause muscle disfunction define genes required to construct a normal muscle cell. The nature of mutant defects provides insights into the functions of the wild-type gene products. Genetic analysis is not restricted to the study of abundant proteins. The genes for even low-abundance proteins are subject to mutation, and if a gene product is required for muscle assembly or function, such mutants will be muscle-defective. Macromolecular complexes provide special opportunities for genetic intervention, because gene products that directly interact in the structure can often be identified. Mutational defects that affect one member of an interacting pair of proteins can be compensated by mutations affecting its interacting partner. Not all genes, however, are directly accessible to genetic analysis. For example, genes whose products are essential for cell or organism viability and genes having more than one functional copy present special problems.

Certain features of the life cycle of the free-living nematode *Caenorhabditis elegans* allow a detailed genetic analysis of striated muscle. *C. elegans* is a diploid animal that reproduces as a self-fertilizing hermaphrodite. Induced mutations, therefore, quickly become homozygous, allowing recessive mutant phenotypes to be identified. Self-fertilization is especially important for analysis of muscle. Even severely paralyzed mutants are viable, because individual animals do not have to mate to reproduce. In addition to self-fertilization, *C. elegans* has a sexual cycle that permits all standard complementation and recombination tests to be performed. The life cycle is quick, and genetic analysis of *C. elegans* is quite well developed.

C. elegans mutants defective for the synthesis, assembly, or function of muscle are readily available, and steady progress is being made in relating mutant genes to specific polypeptide products. Muscle-defective mutants are defined as those in which muscle ultrastructure is abnormal. This usually leads to defective movement of the animal, although the precise phenotype depends on the gene involved. To date, approximately two dozen *C. elegans* genes necessary for normal muscle function have been identified by the isolation of several hundred mutants (Waterston et al. 1980; Zengel and Epstein 1980). The protein products of three of these genes are known: The *unc-54* gene encodes one of two myosin heavy-chain isozymes expressed in body-wall muscle cells (Epstein et al. 1974); the *unc-15* gene encodes the thick-filament component paramyosin (Waterston et al. 1977); and the *unc-92* gene (now called the *act-1*, *act-2*, and *act-3* gene cluster) encodes a muscle actin (Krause and Hirsh, this volume.).

Clearly, our knowledge of the genetic basis for *C. elegans* muscle synthesis and function is incomplete. The protein products of many genes identified with mutants are unknown. Several genes have been defined only by the isolation of a single mutant allele. This indicates that not all genes have yet been identified. The structural genes for many proteins known from biochemical analyses to be components of muscle have not been identified genetically. We need to know how many genes are required for sarcomere assembly and function, what the gene products are. and how they function during assembly and contraction.

We discuss here two techniques that should provide answers to some of these important questions. The first technique is a method for selecting muscle-defective mutants of *C. elegans*. This allows us to isolate many more muscle-defective mutants and to identify new genes. The second technique is a method for identifying recombinant DNA clones that correspond to these genes, even though the protein products are unknown. DNA clones offer realistic approaches to identifying the gene products.

DISCUSSION

Selection for Muscle-defective Mutants

Wild-type Motility

Wild-type *C. elegans* animals are freely motile. The controlled contraction/relaxation of body-wall muscle cells generates a wave in the body form that propagates down (or up) the length of an animal and propels it forward (or backward). Mutants having defective body-wall muscle are less motile, but the precise phenotype depends on the gene involved. Muscle-defective mutants are relatively rare (even after chemical mutagenesis), and they can be difficult to identify among nondefective (wild-type) animals. Enrichment or selection techniques would greatly improve the methods of isolation. Park and Horvitz have recently developed genetic techniques that select directly for mutations affecting the muscle-related genes *unc-22* and *unc-54* (E. Park and R. Horvitz, pers. comm.). We have discovered that these genetic techniques can also be used to select for other classes of muscle-defective mutants. This selection is based on properties of the mutant *unc-105(n490)*.

unc-105(n490) *and Its Revertants*

The mutant strain *unc-105(n490)* was isolated and characterized by E. Park (pers. comm.). The product of the *unc-105* gene is unknown, but mutant animals containing the *unc-105(n490)* allele of this gene are extremely paralyzed. The reason for this appears to be due, at least in part, to a contractile defect of body-wall muscle cells. Body-wall muscle of *unc-105(n490)* animals is abnormal. When viewed with polarized light optics (a rapid method to determine gross sarcomere organization), birefringent striations, although quite strong, are diffuse and irregular, having no consistent periodicity. Occasional patches of increased birefringence are found in these cells. *unc-105(n490)* animals are short, attaining less than one-half normal body length. The nature of *unc-105(n490)* partial revertants (see below) suggests that the primary defect in *unc-105(n490)* leads to sustained contraction ("hypercontraction") of body-wall muscle cells.

Motile revertants of *unc-105(n490)* are easily selected. Partial revertants (those that are intermediate between mutants and wild type) usually contain mutations unlinked to *unc-105(n490)* that suppress the hypercontractive paralysis. These suppressors, when backcrossed and segregated in the absence of *unc-105 (n490)*, are usually muscle-defective mutants themselves. The suppressors, therefore, are epistatic to *unc-105(n490)*, and this is the basis for their selection. Most suppressors are mutations in either the *unc-22* or *unc-54* gene (E. Park, pers. comm.). Figure 1 shows an example of an *unc-54* myosin heavy-chain mutant isolated as a revertant of *unc-105(n490)*. Although the defect(s) underlying *unc-105(n490)* paralysis are not understood, the isolation of its suppressors appears simple: Muscle-defective mutants are unable to generate the force necessary for the hypercontractive rigor of *unc-105(n490)*. Although many muscle-defective mutants would normally be classified as "paralyzed," even the slight motility of these mutants is sufficient to select them from their *unc-105(n490)* paralyzed siblings.

We have isolated over 300 muscle-defective revertants of *unc-105(n490)*. Table 1 summarizes our genetic analysis of these mutants. Eight genes known from previous work to be necessary for normal muscle function are represented in our collection. In addition, three mutations complement all existing muscle-defective mutants (including each other); therefore, they identify three new genes. The distribution of mutations among the 11 genes in Table 1 is highly biased, but in most of the reversion experiments contributing to Table 1, we were exclusively retaining *unc-54* mutants (see below). We hope to saturate the nematode genome for mutants that are muscle-defective using this selective technique. Such mutants provide a minimal estimate of the genetic complexity of muscle and are a resource for biochemical analysis of mutant gene products.

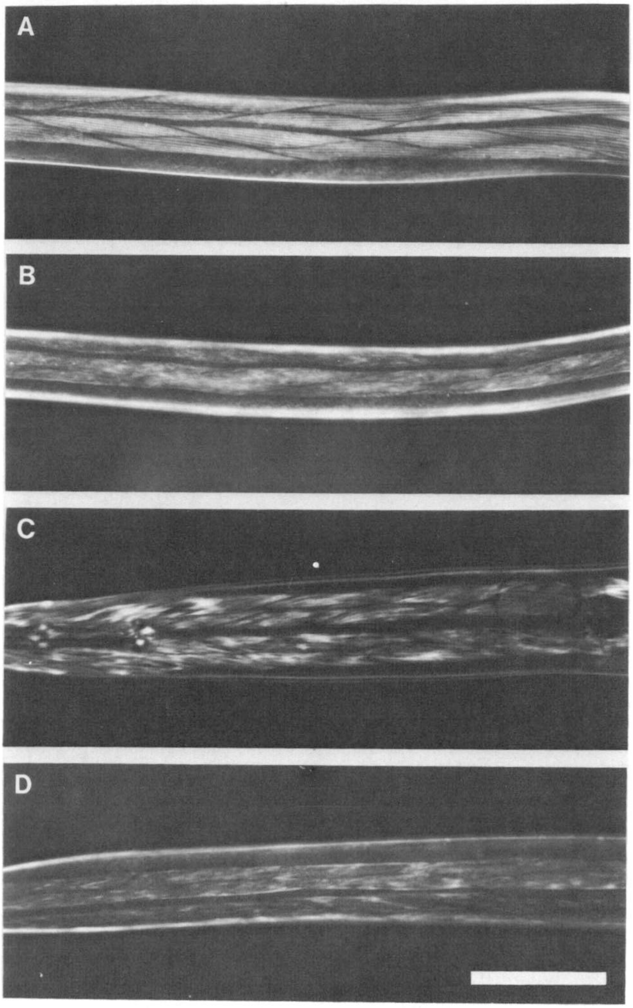

Figure 1
Polarized light micrographs of muscle-defective mutants and control strains. (*A*) Wild-type muscle when viewed using polarized light optics. *C. elegans* contains 95 mononucleate, obliquely striated body-wall muscle cells. (*B*) Mutant muscle from the control strain *unc-54(e1092)*. This strain contains a null allele of the *unc-54* myosin heavy-chain gene. (*C*) Muscle of the paralyzed, hypercontracted strain *unc-105(n490)*. (*D*) Muscle of *unc-105(n490); unc-54(r325)*, a spontaneous partial revertant of *unc-105(n490)*. This strain has a muscle phenotype of typical *unc-54* mutants. Bar, 100 μm.

Use of the Transposable Element Tc*1* to Identify Gene Clones

In many different organisms, spontaneous mutations (those that arise in the absence of any known mutagenic treatment) are often caused by insertion of transposable elements into or nearby a structural gene (for review, see Shapiro 1983).

Table 1
Summary of Muscle-defective Mutants Selected as Suppressors of *unc-105* (*n490*) Hypercontraction

Gene	Gene product	Number of alleles
unc-54	myosin heavy chain	186
unc-22	unknown	100
unc-15	paramyosin	20
unc-52	unknown	1
unc-60	unknown	1
unc-82	unknown	1
unc-87	unknown	1
unc-89	unknown	1
New genes (3)	unknown	1 each
Total		314

Mutations caused by transposon insertion provide a convenient method for cloning the DNA sequences that surround the insertion site (Kleckner et al. 1977; Bingham et al. 1981). In the case of eukaryotic genes, clones can be identified by hybridization with the transposable element; flanking DNA represents the gene into which the transposon was inserted.

Structure of Spontaneous unc-54 *Mutations*

The selective techniques described above are sufficiently strong that spontaneous muscle-defective mutants can be isolated. We have studied the gene structure of 114 spontaneous mutations affecting the *unc-54* myosin heavy-chain gene. *unc-54* encodes one of two myosin heavy-chain isozymes expressed in body-wall muscle cells (Epstein et al. 1974). We chose *unc-54* for analysis because recombinant DNA clones are available and a detailed restriction map of the wild-type gene is known (MacLeod et al. 1981).

We analyzed spontaneous *unc-54* mutants isolated in the genetic backgrounds of three different varieties of *C. elegans*. These varieties were chosen because each represents an independent natural isolate of *C. elegans*, and we did not know which, if any, of them contained genetically active transposable elements. We find that in one (and only one) of these varieties, most spontaneous *unc-54* mutations are due to insertion of a transposable element, designated Tc*1*, within the gene.

*Transposable Element Tc*1

The transposon Tc*1* was first identified as a repetitive DNA sequence dispersed throughout the *C. elegans* genome (Emmons et al. 1983; Liao et al. 1983). Tc*1* has structural features characteristic of a transposable element, but it had not been observed to transpose prior to this work. Most wild isolates of *C. elegans* contain approximately 25 copies of Tc*1* within their genomes, although certain varieties contain many more copies (Emmons et al. 1983; Liao et al. 1983). Of the 18 spontaneous *unc-54* mutations we isolated in the Bergerac variety of *C.*

elegans, 10 of them are insertion mutations. Using a combination of hybridization and gene-cloning techniques, we have shown that each of these mutations is caused by insertion of Tc*1*. The sites of ten independent Tc*1* insertions within *unc-54* are shown in Figure 2.

Genetic Properties of unc-54⁻::*Tc*1 *Mutants*

Each of the *unc-54*⁻::Tc*1* insertion mutants has a phenotype typical of *unc-54* mutants. These mutants are unusual, however, because most of them are genetically unstable. They revert spontaneously to *unc-54*⁺ at relatively high frequencies. For example, strain TR445 (*unc-54(r322)*::Tc*1*) reverts spontaneously to *unc-54*⁺ at a frequency of approximately 4×10^{-6}. Revertants are detected as wild-type (motile) animals spontaneously occurring in cultures of the paralyzed mutants. Revertants of TR445 no longer contain Tc*1* inserts within *unc-54*. In each case, the excisions of Tc*1* are precise or nearly precise (± 50 bp).

*Utility of Tc*1

These observations suggest a general method for cloning *C. elegans* genes. Spontaneous mutations that are isolated in the Bergerac variety of *C. elegans* and that are genetically unstable likely contain insertions of Tc*1* within the affected gene. Tc*1* insertion sites can be cloned and the flanking sequences can be obtained. A comparison of wild-type, insertion-mutant, and revertant strains identifies which of the many Tc*1* copies corresponds to the gene of interest. Insertion mutations can be substituted into any genetic background, and the copy number of Tc*1* is sufficiently low in most *C. elegans* varieties that this is a realistic approach.

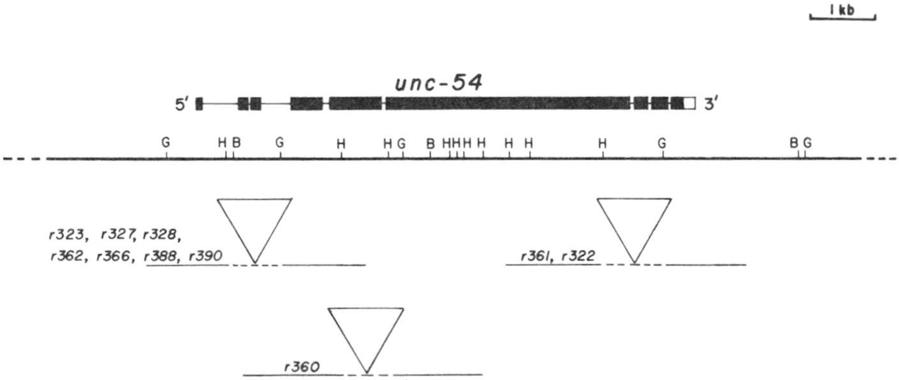

Figure 2
Tc*1* insertions affecting *unc-54*. The position of each Tc*1* insertion within *unc-54* is shown relative to the restriction map of this region. Dashed line indicates the limits of precision for mapping data; each insertion resides somewhere within the dashed region. The *unc-54* gene spans over 7.2 kb and is interrupted by eight short introns (Karn et al. 1983). The organization of *unc-54* coding sequences is indicated above its restriction map. (■) Exons; (□) untranslated 3′ trailer sequence. Positions of the 5′ and 3′ termini of *unc-54* mRNA have not been precisely defined. The limits of *unc-54* shown here are the AUG translational initiation codon (5′) and the AAUAAA polyadenylation signal (3′). The precise termini are thought to be nearby (Karn et al. 1983). (B) *Bam*HI; (G) *Bgl*II; (H) *Hin*dIII.

SUMMARY

C. elegans is one of few organisms for which a classical genetic analysis of striated muscle is possible. The components of muscle are defined by the isolation and genetic analysis of muscle-defective mutants. We have described genetic techniques that select directly for muscle-defective mutants, and we have identified several new genes required for normal muscle function. Our next major challenge is to identify the protein products for these and other muscle-related genes.

Properties of the transposable element Tc*1* may provide a method to identify the products of genes that have been defined only through the phenotypic effects of mutations. In the Bergerac variety of *C. elegans,* most spontaneous mutations affecting the *unc-54* myosin heavy-chain gene are due to insertion of Tc*1*. This suggests that any gene for which spontaneous mutations are available can be cloned using "transposon-tagging" techniques. Because of the strong selective techniques available, genes involved in muscle function are especially well suited to this approach.

REFERENCES

Bingham, P.M., R. Levis, and G.M. Rubin. 1981. Cloning of DNA sequences from the *white* locus of *D. melanogaster* by a novel and general method. *Cell* **25:** 693.

Emmons, S.W., L. Yesner, K. Ruan, and D. Katzenberg. 1983. Evidence for a transposon in *Caenorhabditis elegans. Cell* **32:** 55.

Epstein, H.F., R.H. Waterston, and S. Brenner. 1974. A mutant affecting the heavy chain of myosin in *Caenorhabditis elegans. J. Mol. Biol.* **90:** 291.

Karn, J., S. Brenner, and L. Barnett. 1983. Protein structural domains in the *Caenorhabditis elegans unc-54* myosin heavy chain gene are not separated by introns. *Proc. Natl. Acad. Sci.* **80:** 4253.

Kleckner, N., J.R. Roth, and D. Botstein. 1977. Genetic engineering *in vivo* using translocatable drug resistance elements. *J. Mol. Biol.* **116:** 126.

Liao, L.W., B. Rosenzweig, and D. Hirsh. 1983. Analysis of a transposable element in *Caenorhabditis elegans. Proc. Natl. Acad. Sci.* **80:** 3585.

MacLeod, A.R., J. Karn, and S. Brenner. 1981. Molecular analysis of the *unc-54* myosin heavy chain gene of *Caenorhabditis elegans. Nature* **291:** 386.

Shapiro, J.A., ed. 1983. *Mobile genetic elements*. Academic Press, New York.

Waterston, R.H., R.M. Fishpool, and S. Brenner. 1977. Mutants affecting paramyosin in *Caenorhabditis elegans. J. Mol. Biol.* **117:** 679.

Waterston, R.H., J.N. Thomson, and S. Brenner. 1980. Mutants with altered muscle structure in *Caenorhabditis elegans. Dev. Biol.* **77:** 271.

Zengel, J.M. and H.F. Epstein. 1980. Identification of genetic elements associated with muscle structure in the nematode *Caenorhabditis elegans. Cell Motil.* **1:** 73.

Nematode Thick-filament Structure and Assembly

Henry F. Epstein, Irving Ortiz,
Gary C. Berliner, and David M. Miller III
Department of Neurology
Baylor College of Medicine
Houston, Texas 77030

Myosins from slime molds to brain cells show a remarkable commonality of general molecular properties. These characteristics include two globular domains or heads that contain ATPase and actin-binding sites and the fibrous, coiled-coil α-helical rod that interacts with other molecules in assembly. Two heavy chains (m.w. ~200,000) contribute to both heads, whereas two kinds of light chains bind to each head. In this paper, we consider striated muscles and their myosins. The phylogenetically distant nematode body-wall muscles and rabbit fast skeletal muscles produce myosin heavy chains, with about 47% of the amino acid sequences in the heads and about 37% of the amino acid sequences in the rod being identical (Karn et al. 1984). Myosin heavy chains are therefore highly conserved proteins.

Contrasting with the phylogenetic conservation of myosin structure and sequence is the diversity of supramolecular arrangements of myosin assemblies in striated muscles, the so-called thick filaments. The lengths of thick filaments range from 1.55 μm in vertebrates, 2–4 μm in insect flight muscles, 10 μm in the nematode to 40 μm in certain mollusks. The average diameters of these filaments range from about 15 nm in vertebrates, 20 nm in insects, 25 nm in nematodes to 50–100 nm in some molluscan muscles. The surface arrangements of the myosin heads also vary in these different species. The lattice arrangements between thick filaments and the interdigitating, actin-containing thin filaments differ in terms of symmetry and thick:thin stoichiometry between these muscles. It appears likely that other protein components of these muscles interact with the very similar myosins to produce this structural diversity. The relatively subtle differences between myosin isoforms may also be important in these interactions.

We define isoform in the case of myosin, for example, as a protein that is defined as a myosin by biochemical criteria but that can be distinguished on the basis of intrinsic molecular structure from another myosin within the same organism.

In this paper, we describe experiments suggesting that two genetically different isoforms of myosin play distinct roles in concert with other proteins during the assembly of thick filaments in the nematode, *Caenorhabditis elegans*.

Different Locations of Myosin Isoforms

There are two major types of striated muscles in the nematode, the body-wall muscle and the pharyngeal muscle. Their physiological functions and filament structures are very different. It is not surprising that the myosins of the two muscles are also distinct. However, the observation that each muscle type contains two isoforms of myosin heavy chain was enigmatic (Epstein et al. 1974).

More recently, we have determined that the two body-wall isoforms, A and B, are located in different positions within the same cells (Miller et al. 1983). These experiments utilized monoclonal antibodies specific to the body-wall isoforms. Monoclonal antibody 28.2 is specific to the B heavy chain, and monoclonal antibody 5.6 is specific to the A heavy chain. In immunofluorescence experiments, each of the antibodies reacts separately with all of the 95 body-wall muscle cells. The anti-B antibody reacts with all but a central region of the thick-filament-containing A bands, whereas the anti-A antibody reacts only with the central region. These observations are confirmed by double labeling the same muscle cells with the two antibodies. The reactions of 28.2 with the B isoform and of 5.6 with the A isoform appear complementary (Fig. 1). Thus, two myosin heavy chains coexist in the same muscle cells and the same sarcomeres, confirming previous work with affinity-purified polyclonal rabbit antibodies (Mackenzie et al. 1978). Furthermore, the B isoform appears to be restricted to the flanking regions of the thick filaments, whereas the A isoform appears localized to the central portions of the thick filaments.

To confirm the above results and extend them to the level of individual thick filaments, the monoclonal antibodies were reacted separately with isolated filaments (Mackenzie and Epstein 1980, 1981). The reactions were identified by secondary goat anti-mouse antibodies, and the preparations were negatively stained with uranyl acetate before electron microscopy. All thick filaments react with either antibody, indicating that both forms of the myosin heavy chain coexist in the same filaments. The A form is localized to the central 1.8 μm, whereas the B form is found in the polar 4.4 μm on either side (Fig. 2). The localizations suggest that the surfaces of nematode thick filaments are differentiated by myosin isoform content into five zones.

This model is important because it explains previous observations regarding nematode body-wall myosins and because it implies distinct roles for the two isoforms during assembly and in the final structure. Previous studies of nematode myosins indicated that native molecules are homodimers of either heavy-chain isoform; no heterodimers are detected (Schachat et al. 1977, 1978). In the present model, the two heavy-chain isoforms would be distributed into different myosins, as well. The ratio of the heavy-chain isoforms appears constant during the postembryonic stages of the nematode life cycle despite a 40-fold increase

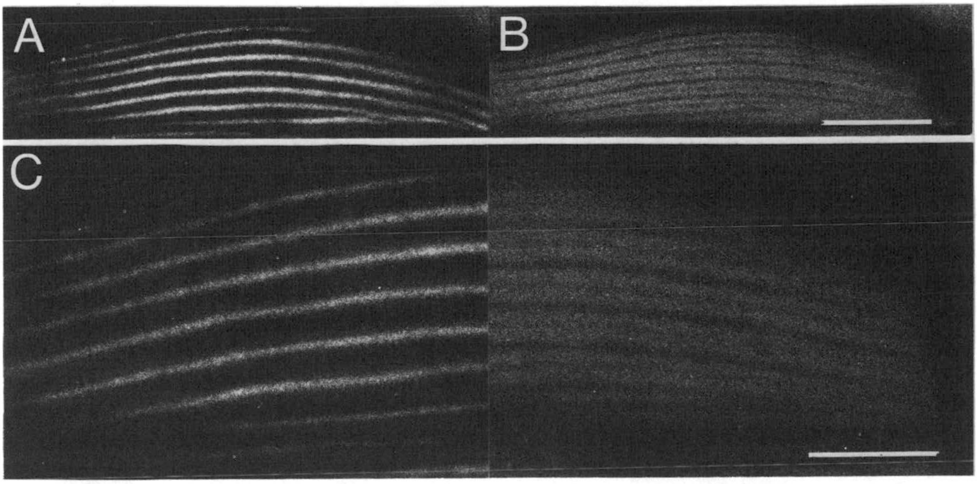

Figure 1
Simultaneous immunofluorescence labeling of a body-wall muscle cell with antibodies 28.2 and 5.6. Fluorescent images of a single muscle cell double-labeled with rhodamine-coupled antibody 5.6 (1.7 μg/ml) (*A*) and antibody 28.2 (10 μg/ml) plus fluorescein-coupled goat anti-mouse IgG (*B*). Bar, 10 μm. (*C*) Cutaway construction of images in *A* and *B*. (*Left*) Cell in *A*; (*right*) cell in *B*. Bar, 5 μm. Note that the bright lines of the antibody-5.6 reaction match with the nonfluorescent gaps in the middle of the A bands of the antibody-28.2 image. (Reprinted, with permission, from Miller et al. 1983.)

in total myosin content (Garcea et al. 1978). This fixed ratio is consistent with the relative distributions of the two forms in the thick filaments.

The different locations of the two myosins have several structural implications. The central regions that contain myosin A only are the regions in which myosins pack in an antiparallel fashion to form the bare zone. This zone is believed to be the site of initiation for myosin assembly. The same zone is where the cross-linking structures of the M line interact with the thick filaments to produce the ordered lattice of the A band. All of these interactions involve myosin A. In contrast, B myosins are located in the primary cross-bridge-bearing regions of the thick filaments that interact with thin filaments to produce movement. The myosin-B molecules pack only in a parallel fashion. The elongation and termination reactions during assembly involve myosin B.

Discovery of a New Core Structure

Nematode thick filaments were treated with solutions of increasing KCl concentration in order to sequentially dissociate myosins B and A from the polar ends and confirm their differential locations. Similar treatments of mammalian thick filaments yield the antiparallel assemblage of myosin about the central bare zone and solubilize the more distal myosins (Trinick and Cooper 1980; Niederman and Peters 1982). Thick filaments from nematodes, like other invertebrates, contain paramyosin as a major protein component internal to the myosin (Waterston et al. 1974, 1977; Harris and Epstein 1977). With progressively increasing KCl con-

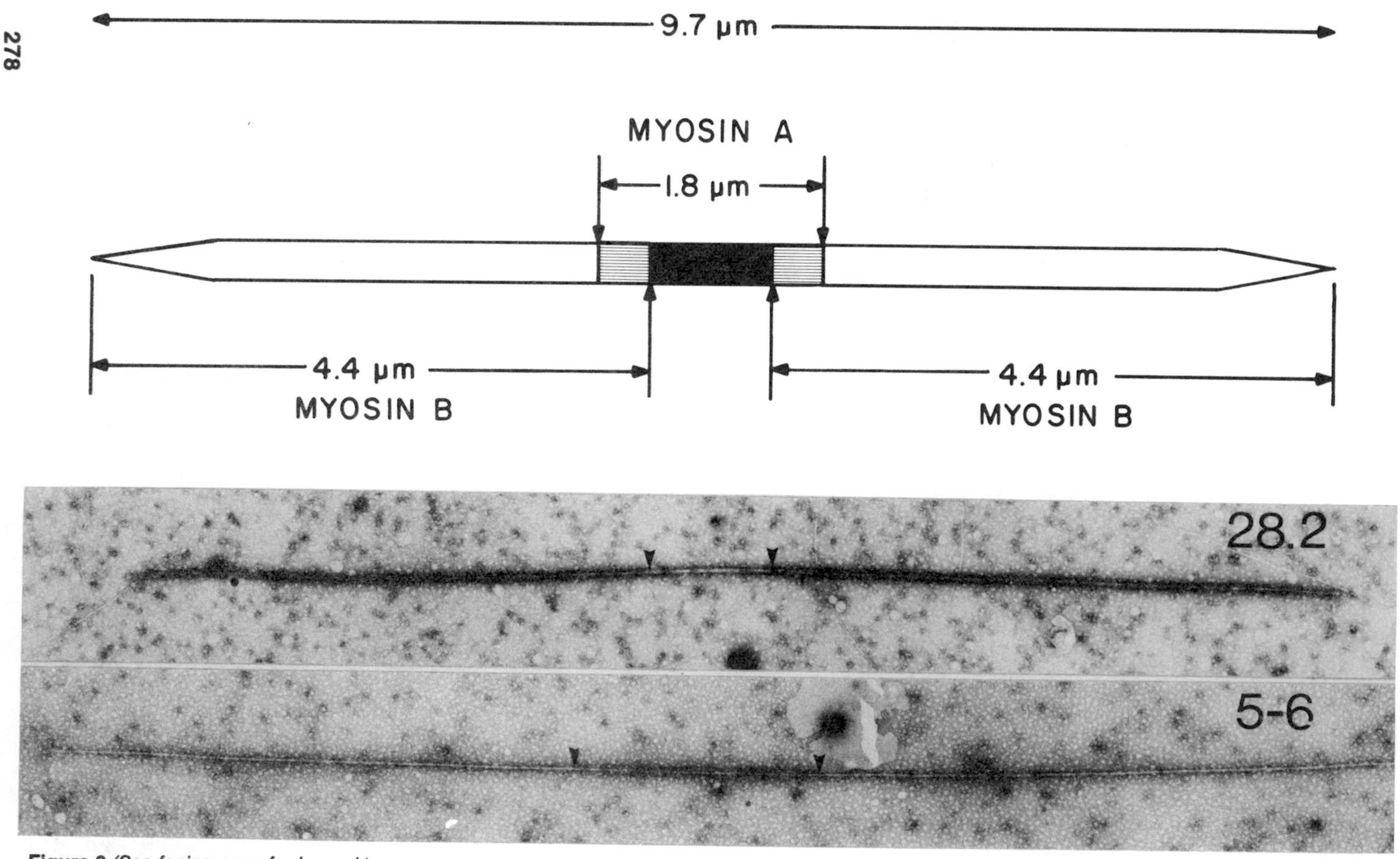

Figure 2 *(See facing page for legend.)*

centrations at pH 6.35, up to 80% of total myosin heavy chain and 95% of the paramyosin are solubilized (Fig. 3). When the amounts of these proteins remaining in sedimentable form are plotted as a function of KCl concentration, the resulting functions are very similar to one another, indicating that myosin and paramyosin are dissociated together. The sharp transitions in these functions resemble the cooperative equilibria observed in other confirmational transitions such as protein folding and DNA melting. Myosin A still remains in the central 1.8 μm at 0.35 M KCl and in the central 0.9 μm at 0.45 M KCl. Myosin B is not detected in the central 0.9 μm at either 0.35 M or 0.45 M KCl but is in the remainder of the central 1.8 μm at 0.35 M KCl. The lengths of the remaining myosin-containing regions, as judged from electron microscopy of antibody-treated material, follow a function of salt concentration similar to that indicated by biochemical measurements. However, close examination of the thick-filament remnants indicates that 15-nm-wide filamentous structures emanate from the ends of the myosin-containing regions (Fig. 4). At 0.35 M and 0.45 M KCl, these 15-nm structures have variable lengths up to several micrometers. It is likely that they extend the entire 10-μm length of native filaments; at 0.25 M KCl, full-length filaments are stripped of myosin and paramyosin at their ends, and 15-nm core structures are revealed.

These new core structures do not react with 28.2 monoclonal antibodies specific to myosin B, 5.6 monoclonal antibodies specific to myosin A, or 5-9 and 5-23 monoclonal antibodies specific to paramyosin. Similar structures are observed in dissociated filaments from the myosin-B-deficient mutant E190 and the paramyosin-deficient mutant E1214. Both genetic and immunological evidence indicate that these core structures are composed of proteins other than myosin and paramyosin.

The core structures are probably tubes that are at least partially hollow. Negative stain can penetrate part of the length of isolated core structures. In cross sections of nematode body-wall and pharyngeal muscles, the polar ends of the thick filaments appear hollow, whereas the central regions in the H zone appear solid (Fig. 5). In the body-wall thick filaments, there appears to be a gradient of density in the cores from the central region to the polar ends. The change in core density correlates with the tapering of thick filaments that have diameters of 37.5 nm in the central region and 15 nm at the polar ends. The diameters of the polar ends are about the same dimensions as the dissociated core structures. Thus, the myosin and paramyosin domains must be changing in their molecular packing and number of molecules per unit length from the central to the polar regions. The core structure appears to be relatively constant in diameter.

Figure 2
(*Top*) Locations of myosins A and B. Model of nematode thick filament depicting distribution of myosin isoforms suggested by monoclonal antibody reactions (*middle*). Electron micrograph of thick filaments reacted with antibody 28.2 (10 μg/ml) plus goat anti-mouse IgG (20 μg/ml), then negatively stained (*bottom*). Electron micrograph of thick filament reacted with antibody 5-6 (1 μg/ml) plus goat anti-mouse IgG (20 μg/ml), then negatively stained. All three panels are at the horizontal scale in micrometers. (Reprinted, with permission, from Miller et al. 1983.)

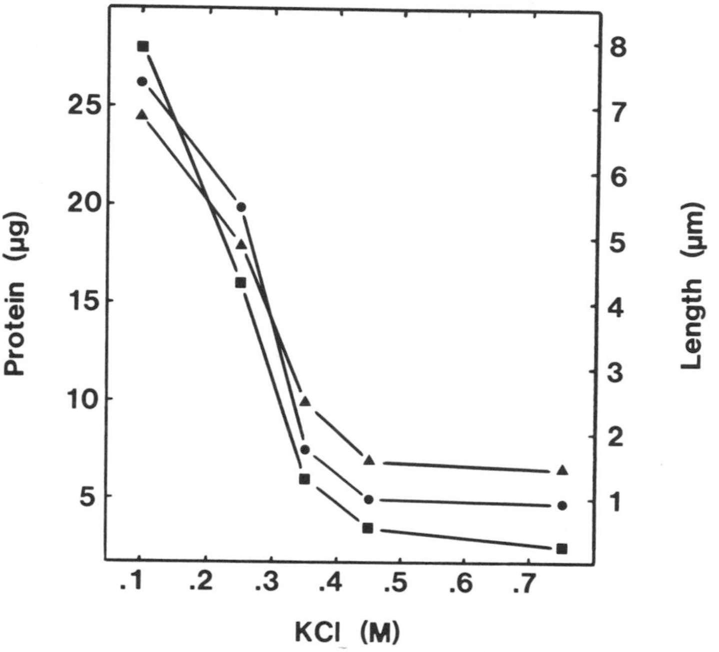

Figure 3
Dissociation of thick filaments by KCl. Equal amounts of filaments were incubated in equal volumes of the corresponding KCl solutions. After 30 min, the solutions were centrifuged at 100,000*g* for 60 min. Protein analyses were carried out by SDS-polyacrylamide (7.5%) gel electrophoresis, staining by Coomassie brilliant blue, and densitometry. Length analysis was performed on filaments after reaction with antibody 28.2 and electron microscopy as in Fig. 2. (■) Paramyosin content; (▲) myosin heavy-chain content; (●) myosin-containing region length.

The qualitative aspects of thick-filament tapering and changes in core density that we find in *C. elegans* have been observed in other nematodes (Hirumi et al. 1971), in lobsters (Hayes et al. 1971), and in various insects (Goode 1972; Reedy et al. 1981). As with previous observations in *C. elegans* (Waterston et al. 1974, 1977), the results with other invertebrate muscles were interpreted in terms of a pure myosin filament or myosin-paramyosin cofilament. Dissociation experiments of the kind reported here were not done previously in these systems. Thus, it appears likely that the core structure and its internal components as well as some tapering of the myosin and paramyosin domains may be general elements in invertebrate thick filaments. Theoretical considerations of myosin structure and content in thick filaments led Squire (1971) to propose a separate core structure.

Model of Nematode Thick-filament Assembly

The structural observations above suggest a model for the structure of nematode thick filaments and their assembly. The body-wall thick filaments are composed

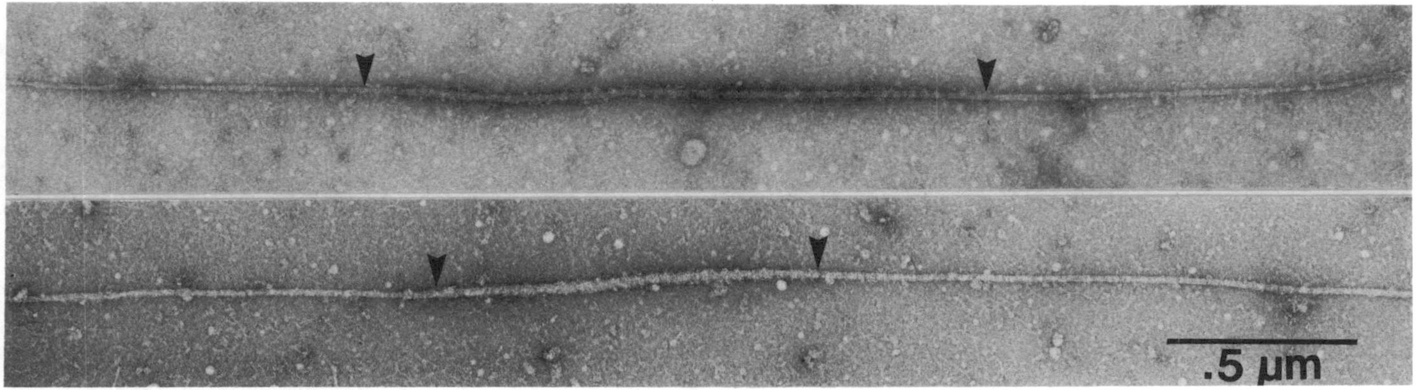

Figure 4
Electron micrographs of dissociated thick filaments. The upper and lower figures are typical negatively stained thick filaments after treatment with 0.35 M and 0.45 M KCl at pH 6.35, respectively. Arrowheads denote the myosin-containing regions that are longer at 0.35 M than at 0.45 M KCl (see Fig. 3). Extending from these regions are the newly identified core structures.

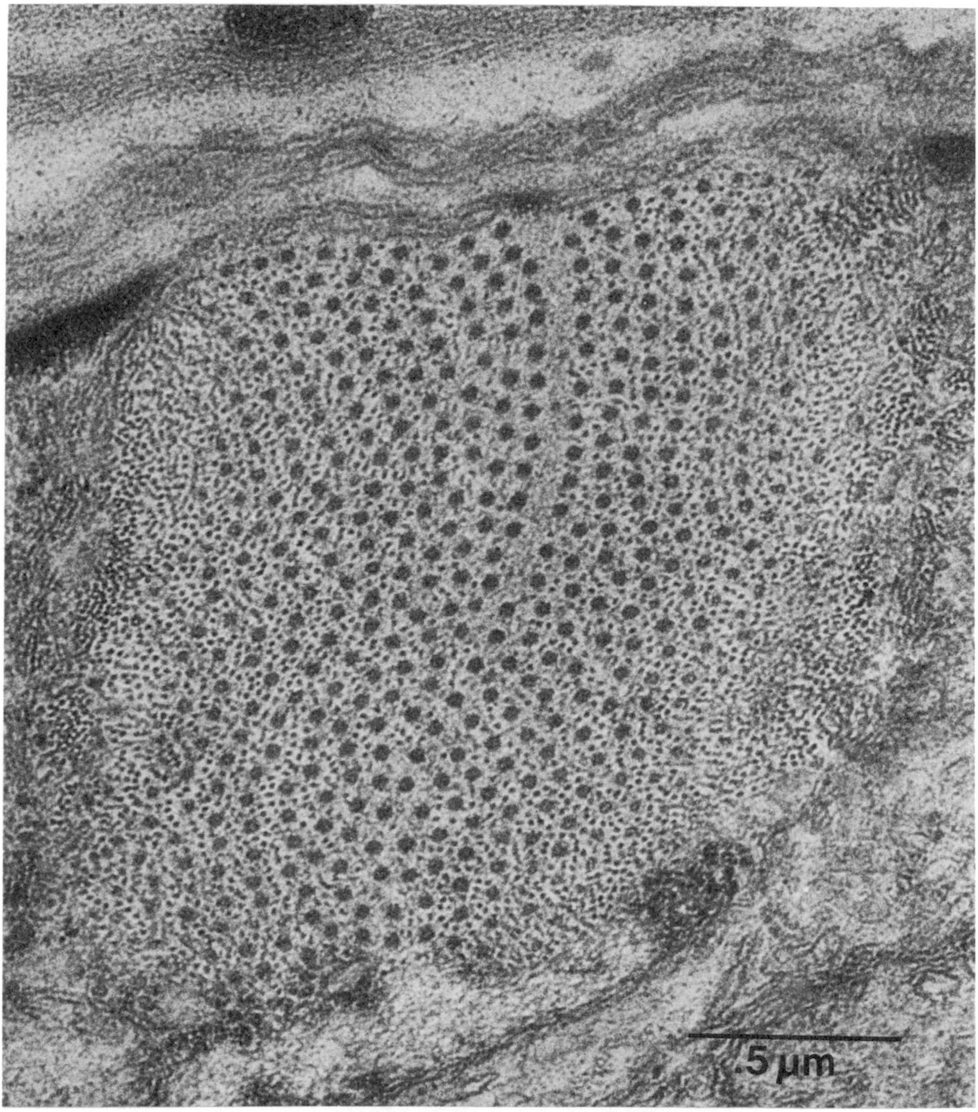

Figure 5

Electron micrograph of body-wall muscle. Cross section of body-wall muscle prepared according to the method of Epstein et al. (1974). Note that the diameters of thick filaments in the center of the array are significantly wider than the filament diameters at the periphery. The cores of the central filaments are more electron dense than the more peripheral filaments. Because these muscles are obliquely striated, the central filaments in cross section represent thick filaments sectioned in their central regions, whereas the peripheral filaments represent thick filaments sectioned in their polar regions. Thus, the cross section represents a sampling of thick filaments along their entire length from polar ends to polar ends.

of three coaxial domains: the myosin domain at the filament surface, the paramyosin domain in an intermediate volume, and the core domain (Fig. 6). The appearance and position of the paramyosin domain are consistent with previous studies of molluscan thick filaments in which myosin was solubilized (Szent-Györgyi et al. 1971). We do not know the complexity of the core domain in terms of molecular composition; there are at least two distinct structures: the core tube and the core internal structure. There may be additional proteins in each of these structures. The myosin domain is clearly subdivided into five subdomains based on the localization of myosins A and B (Fig. 1). The other domains may also be differentiated molecularly and structurally.

The various structures obtained by dissociation of wild-type filaments and studies of specific mutants affecting myosin and paramyosin suggest the following model for the assembly of thick filaments in the body-wall muscles of *C. elegans* (Fig. 7). Structure I represents a nucleation structure whose formation is rate-limiting for assembly as a whole. Myosin A is necessary for the formation of this structure. There may be central internal components in this structure other than paramyosin, since myosin A does form 1.5-μm bipolar structures in the paramyosin-deficient mutant E1214 (Mackenzie and Epstein 1980). The polar-core template is the material of varying density at the center of the cores. In this model, this template would be the ultimate determinant of thick-filament length in a fashion analogous to the role of RNA in tobacco mosaic virus.

Structure II represents the additional formation of the polar-core tube that emanates from the central myosin assemblages in the salt-dissociation experiments. We have recently obtained structures very similar in appearance to structure II from the paramyosin-deficient *unc-15* mutant E1214. Structure III represents the polymerization of paramyosin about the polar core. We observe paracrystalline paramyosin in rare partially dissociated thick filaments at 0.25 M KCl. *unc-15* mutants in which paramyosin is present but presumably altered (such as E73) produce paracrystalline paramyosin structures that resemble structure III. Further analysis of these mutants is in progress to determine whether the E73 paracrystals contain myosin A and core structures. Structure IV is the final thick filament after polymerization of myosin B about paramyosin.

This model for assembly is presented as a plausible basis for further experiments. One set of observations that supports its plausibility concerns the myosin-B-deficient mutant E190. In the body-wall muscles of this mutant, only 20% of the normal number of thick filaments are present, and this number correlates with the fraction of body-wall myosin that is myosin A normally (MacLeod et al. 1977). No structures resembling any of the intermediates are observed. We have shown that these filaments are of normal length. In contrast to wild-type filaments, myosin A is present along the entire length. These observations are consistent with our assembly model. The nucleation is critically dependent on myosin-A concentration. Once a specific structure is nucleated, assembly continues until completion. In this case, most of the myosin A is utilized in forming the polar myosin subdomain of initiated filaments, and fewer additional structures become nucleated because the available myosin-A concentration becomes too low.

Further experiments are required to establish whether this pathway requires the presence of other components or kinetic compartments to ensure the stepwise addition of myosin A, polar-core protein, paramyosin, and myosin B. The

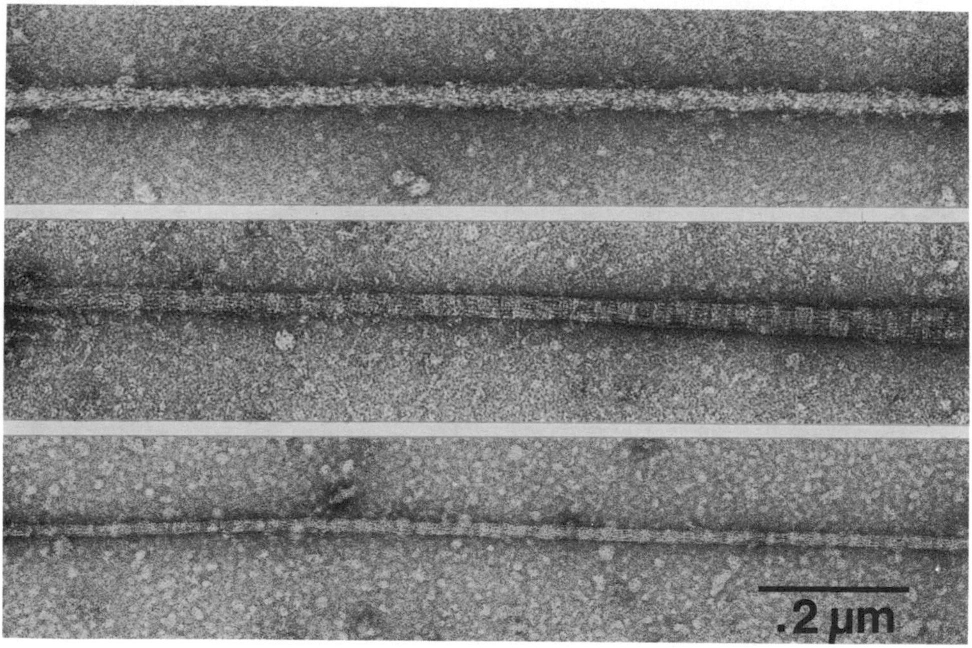

Figure 6
Electron micrographs of three domains of thick filaments. (*Top*) Myosin surface domain of native thick filament at 0.1 M KCl. (*Middle*) Paramyosin intermediate domain at right and myosin surface domain at left of thick filament at 0.25 M KCl. Note paracrystalline "checkerboard" arrangement of paramyosin (see Szent-Györgyi et al. 1971). (*Bottom*) Core structure of thick filament at 0.35 M KCl.

study of mutants in other genes of *C. elegans* and the identification and localization of additional components will be useful in the confirmation of this model.

SUMMARY AND CONCLUSIONS

The body-wall muscles of the nematode *C. elegans* contain two myosin isoforms, A and B. By specific monoclonal antibodies, myosins A and B are localized to the central and polar subdomains, respectively, of body-wall thick filaments. Salt dissociation of the filaments shows that the myosins occupy a surface domain, paramyosin constitutes an intermediate domain, and a new core structure exists. These findings suggest a model for assembly in which the myosin-A subdomain, the core structure, paramyosin domain, and myosin-B subdomain are formed in sequential steps. Relevant information from specific nematode mutants supports this model, and data from additional mutants may confirm it.

ACKNOWLEDGMENTS

We thank Drs. Sandra Honda, Roger Kornberg, Tony Otsuka, and Mathoor Sivaramakrishnan for discussions. This research was supported by grants from the

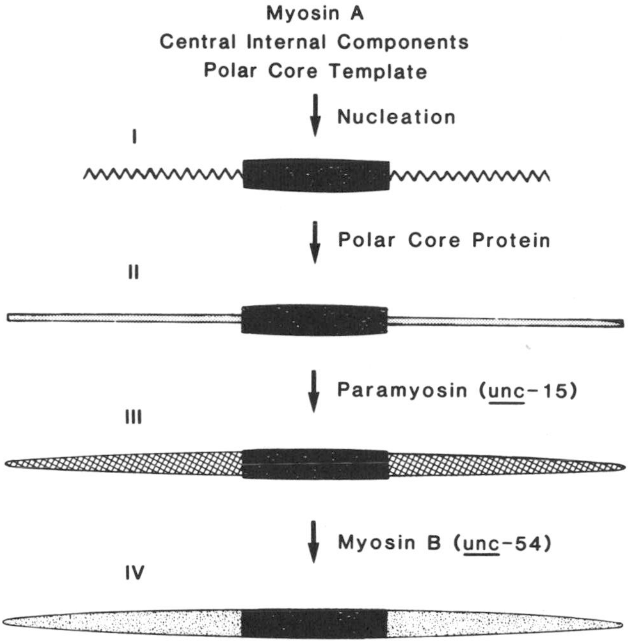

Figure 7
Model of thick-filament assembly in body-wall muscles of the nematode, *C. elegans*.

National Institute of Aging, National Institute of General Medical Sciences, and the Muscular Dystrophy Association. H.F.E. held a research career development award from the National Institute of Child Health and Human Development, and D.M.M. was a Muscular Dystrophy Association fellow during part of this research.

REFERENCES

Epstein, H.F., R.H. Waterston, and S. Brenner. 1974. A mutant affecting the heavy chain of myosin in *Caenorhabditis elegans. J. Mol. Biol.* **90:** 291.

Garcea, R.L., F. Schachat, and H.F. Epstein. 1978. Coordinate synthesis of two myosins in wild-type and mutant nematode muscle during larval development. *Cell* **15:** 421.

Goode, M.D. 1972. Ultrastructure and contractile properties of isolated myofibrils and myofilaments from *Drosophila* flight muscle. *Trans. Am. Microsc. Soc.* **91:** 182.

Harris, H.E. and H.F. Epstein. 1977. Myosin and paramyosin of *Caenorhabditis elegans*: Biochemical and structural properties of wild-type and mutant proteins. *Cell* **10:** 709.

Hayes, D., M. Huang, and C.R. Zobel. 1971. Electron microscope observations on thick filaments in striated muscle from the lobster *Homarus americanus. J. Ultrastruct. Res.* **37:** 17.

Hirumi, H., D.J. Raski, and N.O. Jones. 1971. Primitive muscle cells of nematodes: Morphological aspects of platymyarian and shallow coelomyarian muscles in two plant parasitic nematodes, *Trichodurus christei* and *Longidorus elongatus. J. Ultrastruct. Res.* **34:** 517.

Karn, J., N.J. Dibb, and D.M. Miller. 1984. Cloning namatode myosin genes. In *Cell and muscle motility* (ed. R.N. Dawben and J.W. Shay). (In press.)

Mackenzie, J.M., Jr. and H.F. Epstein. 1980. Paramyosin is necessary for determination of nematode thick filament *in vivo*. *Cell* **22:** 747.

———. 1981. Electron microscopy of nematode thick filaments. *J. Ultrastruct. Res.* **76:** 277.

Mackenzie, J.M., Jr., F. Schachat, and H.F. Epstein. 1978. Immunocytochemical localization of two myosins within the same muscle cells in *Caenorhabditis elegans*. *Cell* **15:** 413.

MacLeod, A.R., R.H. Waterston, R.M. Fishpool, and S. Brenner. 1977. Identification of the structural gene for a myosin heavy chain in *Caenorhabditis elegans*. *J. Mol. Biol.* **114:** 133.

Miller, D.M., III, I. Ortiz, G.C. Berliner, and H.F. Epstein. 1983. Differential localization of two myosins within nematode thick filaments. *Cell* **34:** 477.

Niederman, R. and L.K. Peters. 1982. Native bare zone assemblage nucleates myosin filament assembly. *J. Mol. Biol.* **161:** 505.

Reedy, M.K., K.R. Leonard, R. Freeman, and T. Arad. 1981. Thick myofilament mass determination by electron scattering measurements with the scanning transmission electron microscope. *J. Muscle Res. Cell Motil.* **2:** 45.

Schachat, F.H., R.L. Garcea, and H.F. Epstein. 1978. Myosins exist as homodimers of heavy chains: Demonstration with specific antibody purified by nematode mutant myosin affinity chromatography. *Cell* **15:** 405.

Schachat, F.H., H.E. Harris, and H.F. Epstein. 1977. Two homogeneous myosins in body-wall muscle of *Caenorhabditis elegans*. *Cell* **10:** 721.

Squire, J.M. 1971. General model for the structure of all myosin-containing filaments. *Nature* **233:** 457.

Szent-Györgyi, A.G., C. Cohen, and J. Kendrick-Jones. 1971. Paramyosin and the filaments of molluscan "catch" muscles. II. Native filaments: Isolation and characterization. *J. Mol. Biol.* **56:** 239.

Trinick, J. and J. Cooper. 1980. Sequential disassembly of vertebrate thick filaments. *J. Mol. Biol.* **141:** 315.

Waterston, R.H., H.F. Epstein, and S. Brenner. 1974. Paramyosin of *Caenorhabditis elegans*. *J. Mol. Biol.* **90:** 285.

Waterston, R.H., R.M. Fishpool, and S. Brenner. 1977. Mutants affecting paramyosin in *Caenorhabditis elegans*. *J. Mol. Biol.* **117:** 825.

Actin Gene Expression in *Caenorhabditis elegans*

Michael Krause and David Hirsh
Department of Molecular, Cellular, and Developmental Biology, University of Colorado Boulder, Colorado 80309

The switching on or off of specific genes is a fundamental aspect of cellular differentiation during metazoan development. The molecular events involved in this switching are not yet understood, but they are now subject to analysis with the current technology available in molecular biology.

Much of the work directed toward the understanding of developmental gene regulation has focused on the genes encoding the protein actin. Actin is the major thin-filament protein in both muscle and nonmuscle cells (Pollard and Weihing 1974; Clarke and Spudich 1977). The protein sequences of actins from a variety of tissues in several organisms have been determined (Collins and Elzinga 1975; Vandekerckhove and Weber 1978; 1979), and the actin genes from a number of organisms have been isolated and are currently being studied (Buckingham and Minty 1983; Fyrberg et al. 1983; Sanchez et al. 1983). This work has revealed that actins are evolutionarily conserved, are encoded in most species by multigene families, and are differentially regulated, both spatially and temporally, during development.

Caenorhabditis elegans Actin Genes

The actin genes of the nematode *C. elegans* represent an attractive system for studying the developmental regulation of gene expression. The availability of defined cell lineages, genetics, and actin mutants makes *C. elegans* uniquely qualified for such studies.

The four actin genes of *C. elegans* have been cloned and characterized (Files et al. 1983). Actin genes *1, 2,* and *3* are clustered in a 12-kb region of the genome. The exact position of actin gene *4* relative to the cluster is unknown, but it is at least 30 kb away.

All four actin genes have been sequenced (M. Wild, pers. comm.). A comparison of the sequence data shows that genes *1* and *3* are identical at the nucleotide level. Actin genes *1* and *2* differ by 35 silent third-base changes and only three amino acid substitutions. A comparison of genes *1* and *4* shows many nucleotide changes but only one amino acid substitution. A distinction between *C. elegans* cytoplasmic and muscular actin genes cannot be ascertained by sequence, as each gene codes for some amino acids specific to either vertebrate cytoplasmic actin or skeletal-muscle actin.

Wild-type Actin Gene Expression

Gene-specific probes were constructed from the putative 3′-transcribed, but untranslated, region of each gene. These gene-specific probes were used to analyze the transcription of each gene during postembryonic development with the following results.

All four actin genes are transcribed (Fig. 1). Three transcript size classes, with lengths of 1450, 1550, and 1650 nucleotides, are observed. At present, we believe that the 1450 class is composed of transcripts from genes *1* and *2*; the 1550- and 1650-size classes are apparently composed of transcripts from genes *3* and *4*, respectively.

The abundance of transcripts from any given gene remains constant during postembryonic development. There is variation, however, when comparing the relative abundance of transcripts among the four genes. Actin gene-*4* transcripts are the most abundant. Gene-*1* and -*3* transcripts occur at equal abundance and at about one-half the level seen for gene *4*. Gene-*2* transcripts are the least abundant class at about one-fifth the level of gene *4*.

Identification of Actin Mutants

A DNA polymorphism between two closely related strains of *C. elegans* is linked to the cluster of three actin genes. This polymorphism served as a phenotypic marker that was genetically mapped to a 2-cM region of the genome (Files et al. 1983). This same 2-cM region contained several tightly linked mutations that were candidates for being actin gene lesions. Animals with dominant mutations at this map position have disorganized thin-filament structure in the body-wall musculature (R.H. Waterston and T.R. Lane, in prep.). One might expect actin mutants to be dominant because recessive loss of function mutations might be masked by compensatory activity from other members of the gene family. Another characteristic of mutants at this locus is a high frequency of reversion. The phenotypes of these revertants are indistinguishable from wild-type animals. It seemed that frequent reversion could be related to recombination or conversion among the members of the cluster.

DNA was isolated from 73 revertants and analyzed for changes in the restriction fragment pattern for the actin genes. Four revertants had detectable abnormalities, indicating that rearrangements had occurred in the actin cluster. We

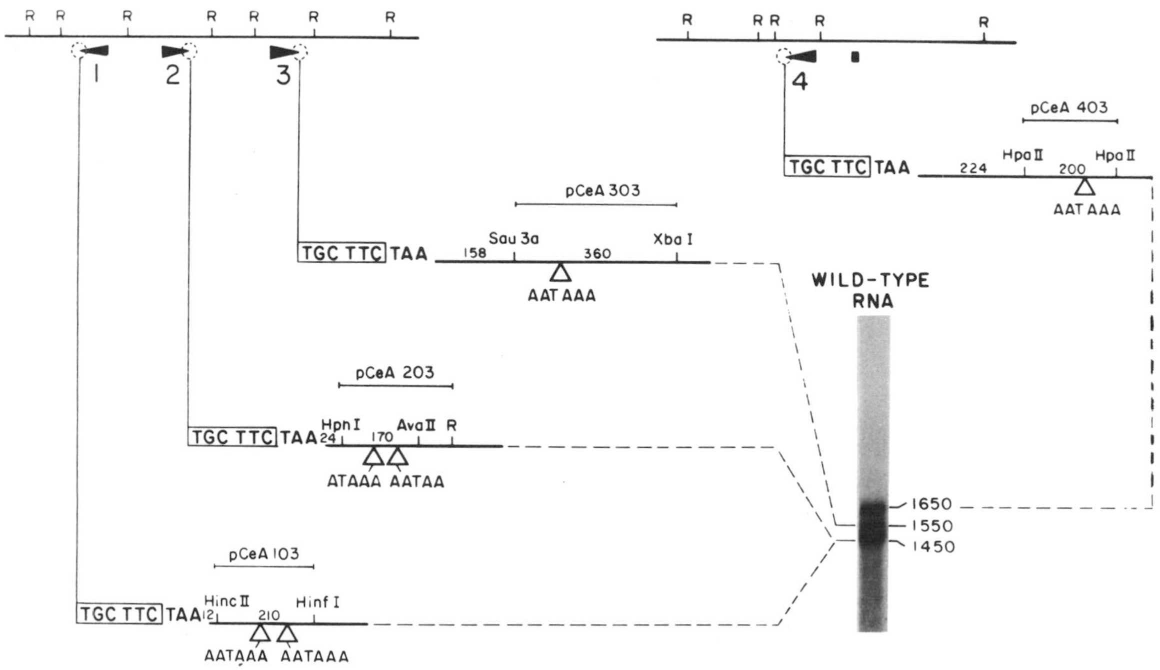

Figure 1
Actin gene-specific probes. The last two codons of each gene are boxed and the translational terminational signals are indicated. Gene-specific probes were subcloned from the 3′-untranslated region of each gene. The region covered by each probe is designated by the lines marked pCeA 103, 203, 303, and 403. These probes were used to correlate the genes with observed transcript size classes. Arrowheads indicate the possible poly(A)-addition signals. Distances between restriction sites are given in base pairs.

assume that these rearrangements are the cause of the reversion and that the original mutations are within the actin genes.

The four revertants having abnormal actin restriction patterns were further analyzed to define the nature of the rearrangement. In two revertants, RW2580 and RW2458, the rearrangement is confined to a single gene (Fig. 2). We assume that the original mutations of these two revertants are alleles of the rearranged genes. The other two reversion events involve more than one gene, and we are unable to determine which gene carried the original mutation.

Revertant Analysis

We were interested in the effect of the revertant DNA rearrangements on the transcription of the actin genes. A comparison of actin gene transcription in wild-type and three of the revertant animals is shown (Fig. 3). Each revertant tested has a developmental pattern of transcription that is different from that seen in wild-type animals. Specifically, RW2246 has two (rather than three) size classes of actin transcripts throughout postembryonic development. RW2458 and

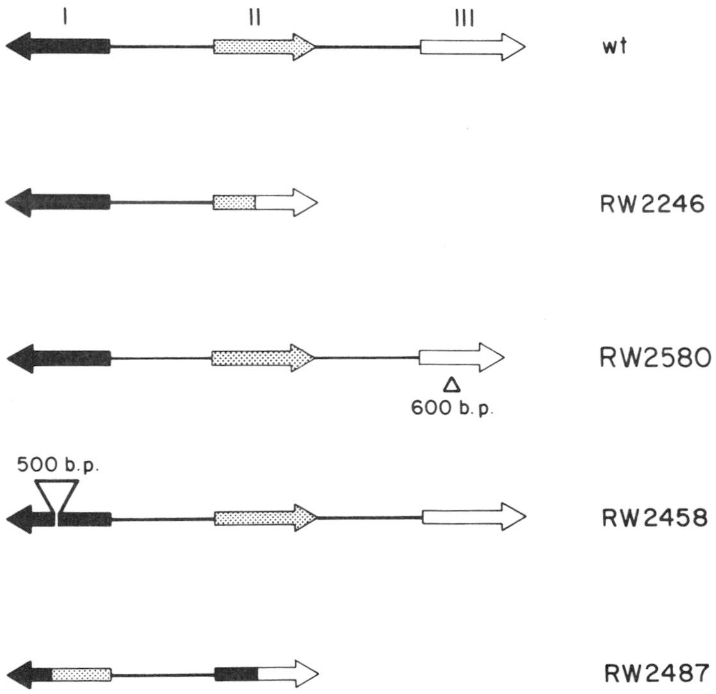

Figure 2
Genomic organization of the actin gene cluster for wild-type (N2) and the revertants RW2246, RW2580, RW2458, and RW2487. Break points for insertions, deletions, and recombination are approximate and are only meant to indicate the region involved in the rearrangements.

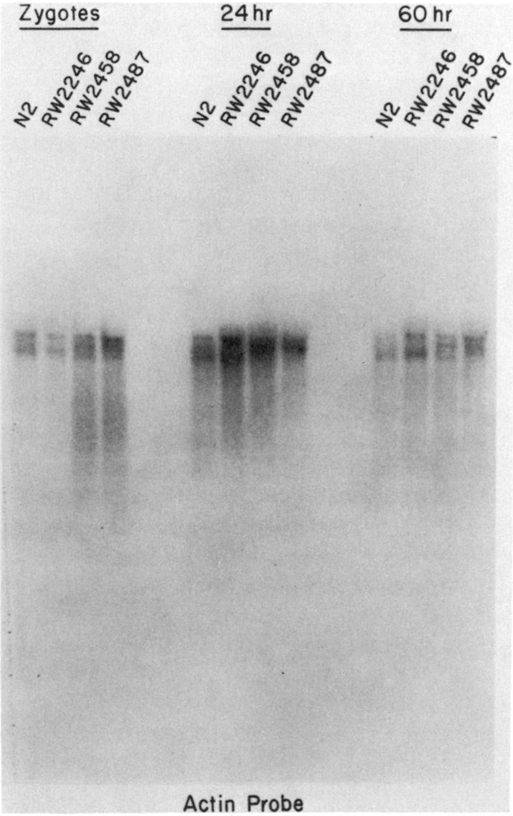

Figure 3
Comparison of wild-type (N2) and RW2246, RW2458, and RW2487 actin transcripts. Total RNA, isolated from staged zygotes, mid-larval (24 hr), and gravid adults, was Northern-blotted and probed with a cross-hybridizing actin probe.

RW2487 both have three size classes of actin message at zygotic and adult stages but only two at the 24-hour stage.

We are currently analyzing these three revertants with gene-specific probes to correlate the revertant transcriptional patterns with their respective DNA rearrangements.

SUMMARY

We have shown that all four actin genes of *C. elegans* are transcribed and that the abundance of actin transcripts is constant during postembryonic development. In addition, we have identified revertants of actin gene mutants that have gross rearrangements in the actin gene cluster, leading to abnormal patterns of actin gene transcription. Using gene-specific probes, we can now correlate the DNA rearrangements with the transcriptional patterns in these and future revertants. We believe that such studies will allow us to identify the physiological role of each actin gene product and to define DNA sequences involved in the developmental regulation of the actin genes during *C. elegans* development.

ACKNOWLEDGMENTS

We thank Dr. Robert Waterston for supplying us with the actin mutant and revertant strains on which much of this work is based. We also thank Dr. Martha Wild for the actin gene sequence data. This work was supported by grants GM-19851 and GM-26515 from the U.S. Public Health Service. M.K. received support from a U.S. Public Health Service training grant (GM-07135).

REFERENCES

Buckingham, M. and A. Minty. 1983. Contractile protein genes. In *Eukaryotic genes: Their structure, activity and regulation* (ed. N. MacLean et al.). (In press.)

Clarke, M. and J.A. Spudich. 1977. Non-muscle contractile proteins: The role of actin and myosin in cell motility and shape determination. *Annu. Rev. Biochem.* **46:** 797.

Collins, J.H. and M. Elzinga. 1975. The primary structure of actin from rabbit skeletal muscle. Completion and analysis of the amino acid sequence. *J. Biol. Chem.* **250:** 5915.

Files, J., S. Carr, and D. Hirsh. 1983. The actin gene family in *Caenorhabditis elegans. J. Mol. Biol.* **164:** 355.

Fyrberg, E.A., J.W. Manaffey, B.J. Band, and N. Davidson. 1983. Transcripts of the six *Drosophila* actin genes accumulate in a stage- and tissue-specific manner. *Cell* **33:** 115.

Pollard, T.J. and R.R. Weihing. 1974. Actin and myosin and cell movement. *Crit. Rev. Biochem.* **2:** 1.

Sanchez, F., S.L. Tobin, V. Rdest, E. Zulauf, and B.J. McCarthy. 1983. Two *Drosophila* actin genes in detail. Gene structure, protein structure and transcription during development. *J. Mol. Biol.* **163:** 533.

Vandekerckhove, J. and K. Weber. 1978. Actin amino acid sequences. Comparison of actins from calf thymus, bovine brain and SV40-transformed mouse 3T3 cells with rabbit skeletal muscle. *Eur. J. Biochem.* **90:** 451.

———. 1979. The amino acid sequence of actin from chicken skeletal muscle actin and chicken gizzard smooth muscle actin. *FEBS Lett.* **102:** 219.

Genetic Analysis of Microtubule Function in *Drosophila*

Elizabeth C. Raff
Institute for Molecular and Cellular Biology and Department of Biology, Indiana University Bloomington, Indiana 47405

Margaret T. Fuller
Department of Molecular, Cellular, and Developmental Biology, University of Colorado Boulder, Colorado 80302

Two hypotheses are generally put forward to explain the enormous functional diversity of microtubules in eukaryotic cells. First, the tubulin dimer may be viewed as a useful ubiquitous building block that imparts to the structures into which it is constructed only the general constraints intrinsic to the size and shape of microtubules. In this view, the morphological and functional specificity of each set of microtubules is thought to be derived from additional structural or regulatory components, which include the diverse set of proteins grouped collectively as microtubule-associated proteins (MAPs). Alternatively, the tubulin subunits may themselves impart some or all of the specificity of assembly into different microtubule organelles by virtue of small differences in their primary structure, derived either through genetic diversity or by posttranslational modifications; i.e., tubulin dimers might be more like Legos than bricks. It seems reasonable to suppose that both models may be true, probably to differing extents in different circumstances. The application of genetics to the study of tubulin gene expression and microtubule function allows a direct experimental approach to this question. Analysis of mutations in tubulin and interacting genes promises to make unique and powerful contributions to our understanding of the mechanisms that regulate microtubule function in vivo.

Spermatogenesis in *Drosophila melanogaster*

We are studying the relationship of differential tubulin gene expression to microtubule function during spermatogenesis in *Drosophila*. The *Drosophila* tubulin gene family consists of four genes each for α- and β-tubulin. At least one gene from each class is prevalently expressed in many tissues, whereas other genes are differentially expressed with respect to developmental time and/or cell type (for review, see Raff 1984). Our genetic studies have centered on analysis of the function of β_2-tubulin, a testis-specific subunit encoded by a single gene on the right arm of the third chromosome (Kemphues et al. 1979, 1980).

Spermatogenesis affords an excellent model system in which to study the complex interactions among structural proteins, resulting in assembly of subcellular organelles. As illustrated in Figure 1, the differentiation of the motile spermatozoa proceeds by a series of dramatic changes in cellular architecture (Tates 1971; Lindsley and Tokuyasu 1980). Many of these changes in morphology are microtubule-mediated events utilizing morphologically different microtubule arrays. Four crucial classes of microtubules are the mitotic spindle, the meiotic spindle, the sperm flagellar axoneme, and the cytoplasmic microtubules involved in nuclear shaping. Although the adult testis contains cysts in all stages of development, differentiation of the testis proceeds coordinately with the development of the body, and the first appearance of each sequential stage in spermatogenesis can be examined by looking at developing testes from the larval through pupal stages to adulthood. Using this approach, Kemphues et al. (1982) examined tubulin expression during spermatogenesis. As shown in Figure 1, during early stages (including the mitotic cell division and growth of the primary spermatocytes), only the ubiquitous subunit β_1-tubulin is present. Synthesis of the testis-specific gene product, β_2-tubulin, first begins shortly before meiosis; thereafter, this form becomes the major β-tubulin component of the testis tubulin pool and is the only β-tubulin detectable in Coomassie-blue-stained gels of proteins in mature motile sperm.

Class I Recessive Mutations in the Gene for β_2-Tubulin: Loss of Microtubule Function

To study β_2-tubulin function in spermatogenesis, we have isolated a series of mutations in the β_2 structural gene. Because the expression of this gene is restricted to the testis, the sole phenotype of such mutations is male sterility; neither the viability of the male nor the fertility or viability of females is affected. The first set of recessive male-sterile mutations in the β_2-tubulin gene that we reported all exhibit the same phenotype in spermatogenesis and are designated class I β_2-tubulin mutations. In these mutations, although the mitotic cell divisions proceed normally, all of the subsequent microtubule-mediated events fail to occur: Neither meiosis, nuclear shaping, nor axoneme assembly takes place. In fact, in the testes of males homozygous for any of the class I mutations, virtually no microtubules are assembled in any of the cells in postmitotic stages of spermatogenesis. The failure of microtubule function in the class I mutations is consistent with the timing of expression of the testis-specific tubulin gene. Analysis of their phenotype suggested that in normal males, the β_2-tubulin subunit is required for assembly of the meiotic spindle, of the flagellar axoneme, and of the cytoplasmic microtubules required for nuclear shaping (Kemphues et al. 1982, 1983; Raff and Kemphues 1983).

As illustrated in Figure 2, the basis for the loss of microtubule function in the class I mutations is the failure to establish a stable functional testis tubulin pool.

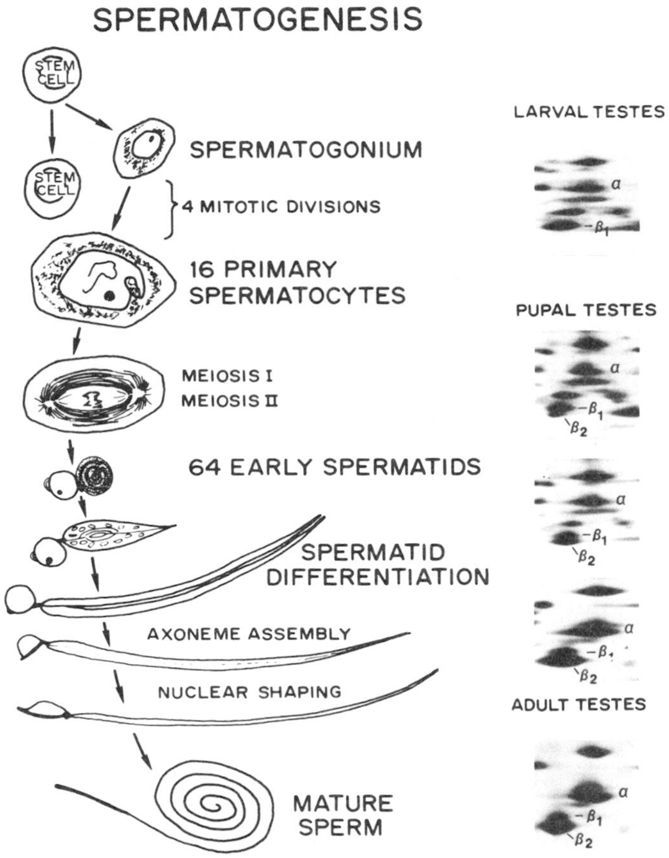

Figure 1

Spermatogenesis in *Drosophila melanogaster*. On the left side of the figure, the stages in spermatogenesis are illustrated diagrammatically, as described by Tates (1971) and as observed in phase-contrast microscopy of living squash preparations of testes (Kemphues et al. 1980, 1982). On the right are portions of autoradiograms of two-dimensional gels showing the tubulins synthesized in testes dissected from males of the developmental stages indicated and labeled in culture with [^{35}S]methionine. The age (at 25°C) of the males from which each set of testes were dissected and labeled is, from top to bottom, 72, 95, 120, 146, and 216 hr. Two-dimensional gel electrophoresis was performed with nonequilibrium pH gradients in the first dimension according to the procedure of O'Farrell et al. (1977) as modified by Waring et al. (1978). The acidic region of the first-dimension gel is at the left; SDS electrophoresis was from top to bottom. (Gel panels from Fig. 1 in Kemphues et al. [1982] were reprinted with permission.)

The class I mutations encode β_2-tubulin variants that are synthesized at approximately normal rates but are then rapidly degraded within the cell (Kemphues et al. 1982, 1983). Recently, K.A. Matthews (unpubl.) isolated a null mutation in which no β_2-tubulin is synthesized and which exhibits a spermatogenic phenotype identical with that of the class I mutations. In both cases, presumably be-

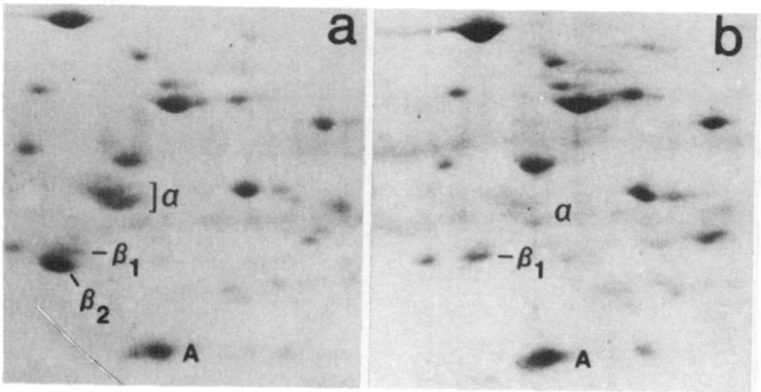

Figure 2
Testis tubulin pools in a wild-type *Drosophila* male (*a*) and a male homozygous for the class I β_2-tubulin mutation *B2t*^{4} (*b*). The panels shown are portions of Coomassie-blue-stained two-dimensional gels displaying total proteins in the testes from seven males. The tubulin subunits present are indicated, as is actin (A), as a reference protein. The wild-type testis protein pool contains significant levels of β_2-tubulin, whereas this protein and much of the α-tubulin are absent from the testes of the male of genotype *B2t*^{4}/*B2t*^{4}. Gel electrophoresis was carried out as described in Fig. 1. (Reprinted, with permission, from Kemphues et al. 1982.)

cause of failure to form a normal α,β-tubulin dimer (the stable form of unpolymerized tubulin in the cell), the normal wild-type α-tubulins are also rapidly degraded, resulting in virtually complete absence of the normal pool of both α-tubulin and β_2-tubulin in the testis.

Class II Recessive Mutations in the Gene for β_2-Tubulin: Specific Defects in Microtubule Function

Analyis of the class I β_2-tubulin mutations demonstrated that although the expression of this tubulin gene product is limited to a single tissue, it participates in a variety of microtubule structures. To define genetically the specific domains in the β_2-tubulin molecule that might be required for only certain aspects of its functions, we set out to isolate mutations in the β_2-tubulin strctural gene that differentially affect the microtubules into which it is assembled. We have now isolated and characterized a number of new recessive male-sterile mutations that fail to complement the class I β_2-tubulin mutations described above and map by recombination to the β_2-tubulin locus but that show a wide range of phenotypes. These new β_2-tubulin mutations have been designated class II mutations.

The class II β_2-tubulin mutations encode β_2-tubulin variants that are stable. In contrast to the class I mutations, the phenotypes of the class II mutations reflect the effects of the presence of the variant subunit in the testis tubulin pool during spermatogenesis. Each of the class II mutations shown in Figure 3 has a unique phenotype and exhibits a different set of defects in microtubule assembly during spermatogenesis. Furthermore, each of these variant β_2-tubulins is at least to some extent capable of participating in normal microtubule assembly, since in males heterozygous for a class II β_2-tubulin allele and wild type, the variant β_2-tubulin is incorporated into mature motile sperm in approximately equal amounts with the normal β_2-tubulin encoded by the wild-type allele. To date, we have not isolated any mutations that encode a stable β_2-tubulin variant that is not under

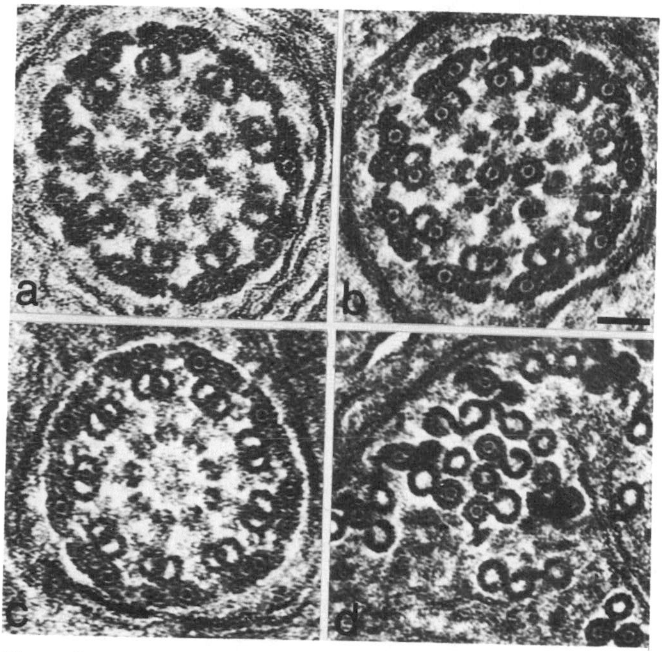

Figure 3
Ultrastructure of the sperm flagellar axonemes of a fertile wild-type *Drosophila* male (*a*) and sterile males homozygous for class II β_2-tubulin mutations (*b–d*). (*a*) A wild-type axoneme, which shows the typical cross-sectional structure of a central pair of microtubules surrounded by nine outer-doublet tubules, plus the nine accessory tubules present in *Drosophila* sperm. Note that in the nearly mature sperm flagellum, as shown here, the central-pair and accessory tubules are filled with an electron-dense core. (*b–d*) Comparable stages of spermatid differentiation in three of the class II mutations. In these males, no motile sperm are produced and the abnormally developing spermatids eventually degenerate. (*b*) Axoneme of a male of genotype *B2t⁶/B2t⁶*. The morphology of the axoneme is essentially identical with that of wild type, except that the lumen of the A tubules of the doublets as well as of the central-pair and accessory tubules become filled with electron-dense material. In the latest stages of spermatid development reached in this mutation, the B tubules of the doublets may also become filled in. (*c*) Axoneme of a male of genotype *B2t⁷/B2t⁷*. The morphology of the axoneme resembles that of wild type except that one or both of the central-pair microtubules are always absent. At least some of the radial spoke structures are present, however. Axonemes in males of this genotype are intact only for very short lengths (less than 100 μm, compared with the 1.8-mm length attained by mature wild-type sperm). (*d*) Axonemelike structure in a male of genotype *B2t⁸/B2t⁸*. The abnormal S- or hook-shaped microtubules are typical of this mutation. We interpret clusters of microtubules such as that shown here to represent abortive axoneme assembly because of their association with membranes and/or mitochondrial derivatives and because of the electron-dense core present in the lumen of many of the tubules, which resembles that normally found in the central-pair and accessory tubules. The magnification is the same in all panels. Bar, 0.05 μm.

some circumstances incorporated into microtubules in vivo. It may be that our sample size is not yet sufficiently large for us to have identified what may be a fairly rare class of β_2-tubulin mutation—a totally afunctional subunit. Another possibility is that the potential for microtubule assembly and ability to form a stable α,β-tubulin dimer are inextricably linked. Thus, any mutation that disrupts the potential for microtubule assembly may also disrupt the ability of β_2-tubulin to dimerize with α-tubulin and therefore result in the class I phenotype. Conversely, even the residual tubulin dimers formed by the unstable class I variant β_2-tubulins participate in microtubule assembly, since, in heterozygous males, very small amounts of the unstable variant may be incorporated into motile sperm along with the wild-type β_2-tubulin subunit (Kemphues et al. 1983).

All of the class II mutations we have isolated to date exhibit multiple defects, including meiotic defects and, as illustrated in Figure 3, defects in axoneme assembly. For example, the mutation $B2t^6$ causes profoundly defective meiotic spindle morphology and function but only a very subtle morphological abnormality in construction of the axoneme (Fig. 3b). In this mutation, the outer-doublet microtubules become filled with an electron-dense core normally characteristic only of the central-pair and accessory tubules in the axoneme of wild-type males (Fig. 3a). In $B2t^7$, spindle morphology appears to be normal and chromosome disjunction occurs normally, although cytokinesis sometimes fails to occur. However, the axonemes that form in $B2t^7$ homozygotes are very short and always lack one or both of the central-pair microtubules (Fig. 3c). Finally, in $B2t^8$, virtually all of the microtubules assembled following expression of the variant β_2-tubulin subunit are morphologically abnormal. The S-shaped microtubules characteristic of this mutation are present in the meiotic spindle, in the cytoplasm, adjacent to the nucleus during nuclear shaping, and in the aberrant axonemes (Fig. 3d). Other mutations in class II also exhibit multiple defects. Thus, although we have not demonstrated whether β_2-tubulin possesses specific functional domains uniquely required for each of the microtubule arrays in which it participates, we have shown that different mutations in the β_2-tubulin structural gene may have very different effects on microtubule assembly and function. It is therefore possible that the factors which specify each set of functions may be separable to some degree.

Specificity of β_2-Tubulin Function: Second-site Noncomplementing Mutations

If the specificity for assembly into different sets of microtubules is determined by factors that are either in whole or in part extrinsic to the testis-specific β_2-tubulin subunit itself, then the specificity may reside either in the other unit of the tubulin dimer, α-tubulin, or in other components (either structural or regulatory) that interact with the tubulin dimer. We have identified several potential candidates for genes that encode such products. In the screen described above for new β_2-tubulin mutations, we isolated several new recessive male-sterile mutations that fail to complement class I β_2-tubulin mutations but that map to other genes. Failure to complement for the phenotype in question is a classic genetic test to demonstrate that two mutations represent different alleles of the same gene. Thus, all of the mutations derived from our screen would be expected to be mutations in the β_2-tubulin gene. However, as shown in Figure 4, the "second-site noncomplementing" mutations characterized to date map to several locations on the third

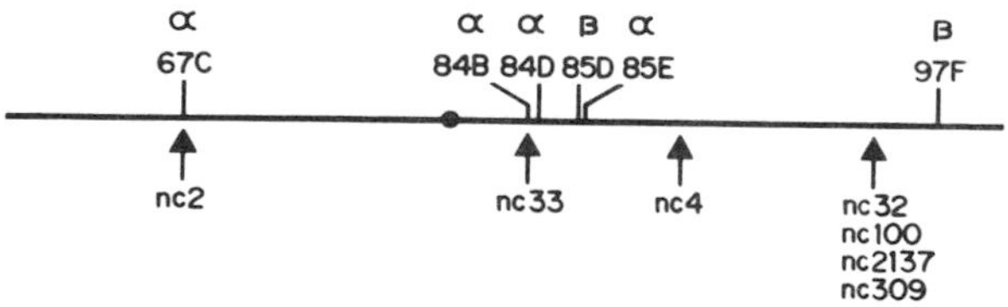

Figure 4
Third-chromosome second-site noncomplementing mutations. The third chromosome is diagramed approximately to scale, giving the cytogenetic locations of the six tubulin genes as determined by in situ hybridization of cloned tubulin sequences to salivary gland polytene chromosomes: α-tubulin genes at 67C4-6, 84B3-6, 84D4-8, and 85E6-10 and β-tubulin genes at 85D4-7 and 97F (for review, see Raff 1984). The remaining two *Drosophila* β-tubulin genes are located on the second chromosome. The recombination map position is known only for the tubulin gene at 85D, which encodes the testis-specific β_2-tubulin subunit (48.49 ± 0.06; determined by Kemphues et al. 1980). The approximate locations of the second-site mutations are indicated, based on their recombination map positions: (*nc2*) 32.1 ± 2.2, mapped between *hairy* (26.5) and *scarlet* (44.0). This recombination map position tentatively places *nc2* within the region 67B-68A based on the cytogenetic localization of the low-molecular-weight heat-shock protein gene cluster *hsp22-27* (29.1) to bands 67B1,2 by Peterson et al. (1979) and on the localization of *rose* (35.0) to bands 68A1,2 by Akam et al. (1978). (*nc33*) 47.7 ± 0.1, mapped between *Kinked* (47.6) and *pink* (48.0). We localized this mutation to region 84B1,2–84B6 by complementation for male fertility using a series of overlapping third-chromosome deletions. (*nc4*) 54.2 ± 0.2, mapped between *crossveinless-c* (54.1) and *stubbloid* (58.2). This recombination position places *nc4* within the region 88C-88F based on the localization of *red* (53.6) within region 88A-C and *c(3)G* (57.4) within region 88F9-89B9. (*nc32*) 82.9 ± 1.1, mapped between *bar-3* (79.1) and *Prickly* (90.0). The other three alleles also map within the same region as *nc32*, and all four alleles are mutually noncomplementing for male fertility in all inter se combinations. No cytogenetic localization is as yet available for this group of mutations, except that they must be distal to band 93, which contains *ebony* (70.7), and proximal to band 99, which contains *claret* (100.7). Stocks used in mapping are described by Lindsley and Grell (1968).

chromosome. Males heterozygous for a class I β_2-tubulin mutation and a second-site mutation are sterile even though they possess one wild-type allele for each of the two genes. Our present hypothesis to explain this rather unexpected finding is that the second-site mutations represent lesions in genes encoding products that directly interact with β_2-tubulin at some point in spermatogenesis. Consistent with this hypothesis (see below), at least one of the second-site mutations maps very close to an α-tubulin gene known to be expressed during spermatogenesis. The sterility of the double heterozygotes may reflect dosage requirements for minimal levels of the product of the interaction between the second-site protein and β_2-tubulin during spermatogenesis. For example, if the second-site mutation encodes a variant gene product that competes randomly with the wild-type product to associate with β_2-tubulin, then in heterozygous combinations with a class I β_2-tubulin allele, only one fourth of the normal level of functional complex can be formed. We conclude that this level is insufficient for normal completion of spermatogenesis but that half the normal level is sufficient (since heterozygotes for the null β_2-tubulin allele alone are fertile). The fertility of most heterozygous combinations of the second-site mutations with class II β_2-tubulin mutations may reflect cases in which the stable β_2-tubulin variants retain some ability to function in the hypothetical complex with the second-site gene product.

If the second-site mutations identify genes encoding proteins that interact with β_2-tubulin and specify its function, then we would expect that different sets of interacting proteins might be used in different microtubule arrays. Consistent with this idea, homozygotes for different second-site mutations exhibit a variety of phenotypes. Of the mutations shown in Figure 4, two (*nc4* and *nc33*) affect spermatid differentiation but do not affect meiosis or earlier stages. The mutation *nc2*, like the β_2-tubulin mutations, exhibits multiple defects, including both abnormal meiosis and defective spermatid differentiation. Finally, the *nc32* group, for which we have isolated several alleles, affects the transition between the early mitotic and cell-growth stages and the onset of meiosis.

Table 1 summarizes the interactions among the two classes of β_2-tubulin mutations and the second-site mutations. All heterozygous combinations of class I β_2-tubulin alleles exhibit the typical phenotype by which this class is defined (Kemphues et al. 1982, 1983). Other interactions within these groups of mutations are more complex. The double heterozygotes for the class I β_2-tubulin mutations and the class II mutations exhibit spermatogenic phenotypes that in some cases are less severe than homozygotes for either mutation alone. This is also true for different combinations of class II alleles. Indeed, the combination *B2t*^{6}/*B2t*^{8} results in fertile males in which the motile spermatozoa produced contain equal parts of the two variant β_2-tubulin subunits and do not contain any wild-type β_2-tubulin. In heterozygotes for class I alleles and a second-site mutation, the typical class I phenotype is ameliorated. It should be noted that one of the second-site mutations, *nc32*, distinguishes by complementation between different class I β_2-tubulin alleles. We do not as yet know the basis for this.

Table 1
Complementation for Male Fertility between β_2-Tubulin Mutations and Second-site Mutations

	β_2-Tubulin mutations									
	class I			*class II*			*Second-site mutations*			
	B2t^{D+R1}	B2t^{D+R2}	B2t^3	B2t^6	B2t^7	B2t^8	nc2	nc4	nc32	nc33
B2t^{D+R1}	ms	ms	ms	ms	ms	ms	ms	ms	ms	ms
B2t^{D+R2}		ms	ms	ms	ms	ms	ms	ms	ms	ms
B2t^3			ms	ms	ms	ms	ms	ms	f	ms
B2t^6				ms	(f)	f	f	f	f	f
B2t^7					ms	ms	ms	f	f	ms
B2t^8						ms	f	f	ms	ms
nc2							ms	f	f	(f)
nc4								ms	f	f
nc32									ms	f
nc33										ms

Complementation tests for male fertility were done by mating and/or by scoring for the presence of motile sperm by dissection of seminal vesicles of 5- to 7-day-old males. Data are given for the most well-characterized mutations in each set. Total number of mutations isolated in each set to date are class I, 5; class II, 7; and second-site, 7. ms, male sterile; f, male fertile; (f), very weakly fertile (a few motile sperm produced).

Specificity of β_2-Tubulin Function: α-Tubulin

Although we have shown that β_2-tubulin is required for assembly of a number of different sets of microtubules, including both singlet and doublet tubules, we do not as yet know the identity of the α-tubulins with which β_2-tubulin is associated. One possibility is that some or all of the specificity for assembly of different microtubule organelles may be determined by means of the α-tubulin component of the tubulin dimer.

As shown in Figure 5, we have found that the most prevalent α-tubulin species synthesized in testis differs in electrophoretic properties from the α-tubulin incorporated into sperm. The steady-state testis tubulin pool revealed by Coomassie blue staining of two-dimensional gels of total testis proteins contains two major α-tubulin species that differ in isoelectric points (Fig. 5a). Autoradiograms of testis proteins labeled with [^{35}S]methionine in culture show that most of the radioactive precursor is incorporated into the more basic α-tubulin species (Fig. 5b,d). However, in mature motile sperm isolated from the seminal vesicle, only the more acidic form of α-tubulin is present (Fig. 5c).

A similar pattern of α-tubulin subunits was observed in *Chlamydomonas* and the related alga *Polytomella* by McKeithan and Rosenbaum (1981), who found that the α-tubulin in the axoneme of the flagellum migrates to a more acidic isoelectric focusing point than the cell-body α-tubulin. More recently, it has been demonstrated that both α-tubulins are translated from the same mRNA species and that the acidic flagellar isoform is derived from the more basic cell-body form by posttranslational acetylation of a lysine residue (Brunke et al. 1982; McKeithan et al. 1983; L'Hernault and Rosenbaum 1983). Although we have not as yet experimentally demonstrated a precursor-product relationship between the α-tubulin isoforms in *Drosophila* spermatogenesis, it is intriguing to speculate that the situation may be analogous to that in *Chlamydomonas*.

We do not as yet know how many different gene products are represented in the α-tubulin populations we observe on two-dimensional gels of testis proteins. Studies by Kalfayan and Wensink (1982) and Natzle and McCarthy (1984) have shown that three of the four *Drosophila* α-tubulin genes are expressed during spermatogenesis. The major α-tubulin transcript in testis mRNA populations is that derived from the α-tubulin gene in the polytene chromosome interval 84B3-6. Significant levels of transcripts from the gene at 85E6-10 are also present. Transcripts from the α-tubulin gene at 84D4-8 were also detected but appear to be much less prevalent than either of the other two gene products.

Two of the second-site noncomplementing mutations shown in Figure 4 map within regions of the third chromosome that contain α-tubulin genes. One of these, *nc33*, was localized by deletion mapping to 84B1,2-6, a small region that contains the major testis α-tubulin. However, both studies of mRNA populations (Kalfayan and Wensink 1982; Mischke and Pardue 1982; Natzle and McCarthy 1984) and our genetic localization (Raff et al. 1982; Raff 1984) have shown that virtually all of the α-tubulin synthesized during embryogenesis is also encoded at the 84B locus. Thus, if *nc33* represents a lesion in the 84B α-tubulin gene, the lesion must be such that α-tubulin function during embryogenesis is not disrupted. A tempting speculation—for which we at present have no experimental evidence—is that the lesion might alter the function of the putative modified α-tubulin discussed above. This is consistent with the fact that in males homozygous for *nc33*, axo-

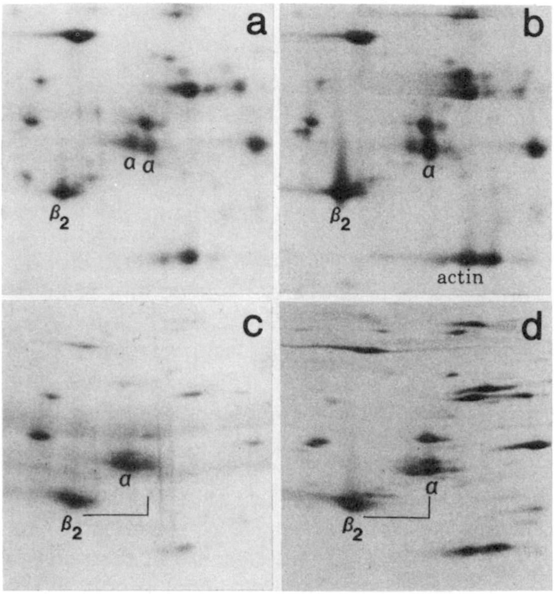

Figure 5
α-Tubulins in the testis and mature sperm of *Drosophila*. All four panels show the tubulin region of two-dimensional gels. Gel electrophoresis was performed as described in Fig. 1, except that the pH gradient was expanded in the pH 5-6 region in order to separate the α-tubulin isoforms. The acidic region of the first-dimension gel is at the left; SDS electrophoresis was from top to bottom. (*a,b*) Experiment in which the total testis proteins from seven fertile males from a stock marked with *red* and *ebony* are displayed. Four of the testes were first labeled in culture for 1 hr with [^{35}S]methionine. (*a*) Coomassie-blue-staining pattern of the gel reveals that the tubulin pool contains approximately equal amounts of two isoforms of α-tubulin. Position of β_2-tubulin is also indicated. (*b*) Autoradiogram of this gel demonstrates that most of the [^{35}S]methionine was incorporated into the α-tubulin which migrates to a more basic position in the pH gradient in the first-dimension gel. (*c,d*) Experiment in which unlabeled mature sperm were coelectrophoresed with a single testis labeled in culture with [^{35}S]methionine. Sperm were dissected from the seminal vesicles of ten wild-type males (strain Oregon R). One testis from a male of the same stock was labeled for 1 hr in culture with [^{35}S]methionine. The two samples were then mixed, and the total proteins were displayed on a two-dimensional gel. (*c*) Coomassie-blue-staining pattern of the resulting gel represents the proteins present in the sperm; (*d*) autoradiogram of the testis proteins synthesized in culture. As in the experiment shown above, most of the [^{35}S]methionine is incorporated into the more basic α-tubulin subunit (compare *b* and *d*). The α-tubulin incorporated into mature sperm, however, is the more acidic of the two α-tubulins (compare *a* and *c*). The major proteins in panel c are sperm proteins; the low levels of actin and other testis proteins that are also visible are due to the single testis in the sample.

neme assembly during late stages of spermatogenesis is defective, but earlier stages are not affected.

Finally, our initial localization of the second-site mutation *nc2* mapped it within a region on the left arm of the third chromosome that contains the 67C α-tubulin gene. However, expression of this gene appears to be limited to oogenesis: Transcripts were found only in ovaries and very early embryos and not in testis (Kalfayan and Wensink 1982; Baum et al. 1983; Natzle and McCarthy 1984). Thus, it seems unlikely that *nc2* represents a mutation in the α-tubulin gene. More accurate recombination mapping and deletion analysis are required to place *nc2* within this region more precisely.

CONCLUSIONS

Our genetic analysis of tubulin function in *Drosophila* has shown that although a specific β-tubulin gene is expressed only in the testis, the product of this gene, β_2-tubulin, is not used exclusively for assembly of the axoneme, an organelle that in *Drosophila* is unique to the testis. Rather, β_2-tubulin is a major structural component of at least three kinds of microtubule arrays: the meiotic spindles, the sperm flagellar axonemes, and the cytoplasmic microtubules involved in nuclear shaping.

We have isolated a number of recessive male-sterile mutations in the structural gene for β_2-tubulin, each of which has different effects on microtubule assembly and function in vivo. Molecular analysis of the variant β_2-tubulin subunits encoded by these mutations may lead toward a better understanding of how alterations in the β_2-tubulin molecule result in changes in its function in vivo, as revealed by the mutant phenotype. Finally, we have identified a set of recessive male-sterile mutations that fail to complement certain β_2-tubulin mutations but that map to other genes. These second-site noncomplementing mutations may provide a novel way to identify genes encoding proteins that interact with β_2-tubulin and that may specify its function in different microtubule arrays.

ACKNOWLEDGMENTS

We thank Joan Caulton for her splendid electron micrographs and Jeffrey Hutchens for unfailing and excellent technical assistance. We also thank Dr. Kathleen Matthews for making available her unpublished data. This work was supported by a National Science Foundation grant (PCM-83-02149) to E.C.R. and by a postdoctoral fellowship from the Jane Coffin Childs Fund for Medical Research and a U.S. Public Health Service grant (1-R01-HD-18127-01) to M.T.F.

REFERENCES

Akam, M.E., D.B. Roberts, G.P. Richards, and M. Ashburner. 1978. *Drosophila*: The genetics of two major larval proteins. *Cell* **13:** 215.

Baum, H.J., Y. Livneh, and P.C. Wensink. 1983. Homology maps of the *Drosophila* α-tubulin gene family: One of the four genes is different. *Nucleic Acids Res.* **11:** 5569.

Brunke, K.J., P.S. Collis, and D.P. Weeks. 1982. Post-translational modification of tubulin dependent on organelle assembly. *Nature* **297:** 516.

Kalfayan, L. and P.C. Wensink. 1982. Developmental regulation of *Drosophila* α-tubulin genes. *Cell* **29:** 91.

Kemphues, K.J., E.C. Raff, and T.C. Kaufman. 1983. Genetic and cytogenetic analysis of *B2t*, the gene for a testis-specific tubulin subunit in *Drosophila melanogaster. Genetics* **105:** 345.

Kemphues, K.J., T.C. Kaufman, R.A. Raff, and E.C. Raff. 1982. The testis-specific β-tubulin subunit in *Drosophila melanogaster* has multiple functions in spermatogenesis. *Cell* **31:** 655.

Kemphues, K.J., E.C. Raff, R.A. Raff, and T.C. Kaufman. 1980. Mutation in a testis-specific β-tubulin in *Drosophila*: Analysis of its effects on meiosis and map location of the gene. *Cell* **21:** 445.

Kemphues, K.J., R.A. Raff, T.C. Kaufman, and E.C. Raff. 1979. Mutation in a structural gene for a β-tubulin specific to testis in *Drosophila melanogaster. Proc. Natl. Acad. Sci.* **76:** 3991.

L'Hernault, S.W. and J.L. Rosenbaum. 1983. *Chlamydomonas* α-tubulin is post-translationally modified in flagella during flagellar assembly. *J. Cell Biol.* **97:** 258.

Lindsley, D.L. and E.H. Grell. 1968. Genetic variations of *Drosophila melanogaster. Carnegie Inst. Wash. Publ. 627.*

Lindsley, D.L. and K.T. Tokuyasu. 1980. Spermatogenesis. In *The genetics and biology of* Drosophila (ed. M. Ashburner and T.R.F. Wright), vol. 2d, p. 226. Academic Press, New York.

McKeithan, T.W. and J.L. Rosenbaum. 1981. Multiple forms of tubulin in the cytoskeletal and flagellar microtubules of *Polytomella. J. Cell Biol.* **91:** 352.

McKeithan, T.W., P.A. Lefebvre, C.D. Silflow, and J.L. Rosenbaum. 1983. Multiple forms of tubulin in *Polytomella* and *Chlamydomonas*: Evidence for a precursor of flagellar α-tubulin. *J. Cell Biol.* **96:** 1056.

Mischke, D. and M.L. Pardue. 1982. Organization and expression of α-tubulin genes in *Drosophila melanogaster.* One member of the α-tubulin gene family is transcribed in both oogenesis and later embryonic development. *J. Mol. Biol.* **156:** 449.

Natzle, J. and B.J. McCarthy. 1984. Regulation of *Drosophila* α- and β-tubulin genes during development. *Dev. Biol.* **104:** 187.

O'Farrell, P.Z., H.M. Goodman, and P.H. O'Farrell. 1977. High resolution two-dimensional electrophoresis of basic as well as acidic proteins. *Cell* **12:** 1133.

Peterson, N.S., G. Moller, and H.K. Mitchell. 1979. Genetic mapping of the coding regions for three heat-shock proteins in *Drosophila melanogaster. Genetics* **92:** 891.

Raff, E.C. 1984. Genetics of microtubule systems. *J. Cell Biol.* **99:** 1.

Raff, E.C. and K.J. Kemphues. 1983. The expression and function in spermatogenesis of a testis-specific tubulin subunit in *Drosophila melanogaster. J. Submicrosc. Cytol.* **15:** 341.

Raff, E.C., M.T. Fuller, T.C. Kaufman, K.J. Kemphues, and R.A. Raff. 1982. Regulation of tubulin gene expression during embryogenesis in *Drosophila melanogaster. Cell* **28:** 33.

Tates, A.D. 1971. "Cytodifferentiation during spermatogenesis in *Drosophila melanogaster*: An electron microscope study." Ph.D. thesis, Drukkerij J.H. Pasmans, 'S-Gravenhage.

Waring, G.J., C.D. Allis, and A.P. Mahowald. 1978. Isolation of polar granules and the identification of polar granule-specific protein. *Dev. Biol.* **66:** 197.

Genetic Analysis of Microtubule Function in CHO Cells

Fernando Cabral and Matthew Schibler
Division of Endocrinology, University of Texas Medical School, Houston, Texas 77025

Ryoko Kuriyama
Department of Molecular Biology
University of Wisconsin
Madison, Wisconsin 53706

Irene Abraham, Carolyn Whitfield, Clark McClurkin, Susan Mackensen, and Michael M. Gottesman
Laboratory of Molecular Biology
National Cancer Institute
National Institutes of Health
Bethesda, Maryland 20205

As is amply supported by many of the other chapters in this volume, the genetic analysis of microtubules has proved to be a highly rewarding approach to understanding the function of microtubules in a variety of organisms. For several years, we have been interested in a genetic analysis of microtubules in mammalian cells and have employed for these studies cultured Chinese hamster ovary (CHO) cells. Because of their rapid doubling time (12 hours), simple and relatively stable karyotype, and the ease with which mutants can be obtained in culture (Siminovitch 1976), CHO cells are frequently the cell line of choice for somatic cell genetic studies. Recently, we have succeeded in optimizing conditions for high-frequency DNA-mediated gene transformation of CHO cells (Abraham et al. 1982; Abraham 1984; Abraham and Gottesman 1984), which, as indicated below, increases the genetic versatility of these cells to an even greater extent.

Our initial strategy involved the isolation of mutants affecting the major microtubule proteins α-tubulin and β-tubulin with the intent of using the properties of these mutants to learn something about the functions of microtubules in intact cells. All of the mutants we have analyzed to date appear to be specifically altered in mitotic spindle function. Because of the ability of genetics to provide cells with a single altered peptide chain, we were able to obtain mutant tubulins shifted in their isoelectric points and hence distinguishable from the wild-type proteins. Thus, we were able to draw some conclusions about the number of different tubulin genes in CHO cells. The ability to mutate a single tubulin subunit among many also allowed us to examine the question of whether different tubu-

  0-87969-174-3/84 $1.00

lins have different functions in CHO cells. Because many of our tubulin mutants have proved to have conditional lethal phenotypes, it was possible to isolate revertant sublines. As shown below, some of these revertants affect tubulin itself, whereas others appear to affect proteins that interact with tubulin in microtubules. Finally, the ability to isolate microtubule mutants in CHO cells has proved that the genetic analysis of a system as complex as the mammalian microtubule system is feasible and offers future promise as a tool to analyze regulatory phenomena involving microtubules, such as control of the synthesis of microtubule proteins and the steps involved in the assembly of the mammalian spindle.

DISCUSSION

Selection and Screening of Microtubule Mutants in CHO Cells

The availability of a variety of drugs that interact specifically with tubulin in mammalian cells allowed us to design a simple, single-step procedure for the isolation of CHO tubulin mutants. Mutagenized cells are exposed to a concentration of an anti-microtubule drug that is approximately two- to threefold higher than the minimum dose that will kill 99.9% of the cells in a population. Drugs used successfully to date to select microtubule mutants in our laboratories include Colcemid, colchicine, griseofulvin, taxol, and maytansine. Mutant clones appear on dishes of mutagenized cells with a frequency of approximately 10^{-5}. These colonies are picked and subcloned in the selecting drug at least twice, and the purified colonies are tested for drug resistance and drug dependence. The drug-resistant colonies are screened for cross-resistance to other drugs such as puromycin, adriamycin, and actinomycin D, which appear to share a permeation pathway with many of the anti-microtubule agents; the cross-resistant mutants are not analyzed further, since they are assumed to be permeability mutants of the kind extensively analyzed by Ling and co-workers (Ling and Thompson 1973).

Mutants that are resistant only to anti-microtubule agents are labeled with [^{35}S]methionine, and whole cell lysates are analyzed on two-dimensional gels for the presence of altered tubulin subunits. Approximately 30–50% of our microtubule drug-resistant mutants have been found to carry isoelectric variant α- or β-tubulins (Cabral et al. 1980; M.J. Schibler and F.R. Cabral, unpubl.). As shown below, mutants with electrophoretically variant tubulins have been studied to determine whether these mutant tubulins are incorporated into microtubules and what the physiological effect of these tubulins is on the cell. Table 1 summarizes the results of several successful selections using anti-microtubule drugs and indicates the range of protein alterations and phenotypes, which have been described in the literature. As can be seen, many of the mutants have conditional lethal phenotypes (cold- or heat-sensitive or drug-dependent). As we will show, these phenotypes can be used to study the physiological effects of the mutant subunits and can be exploited to isolate revertants of these cell lines.

Isolation of Revertants of Tubulin Mutants

The isolation of revertants is an important step in establishing that electrophoretically variant tubulins are responsible for mutant phenotypes. In this way, both

Table 1
Microtubule Mutants in CHO cells

Selection[a]	*Phenotype*[b]	*Reference*
Cmd^R	β-tubulin, ts	Cabral et al. (1980)
	α-tubulin	Ling et al. (1979)
		Keates et al. (1981)
	Cmd-dependent	Whitfield and Gottesman (this paper)
Grs^R	β-tubulin, ts	Cabral et al. (1980)
Tax^R	α-tubulin, ts	Cabral et al. (1981)
	β-tubulin	Cabral and Schibler (this paper)
	taxol-dependent	Cabral (1983)
	cs	Gottesman (this paper)
Ben^R	taxol sensitivity	Warr et al. (1982)
Pod^R	altered 66K protein	Gupta et al. (1982)
Col^R Pod^R	altered 70K protein	Gupta and Gupta (1984)
May^R	α-tubulin	Schilber and Cabral (1983)
Revertants	tr	Cabral et al. (1982)

[a]Cmd, Colcemid; Grs, griseofulvin; Tax, taxol; Pod, podophyllotoxin; Col, colchicine; Ben, benzimidazole carbamate.
[b]ts, temperature sensitive; tr, temperature resistant; cs, cold sensitive.

Tax-1 (Cabral et al. 1981) and Cmd-4 (Cabral et al. 1980) have been established as bona fide tubulin mutants having increased drug resistance. In the better-studied example, revertants of Cmd-4 were isolated using the observation that the mutant cells are temperature sensitive for growth in addition to being drug resistant. Thus, we argued that if a single lesion is responsible for both the temperature-sensitive and drug-resistant phenotypes, then both should revert at high frequency.

To test the prediction that the temperature-sensitive phenotype and drug-resistant phenotype were linked, 24 revertants of Cmd-4 able to grow at the nonpermissive temperature were selected from seven independent nonmutagenized cultures and were then screened for Colcemid resistance and for the presence or absence of the electrophoretically variant β-tubulin (β^*) on two-dimensional gels (see Fig. 1). The revertants fell into three classes (Cabral et al. 1982) as shown in Table 2. The largest class (19/24) consists of cells that retain β^* and

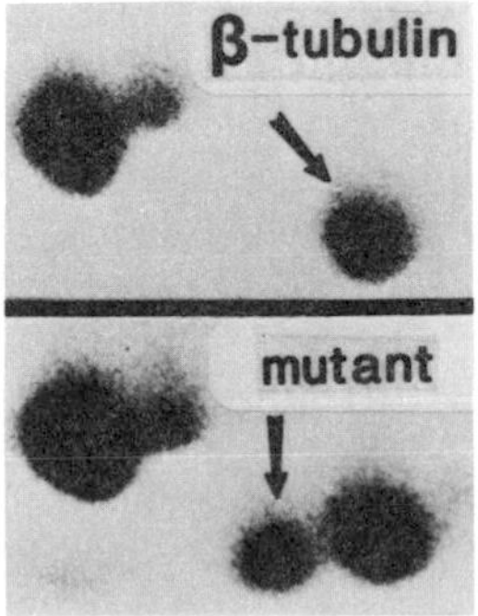

Figure 1
Mutant β-tubulin spot in Cmd-4. (*Upper panel*) Enlargement of a two-dimensional gel of a [^{35}S]methionine-labeled whole-cell extract from wild-type CHO cells. (*Bottom panel*) Same region of the gel from the Colcemid-resistant mutant Cmd-4.

Table 2
Revertants of Mutant Cmd-4, Selected for Growth at the Nonpermissive Temperature

	Number isolated	*Colcemid resistance*[a]	*β-Tubulin species present*[b]
Class I	19	+	$\beta^* \ \beta^{wt}$
Class II	4	–	β^{wt}
Class III	1	–	$\beta^{**} \ \beta^{wt}$

[a] + indicates Colcemid resistant; – indicates Colcemid sensitive.
[b] Since CHO are quasi-diploid cells, mutant Cmd-4 has both mutant (β^*) and wild-type (β^{wt}) β-tubulin species.

are still Colcemid-resistant but have lost their temperature sensitivity. There are several possible explanations for this result, but the one we feel is most exciting postulates the existence of alterations in associated proteins (i.e., microtubule-associated proteins or MAPs) that allow these proteins to interact with β^* in such a way as to stabilize the microtubules to high temperatures without removing the ability of β^* to confer drug resistance on the cells. We are currently pursuing this idea by isolating mitotic spindles from wild-type, mutant, and revertant cells and looking for alterations among the spindle-associated proteins (R. Kuriyama and M.M. Gottesman, in prep.). Success in these studies would allow us to define functional associations between microtubules and associated proteins in vivo.

The second class of revertants (4/24) have lost temperature sensitivity, drug resistance, and β^* in one step and at high frequency (10^{-5}). This is good evidence that a single locus controls all three mutant properties and establishes that the alteration in β^*-tubulin is responsible for the drug resistance and temperature sensitivity of the strain.

The third class of revertants has only a single member and is in many ways the most interesting. This cell line has an alteration in β^* to produce β^{**}, which has yet another electrophoretic mobility. The production of β^{**} from β^* leads to the loss of drug resistance and temperature sensitivity, providing further evidence that β^* in Cmd-4 is responsible for both properties. Analysis of in-vivo-assembled microtubules in this revertant reveals an absence of β^{**}. Thus, β^{**} has an alteration that precludes its assembly into microtubules. From these results, we may infer that β^* in the mutant Cmd-4 is able to confer drug resistance and temperature sensitivity by assembling and subtly altering the structure of the microtubules. We are now trying to map the site of the lesion in β^{**}, since this may define a region in β-tubulin that is involved in assembly (F.R. Cabral and M.J. Schibler, in prep.). We are also using this revertant as a starting point for the isolation of second-site suppressor mutations as a way of identifying interacting proteins in microtubule assembly. A similar approach has previously been used to study phage assembly (Jarvik and Botstein 1975).

The revertants just described were all isolated on the basis of their temperature-sensitive phenotype. We have recently found that some of our tubulin mutants are supersensitive to other antimitotic drugs (see Table 3, below). This property should allow the selection of other classes of microtubule-defective mutants. In addition, some CHO mutants (selected on the basis of cAMP resistance)

Table 3
Relative LD_{50} of Antimitotic Drugs to Tubulin Mutants Compared with Wild Type

	tax	*grs*	*cmd*	*podo*	*VP-16*	*noco*	*DCBT*	*may*	*IPC-Cl*	*gel*
Grs-2	0.5	2.5	1.6	1.4	0.9	1.2	1.3	1.6	0.8	0.7
Tax-1	2.1	1:2	1.0	1.1	1.0	1.1	0.7	1.4	0.5	0.6

tax, taxol; grs, griseofulvin; cmd, Colcemid; podo, podophyllotoxin; noco, nocodazole; DCBT, 2,4-dichlorobenzylthiocyanate; may, maytansine; IPC-Cl, isopropyl *N*-phenyl carbamate-chloride; gel, geldanomycin.

that have altered cAMP-dependent protein kinases have been shown to be supersensitive to certain anti-microtubule agents (Colcemid, colchicine, and geldanomycin). We have used these mutants and these drugs to select cells normally resistant to the drugs. Two classes of these revertants have been detected. The first class consists of revertants of the cAMP-dependent protein kinase phenotype, demonstrating that the supersensitivity of the mutants to the drugs is the result of the kinase defect. The second class still has altered kinases and is presumed to consist of mutant substrates for these kinases in the cell (I. Abraham and M.M. Gottesman, unpubl.). These results argue that microtubule stability in CHO cells is either directly or indirectly dependent on an intact cAMP-dependent protein kinase system and strongly suggests that some element of the spindle is phosphorylated in a functionally significant cAMP-dependent manner.

Biochemical Defects in Microtubule Mutants

As summarized in Table 1, our original selections using Colcemid, colchicine, griseofulvin, and taxol yielded several cell lines with alterations in β-tubulin and one cell line with an alteration in α-tubulin (Cabral et al. 1980, 1981). Figure 1 shows enlargements of two-dimensional gels in which the major spots seen are α-tubulin and β-tubulin. The mutant cell contains wild-type β-tubulin as well as an additional polypeptide spot that is slightly more basic than wild-type β-tubulin. Because the mutant spot is synthesized at approximately 50% of the rate of the wild-type spot, we tentatively concluded from these studies that there may be three functional β-tubulin genes (or alleles) in CHO cells (Cabral et al. 1980). Similar results were found for α-tubulin. Recent analysis of expression of human tubulin genes by Cowan and co-workers (Lewis et al., this volume) suggests that there may be three functional human β-tubulin genes as well.

The presence of an additional spot on a two-dimensional gel does not prove that this is a mutant tubulin or that this mutant spot is responsible for the observed phenotype of the cell. Because of the inability to perform formal genetic analysis with cultured cells, it is essential to use other approaches to be certain that biochemical genotypes and phenotypes are linked. We have proved that the mutant spots are tubulins in the following way: (1) Mutant tubulins and wild-type tubulins copurify (Cabral et al. 1980); (2) protease maps (one-dimensional V8 protease maps and two-dimensional tryptic peptide maps) of mutant and wild-type tubulins are the same (Cabral et al. 1980, 1981); and (3) monoclonal antibodies to tubulin recognize both wild-type and mutant subunits (Cabral et al. 1982 and this paper). In addition, in vitro translation of mRNA from mutant cells indicates that there is an mRNA species in these cells encoding the mutant polypeptide, suggesting that posttranslational modifications do not produce the extra tubulin spots (Cabral et al. 1980). Evidence for the linkage of the drug-resistant phenotypes with tubulin alterations is based on reversion analysis (see above) and on recent experiments in which we have been able to transfer drug resistance via DNA-mediated transformation. This transferred drug resistance appears to be linked to amplification of a genomic tubulin sequence (see below).

Physiological Defects in the Microtubule Mutants

The finding that the mutant tubulin subunits copolymerized with wild-type subunits in the mutant cells suggested that these mutant microtubules might be defective in some function. This surmise was confirmed by the discovery that many of our tubulin mutants are temperature sensitive for growth (Cabral et al. 1981, 1982; Abraham et al. 1983). We examined events surrounding the death of these mutants by a detailed time-lapse video microscopy study. This analysis indicated that at the nonpermissive temperature, the mutant cells look quite normal until mitosis, at which time they round up and remain round for up to 1 hour. Many of these cells are unable to complete cytokinesis, and when they flatten are seen to be binucleate. After two cell cycles, many multinucleate and micronucleate cells can be seen in the population. As also is seen clearly for the taxol-dependent phenotype in the absence of taxol (see below), the cells appear to go through the cell cycle but cannot complete mitosis correctly. A high percentage of spindles from mutant cells grown under nonpermissive conditions are multipolar and incomplete (Abraham et al. 1983). The exact cause of cell death is not known, but in these mutants, death is not cell-cycle-specific, since dead cells can be found in all phases of the cell cycle, as revealed by premature chromosome condensation analysis (Abraham et al. 1983). Presumably, cell death results from massive gene imbalance as a consequence of wholesale nondisjunction events due to the defective spindles. Interphase (cytoplasmic) microtubules are normal in appearance (Cabral et al. 1982) and appear to function normally at the nonpermissive temperature with respect to saltatory motion, secretion, and determination of cell shape.

Spindles isolated from our griseofulvin-resistant, β-tubulin mutant have been found to contain both wild-type and mutant subunits (Fig. 2), as do interphase microtubules, suggesting that the simple hypothesis that these cells have an altered spindle-specific β-tubulin is not correct.

The isolation of a novel class of mutants requiring the continuous presence of taxol for cell growth has greatly simplified physiological studies of microtubule mutants. A representative of this mutant class, Tax-18, illustrates the kind of information these mutants can provide concerning the role of microtubules in mitosis (Cabral 1983; Cabral et al. 1983). Tax-18 grows normally and maintains normal morphology when incubated in the presence of taxol. In the absence of the drug, however, cell division stops and a profound morphological change occurs within 24–48 hours. The cells become larger and flatter and they exhibit oddly shaped nuclei and micronuclei.

Since taxol is known to be a microtubule-stabilizing drug (Schiff and Horwitz 1980), we reasoned that these cells might have hyperlabile microtubules incap-

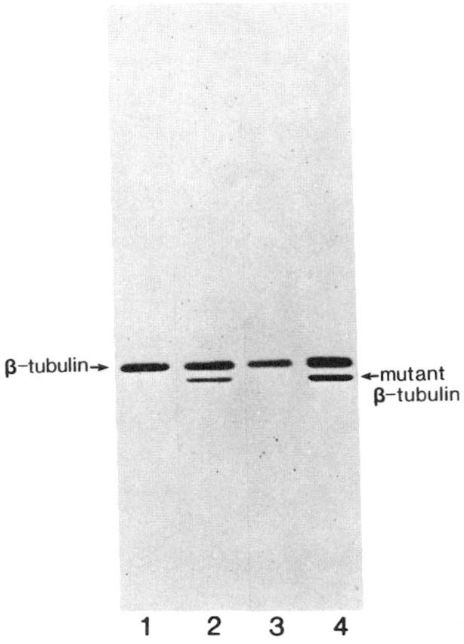

Figure 2
Both wild-type and mutant β-tubulins are present in spindles prepared from a griseofulvin-resistant mutant, Grs-2. (*1*,*2*) Whole-cell extracts from wild-type CHO cells and Grs-2, respectively; (*3*,*4*) spindles purified from wild-type and Grs-2 cells, respectively. Samples were solubilized in boiling SDS sample buffer, electrophoresed on a 10% polyacrylamide gel with the separating gel adjusted to pH 8, and electrophoretically transferred to nitrocellulose paper. β-Tubulin was localized using a mouse monoclonal antibody to Chinese hamster brain β-tubulin (a gift from S. Binder, University of Virginia), and the antibody was detected using immunoperoxidase-conjugated goat anti-mouse antibody (Cappel).

able of carrying out microtubule-mediated functions. The addition of taxol would then stabilize the microtubules and restore their function. To test this idea, the mutant cells grown with or without taxol were examined by tubulin immunofluorescence. To our surprise, the mutant cells elaborated an extensive cytoplasmic microtubule complex. When the cells were tested for their ability to transport intracellular vesicles by saltatory motion and to cap cell-surface molecules, no differences were found in taxol-containing or taxol-deficient media, indicating that these microtubule-related functions are not affected by the mutation. Thus, we concluded that the cytoplasmic microtubules in Tax-18 are morphologically and functionally intact.

Examination of the mitotic cells in the culture, however, revealed a very different situation. Tax-18 grown in taxol is able to display metaphase, anaphase, and telophase stages of mitosis. In the absence of the drug, however, "prometaphaselike" figures accumulate (Fig. 3) and metaphase, anaphase, and telophase figures do not appear. A closer examination of the prometaphaselike figures using electron microscopy reveals the persistence of kinetochore-to-pole microtubules but an apparent absence of the interpolar class of spindle microtubules. The simplest explanation for this result is that the interpolar microtubules are already the most labile in the cell. Thus, any mutation affecting microtubule stability will manifest itself preferentially in these already labile structures.

The consequences of the lesion in Tax-18 are severe. The inability of the spindle assembly to progress beyond the prometaphase stage of mitosis results in abnormal chromosome organization and segregation. Since the cells are unable to organize chromosomes about the spindle poles, nuclear membranes re-form around the scattered chromosomes, resulting in the formation of micronuclei (F. Cabral and A. Pandya, unpubl.). A second consequence of the lesion is the inability of the mutant cell to undergo cytokinesis. This was obvious from growth

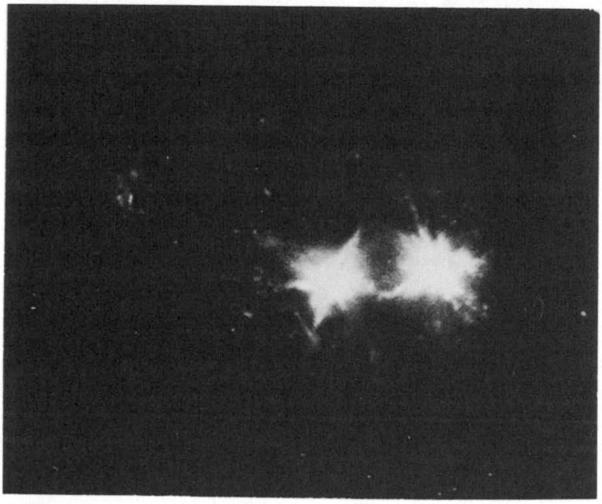

Figure 3
Tubulin immunofluorescence of Tax-18 deprived of taxol for 2 days. The cell is in a prometaphaselike state (see text for details). This photograph was taken in collaboration with B.R. Brinkley (Baylor College of Medicine). Magnification, 600 ×.

curves and morphological observation of the mutant deprived of taxol and was confirmed by time-lapse observation. Although the cells are delayed in a prometaphaselike state, they eventually proceed directly to a "telophaselike" condition of intense membrane activity. During this time, a cleavage furrow appears to form but is unable to completely divide the cytoplasm and relaxes as the cell enters G_1 without having divided into daughter cells. We speculate that the final stages of cytokinesis require the stable interaction of the contractile ring with the interpolar microtubules. Since these microtubules are missing from the mutant, cytokinesis cannot be completed.

In spite of the inability of the cells to assemble the mitotic spindle and divide, the signals for progression through mitosis and interphase are intact and must therefore, be independent of spindle microtubule assembly or function. The cells that reenter G_1 continue to synthesize protein and go on to carry out cell-cycle-specific functions such as chromosome replication and centriole duplication. After traversing the cell cycle several times without dividing, the mutant cells become large multinucleated structures with as many as 16 or more centrosomes and greater than octaploid DNA content per cell. Progression beyond this results in severe chromosome abnormalities (F. Cabral and E. Sutcliffe, unpubl.) and the eventual death of the cells.

The study of Tax-18 has given us a unique view of microtubule involvement in chromosome organization and cytokinesis and has given us a measure of the importance of microtubules in the life cycle of mammalian cells. The results argue that microtubules are not necessary for cell viability in the short term, but if continued cell-cycle progression occurs in the absence of spindle microtubule assembly, the resultant loss of chromosome organization and segregation, combined with the absence of cell division, leads to eventual cell death.

Tax-18 and six other taxol-requiring mutants isolated at the same time have all failed to exhibit any electrophoretic shifts in tubulin. To determine whether alterations in tubulin are capable of producing the taxol-dependent phenotype, we recently isolated 139 taxol-resistant mutants (M.J. Schibler and F. Cabral, unpubl.). Our current analysis of these mutants is summarized in Figure 4. It is apparent that taxol-requiring mutants are common in UV-mutagenized CHO cells, making up approximately half of the taxol-resistant mutants. On the other hand, cells that are cross-resistant to puromycin and that probably represent permeability mutants account for only about 7% of the taxol-resistant cells. This contrasts greatly with selections using other microtubule-active drugs in which permeability mutants account for more than 90% of the drug-resistant cells. Analysis of the taxol-resistant cells by two-dimensional gel analysis reveals that alterations in α-tubulin or β-tubulin may account for the emergence of taxol dependency in CHO cells, but the frequency (10–15%) is low compared to selections with other drugs where electrophoretic alterations in tubulin are seen in 30–50% of the cells that have no cross-resistance to puromycin. This suggests that alterations in other proteins (e.g., MAPs) may also produce the taxol-resistant phenotype. One of our major goals will be to demonstrate this possibility biochemically.

Cross-resistance of Mutants to Antimitotic Drugs

One way to learn something about the functional differences among the tubulin mutants discussed above is to determine cross-resistance and collateral sensitivity to other antimitotic drugs. We have measured the LD_{50} of various drugs for the β-tubulin mutant Grs-2 and the α-tubulin mutant Tax-1 (see Table 3) (I. Abraham et al., unpubl.). From this analysis, we can also classify the drugs according to which mutants are resistant to them.

It is clear from these data that drug resistance or sensitivity is not subunit-dependent since, for example, both an α-tubulin mutant and β-tubulin mutants show resistance to maytansine (Table 3, column 8). Since griseofulvin depolymerizes microtubules and taxol increases the stability of microtubules, one might

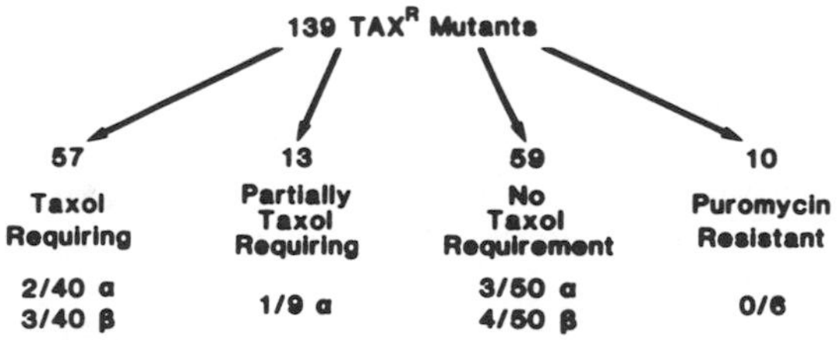

Figure 4
Analysis of taxol-resistant cells. This figure summarizes the current status of the characterization of taxol resistance in CHO cells. Taxol dependency was assayed by observing the morphology of the cells 2–5 days after taxol deprivation. Cross-resistance to puromycin was established by determining the plating efficiency of the mutant and wild-type cells in puromycin. Alterations in α- or β-tubulin were identified as electrophoretic shifts in the proteins on two-dimensional gels.

expect that mutants selected with these drugs would show opposing sensitivities and resistances to new drugs (Table 3, column 1). However, as this is not always the case (e.g., compare the sensitivities of both Grs-2 and Tax-1 to geldanomycin, Table 3, column 10), this simple hypothesis may not be adequate to predict the functional behavior of all of the mutants. An additional problem with analyzing these data is the phenotypic drift that occurs in tissue culture with both wild-type and mutant cell lines. This can especially pose a problem with the interpretation of small differences in drug sensitivity or resistance.

These results show that although we cannot yet predict the resistance or sensitivity of our mutants to new antimitotic drugs on the basis of the selection conditions alone, the patterns of cross-resistance or sensitivity of each mutant are unique and define a "fingerprint" for that mutant. In addition, we have shown that the altered tubulin molecules apparently affect the structure and function of the microtubules, since antimitotic drugs that are not thought to have tubulin-binding sites (griseofulvin, IPC-Cl) have altered effects on the tubulin mutants as compared with the wild type.

Many of these drugs will be useful for selecting new mutants from wild-type cells. The drugs to which the mutants are sensitive will also be useful for the selection of intra- and extragenic suppressors of the mutants. Finally, a new drug, DCBT (R. Dion and E. Hamel, unpubl.), is interesting in that it is "griseofulvinlike" by our drug testing (Table 3, column 7), and we have shown by immunofluorescence that it drastically alters the morphology of microtubules in the cell (I. Abraham and M.M. Gottesman, unpubl.).

DNA-mediated Transfer of the Colcemid-resistant Phenotype

Recent experiments suggest that the Colcemid-resistant phenotype can be transferred using genomic DNA from the Cmd-4 mutant mixed with pSV$_2$*neo* plasmid DNA.Transformants were selected with the neomycin analog G418 followed by Colcemid. The initial selection with G418 is expected to enrich for cells that have taken up the transferred DNA and to reduce the spontaneous background of Colcemid resistance. The frequency of Colcemid-resistant colonies was about 5 in 10^7 cells, and no significant difference in frequency was observed between wild-type and mutant DNA. One interesting cell line from the control experiment with wild-type DNA was dependent on Colcemid for growth (C. Whitfield and M.M. Gottesman, unpubl.). Another cell line arising from the transfer of mutant DNA and the pSV$_2$*neo* plasmid exhibited the properties expected for a transformant. This cell line (11801) was Colcemid-resistant and puromycin-sensitive to the same extent as the original Cmd-4 mutant. In addition, the putative transformant was neomycin-resistant, indicating that it had acquired the transferred DNA.

We analyzed this putative transformant by restriction digestion, electrophoresis, and Southern blotting to nitrocellulose, expecting to see a new or amplified band hybridizing to human β-tubulin probe obtained from N.J. Cowan (New York University School of Medicine). Multiple bands were observed in CHO cells with the human β-tubulin probe, suggesting the presence of multiple functional and nonfunctional genes (Fig. 5). A band was amplified three- to fivefold in the transformant (11801), compared with the wild type, Cmd-4 mutant, or control with wild-

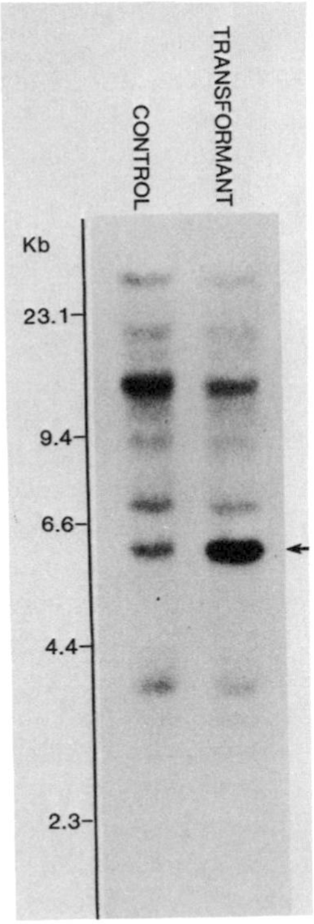

Figure 5
Amplification of a β-tubulin-containing restriction fragment in a Colcemid-resistant transformant. DNA from the Colcemid-resistant mutant Cmd-4 was used to transform wild-type CHO cells as described in the text, and a Cmd-resistant transformant was picked. The figure shows Southern blots of *Bam*HI-digested DNA using a human β-tubulin probe (a gift from N.J. Cowan, New York University School of Medicine). Arrow points to the restriction fragment that is amplified severalfold in the Colcemid-resistant transformant compared with the control transformant (first lane), wild-type recipient cells, or Cmd-4 (data not shown).

type DNA in the DNA transfer. Data were quantitated by densitometry and normalized to an unaffected band. Since no alterations in Southern blotting have been observed with any of our tubulin mutants or with the controls using wild-type DNA in the DNA transfer, we believe that the Colcemid resistance of 11801 is due to transfer and amplification of the mutant β-tubulin gene. Amplification of a functional β-tubulin gene in the recipient cannot be ruled out at this time. However, in other experiments with the herpes simplex virus thymidine kinase gene (Abraham et al. 1982) or genomic transfer of a mutant protein kinase (I. Abraham et al., unpubl.), gene transfer into CHO cells has been associated with the presence of extra copies of the transferred gene. In conclusion, DNA-mediated transfer of a mutant gene is a unique way to identify a functional gene in a multigene family. The construction of this amplified transformant will facilitate the cloning of the mutant gene. These data also demonstrate that the Colcemid-resistant phenotype is linked to expression of a specific tubulin gene, in support of the biochemical data outlined above.

SUMMARY AND CONCLUSIONS

The study of microtubules in CHO cells has already yielded important information concerning the role of microtubules in mitotic events, the way in which microtubules affect cell viability, the number of expressed tubulin loci in CHO cells, mechanisms of drug resistance, and the improbability that specific tubulin subunits perform specific functions in these cells. One outstanding problem that remains is the identification of microtubule-associated proteins by second-site reversion analysis. This identification would provide genetic proof that MAPs play a role in microtubule assembly or stability in vivo. Although considerable in vitro evidence for MAP involvement in microtubule assembly exists, their in vivo significance remains to be established. We believe that we may already have alterations in MAPs among our mutants and revertants and are working to purify and identify these proteins in CHO cells. As often happens, the genetics is well ahead of the biochemistry in these cells.

Another important area for future study involves the mapping of alterations in mutant tubulin subunits or their genes. This information is especially important in mutants with increased or decreased microtubule stability, since it is likely that the affected sites represent regions of interaction between the mutant tubulin and other subunits in the microtubule structure. An extreme case of this is the revertant of Cmd-4, which has a nonassembling β-tubulin (Cabral et al. 1982). Mapping the alteration in this subunit should reveal which part of the molecule is involved in the initial phases of microtubule assembly. These studies will ultimately have to be combined with physical studies to determine whether the mutation affects assembly by altering the gross conformation of the molecule or by causing more subtle changes in the area of subunit-subunit interaction.

Many of the studies just mentioned would be greatly aided by the isolation of cloned cDNAs to the CHO tubulins. As reported at this meeting, three unique full-length cDNA clones to CHO α-tubulin have recently been isolated (E.M. Elliott and V. Ling, pers. comm.), and we hope to isolate similar probes for β-tubulin in the near future.

The outlook for the ability of somatic cell genetics to make an impact on microtubule research is bright. Although formal genetics is difficult if not impossible with these cells, we and others have been able to isolate and analyze mutants capable of addressing significant questions in microtubule structure, physiology, and regulation. With the introduction of techniques in molecular biology and gene transfer, we foresee an even greater ability to analyze the role of microtubules in mammalian cells.

REFERENCES

Abraham, I. 1984. DNA-mediated gene transfer. In *Molecular cell genetics* (ed. M.M. Gottesman). Wiley, New York. (In press.)

Abraham, I. and M.M. Gottesman. 1984. Gene transfer in CHO cells. *Prog. Cancer Res. Ther.* **30:** 31.

Abraham, I., J.S. Tyagi, and M.M. Gottesman. 1982. Transfer of genes to Chinese hamster ovary cells by DNA-mediated transformation. *Somatic Cell Genet.* **8:** 23.

Abraham, I., M. Marcus, F. Cabral, and M.M. Gottesman. 1983. Mutations in α- and β-tubulin affect spindle formation in Chinese hamster ovary cells. *J. Cell Biol.* **97:** 1055.

Cabral, F. 1983. The isolation of CHO mutants requiring the continuous presence of taxol for cell division. *J. Cell Biol.* **97:** 22.

Cabral, F., I. Abraham, and M.M. Gottesman. 1981. Isolation of a taxol-resistant CHO cell with an alteration in α-tubulin. *Proc. Natl. Acad. Sci.* **78:** 4388.

———. 1982. Revertants of a Chinese hamster ovary cell mutant with an altered β-tubulin:

Evidence that the altered tubulin confers both Colcemid resistance and temperature sensitivity on the cells. *Mol. Cell. Biol.* **2:** 720.

Cabral, F., M.E. Sobel, and M.M. Gottesman. 1980. CHO mutants resistant to colchicine, Colcemid or griseofulvin have an altered β-tubulin. *Cell* **20:** 29.

Cabral, F., L. Wible, S. Brenner, and B.R. Brinkley. 1983. A taxol requiring mutant of CHO cells with impaired mitotic spindle assembly. *J. Cell Biol.* **97:** 30.

Gupta, R.S. and R. Gupta. 1984. Mutants of Chinese hamster ovary cells affected in two different microtubule-associated proteins. *J. Biol. Chem.* **259:** 1882.

Gupta, R.S., T.K.W. Ho, M.R.K. Moffat, and R. Gupta. 1982. Podophyllotoxin-resistant mutants of Chinese hamster ovary cells. Alteration in a microtubule-associated protein. *J. Biol. Chem.* **257:** 1071.

Jarvik, J. and D. Botstein. 1975. Conditional-lethal mutants that suppress genetic defects in morphogenesis by altering structural proteins. *Proc. Natl. Acad. Sci.* **72:** 2738.

Keates, R.A.B., F. Sarangi, and V. Ling. 1981. Structural and functional alterations in microtubule protein from Chinese hamster ovary cell mutants. *Proc. Natl. Acad. Sci.* **78:** 5638.

Ling, V. and L.H. Thompson. 1973. Reduced permeability of CHO cells as a mechanism of resistance to colchicine. *J. Cell. Physiol.* **83:** 103.

Ling, V., J.E. Aubin, A. Chase, and F. Sarangi. 1979. Mutants of Chinese hamster ovary (CHO) cells with altered Colcemid-binding activity. *Cell* **18:** 423.

Schibler, M.J. and F.R. Cabral. 1983. Maytansine resistant mutants of Chinese hamster ovary cells with an alteration in α-tubulin. *J. Cell Biol.* **97:** 218a.

Schiff, P.B. and S.B. Horwitz. 1980. Taxol stabilizes microtubules in mouse fibroblastic cells. *Proc. Natl. Acad. Sci.* **77:** 1561.

Siminovitch, L. 1976. On the nature of hereditable variation in cultured cells. *Cell* **7:** 1.

Warr, J.R., D.J. Flanagan, and M. Anderson. 1982. Mutants of Chinese hamster ovary cells with altered sensitivity to taxol and benzimidazole carbamates. *Cell Biol. Int. Rep.* **6:** 455.

CYTOSKELETAL GENES: STRUCTURE, EXPRESSION, AND REGULATION

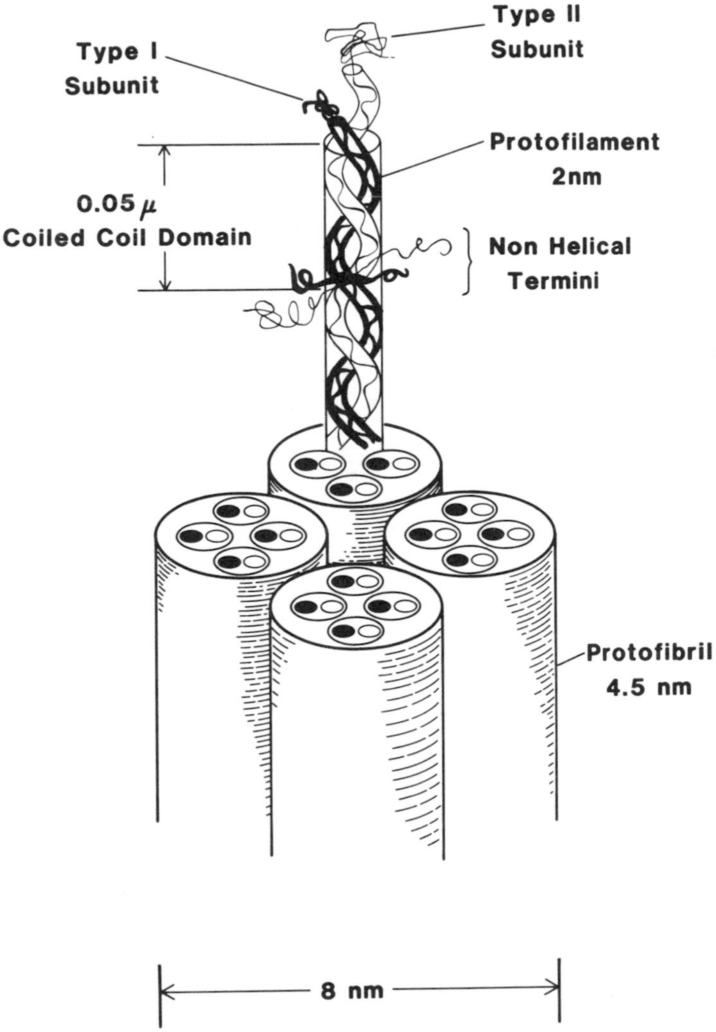

Recto: Structural model of the 8-nm keratin filament. The model is based on data compiled from X-ray diffraction, electron microscopic, physicochemical, and cDNA sequencing studies. The model is intended to summarize the findings of a number of laboratories in the field. (Figure courtesy of E. Fuchs, The University of Chicago.)

Primary Structure and Expression of a Vertebrate β-Tubulin Gene Family

Kevin F. Sullivan, Jane C. Havercroft, and Don W. Cleveland
Department of Biological Chemistry
The Johns Hopkins University School of Medicine, Baltimore, Maryland 21205

Microtubules are ubiquitous components of eukaryotic cells and constitute a major structural element of the cytoskeleton, as well as a number of specialized structures such as the mitotic spindle and flagella. Although the ultrastructure of microtubules in different cells and organisms is, with a few notable exceptions (Fujiwara and Tilney 1975; Chalfie and Thompson 1982), indistinguishable, it is clear from a number of studies that microtubules from different microtubule systems even within single cells can differ in their molecular properties (see, e.g., Behnke and Forer 1967; Brinkley and Cartwright 1975; Brady et al. 1984; Cumming et al. 1984; Thompson et al. 1984). One potential mechanism for the establishment of microtubules of different biochemical and functional properties within cells is at the level of the α/β-tubulin heterodimer which constitutes the major structural subunit of the microtubule. Using a variety of analytical protein chemistry techniques, several investigators have shown that multiple forms of α- and β-tubulins exist in cells (see, e.g., Gozes and Littauer 1978; Marotta et al. 1978; George et al. 1981; Murphy and Wallis 1983a; Sullivan and Wilson 1984), thus encouraging speculation that different tubulins do in fact play a role in the establishment of distinct microtubule systems within cells.

With the advent of molecular cloning technology, it rapidly became apparent that most eukaryotes possess small multigene families that encode α- and β-tubulins (Cleveland et al. 1980; Sanchez et al. 1980; Cowan et al. 1981; Kalfayan and Wensink 1981; Silflow and Rosenbaum 1981; Wilde et al. 1982; Alexandraki and Ruderman 1983; Toda et al. 1984). Among vertebrates, both protein and DNA sequence analyses have established that multiple tubulin genes encode structurally distinct polypeptides within individual species (Krauhs et al. 1981;

Ponstingl et al. 1981; Cowan et al. 1983; Hall et al. 1983). However, analysis of the structure and expression of functional tubulin genes in the large tubulin multigene families of higher eukaryotes has been hampered by the finding that many, if not most, of the sequences within mammalian genomes that contain strong homology with a cloned β-tubulin mRNA are nonfunctional pseudogenes (Gwo-Shu Lee et al. 1983; N. Cowan, pers. comm.). Consequently, the task of identifying functional genes in these genomes is a formidable one indeed.

We have chosen to analyze the tubulin gene families of the chicken, a vertebrate that fortuitously lacks the extensive accumulation of nonfunctional pseudogenes characteristic of mammalian genomes and is thus a uniquely tractable higher eukaryotic system in which to study the structure, expression, and, ultimately, the function of multiple tubulin genes and their protein products.

RESULTS AND DISCUSSION

Primary Structure of the Chicken β-Tubulin Gene Family

We have previously shown (Cleveland et al. 1980) that the chicken genome possesses four segments with detectable hybridization to both the aminoterminal and carboxyterminal coding portions of a cloned chicken brain β-tubulin cDNA (Fig. 1A). Each of these sequences has been isolated by molecular cloning methods and has been designated β_1, β_2, β_3, and β_4, respectively, in order of increasing size of the genomic *Eco*RI fragment on which they are found (Lopata et al. 1983). We now report that each of these four genes has been subjected to DNA sequence analysis using M13/dideoxy sequencing techniques (Sanger et al. 1977; Messing and Vieira 1982). Complete sequences of three of these genes and 72% of the coding sequence of the fourth gene have been determined.The structure of each gene, as determined by comparison with the sequence of the chick brain β-tubulin cDNA pT2, is diagramed in Figure 1B. Each gene consists of four coding-sequence exons interrupted by three intervening sequences at codons 19, 56, and 93. This gene organization is identical with that previously reported by Gwo-Shu Lee et al. (1983) for the human β-tubulin gene M40, thus demonstrating that the structural organization of β-tubulin genes has been conserved during evolution of higher eukaryotes.

In addition, we report that a fifth segment of chicken DNA (designated β_5 in Fig. 1A), which displays weak, but detectable, homology only with aminoterminal coding regions of a chick brain β-tubulin cDNA, has now also been cloned (K. Murphy and D. Cleveland, unpubl.). Partial DNA sequence data (53% of the coding region) for this putatively functional gene has also been obtained. Two additional DNA segments with detectable homology with β-tubulin are also evident in the intentionally overexposed blot pattern in Figure 1A. The first of these corresponds to a large DNA fragment migrating above β_4; the second (which hybridizes very weakly) migrates just below β_3. Neither of these segments has yet been cloned.

From the β-tubulin gene DNA sequence data, we have deduced the bulk of the sequences of the proteins encoded by these five chicken β-tubulin genes. These studies have defined four distinct isotypes of β-tubulin that differ markedly in their primary sequences. These isotypic classes, represented by β_1/β_2, β_3, β_4, and β_5,

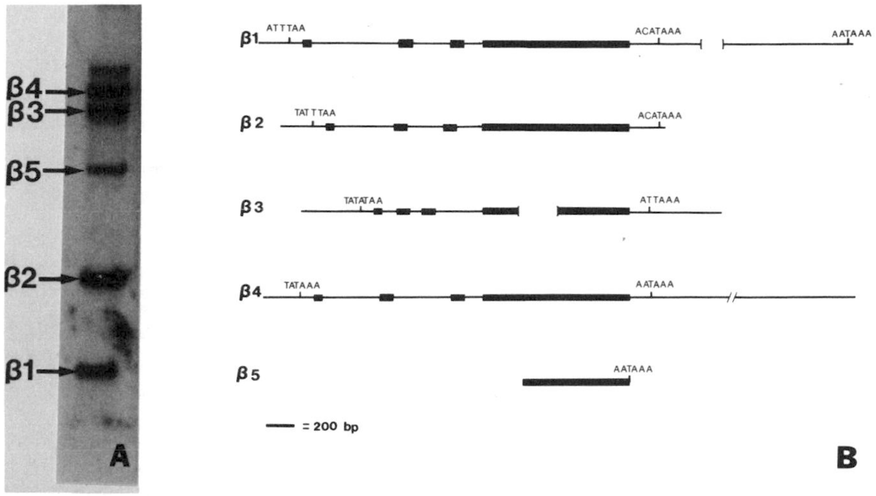

Figure 1
Identification and analysis of the chick β-tubulin gene family. (*A*) Genomic chicken DNA was digested to completion with *Eco*RI restriction endonuclease, electrophoresed on an agarose gel, and transferred to nitrocellulose. To identify DNA segments homologous to β-tubulin, the resultant filter was hybridized to a ^{32}P-labeled probe carrying the entire coding sequence for a chicken brain β-tubulin. An autoradiogram of the hybrids formed is shown. Arrows indicate the positions of individual β-tubulin segments. (*B*) Gene organization of five isolated β-tubulin genes as determined by DNA sequencing. Only regions of known sequence are depicted. Coding-region (exon) sequences are indicated by the raised regions. Positions of signal sequences for polyadenylation (AATAAA, ATTAAA, or ACATAAA) are marked as are putative "TATA" promoter signals.

are demonstrated in Figure 2, in which we have aligned available protein sequences predicted from the five chicken genes and compared them with each other and with all other published metazoan β-tubulin sequences. The sequence used for reference is that of the β_2-coding region, which was originally determined from a chicken brain cDNA clone, pT2 (Valenzuela et al. 1981). Several noteworthy points are illustrated in this figure. First, genes β_1 and β_2 encode essentially identical polypeptides that differ only in two conservative amino acid substitutions at residues 55 and 187. Indeed, these two genes are 99.1% identical at the nucleotide level in the coding region and retain significant homology even within corresponding intron regions. Moreover, the 5′-flanking regions up to and including a putative "TATA" promoter region also show 80% conservation. Nonetheless, as described below, the program of expression of these two genes that encode a single isotypic class of β-tubulin polypeptide is markedly different. Second, the polypeptide isotype encoded by gene β_4 is remarkably *divergent* in its primary sequence from that of β_1/β_2. This is particularly apparent in the carboxyterminal region, which is in fact four amino acids longer than the β_1/β_2 and β_3 isotypes. Third, the isotype represented by gene β_3 is of intermediate sequence divergence with respect to β_1/β_2 but still differs in sequence at 16 of the 320 known residue positions. Finally, the fourth class for which we have only 53% of the polypeptide sequence also displays the extreme divergence charac-

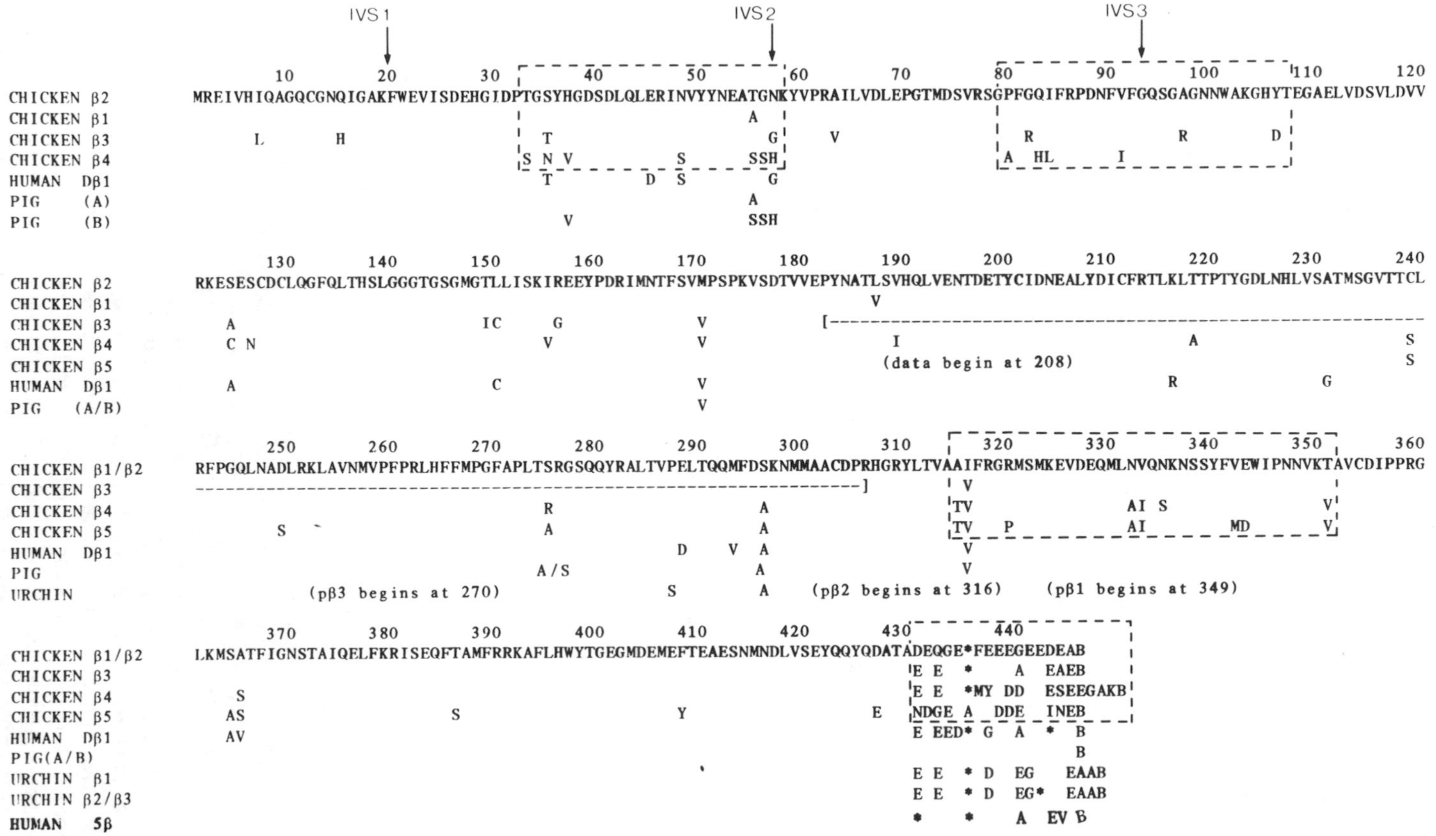
IVS1
IVS2
IVS3
CHICKEN β2
CHICKEN β1
CHICKEN β3
CHICKEN β4
HUMAN Dβ1
PIG (A)
PIG (B)
MREIVHIQAGQCGNQIGAKFWEVISDEHGIDPTGSYHGDSDLQLERINVYYNEATGNKYVPRAILVDLEPGTMDSVRSGPFGQIFRPDNFVFGQSGAGNNWAKGHYTEGAELVDSVLDVV
CHICKEN β5
PIG (A/B)
RKESESCDCLQGFQLTHSLGGGTGSGMGTLLISKIREEYPDRIMNTFSVMPSPKVSDTVVEPYNATLSVHQLVENTDETYCIDNEALYDICFRTLKLTTPTYGDLNHLVSATMSGVTTCL
(data begin at 208)
CHICKEN β1/β2
PIG
URCHIN
RFPGQLNADLRKLAVNMVPFPRLHFFMPGFAPLTSRGSQQYRALTVPELTQQMFDSKNMMAACDPRHGRYLTVAAIFRGRMSMKEVDEQMLNVQNKNSSYFVEWIPNNVKTAVCDIPPRG
(pβ3 begins at 270)
(pβ2 begins at 316)
(pβ1 begins at 349)
PIG(A/B)
URCHIN β1
URCHIN β2/β3
HUMAN 5β
LKMSATFIGNSTAIQELFKRISEQFTAMFRRKAFLHWYTGEGMDEMEFTEAESNMNDLVSEYQQYQDATADEQGE*FEEEGEEDEAB

Figure 2 *(See facing page for legend.)*

teristic of the carboxyl terminus, in addition to numerous single amino acid substitutions in a cluster between residues 315 and 351.

Furthermore, it is clear in Figure 2 that the amino acid substitutions between isotypes are not randomly distributed throughout the molecule but rather are localized into four primary clusters. The carboxyterminal region between residues 431 and 445 is extremely variable, showing up to 60% divergence between isotypes, as well as variation in the length of the encoded polypeptide. As also noted in the figure, similar divergence has been documented by Hall et al. (1983) in the analysis of two human β-tubulins. Two additional clusters of heterogeneity are observed between residues 33–57 and 80–106. Together, these two regions account for as much as 30% of the substitution positions between chicken isotypes β_2 and β_4, although they represent only 13% of the polypeptide. Region 33–57 is encoded almost entirely on exon 2, whereas the majority of the variable residue positions of region 80–106 are encoded on exon 3 of the genes, indicating that these regions may comprise discrete, functionally differentiable domains within the structure of β-tubulin. A fourth cluster is observed in exon 4 in the region between residues 315 and 351. In addition, not only are clusters of divergence present, regions of high sequence conservation are also evident. One 65-amino-acid region, between residues 366 and 431, shows no variation in any of the vertebrate sequences shown in Figure 2, except for chicken β_5, which has not yet been definitively identified as a functional gene. Also, the region from the amino terminus to residue 32 is very highly conserved. This level of conservation indicates that these regions must be highly constrained during evolution and, therefore, that they may play an essential role in conserved functions of β-tubulin such as GTP binding, α/β dimer formation, and microtubule polymerization.

Although at present we cannot directly ascribe an individual function to any particular β-tubulin isotype (however, see Murphy et al., this volume), comparison of the β-tubulins of different species strongly suggests that these variable sequences may have a role in establishment of isotype-specific β-tubulin function. For example, the sequence of chicken β_2, which encodes the dominant neural β-tubulin (see Table 1), is virtually identical with that of one of two pig brain β-tubulins (Krauhs et al. 1981) even in the highly variable carboxyterminal region. In addition, several of the variant residues in the most aminoterminal substitution cluster are identical in chicken β_4 and the second porcine brain β isotype (Krauhs et al. 1981). In view of the *intraspecies* divergence of these regions in different β isotypes, this *interspecies* conservation of specific sequence variants must be a result of specific evolutionary selection pressures. We interpret these data as strongly suggestive of divergent functional roles for the different isotypes of β-tubulin.

Figure 2

Comparative protein sequences for currently available metazoan β-tubulins. Sequence data determined either by direct protein sequencing techniques or deduced from DNA sequence data are shown for five chicken β-tubulins, two human β-tubulins, and two hog β-tubulins. The one-letter amino acid code has been utilized. Areas of sequence divergence among these polypeptides are boxed. The region of known chicken β_5 sequence begins at residue 208 (indicated by left-hand bracket). Sequences for the human Dβ_1 and 5β are from Hall et al. (1983), the porcine sequences are from Krauhs et al. (1981), the urchin sequences are from Alexandraki and Ruderman (1983).

Table 1
Differential Expression of β-Tubulin Genes in Chicken Cells and Tissues

	β_1	β_2	β_3	β_4'	*Developmental stage*[a]
Cells					
fibroblasts	+	+ +	+	+	E
chondroblasts	+	+	+	+ +	E
skeletal muscle	+ +	+ +	+	+	E
smooth muscle	+	+ +	+	+	E
neurons	+	+ + +	+	+	E
glia	+	+ +	+	+	E
hepatocytes			+	+ + +	A
primitive red blood cells			+	+ + +	E
Tissues					
lung	+	+ +	+	+	A
esophagus	+	+	+ +	+ +	A
oviduct	+		+ +	+ +	A
liver			+	+ + +	A
brain (adult)	+	+ +	+	+	A
brain (embryo)	+	+ + + +		+	E
intestine			+ +	+ + +	A
spleen		+	+	+ + +	A
bursa			+ +	+ + +	A
thymus			+ +	+ + +	A
testis			+ + + + + +	+	A

Data from J. C. Havercroft and D. W. Cleveland (in press).
[a]E indicates that the cell or tissue was isolated from an embryo; and A indicates that the cell or tissue was isolated from an adult.

Differential Programmed Expression of the Chicken β-Tubulin Gene Family

To begin to determine the biological potential for functional distinction of the multiple β-tubulin genes, β_1, β_2, β_3, and β_4, of the chicken, we have determined the program of expression of each gene by analysis of its respective steady-state mRNA level in a number of chicken tissues and cell types and as a function of chicken development. To do this, it was initially necessary to identify regions of each gene that were specific to that particular gene and that did not detect the mRNAs derived from other β genes. This was achieved for genes β_1, β_2, and β_3 by construction of subclones that contain only sequences corresponding to their putative 3′-untranslated regions extending from the translation-termination codon through the site of poly(A) addition. Each of these three probes hybridizes uniquely to its parent gene. For gene β_4, the initial probe utilized began 480 bases 5′ to the translation-initiation codon, extended through the entire coding region, and included 2.5 kb of flanking 3′ DNA sequence. (The pattern of β_5 expression has not yet been investigated.)

In an initial search for transcripts derived from the β_1–β_4 genes, smooth-muscle mRNA was prepared and examined by blot analysis (see Fig. 3 and Table 1).

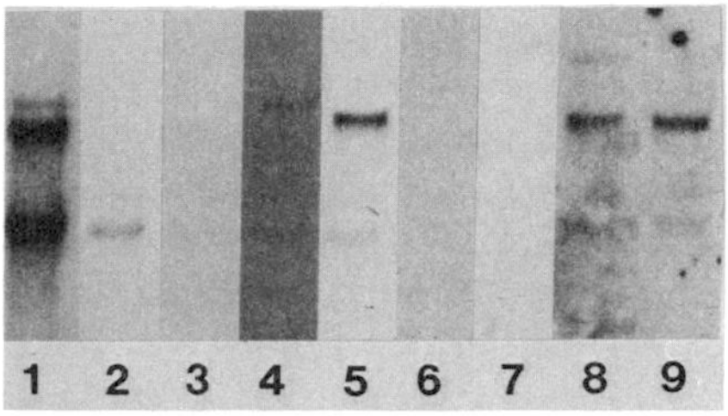

Figure 3
Identification of the RNAs encoded by the β_1, β_2, β_3, and β_4 genes. Poly(A)-containing RNA isolated from chick smooth muscle was analyzed by RNA blot analysis following separation on denaturing gels. Replicate blots were probed with a variety of probes derived from various regions of the β_1, β_2, β_3, or β_4 genes. (*1*) Probe from pT2, a full-length β-tubulin cDNA; (*2*) β_2 3′-specific probe (see text); (*3*) β_3 3′-specific probe (see text); (*4*) β_1 3′ probe (see text); (*5*) β_4 probe spanning the entire coding sequence and including some 5′- and 3′-flanking sequences; (*6*) β_4 probe containing 652 bases of flanking 3′ sequences and only seven nucleotides of coding-region sequence; (*7*) β_4 5′ probe starting at the translation-initiation codon and extending 480 bases into the flanking 5′ sequences; (*8*) same β_4 probe as in lane *7* except including the first 57 bases of aminoterminal coding-region sequence; (*9*) β_4 probe containing the 217 most 3′ coding-region sequences in addition to 2.6 kb of 3′-flanking region.

Using a conserved coding-sequence probe that identifies all four genes (constructed from the β-tubulin cDNA clone pT2), three major size classes of mRNA (4000, 3500, and 1800 bases) are seen (Fig. 3, lane 1). The probes specific for β_2 and β_3 hybridize uniquely to the 1800-base class (Fig. 3, lanes 2 and 3, and Fig. 4B,C), whereas the β_1 probe hybridizes to both the 4000-base and 1800-base species (Fig. 3, lane 4). For β_4, strong preferential hybridization to the 3500-base species is seen (Fig. 3, lane 5), along with much weaker detection of the other size classes. This weaker identification of all β-tubulin RNA classes was not unexpected, since this initial β_4 probe contains the entire β_4-coding region. For a more refined analysis of the patterns of β_4 gene expression, a specific 3′-untranslated region probe, extending from 7 bp 5′ to the translation-termination codon through 652 bp of 3′-flanking sequence, was prepared. In the face of the markedly preferential hybridization to the 3500-base mRNA species, we were more than a little surprised to find that this 3′ β_4-specific probe did *not* hybridize to the 3500-base mRNA (Fig. 3, lane 6). Similarly, a probe from β_4 containing sequences 5′ to the translation-initiation codon also failed to detect any mRNA species (Fig. 3, lane 7). However, probes containing small amounts of either 5′ or 3′ β_4-coding sequence hybridize with strong preference (Fig. 3, lanes 8 and 9). These findings, along with a number of other data (not shown), indicate that the dominant 3500-base β-tubulin RNA species cannot be a transcript of the β_4 gene, but rather probably derives from a very closely related sister gene that we have not yet isolated. We note that precisely such a situation is known to exist for the β_1 and β_2 genes. In this instance, the β_1 and β_2 genes share more than 99% nucleotide homology in the coding regions and extensive homology in the 5′-untranslated regions and introns but are completely nonhomologous in their 3′-untranslated regions. Despite the present ambiguity as to the progenitor gene for the 3500-base transcript (which we designate β_4'), we would stress that we have demonstrated by in vitro translation that this RNA species does encode a true β-tubulin polypeptide (Lopata et al. 1983) and that we can identify this transcript by its size and selective hybridization to the β_4 gene.

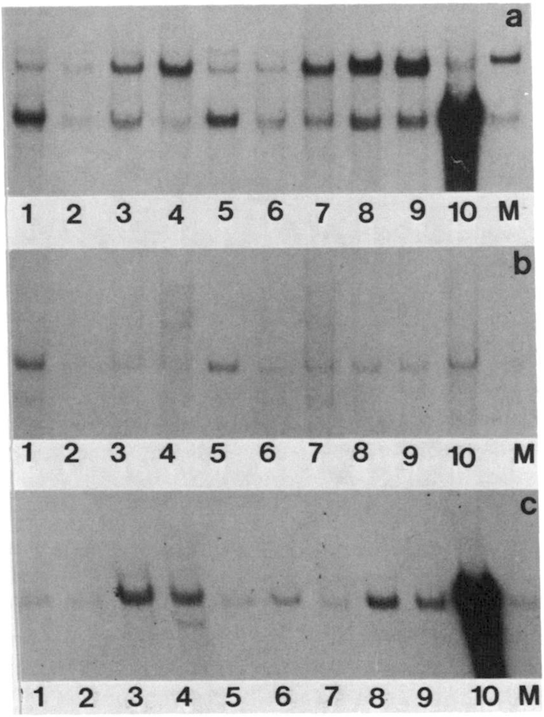

Figure 4
Expression of β-tubulin mRNAs in isolated chicken tissues. Poly(A)-containing RNA was isolated from a number of adult chicken tissues and analyzed by RNA blot analysis with a complete β-tubulin coding region from pT2 (*A*), the β_2 3′-specific probe (*B*), and the β_3 3′-specific probe (*C*). Tissues from which RNA was prepared: (*1*) lung; (*2*) esophagus; (*3*) oviduct; (*4*) liver; (*5*) brain; (*6*) intestine; (*7*) spleen; (*8*) bursa; (*9*) thymus; (*10*) testis. M, a cloned chick T lymphocyte cell line, MSB.

Having established probes with which we can specifically identify β_1, β_2, β_3, and β_4' RNAs, we next sought to determine the pattern of differential β-tubulin gene expression in different tissues isolated from adult chickens. Results of that investigation are shown in Figure 4, A through C, respectively, for the β_1, β_2, β_3, and β_4' mRNAs. Markedly different levels of the various transcripts are readily apparent in different tissues. Transcripts of the β_1 gene are present at low or undetectable levels in all tissues. The β_3 gene is transcribed at varying levels in all adult tissues examined but is overwhelmingly represented in testes. The β_2 gene, on the other hand, is expressed at high levels only in brain and lung tissues. The β_4' transcript is expressed at moderate to high levels in all tissues examined and is particularly significant in liver, spleen, bursa, and thymus.

Although these experiments demonstrate that the β-tubulin genes are expressed in a complex pattern as a function of differentiation, it is difficult to interpret the biological significance of results determined with tissue samples because many of the tissues used are composed of multiple cell types. To refine these studies, experiments were performed using several isolated cell types. A comparison of neurons and glial cells (Table 1) demonstrates that the high level of β_2 transcripts in the brain is due to the fact that it is the major neuronal β-tubulin, whereas the glial cells show a more balanced expression of all four genes. The β_1 transcript is present at low levels in all cell types except for skeletal-muscle myotubes, where it represents the major β-tubulin transcript. β_3 is present at low levels in all cell types, whereas β_4' is a significant transcript in most cells and is abundant in hepatocytes and primitive red blood cells.

These and many additional experiments are summarized in Table 1. Collectively, they demonstrate two important points. First, each gene is expressed in a specific, but complicated, program during differentiation and certain genes are expressed at high levels in one or a few cell types. Thus, β_2 appears to be the dominant neuronal β-tubulin, whereas β_3 is the overwhelming species in testis and β_1 is expressed at high levels only in skeletal-muscle myotubes. β_4', on the other hand, is expressed in relatively high abundance in almost all cells and tissues and thus appears to encode an essential "housekeeping" β-tubulin. Another significant conclusion is that, without exception, all cells and tissues examined express more than one isotype of β-tubulin. Thus, although we observe overwhelmingly preferential expression of particular genes in some tissue/cell types, it appears that no individual β-tubulin isotype is sufficient to constitute a complete cellular complement of β-tubulin subunits but must be present in cells in concert with other isotypes.

Functions of Multiple Genes Encoding Multiple β-Tubulins

We have demonstrated that the chicken genome encodes at least three different isotypes of β-tubulin that are differentially expressed in a complex pattern as a function of development and differentiation. Two general hypotheses have been advanced regarding the possible functions of multiple tubulin genes/polypeptides. The first of these, a "structural" hypothesis, posits that structurally *distinct* tubulin subunits exist for functional specialization or differentiation of particular microtubule systems. This hypothesis was initially advanced based on the apparent detection of chemically distinct subunits within cells (see, e.g., Fulton and Simpson 1976; Stephens 1978). An alternative proposal, a "regulatory" hypothesis, proposes that multiple tubulin genes may encode functionally *equivalent* tubulin polypeptides but that the genes are present in different regulatory environments within the genome to independently coordinate tubulin synthesis with different programs of cellular differentiation. This hypothesis has received substantial experimental support from the work of Raff and co-workers (Kemphues et al. 1982) in which they have documented a multifunctional testis-specific β-tubulin subunit. Our present findings with the chicken β-tubulin gene family strongly suggest that, in fact, both scenarios govern the structure and expression of at least the β-tubulin genes of vertebrates. Our results demonstrate, as we and others have previously suggested by protein chemistry approaches (Nelles and Bamburg 1979; Sullivan and Wilson 1984), that different forms of β-tubulin are encoded by the chicken genome. Furthermore, these different forms are independently regulated and expressed in a tissue-dependent fashion. However, our findings with the β_1/β_2 gene pair, which represent different but highly homologous genes encoding essentially identical proteins, indicate that a single isotype class of β-tubulin polypeptide is regulated independently by the expression of these two genes. Such a situation may also apply to the β_4 gene and its putative sister gene, β_4'.

What remains unclear from these studies is the molecular *function* of the isotypic differences among β-tubulins. The pattern of substitutions within the primary structure suggests that regions of heterogeneity are embedded within a highly conserved structural framework. It is thus tempting to speculate that the conserved regions of the primary structure are involved in the conserved functions of β-tubulin, such as dimer formation and microtubule polymerization. From the structure of microtubule polymers, we would expect that these conserved regions are found at and around the contact sites between subunits and thus in the protein interior of the polymer. The variable regions, on the other hand, probably could not be accommodated at protein interfaces because of the extensive structural variations that characterize them. We would thus expect to find these areas of the molecule on the outside of the polymer, where they may be involved in establishing a *surface* topology of the microtubule which would then be isotype-specific. Alternatively, from our finding that multiple β-tubulin isotypes are coexpressed in cells, subunits may be present in random or pseudorandom distribution within cellular microtubules. The presence of different isotypes within a microtubule could then provide a stochastic or multifunctional surface topology.

It is, of course, well known that microtubules interact with a variety of cellular components. For example, whereas flagellar microtubules interact primarily with other microtubules through specific associations with flagellar proteins such as dynein (Gibbons and Gibbons 1973), cytoplasmic microtubules interact with different components including intermediate filaments (see, e.g., Blose et al. 1984) and, in neuronal cells, neurofilaments (Leterrier et al. 1982; Runge and Williams 1982). Many of these interactions are likely to be mediated by different kinds of microtubule-associated proteins (see, e.g., Vallee et al., this volume). Thus, one important function of tubulin heterogeneity may be, in concert with the expression of specific microtubule-associated proteins, to specify the kinds of interactions in which a microtubule may participate in different cellular events. Other possible functions of isotypic divergence may also include subtle control of microtubule polymerization properties and/or stability, as has been clearly shown by the work of Murphy and Wallis (1983a,b; Murphy et al., this volume) on the chicken erythrocyte-specific β-tubulin isotype. (The gene encoding this remarkably divergent subunit has not yet been isolated, but, definitively, it is not encoded by genes β_1–β_5 [J. Havercroft et al., unpubl.].)

CONCLUSION

Although many questions remain regarding the roles of multiple tubulins in animal cells, it is becoming increasingly clear that subtle variations in the structures of the highly conserved cytoskeletal proteins may be important for establishing differentiated function. Experiments conducted at the level of the gene are beginning to solidify our understanding of these important protein differences and continue to offer new and well-defined approaches to determining the dynamic roles that the cytoskeleton plays in the life of the cell.

ACKNOWLEDGMENTS

This work has been supported by grants from the National Institutes of Health and March of Dimes to D.W.C., who is the recipient of a National Institutes of Health Research Career Development Award. K.F.S. is the recipient of a National Institutes of Health postdoctoral fellowship.

REFERENCES

Alexandraki, D. and J.V. Ruderman. 1983. Evolution of α and β tubulin genes as inferred by the nucleotide sequences of sea urchin cDNA clones. *J. Mol. Evol.* **19:** 397.

Behnke, O. and A. Forer. 1967. Evidence for four classes of microtubules in individual cells. *J. Cell Sci.* **2:** 169.

Blose, S.H., D.I. Meltzer, and J.R. Feramisco. 1984. 10-nm Filaments are induced to collapse in living cells microinjected with monoclonal and polyclonal antibodies against tubulin. *J. Cell Biol.* **98:** 847.

Brady, S., M. Tytell, and R.J. Lasek. 1984. Axonal tubulin and axonal microtubules: Biochemical evidence for cold stability. *J. Cell Biol.* (in press).

Brinkley, B.R. and J. Cartwright. 1975. Cold-labile and cold-stable microtubules in the mitotic spindle of mammalian cells. *Ann. N.Y. Acad. Sci.* **253:** 428.

Chalfie, M. and J.N. Thompson. 1982. Structural and functional diversity in the neuronal microtubules of *Caenorhabditis elegans. J. Cell Biol.* **93:** 15.

Cleveland, D.W., M.A. Lopata, R.J. MacDonald, N.J. Cowan, W.J. Rutter, and M.W. Kirschner. 1980. Number and evolutionary conservation of α and β tubulin and cytoplasmic β and γ actin genes using specific cloned cDNA probes. *Cell* **20:** 95.

Cowan, N.J., P.R. Dobner, E.V. Fuchs, and D.W. Cleveland. 1983. Expression of human α-tubulin genes: Interspecies conservation of 3′ untranslated regions. *Mol. Cell. Biol.* **3:** 1738.

Cowan, N.J., C.D. Wilde, L.T. Chow, and F.C. Wefald. 1981. Structural variation among human β tubulin genes. *Proc. Natl. Acad. Sci.* **78:** 4877.

Cumming, R., R.D. Burgoyne, and N.A. Lytton. 1984. Immunocytochemical demonstration of α tubulin modification during axonal maturation in the cerebellar cortex. *J. Cell Biol.* **98:** 347.

Fujiwara, K. and L.G. Tilney. 1975. Structural analysis of the microtubule and its polymorphic forms. *Ann. N.Y. Acad. Sci.* **253:** 27.

Fulton, C. and P.A. Simpson. 1976. Selective synthesis and utilization of flagellar tubulin. The multi-tubulin hypothesis. *Cold Spring Harbor Conf. Cell Proliferation* **3:** 987.

George, H.J., L. Misra, D.J. Field, and J.C. Lee. 1981. Polymorphism of brain tubulin. *Biochemistry* **20:** 2402.

Gibbons, B.H. and I.R. Gibbons. 1973. The effect of partial extraction of dynein arms on the movement of reactivated sea urchin spermatozoa. *J. Cell Sci.* **13:** 337.

Gozes, I. and U.Z. Littauer. 1978. Tubulin microheterogeneity increases with rat brain maturation. *Nature* **276:** 411.

Gwo-Shu Lee, M., S.A. Lewis, C.D. Wilde, and N.J. Cowan. 1983. Evolutionary history of a multigene family: An expressed human β tubulin gene and three processed pseudogenes. *Cell* **33:** 477.

Hall, J.L., L. Dudley, P.R. Dobner, S.A. Lewis, and N.J. Cowan. 1983. Identification of two human β tubulin isotypes. *Mol. Cell. Biol.* **3:** 854.

Havercroft, J.C. and D.W. Cleveland. 1984. Programmed expression of β tubulin genes during development and differentiation of the chicken. *J. Cell Biol.* (in press).

Kalfayan, L. and P.C. Wensink. 1981. α Tubulin genes of *Drosophila. Cell* **24:** 97.

Kemphues, K., T.C. Kaufman, R.A. Raff, and E.C. Raff. 1982. The testis specific β tubulin subunit in *Drosophila melanogaster* has multiple functions in spermatogenesis. *Cell* **31:** 655.

Krauhs, E., M. Little, T. Kempf, R. Hofer-Warbinek, W. Ade, and H. Ponstingl. 1981. Complete amino acid sequence of β tubulin from porcine brain. *Proc. Natl. Acad. Sci.* **78:** 4156.

Leterrier, J.F., K.H. Liem, and M.L. Shelanski. 1982. Interactions between neurofilaments and microtubule associated proteins: A possible mechanism for intraorganellar bridging. *J. Cell Biol.* **95:** 982.

Lopata, M.A., J.C. Havercroft, L.T. Chow, and D.W. Cleveland. 1983. Four unique genes required for β tubulin expression in vertebrates. *Cell* **32:** 713.

Marotta, C.A., J.L. Harris, and J.M. Gilbert. 1978. Characterization of multiple forms of brain tubulin subunits. *J. Neurochem.* **30:** 1431.

Messing, J. and J. Vieira. 1982. A new pair of M13 vectors for selecting either strand of double-digested restriction fragments. *Gene* **19:** 269.

Murphy, D.B. and K.T. Wallis. 1983a. Brain and erythrocyte microtubules from chicken contain different β tubulin polypeptides. *J. Biol. Chem.* **258:** 7870.

———. 1983b. Isolation of microtubule protein from chicken erythrocytes and determination of the critical concentration for tubulin polymerization in vitro and in vivo. *J. Biol. Chem.* **258:** 8357.

Nelles, L.P. and J.R. Bamburg. 1979. Comparative peptide mapping and isoelectric focusing of isolated subunits from chick embryo brain tubulin. *J. Neurochem.* **32:** 477.

Ponstingl, H., E. Krauhs, M. Little, and T. Kempf. 1981. Complete amino acid sequence for α tubulin from porcine brain. *Proc. Natl. Acad. Sci.* **78:** 2757.

Runge, M.S. and R.C. Williams. 1982. Formation of an ATP-dependent microtubule-neurofilament complex in vitro. *Cold Spring Harbor Symp. Quant. Biol.* **46:** 483.

Sanchez, F., J.E. Natzle, D.W. Cleveland, M.W. Kirschner, and B.J. McCarthy. 1980. A dispersed multigene family encoding tubulin in *Drosophila melanogaster. Cell* **22:** 845.

Sanger, F., S. Nicken, and A.R. Coulsen. 1977. DNA sequencing with chain terminating inhibitors. *Proc. Natl. Acad. Sci.* **74:** 5463.

Silflow, C.D. and J.L. Rosenbaum. 1981. Multiple α and β tubulin genes in *Chlamydomonas* and regulation of tubulin mRNA levels after deflagellation. *Cell* **24:** 81.

Stephens, R.E. 1978. Primary structural differences among tubulin subunits from flagella, cilia and the cytoplasm. *Biochemistry* **17:** 2882.

Sullivan, K.F. and L. Wilson. 1984. Developmental and biochemical analysis of chick brain tubulin heterogeneity. *J. Neurochem.* **42:** 1363.

Thompson, W.C., D.J. Asai, and D.H. Carney. 1984. Heterogeneity among microtubules of the cytoplasmic microtubule complex detected by a monoclonal antibody to α tubulin. *J. Cell Biol.* **98:** 1017.

Toda, T., Y. Adachi, Y. Hirakoka, and M. Yanagida. 1984. Identification of the pleotropic cell division cycle gene NDA2 as one of the two different α tubulin genes in *Schizosaccharomyces pombe. Cell* **37:** 233.

Valenzuela, P., M. Quiroga, J. Zaldivar, W.J. Rutter, M.W. Kirschner, and D.W. Cleveland. 1981. Nucleotide and corresponding amino acid sequences encoded by α and β tubulin mRNAs. *Nature* **289:** 650.

Wilde, D.C., C.E. Crowther, T.P. Cripe, M. Gwo-Shu Lee, and N.J. Cowan. 1982. Structure of two human α tubulin genes. *Proc. Natl. Acad. Sci.* **79:** 96.

Differential Expression of the Rat β-Tubulin Multigene Family

Stephen R. Farmer, Julian F. Bond, Gregory S. Robinson, David Mbangkollo, Matthew J. Fenton, and Ellen M. Berkowitz
Department of Biochemistry
Boston University Medical School
Boston, Massachusetts 02118

Microtubules are a major cytoskeletal component involved in many diverse functions including mitosis, meiosis, maintenance of cell shape, and motility (Stephens and Edds 1976). They are particularly important in the nervous system, playing crucial roles in neurite extension and axoplasmic transport. The diversity of microtubule function has stimulated speculation that different tubulins may be synthesized to form microtubules that are functionally distinct (Fulton and Simpson 1976). In this regard, several recent studies have revealed an extensive microheterogeneity of both the α- and β-tubulin proteins, which is especially apparent in the brain (Roberts and Hyams 1979). These different tubulins may result from posttranslational modifications of the same polypeptide or they could be the products of different genes.

The eukaryotic genome does contain multiple copies of DNA sequences complementary to β-tubulin cDNA probes. Most organisms studied to date have large tubulin multigene families, ranging from 10 to 20 unique DNA fragments per haploid genome. However, within mammals, many of these sequences are pseudogenes (Cowan et al. 1981; Wilde et al. 1982a,b; Gwo-Shu Lee et al. 1983). Some genomes contain a significantly less complex β-tubulin multigene family containing two or four unique DNA fragments (Silflow and Rosenbaum 1981; Kemphues et al. 1979, 1980; Lopata et al. 1983). In the case of these simpler gene families, it appears that all of the genes are transcriptionally active, and, in some cases, the different genes are expressed in a complicated program in differentiated cells (Lopata et al. 1983). It is still not known to what extent the separate mRNAs code for polypeptides of distinct function, except for one *Drosophila* gene that codes for a developmentally regulated testes-specific β-tubulin (Kemphues et al. 1979, 1980).

In an attempt to elucidate the tissue-specific expression of the rat β-tubulin multigene family, we have isolated and characterized three different cDNA clones. These have been sequenced and shown to correspond to three different β-tubulin isotypes. Hybridization studies with 3′-untranslated-region subclones of these cDNAs reveal that two of the mRNAs are produced only in neural cells and their expression is differentially regulated during rat brain development.

RESULTS

Our recent studies have demonstrated that β-tubulin mRNA production is regulated during rat brain development. This regulation is characterized by a dramatic decrease in the amount of both a 1.8-kb mRNA and a 2.9-kb mRNA at a time when cessation of cell division and neurite extension is initiated. Coincident with this response is an increased production of a new 2.5-kb mRNA species (Fig. 1) (Bond and Farmer 1983). To address the question of whether these β-tubulin mRNA species are transcribed from different genes, we have isolated and characterized several different cDNA clones corresponding to β-tubulin mRNAs (Fig. 2). The initial cDNA library corresponded to mRNA sequences isolated from a 2-day-old rat brain. To identify a rat β-tubulin cDNA, we screened this library with a *Chlamydomonas* β-tubulin cDNA plasmid, kindly given to us by C. Silflow (University of Minnesota) and J. Rosenbaum (Yale University). One of the plasmids (RβT.1) was chosen for further analysis. Hybridization selection of mRNA followed by in vitro translation confirmed that this plasmid contained sequences complementary to a rat brain β-tubulin mRNA. To isolate cDNAs corresponding to the other β-tubulin mRNAs, we constructed two recombinant DNA libraries complementary to mRNA isolated from 32-day-old rat cerebellum and 5-day-old rat spleen.

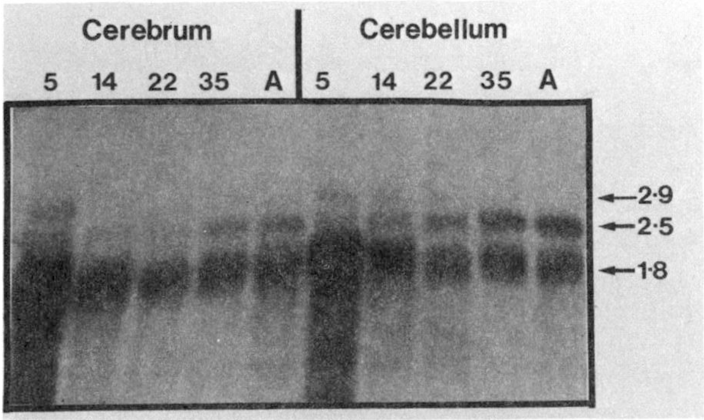

Figure 1
Overall changes in β-tubulin mRNA expression during rat brain development. Poly(A)+ RNA was isolated either from the cerebrum or from the cerebellum at the different ages shown, as described previously (Bond and Farmer 1983; Bond et al. 1984). Equal amounts of RNA (750 ng) were analyzed by RNA gel electrophoresis–filter hybridization (Northern blot) by hybridizing to the entire RβT.1 clone (see Fig. 2). Numbers correspond to ages in days, and A refers to an adult rat (~6 months).

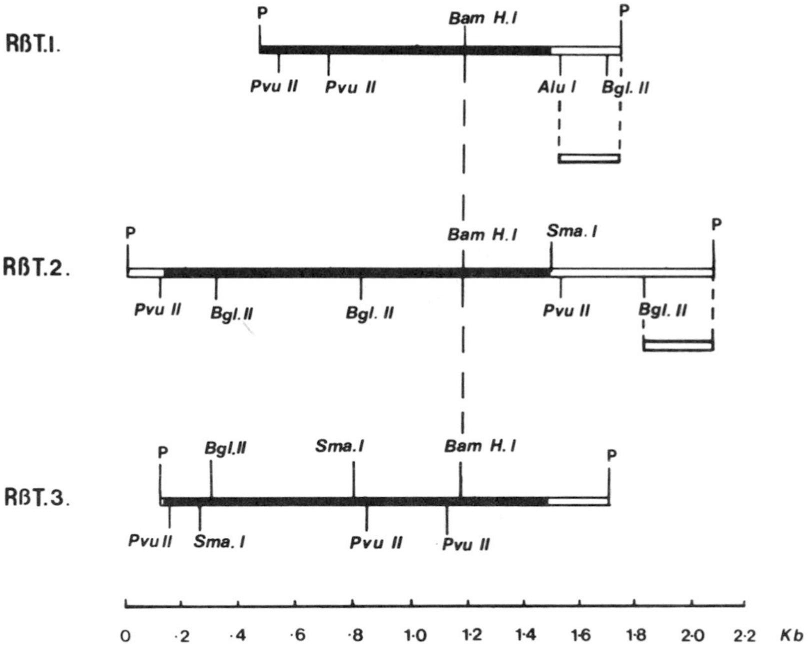

Figure 2
Restriction map of the rat β-tubulin cDNA clones. RβT.1 was isolated from a cDNA library corresponding to 2-day-old rat total brain mRNA, using as a hybridization probe a *Chlamydomonas* β-tubulin cDNA plasmid. RβT.2 was isolated from a 33-day-old cerebellum cDNA library and RβT.3, from a 5-day-old rat spleen cDNA library. Shaded region represents segments of homology among all three mRNAs and corresponds to coding regions. Unshaded region indicates no homology at noncoding areas. Also shown are the positions of 3′-untranslated regions that were subcloned and used for specific filter hybridizations. The size scale (in kilobases) is positioned relative to the largest sequence, RβT.2. (Reprinted, with permission, from Bond et al. 1984.)

To isolate a cDNA complementary to the newly expressed 2.5-kb species expressed in the adult rat brain (Fig. 1), we screened the 32-day-old rat cerebellum library with the entire RβT.1 clone. Several plasmids gave positive hybridization signals and were isolated for further analysis. The largest of the cDNAs that was unique to the adult brain was subjected to extensive analysis and contains all of the coding region. This plasmid was called RβT.2.

We screened the spleen library to obtain plasmid sequences complementary to the 1.8-kb and 2.9-kb mRNAs, which appear to be expressed in all tissues (Fig. 3). A plasmid was identified that was shown to be complementary to a rat β-tubulin mRNA by hybridization to the RβT.1-coding region and by hybridization selection-translation. This cDNA (Fig. 2, RβT.3) is different from RβT.1 and RβT.2, since their 3′-untranslated regions do not cross-hybridize (data not shown). The restriction maps of these three β-tubulin cDNAs, RβT.1, RβT.2, and RβT.3, are shown in Figure 2. Even though the three cDNA inserts are of different lengths,

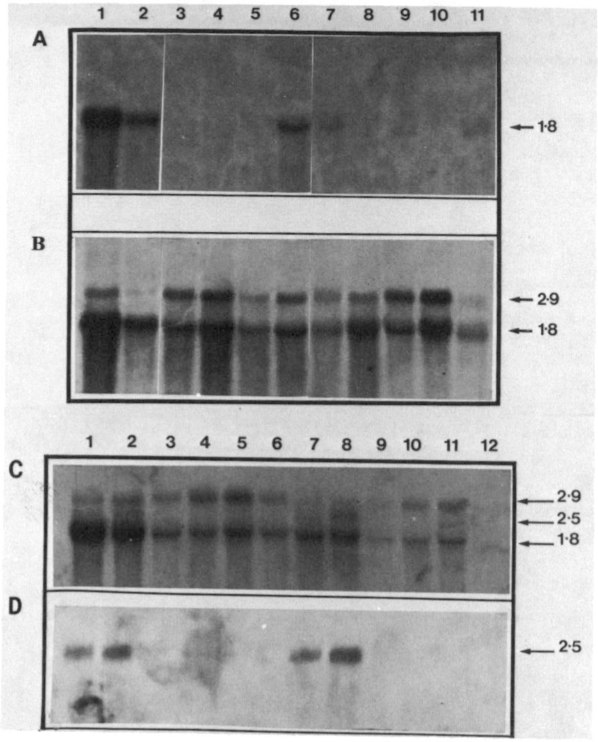

Figure 3
Identification of different β-tubulin mRNAs in a variety of rat tissues and cells. Total RNA was isolated from 5-day (*A,B*), 15-day, and adult (*C,D*) rat tissues or cells and electrophoresed on formaldehyde-agarose gels, followed by transfer to nitrocellulose paper. The filters were hybridized to the 3′-untranslated regions of either RβT.1 (*A*) or RβT.2 (*D*), followed by exhaustive washing and reprobing with either the entire RβT.1 (*B*) or RβT.2 (*C*). RNA samples (25 μg/lane; except lane 2 in *A* and *B*): (*A,B*) (*1–5*) 5-day rat tissues; (*1*) brain; (*2*) brain (5 μg); (*3*) lung; (*4*) spleen; (*5*) kidney; (*6*) B103 neuroblastoma cells; (*7*) PC12 pheochromocytoma cells; (*8*) R1 fibroblasts; (*9*) K16 epithelial cells; (*10*) W8-transformed epithelial cells; (*11*) B50 neuroblastoma cells. (*C,D*) (*1–6*) 15-day-old rat tissues; (*7–12*) adult rat tissues; (*1,7*) cerebrum; (*2,8*) cerebellum; (*3,9*) heart; (*4,10*) lung; (*5,11*) spleen; (*6–12*) kidney. mRNA sizes (kilobases) are indicated at the right. (Reprinted, with permission, from Bond et al. 1984.)

we have aligned them relative to the common *Bam*HI site shown to be present in all β-tubulin cDNAs isolated to date.

Tissue Specificity of the Different β-Tubulin cDNAs

DNA sequence analysis (data not shown) revealed that the specificity of each mRNA sequence resides within its 3′-untranslated region. Therefore, in order to study the tissue-specific expression of these mRNAs, we subcloned these specific regions and hybridized them to equal amounts of mRNA isolated from different rat tissues and cells.

Equal amounts of mRNA were hybridized to a radiolabeled subclone of the *Alu*I-*Pst*I restriction fragment containing the 3′-untranslated region of RβT.1 (Fig. 3A). The subclone was washed off, and the entire RβT.1 cDNA was hybridized to the same filter (Fig. 3B). As expected from the conservation of the coding region, the entire RβT.1 clone hybridizes to both a 1.8-kb mRNA and a 2.9-kb mRNA present in every RNA sample. However, the 3′-untranslated-region subclone of RβT.1 hybridized only to a 1.8-kb mRNA present in brain, B103 and B50 neuroblastoma cells, and PC12 pheochromocytoma cells (Fig. 3A). These data strongly suggest that RβT.1 cDNA corresponds to an mRNA present only in cells that originate from the neural crest.

To characterize the cerebellum-derived RβT.2 cDNA, we analyzed RNA isolated from various tissues of 15-day-old and 6-month-old rats (Fig. 3C,D). The entire RβT.2 clone hybridizes to a 1.8-kb and a 2.9-kb species in all tissues at both ages and to the 2.5-kb species in adult cerebellum, whereas the subclone of the *Bgl*II-*Pst*I restriction fragment containing the 3′-untranslated region of RβT.2 only hybridizes to a 2.5-kb mRNA present in the brain. The amount of this 2.5-kb mRNA was greatest in the adult cerebellum (Fig. 3D).

Expression of Neural-cell-specific β-Tubulin mRNAs during Rat Brain Development

Using the specific 3′-untranslated-region subclone of RβT.1, we failed to detect any complementary sequences present in a 32-day-old cerebellum library. Therefore, we theorized that this particular mRNA sequence is expressed only at very early stages of postnatal brain development. To demonstrate this, we hybridized the 3′-untranslated-region subclone of RβT.1 to mRNA isolated from cerebellum at different stages of postnatal development (Fig. 4B). This Northern blot filter was then washed and reprobed with the entire RβT.1 cDNA clone (previously shown to cross-hybridize with all β-tubulin mRNA species). As demonstrated in Figure 1, hybridization with the entire clone revealed the developmental pattern of overall β-tubulin mRNA expression (Fig. 4A). However, the subclone of RβT.1 only hybridized to a 1.8-kb sequence present at the early stages of cerebellar development.

Our initial studies analyzing changes in overall β-tubulin mRNA expression demonstrated the appearance of a 2.5-kb mRNA species at later stages of development. To confirm that RβT.2 corresponded to this species and to determine more precisely its time of appearance, we hybridized cerebellar mRNA, isolated during different stages of postnatal development, to the specific subclone of RβT.2. This fragment only hybridized to the 2.5-kb mRNA species (Fig. 4C). The time of its appearance is between birth and 3 days, but it is not significantly abundant until after 17 days of development. The appearance of this late neural species coincides with the decrease of the early neural species.

Methylation State of the 3′-untranslated Region of the RβT.1 Gene

To demonstrate a relationship between DNA methylation and neural-cell-specific expression of the RβT.1 gene, we utilized the isoschizomers *Hpa*II and *Msp*I.

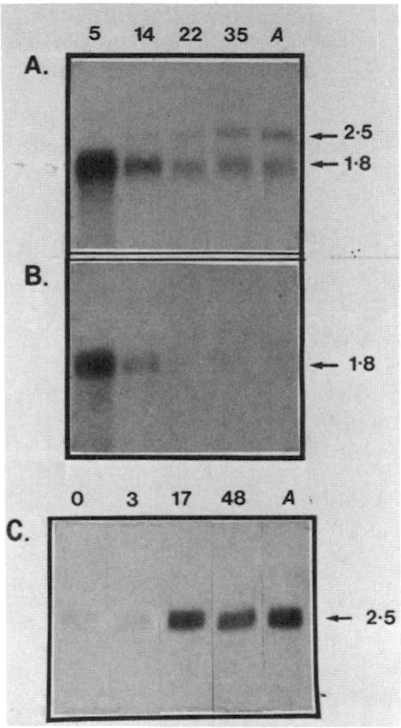

Figure 4
Expression of neural-cell-specific mRNAs during cerebellum development. Poly(A)$^+$ RNA was isolated from the cerebellum at the different ages indicated at the top of the gel. Equal amounts of RNA were analyzed by RNA gel electrophoresis–filter hybridization (Northern blot). One filter was hybridized to the entire RβT.1 clone (*A*), then exhaustively washed and rehybridized to the 3′-untranslated region of RβT.1 (*B*). Another filter (*C*) was hybridized to the 3′-untranslated region of RβT.2. Numbers correspond to ages in days; A refers to an adult rat. (Reprinted, with permission, from Bond et al. 1984.)

These restriction enzymes both recognize the sequence CCGG, containing the predominant methylation site, CpG, in higher eukaryotes. *Hpa*II will cleave this sequence only if the internal C is unmodified on both strands, whereas *Msp*I will cleave it regardless of its state of methylation.

Genomic DNA was isolated from various organs of 6-day-old or 2-month-old rats and digested with either *Hpa*II or *Msp*I. The restriction fragments were then analyzed by the Southern blot technique, using the 3′-untranslated-region subclone of RβT.1 as a hybridization probe. This probe hybridizes to a single *Msp*I fragment in all tissues at both ages (Fig. 5c,d). The presence of a 1.8-kb *Hpa*II fragment only in the cerebrum and cerebellum indicates hypomethylation within the 3′-untranslated region of this gene in the brain and correlates with the expression of mRNA sequences only in neural cells.

Comparison of Carboxyterminal Amino Acids from Different Vertebrate β-Tubulin Isotypes

The three rat β-tubulin cDNAs (RβT.1, RβT.2, and RβT.3) have been sequenced by the Sanger dideoxy chain terminator method. A comparison of the nucleotide sequences with those of other β-tubulins reveals a high degree of homology within the coding regions. This is consistent with the evolutionary conservation of the tubulins. The predicted amino acid sequences have also been compared and there is some divergence. The last 15 amino acids of the carboxyl termini appear to be the region of greatest divergence, particularly among different isotypes. In analyzing these rat carboxyterminal sequences and those of other select vertebrate β-tubulins, it was apparent that we could compile three different isotype

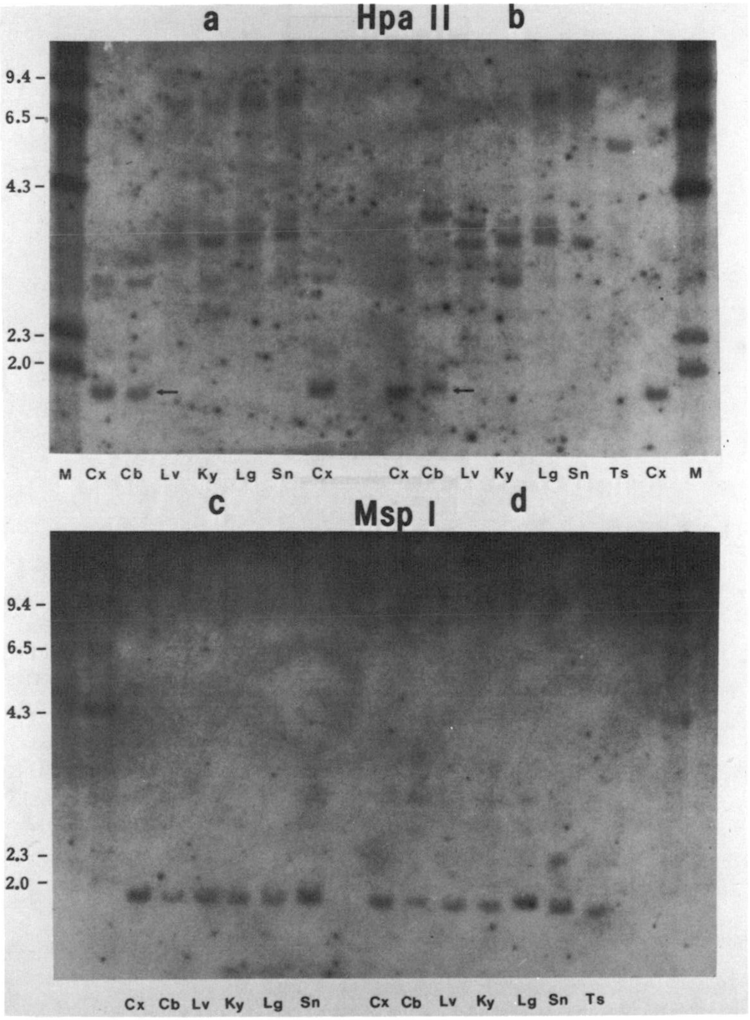

Figure 5

Tissue specificity of *Hpa*II methylation profiles for the RβT.1 β-tubulin sequences. DNA was isolated from each organ of 6-day-old (*a,c*) and 2-month-old (*b,d*) rats, restricted with either *Hpa*II or *Msp*I, electrophoresed on a 1% agarose gel, and transferred to nitrocellulose for hybridization with the 3′-untranslated region of RβT.1. (Cx) Cerebrum; (Cb) Cerebellum; (Lv) liver; (Ky) kidney; (Lg) lung; (Sn) spleen; (Ts) testis; (M) size markers, *Hin*dIII-digested bacteriophage λ DNA.

groups, in which RβT.1, RβT.2, and RβT.3 were members of the different groups (Fig. 6).

The carboxyl terminus of the early neural-cell-specific RβT.1 is almost identical to a chicken C.B2 sequence and a pig sequence. All three of these β-tubulins were derived from brain tissue. We propose that these three sequences are ev-

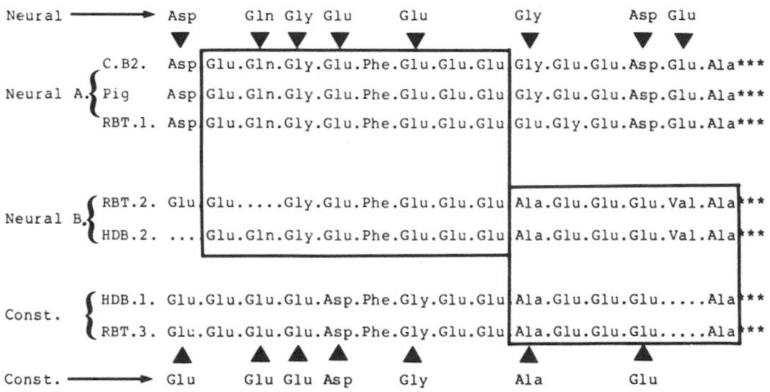

Figure 6
Comparison of the last 15 amino acids of the carboxyl termini of different vertebrate β-tubulins. Data were derived from DNA sequencing of RβT.1, RβT.2, and RβT.3 (data not shown) and the studies of Hall et al. (1983) (human, HDB.1, and HDB.2), Krauhs et al. (1981) (pig), and Valenzuela et al. (1981) (chicken, C.B2).

olutionarily related and are the same neural cell isotype (Fig. 6, Neural A). The carboxyl terminus of RβT.3 is significantly diverged from that of RβT.1 in that 9 of 15 amino acids are different. This sequence is identical with that of human β-tubulin HDB.1, and both mRNAs are expressed in many tissues. We therefore refer to these as constitutive isotypes.

Finally, the carboxyl terminus of the late neural-cell-specific RβT.2 β-tubulin differs somewhat from both of the neural A and the constitutive groups; however, it is closely related to another human β-tubulin, HDB.2. This suggests that HDB.2 is also a neural-cell-specific isotype and that these two isotypes probably constitute another evolutionarily conserved gene (Fig. 6, Neural B). The carboxyterminal sequences of this isotype appear to be a combination of both neural A and constitutive amino acids, implying an interesting evolutionary process within the β-tubulin multigene family.

DISCUSSION

To understand the relationship between tubulin isotypes synthesized and differential cellular functions of microtubules, we have embarked upon a study of the expression of the β-tubulin multigene family during rat development. Our intention is to characterize tissue-specific mRNAs and to define the mechanisms of regulating the differential expression of β-tubulin genes. In this study, we present strong evidence that there is differential expression of multigene β-tubulin mRNAs during rat brain development. Our conclusions are derived from the identification of two different neural-cell-specific β-tubulin mRNAs. One of these mRNAs (corresponding to the RβT.1 cDNA) is expressed in high abundance early in development and decreases dramatically with brain maturation. Thus, the RβT.1 mRNA appears to be restricted in expression to the cell-proliferation stage of this tissue. Its occurrence at this stage may be related to a stockpiling of specific tubulin subunits necessary for later incorporation into neurite microtubules.

The mRNA corresponding to the cDNA clone RβT.2, which also appears to be neural-cell-specific, has a distinctly different pattern of expression in the brain compared with the early neural (RβT.1) form. The RNA product of this gene increases with brain maturation and is apparently more abundant in the cerebellum. This mRNA species may also be a cell-type-specific message that is expressed in a population of cells that increases with later brain development. The greater abundance of this form in cerebellar tissue could then reflect a larger proportion of these cells in this tissue.

The β-tubulin clone RβT.3, which was generated from 5-day-old rat spleen mRNA, is presently being characterized in our laboratory. From the hybridization data obtained using this cDNA as a probe, it appears that this β-tubulin may be a ubiquitous form. Furthermore, from hybridization analysis, it appears that this clone is closely related to the larger 2.9-kb mRNA.

The relationship between DNA methylation and gene expression is somewhat complex, but in many cases, there is a correlation between specific transcription and hypomethylation of CpG sequences within the 5′ regions of genes (Riggs and Jones 1983). We have found that tissue-specific expression of at least one β-tubulin gene (RβT.1) is correlated with CpG hypomethylation in its 3′-untranslated region. Loss of methylation appears to be most extensive in the cerebrum, and the pattern of limited methylation differs between cerebrum and cerebellum. As has been found in other studies, termination of transcription (i.e., during late brain development) does not alter the state of hypomethylation of the RβT.1 gene.

At present, no predictions can be made on the possible functional significance of these different β-tubulin isotypes. However, the fact that the major differences in amino acid sequence reside within the last 15 amino acids of the carboxyl terminus implies that this region of the protein must be involved in some important function within microtubules. This carboxyl terminus is acidic and adopts an α-helical conformation (Krauhs et al. 1981). The amino acid differences between isotypes may change the conformation within this structure, thereby altering the affinity for binding to different cellular components, such as secretory vesicles (Sherline et al. 1977), coated vesicles (Pfeffer et al. 1983), microtubule-associated proteins (Bloom and Vallee 1983), microfilaments (Griffith and Pollard 1982), intermediate filaments (Ball and Singer 1981), and calmodulin (Lee and Wolf 1984).

To determine whether there is specific segregation of these neural sequences to different cell types, future studies will involve in situ hybridization in brain slices, using the radiolabeled subclones as probes. Antibodies to distinct regions of the β-tubulin proteins, in particular the carboxyl termini from the two neural gene products, may provide valuable information on the functional importance of differential mRNA expression.

ACKNOWLEDGMENTS

We acknowledge the help of Laura Dike during the latter stages of this work. We also appreciate the assistance of V. Renee Montgomery in the preparation of the manuscript. This work was supported by U.S. Public Health Service grants AG-00001 and GM-29630 from the National Institutes of Health.

REFERENCES

Ball, E.M. and S.J. Singer. 1981. Association of microtubules and intermediate filaments in normal fibroblasts and its disruption upon transformation by a temperature-sensitive mutant of Rous sarcoma virus. *Proc. Natl. Acad. Sci.* **78:** 6986.

Bloom, G.S. and R.B. Vallee. 1983. Association of microtubule associated protein 2 (MAP2) with microtubules and intermediate filaments in cultured brain cells. *J. Cell Biol.* **96:** 1523.

Bond, J.F. and S.R. Farmer. 1983. Regulation of tubulin and actin mRNA production in rat brain. Expression of a new β-tubulin mRNA with development. *Mol. Cell. Biol.* **3:** 1333.

Bond, J.F., G.S. Robinson, and S.R. Farmer. 1984. Differential expression of two neural cell-specific β-tubulin mRNAs during rat brain development. *Mol. Cell. Biol.* **4:** 1313.

Cowan, N.J., C.D. Wilde, L.T. Chow, and F.C. Wefald. 1981. Structural variation among human β-tubulin genes. *Proc. Natl. Acad. Sci.* **78:** 4877.

Fulton, C. and P.A. Simpson. 1976. Selective synthesis and utilization of flagellar tubulin. The multi-tubulin hypothesis. *Cold Spring Harbor Conf. Cell Proliferation* **3:** 987.

Griffith, L.M. and T.D. Pollard. 1982. The interaction of actin filaments with microtubules and microtubule-associated proteins. *J. Biol. Chem.* **257:** 9143.

Gwo-Shu Lee, M., S.A. Lewis, D.C. Wilde, and N.J. Cowan. 1983. Evolutionary history of a multigene family: An expressed human β-tubulin gene and three processed pseudogenes. *Cell* **33:** 477.

Hall, J.L., L. Dudley, P.L. Dobner, S.A. Lewis, and N.J. Cowan. 1983. Identification of two human β-tubulin isotypes. *Mol. Cell. Biol.* **3:** 854.

Kemphues, K.J., R.A. Raff, T.C. Kaufman, and E.C. Raff. 1979. Mutation in a structural gene for a β-tubulin specific to testis in *Drosophila melanogaster. Proc. Natl. Acad. Sci.* **76:** 3991.

Kemphues, K.J., E.C. Raff, R.A. Raff, and T.C. Kaufman. 1980. Mutation in a testis-specific β-tubulin in *Drosophila*: Analysis of its effects on meiosis and map location of the gene. *Cell* **21:** 445.

Krauhs, E., M. Little, T. Kempf, R. Hofer-Warbinek, W. Ade, and H. Ponstingl. 1981. Complete amino acid sequence of β-tubulin from porcine brain. *Proc. Natl. Acad. Sci.* **78:** 4156.

Lee, Y.C. and J. Wolff. 1984. Calmodulin binds to both microtubule-associated protein 2 and τ-proteins. *J. Biol. Chem.* **259:** 1226.

Lopata, M.A., J.C. Havercroft, L.T. Chow, and D.W. Cleveland. 1983. Four unique genes required for β-tubulin expression in vertebrates. *Cell* **32:** 713.

Pfeffer, S.R., D.G. Drubin, and R.B. Kelly. 1983. Identification of three coated vesicle components as α- and β-tubulin linked to a phosphorylated 50,000 dalton polypeptide. *J. Cell Biol.* **97:** 40.

Riggs, A.D. and P.A. Jones. 1983. 5-Methylcytosine, gene regulation and cancer. *Adv. Cancer Res.* **40:** 1.

Roberts, K. and J.S. Hyams. 1979. *Microtubules*. Academic Press, New York.

Sherline, P., Y. Lee, and L.S. Jacobs. 1977. Binding of microtubules to pituitary secretory granules and secretory granules membranes. *J. Cell Biol.* **72:** 380.

Silflow, C.D. and J.L. Rosenbaum. 1981. Multiple α- and β-tubulin genes in *Chlamydomonas* and regulation of tubulin mRNA levels after deflagellation. *Cell* **24:** 81.

Stephens, R.E. and K.T. Edds. 1976. Microtubule structure, chemistry and function. *Physiol. Rev.* **56:** 709.

Valenzuela, P., M. Quiroga, J. Zaldivar, W.J. Rutter, M.W. Kirschner, and D.W. Cleveland. 1981. Nucleotide and corresponding amino acid sequences encoded by α and β tubulin mRNAs. *Nature* **289:** 650.

Wilde, C.D., C.E. Crowther, and N.J. Cowan. 1982a. Isolation of a multigene family containing human β-tubulin sequences. *J. Mol. Biol.* **155:** 533.

———. 1982b. Diverse mechanisms in the generation of human β-tubulin pseudogenes. *Science* **217:** 549.

Microtubule-associated τ Protein Induction by Nerve Growth Factor during Neurite Outgrowth in PC12 Cells

David Drubin and Marc Kirschner
Department of Biochemistry
University of California
San Francisco, California 94143

Stuart Feinstein
Department of Neurobiology, Stanford University, Stanford, California 94305

Microtubule-associated proteins (MAPs) were first identified as proteins that coassembled with microtubules and dramatically promoted the otherwise poor assembly of purified tubulin in vitro. They include proteins with molecular weights of approximately 300,000, MAP 1 and MAP 2 (Keates and Hall 1975; Murphy and Borisy 1975; Sloboda et al. 1975), MAPs with molecular weights of 120,000 and 210,000 (Bulinski and Borisy 1979), and τ proteins with molecular weights of 55,000–68,000 (Weingarten et al. 1975; Cleveland et al. 1977). Additional MAPs with molecular weights of 69,000 and 80,000 were identified by their association with microtubules in vivo (Pallas and Solomon 1982). Despite the demonstration that MAPs can promote microtubule assembly in vitro, evidence for the function of MAPs in cells is lacking. Thus, although MAPs bind specifically to microtubules in vivo (Connolly et al. 1977, 1978; DeBrabander et al. 1981; Bloom et al. 1984) and in vitro and promote spontaneous microtubule assembly with purified tubulin or increase the rate and extent of nucleated assembly, in no case has it been shown that these proteins function to promote microtubule assembly in vivo. In fact, for some large MAPs, only a small region interacts with microtubules (Vallee and Borisy 1977), perhaps for anchorage, and it may be supposed that the other part of the molecule plays some other role.

To investigate whether some MAPs regulate microtubule polymerization, we sought an in vivo system in which there are major changes in the extent of microtubule polymerization. One system that is easily studied under controlled conditions in culture is the extension of neurites by rat pheochromocytoma (PC12) cells in response to nerve growth factor (NGF) (Greene and Tischler 1976). Since it was known that these neurites are extensively filled with microtubules (Luckenbill-Edds et al. 1979), we thought that extensive changes in microtubule polymer mass might be occurring. In this paper, we demonstrate that neurite outgrowth involves microtubule assembly, and we present some of our recent findings on changes in MAPs, in particular, τ protein, during neurite outgrowth.

DISCUSSION

Microtubule Polymer Levels Increase during PC12 Differentiation

Undifferentiated PC12 cells display a rounded morphology and are about 9 microns in diameter. When these cells are cultured in NGF for 1 week, many neurite processes appear that can be over 500 microns in length. The maintenance of the neurite network is dependent on the continued presence of NGF and requires intact microtubules. When NGF is removed from the culture for 1–2 days, or when microtubule-depolymerizing drugs are added to the culture for 2 hours, the neurite network disappears.

We might imagine that the formation of microtubule-filled processes could be caused either by a rearrangement of existing microtubules or by the polymerization of new microtubules. We can distinguish these possibilities by determining whether there is a change in microtubule polymer mass during neurite outgrowth. To measure the polymer mass, free of the soluble tubulin, detergent-extracted cytoskeletons were prepared under microtubule-stabilizing conditions, and the tubulin content of the cytoskeletons was determined by immunoblotting with antibodies against tubulin. Figure 1 shows a Western blot (Burnette 1981) of detergent-extracted PC12 cytoskeletons stained with a polyclonal tubulin antiserum. The center lane shows the polymer tubulin in undifferentiated PC12 cells. The tubulin signal is barely visible. In contrast, the first lane shows the polymer tubulin from a parallel culture treated for 1 hour with taxol, a drug that drives free tubulin into microtubules. Clearly, much more tubulin is now in the cytoskeleton fraction, implying that a large pool of unpolymerized tubulin that is normally extracted exists in undifferentiated PC12 cells. The last two lanes show the polymer tubulin in PC12 cells grown for 5 and 10 days in the presence of NGF. A massive and progressive increase in polymer tubulin has occurred. Similar, although less pronounced, increases in microtubule polymer mass were reported for neuroblastoma differentiation (Olmstead 1981). These shifts represent either increases in microtubule polymer levels inside the cell or less likely, but possibly, an increase in the stability of the polymer to the extraction conditions. We consider it extremely unlikely that the extraction conditions could induce polymerization because we perform a rapid, virtually infinite dilution of the free tubulin. Thus, NGF either increases the stability of existing microtubules or promotes tubulin assembly, or both. We shall refer to this loosely as an increase in polymer level, although it remains to be seen whether it is mostly a change in polymer stability.

Several possible mechanisms could act to increase microtubule polymer mass during PC12 differentiation. Increased levels of tubulin could drive the assembly of new microtubules. Alternatively, increased levels of MAPs could drive the large tubulin monomer pool into the polymer state. Finally, posttranslational activation of tubulin and MAPs or physiological changes may contribute to increased micro-

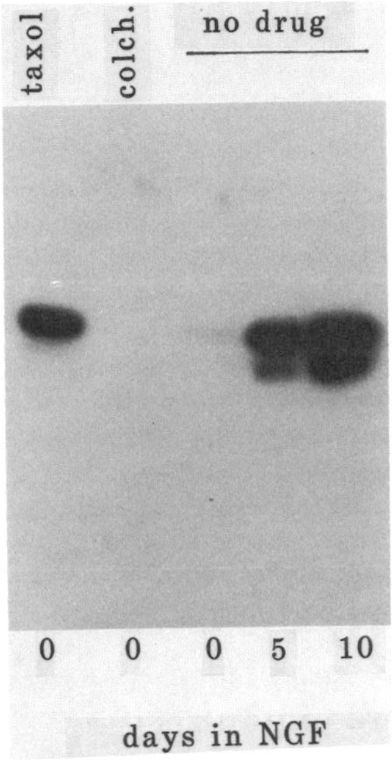

Figure 1
Microtubule polymer levels during NGF-induced PC12 neurite outgrowth. Microtubule polymer levels were measured according to the method of J. Caron (unpubl.) in which PC12 cells were extracted twice for 8 min with microtubule-stabilizing buffer (2.5 M glycerol, 0.1 M PIPES [pH 6.7], 1 mM $MgSO_4$, 1 mM EGTA, plus six protease inhibitors) containing 0.1% Triton X-100, and 15 μg of the resulting cytoskeletal proteins (which were solubilized in SDS-containing buffer) was analyzed with tubulin antiserum on a Western blot. (Lane *3*) Polymeric tubulin in undifferentiated PC12 cells. (Lanes *1,2*) Polymeric tubulin in undifferentiated PC12 cells treated with 10 μM taxol or 10 μM colchicine, respectively. (Lanes *4,5*) Polymeric tubulin in cells cultured in NGF for 5 and 10 days, respectively.

tubule polymer levels. In the following sections, we look specifically at tubulin and associated proteins during neurite outgrowth.

Characterization of PC12 τ Protein

τ Protein from hog brain consists of a collection of four polypeptides with molecular weights of 55,000–68,000 that promote the assembly of tubulin into mirotubules in vitro (Cleveland et al. 1977). Peptide maps, amino acid composition, and the physical properties of the four polypeptides indicate that they are very closely related to one another (Cleveland et al. 1977, 1979). τ Proteins from PC12 cells were analyzed by using an affinity-purified τ antiserum (Pfeffer et al. 1983; Drubin et al. 1984) to immunoprecipitate τ protein from total PC12 protein labeled in vivo with [^{32}P]phosphate (Fig. 2, left). The three lower species migrate in a molecular-weight range characteristic of brain τ protein. The upper band (m.w. 125,000) is much larger than any previously characterized τ polypeptide. We have also detected putative τ proteins in this size range in low abundance in brain tissue and in high abundance in spinal chord. The peptide maps in Figure 2 (right) show that all four PC12 τ proteins are closely related.

Having first characterized the proteins by their structural similarity, we next wanted to determine whether all four ^{32}P-labeled PC12 τ peptides will bind to microtubules in vitro. Unlabeled beef brain tubulin was incubated under polymerizing conditions with ^{32}P-labeled PC12 extracts, and microtubules were collected by centrifugation. Figure 3A shows by immunoprecipitation that the starting τ proteins (Fig. 3c) are quantitatively cleared from the supernatant (Fig. 3d) and end up in the microtubule pellet (Fig. 3e). Thus, all of the τ proteins bind to

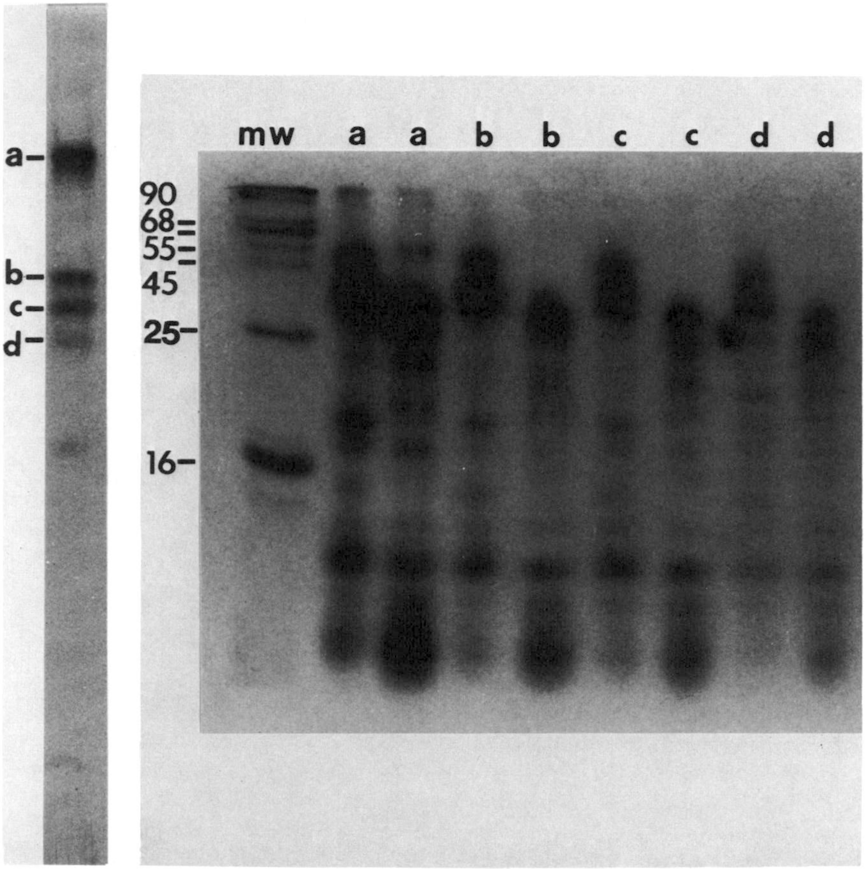

Figure 2
Analysis of immunoprecipitated PC12 τ protein labeled for 30 min in vivo with [^{32}P]-phosphate. Affinity-purified τ antiserum immunoprecipitated four τ proteins which were analyzed on an 8.5% polyacrylamide SDS gel (*left*, bands *a–d*). Each putative τ peptide was excised from the gel, subjected to partial proteolysis with *Staphylococcus aureus* V8 protease (*right*). Letters at the top of each lane designate the immunoprecipitated PC12 τ peptide mapped in that lane, with 10 ng and 100 ng of *S. aureus* protease, respectively.

microtubules in vitro. We next sought to determine whether the τ proteins also bind to microtubules in PC12 cells. Detergent-extracted cytoskeletons were prepared from differentiated PC12 cells under microtubule-stabilizing conditions, and the resulting proteins were analyzed with τ antiserum on Western blots. Figure 3B shows that τ proteins are in fact associated with the cytoskeletal (polymer) protein and not the extracted (monomer) protein. However, when the PC12 cells are pretreated with colchicine to depolymerize the microtubules, the τ proteins become detergent-extractable. We conclude that all four τ species associate with microtubules in PC12 cells.

The close relatedness of the four PC12 τ proteins might suggest that the three smaller species could be derived proteolytically from the 125,000-molecular-

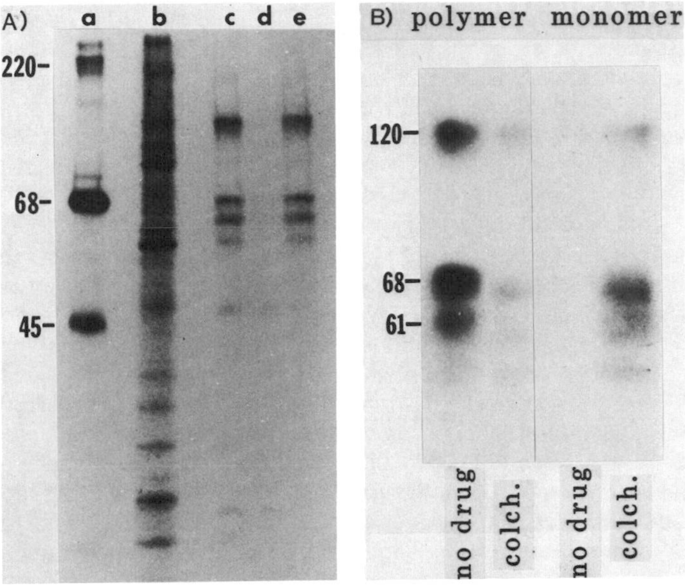

Figure 3
(*A*) Binding of PC12 τ proteins to microtubules in vitro. (*a*) Molecular-weight standards; (*b*) total extract from differentiated PC12 cells labeled with [^{32}P]phosphate; (*c*) immunoprecipitated τ protein from total extract; (*d*) immunoprecipitated τ protein from total extract supernatant after assembling 200 μg of pure tubulin into microtubules in the extract with 10 μg/ml taxol and collecting the microtubules by centrifugation; (*d*) immunoprecipitated τ protein from a resuspension of the microtubule pellet collected from the ^{32}P-labeled PC12 extract. (*B*) Binding of PC12 τ proteins to microtubules in vivo. Detergent-extracted cytoskeletons were prepared from differentiated PC12 cells as described in Fig. 1, and nonextracted or extracted τ proteins were analyzed by Western blotting (adapted from the method of Solomon et al. 1979). Left two lanes show τ proteins in PC12 cytoskeletons from untreated cells or cells treated for 35 min with 10 μM colchicine to depolymerize the microtubules, respectively. Right two lanes show the extracted τ proteins from nontreated and colchicine-treated PC12 cells, respectively.

weight species. However, in vitro translation experiments show that the different brain τ proteins are not proteolytically derived from one precursor, but rather are separate translation products (Ginzburg et al. 1982; Drubin et al. 1984). Another possibility is that the 125,000-molecular-weight τ is a covalent dimer of two lower-molecular-weight τ proteins. This, however, is unlikely, since the 125,000-molecular-weight protein is significantly more basic than the smaller τ species when analyzed on two-dimensional gels (not shown).

Changes in Microtubule Protein Levels during PC12 Differentiation

To determine whether changes in MAP or tubulin levels could account for the dramatic increase in microtubule polymer levels during NGF-induced neurite outgrowth in PC12 cells, the levels of these proteins were determined by Western blotting of extracts prepared at various times during PC12 differentiation. Figure 4A shows that τ-protein levels increase 30–40-fold during PC12 differentiation, with the major increase occurring at 4–7 days when most of the neurite growth occurs (see below). When NGF is removed, there is a sharp drop-off in cells with neurites and in τ-protein levels. The result with τ protein is quite dramatic relative to that seen with tubulin. Only a slight increase and decrease in tubulin levels result from NGF addition and removal (Fig. 4B). Similarly, no change in tubulin levels was observed during mouse N115 neuroblastoma differentiation (not shown), but in this case, the putative 125,000-molecular-weight τ species increased dramatically in abundance (Fig. 5).

Figure 6 summarizes the changes in neurite length, microtubule polymer levels, and amounts of tubulin and associated proteins observed in several experiments where NGF has been added to or removed from PC12 cells. The polymer mass changes very little during the first 3 days, a latent period during which very little or no neurite growth is seen. Removal of NGF causes the disappearance of most neurites and an 80% reduction in polymer mass within 2 days. The τ-protein level increases dramatically and in parallel with both the increase in microtubule polymer level and mean neurite length, and, like these two parameters, drops off sharply when NGF is removed. In contrast to the changes in polymer levels and τ, tubulin changes only slightly and with a different time course. Overall, in 7 days, tubulin levels increase only a fewfold, with the increase occurring steadily during the 3-day latent period. When NGF is removed for 3 days, tubulin levels decrease only 40%, whereas τ levels decrease over 95%.

We have also examined the high-molecular-weight-associated proteins. As

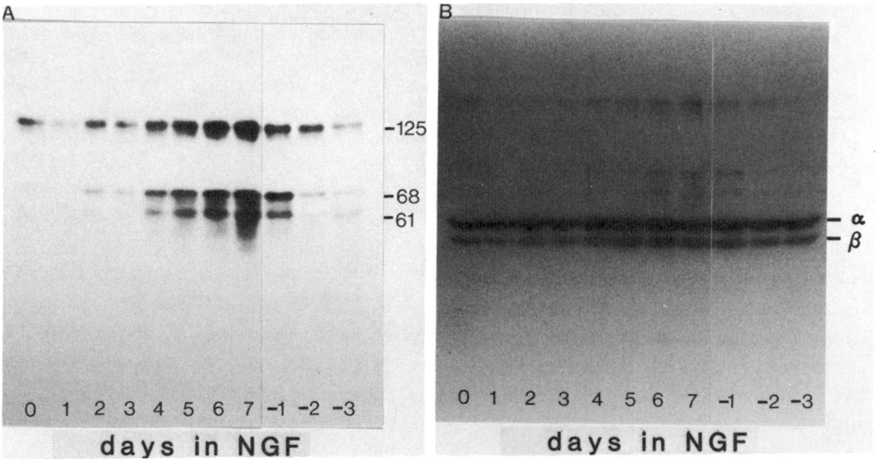

Figure 4

Effect of NGF on total τ and tubulin levels in PC12 cells. Total τ and tubulin levels were analyzed on the same Western blot stained first with an affinity-purified τ antiserum (*A*) and then with an affinity-purified tubulin antiserum (*B*). PC12 extracts were prepared from cells cultured in NGF for 0–7 days. After 7 days, NGF was withdrawn from the cells, and extracts were prepared after 1–3 days of withdrawal (−1 to −3); 30 μg of protein was loaded in each lane.

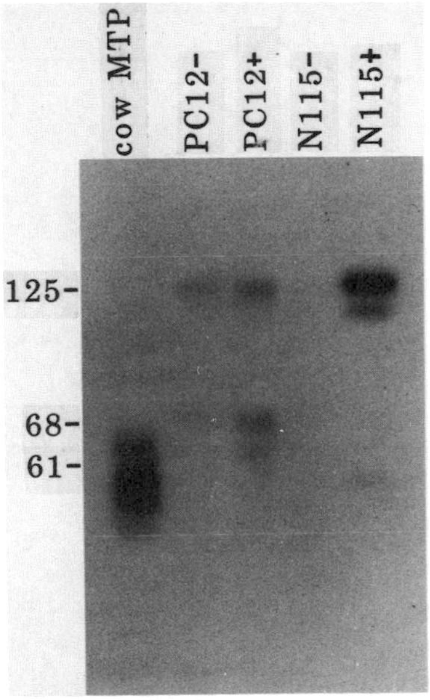

Figure 5
Comparison of τ species in differentiated and undifferentiated PC12 cells and mouse N115 neuroblastoma cells. Extracts from PC12 cells cultured without (−) or with (+) NGF were compared to extracts of undifferentiated N115 cells (−) or cells differentiated by serum starvation (+), on a Western blot stained with an affinity-purified τ antiserum. First lane shows the τ protein present in cow microtubule protein.

previously reported, MAP 1 levels are elevated in differentiated PC12 cells, whereas MAP 2 is undetectable in PC12 cells (Greene et al. 1983). The MAP 1 levels, like τ, show kinetics related to neurite outgrowth. They increase sharply at about 4–7 days and, like the neurites themselves, are completely dependent on the continued presence of NGF.

Control of Microtubule Protein Levels during PC12 Differentiation

We would like to understand how the protein levels of τ, MAP 1, and tubulin are regulated during PC12 cell differentiation. Since tubulin cDNA clones exist (Cleveland et al. 1980) and since we recently have isolated τ cDNA clones (Drubin et al. 1984), we can ask whether τ and tubulin mRNA levels change during this process. Northern blots containing poly(A)$^+$ RNA from various times during neurite outgrowth were analyzed with τ and tubulin cDNA probes (not shown). We found no dramatic changes in τ or tubulin mRNA levels. Although the twofold changes in tubulin levels could be due to slight changes in mRNA levels, another explanation must be sought for the large increases in τ protein. It is also worth noting the lack of correspondence between τ mRNAs and protein during brain development (Ginzburg et al. 1982; Drubin et al. 1984), indicating that the τ-protein levels are under posttranscriptional control.

One possible means of regulating protein levels is to control protein turnover rates. We wanted to determine whether the turnover of τ protein depended on its association with microtubules. To test this, we treated differentiated PC12 cells with appropriate pharmacological agents that affect protein synthesis and micro-

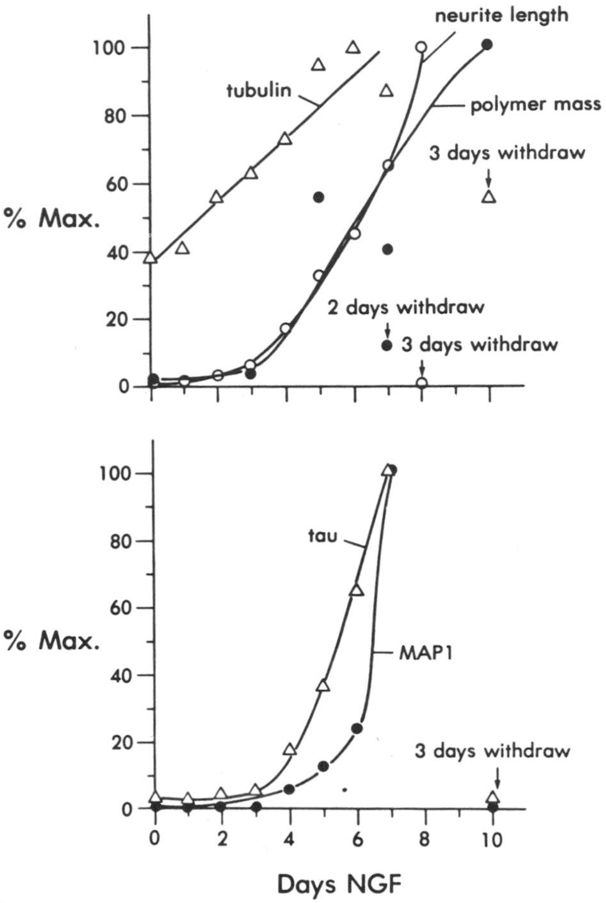

Figure 6
PC12 neurite length, microtubule polymer mass, and microtubule protein levels as a function of culture time in NGF. (*Top*) Neurite length (○) was determined by measuring 200 neurites each day in photographs of random fields of PC12 cells cultured in NGF on three separate plates. Tubulin levels (△) were determined by densimetric analysis of the blot shown in Fig. 4B. Polymer mass (●) was determined by densimetric analysis of blots like the one shown in Fig. 1. (*Bottom*) τ levels (△) were determined by densimetric analysis of the blot shown in Fig. 4A. MAP-1 levels (●) were determined by densimetric analysis of a Western blot (not shown) stained with a monoclonal MAP-1 antiserum (antiserum described by D.J. Asai et al., in prep.). Arrows indicate data points determined after withdrawal from NGF for the indicated number of days.

tubule assembly. Figure 7 is a Western blot showing total τ-protein levels in PC12 cells treated with various drugs. Colchicine treatment for 6 hours, which should create a pool of both free tubulin and τ protein, does not affect total τ levels. Cycloheximide treatment for 2 hours to block new protein synthesis similarly has no effect on τ levels. However, when the two agents are both employed, there is a significant and reproducible decrease in total τ-protein levels. Additional data show that MAP-1 levels also decrease under these conditions. We conclude from

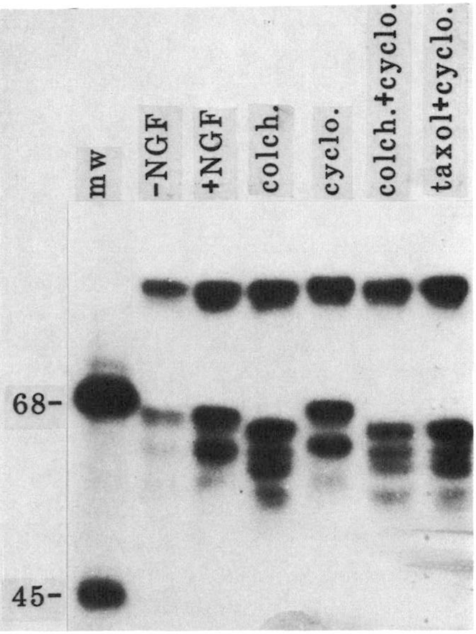

Figure 7
Effect of colchicine and taxol on τ form and stability. Extracts prepared from PC12 cells cultured in various drugs were analyzed by Western blot analysis with an affinity-purified τ antiserum. NGF-treated cells show the typical PC12 τ pattern. When NGF-treated PC12 cells were incubated in 10 μM colchicine for 6 hr to depolymerize the microtubules, a mobility shift in the lower two τ species occurred. Incubation of these cells in 10 μg/ml of cycloheximide for 2 hr to block protein synthesis had no affect. Incubation with colchicine for 6 hr and cycloheximide for the last 2 of those 6 hr resulted in decreased levels of τ protein. Taxol (10 μM) treatment to drive microtubule assembly to completion for 6 hr, with cycloheximide added for the last 2 of those 6 hr, did not result in a change in τ levels.

these experiments that free τ protein and MAP 1 are less stable than τ and MAP 1 associated with microtubules. Consistent with these interpretations, taxol treatment, which drives microtubule assembly to completion, in the presence of cycloheximide does not alter τ-protein levels (Fig. 7). Confirmation of these interpretations can be obtained by directly measuring the rates of τ synthesis and degradation under these various conditions.

Also evident in Figure 7 is the striking effect that colchicine and taxol treatment have on altering the mobilities of some of the τ species on SDS gels. Nocodazole, which, like colchicine, causes microtubules to depolymerize, and vinblastine, which causes tubulin to form paracrystalline aggregates, also cause these same increases in the motility of the τ species on gels (not shown). We have yet to determine the nature of the change in τ species, but since τ mobility on gels is affected by the extent to which the species are phosphorylated (Lindwall and Cole 1984), we suspect that the different mobility groups represent τ proteins with different degrees of phosphorylation. The effect of microtubule drugs on τ protein is exciting for two reasons. The fact that τ-protein changes occur when drugs affecting microtubules are applied to living cells provides strong evidence that τ associates with microtubules in cells. Also, the existence of different states of τ in living cells suggests that posttranslational modifications of τ protein may be important in regulating microtubule assembly and/or function in vivo.

SUMMARY AND CONCLUSIONS

Neurite outgrowth is dependent on the presence of intact microtubules. During PC12 neurite outgrowth, microtubule polymer mass increases dramatically despite only very slight changes in total tubulin levels. Most importantly, τ and MAP-

1 levels increase 30–40-fold in the presence of NGF, and the kinetics of these inductions correlate precisely with the increase in tubulin polymer mass and neurite length. In marked contrast, the kinetics of the slight changes in total tubulin levels are very different. Furthermore, withdrawal of NGF from differentiated PC12 cells and the resulting rapid loss of neurites are accompanied by dramatic reductions in τ- and MAP-1-protein levels as well as tubulin polymer mass, but only modest changes in total tubulin levels. These data suggest that regulation of MAP levels, but not tubulin levels, is fundamental to increasing microtubule polymer levels during neurite outgrowth in NGF-treated PC12 cells. These results are consistent with the observation made in brain tissue (Mareck et al. 1980) and neuroblastoma cells (Seeds and Maccioni 1978) that during neuronal differentiation, microtubule-polymerization-promoting activity increases while tubulin levels remain relatively constant. A more direct test of the role of these proteins in promoting microtubule assembly would be to utilize antisera against τ and MAP 1 to remove them selectively from PC12 extracts to assess their contributions to the assembly-promoting activity in in vitro polymerization assays.

Direct demonstration of the in vivo function of MAPs will require new approaches. Ideally, genetic mutations or drugs that specifically interfere with MAP function could be utilized. Unfortunately, no such mutants or drugs have been identified, but functionally analogous experiments are possible. Microinjection of antisera directed against τ or MAP 1 into PC12 cells might interfere with the function of these proteins. Perhaps more promising is the new technique of transfecting cells with expression vectors that synthesize antisense RNA for a gene of interest and interfere with the translation of the sense mRNA, presumably by in vivo formation of RNA-mRNA hybrids (Izant and Weintraub 1984). We predict that if τ protein drives microtubule assembly during neurite outgrowth, transfection of PC12 cells with an antisense expression vector containing our τ cDNA will inhibit neurite outgrowth.

Understanding how τ and MAP-1 levels are regulated in PC12 cells will be important to understanding how microtubule polymer levels are controlled. The dramatic induction of τ protein in response to NGF cannot be accounted for by an increase in τ mRNA. Other mechanisms must be involved. Importantly, it appears that both τ and MAP 1 are stabilized when they are bound to microtubules. This effect could contribute to the accumulation of these proteins during neurite outgrowth and their rapid disappearance during neurite retraction. This raises the question as to whether polymer assembly drives τ and MAP-1 accumulation or whether τ and MAP-1 accumulation drive polymer assembly. The question therefore remains, What is the primary signal that causes microtubules to assemble and disassemble when NGF is added to or removed from PC12 cells?

Another intriguing observation is that τ proteins change their electrophoretic mobility when drugs affecting microtubules are used on PC12 cells. This has two major implications. First, it strongly suggests that τ does in fact associate with microtubules in living cells. Furthermore, since different forms of τ protein can exist in living cells, it raises the possibility that posttranslational modifications of τ are important for regulating microtubule assembly and/or function in cells. In vitro experiments have shown that unphosphorylated τ protein is more active than phosphorylated τ protein in promoting microtubule assembly (Lindwall and Cole 1984). It is known that dibutyryl cAMP stimulates transient neurite outgrowth in PC12 cells and acts synergistically with NGF to promote neurite outgrowth (Gunning et al. 1981). Perhaps cAMP alters the activity of a τ kinase or phosphatase. The resulting dephosphorylated τ could more effectively promote microtubule assembly. Slight increases in polymer/monomer could lead to further binding of τ and MAP 1, leading to their stabilization and further accumulation. Interestingly, both β-tubulin and the 69,000- and 80,000-molecular-weight MAPs have higher levels of phosphate incorporated when they are associated with microtubules in living cells (Pallas and Solomon 1982; Gard and Kirschner 1984), suggesting that modification of these proteins may be important for regulating microtubule assembly. It will be important to characterize the in vivo phosphorylation sites on MAPs and tubulin and to test the in vitro effect of changes in these sites on stimulation of tubulin polymerization.

The PC12 cells provide a single cell system to study microtubule assembly and the function of associated proteins in response to a natural developmental regulatory molecule, NGF. This system is not simple. Although control does not seem to be exerted on a transcriptional level, there is clearly posttranslational modification and posttranslational control of synthesis or degradation. In addition, both MAP-1 proteins (Bloom et al. 1984) and τ proteins show considerable heterogeneity. We suspect that the different isoforms may be important for specifying the complex organization of microtubules within a given cell. For that reason, we are actively pursuing our studies of τ-protein heterogeneity through analysis of cloned cDNAs. We hope that an understanding of the structural basis of this heterogeneity, coupled with the function of posttranslational modifications, and control of τ-protein turnover will help explain the function of these proteins in PC12 differentiation and contribute to our overall understanding of how microtubules function in cell differentiation.

ACKNOWLEDGMENTS

We thank Eric Shooter in whose laboratory some of this work was done. We are grateful to David Asai for providing MAP-1 antiserum; Peter Nose for providing ^{32}P-labeled PC12 extracts; David Gard for providing N115 extracts and for his surgical and anatomical expertise, which allowed us to determine which parts of a mouse contain τ protein; Rivka Sherman-Gold for her participation in initial experiments with PC12 cells; and Joan Caron for her advice on the monomer/polymer studies. We thank Cynthia Cunningham-Hernandez for her assistance in preparing the manuscript. This work was supported by grants to M.K. from the National Institutes of Health and the American Cancer Society and by a grant from the Muscular Dystrophy Association to S.F.

REFERENCES

Bloom, G.S., T.A. Schoenfeld, and R.B. Vallee. 1984. Widespread distribution of the major polypeptide component of MAP 1 (microtubule-associated protein 1) in the nervous system. *J. Cell Biol.* **98:** 320.

Bulinski, J.C. and G.G. Borisy. 1979. Self-assembly of microtubules in extracts of cultured HeLa cells and the identification of HeLa microtubule-associated proteins. *Proc. Natl. Acad. Sci.* **76:** 293.

Burnette, W.N. 1981. "Western Blotting": Electrophoretic transfer of proteins from sodium dodecyl sulfate-polyacrylamide gels to unmodified nitrocellulose and radiographic detection with antibody and radioiodinated protein A. *Anal. Biochem.* **112:** 195.

Cleveland, D.W., S.Y. Hwo, and M.W. Kirschner. 1977. Physical and chemical properties of purified tau factor and the role of tau in microtubule assembly. *J. Mol. Biol.* **116:** 227.

Cleveland, D.W., B.M. Spiegelman, and M.W. Kirschner. 1979. Conservation of microtubule associated proteins. *J. Biol. Chem.* **254:** 12670.

Cleveland, D.W., M.A. Lopata, R.J. Lopata, N.J. Cowan, W.J. Rutter, and M.W. Kirschner. 1980. Number and evolutionary conservation of α and β tubulin and cytoplasmic β and α actin genes using specific cloned cDNA probes. *Cell* **20:** 95.

Connolly, J.A., V.I. Kalnins, D.W. Cleveland, and M.W. Kirschner. 1977. Immunofluorescent staining of cytoplasmic and spindle microtubules in mouse fibroblasts with antibody to τ protein. *Proc. Natl. Acad. Sci.* **74:** 2437.

———. 1978. Intracellular localization of the high molecular weight microtubule accessory protein by indirect immunofluorescence. *J. Cell Biol.* **76:** 781.

DeBrabander, M., J.C. Bulinski, G. Geuens, J. DeMay, and G.G. Borisy. 1981. Immunoelectron microscopic localization of the 210,000 mol wt microtubule associated protein in cultured cells of primates. *J. Cell Biol.* **91:** 438.

Drubin, D.G., D. Caput, and M.W. Kirschner. 1984. Studies on the expression of the microtubule-associated protein, tau, during mouse brain development, with newly isolated complementary DNA probes. *J. Cell Biol.* **98:** 1090.

Gard, D.L. and M.W. Kirschner. 1984. Polymer-dependent phosphorylation of β-tubulin accompanies differentiation of a mouse neuroblastoma cell line. *J. Cell Biol.* (in press).

Ginzburg, I., T. Scherson, D. Giveon, L. Behar, and U.Z. Littauer. 1982. Modulation of mRNA for microtubule-associated proteins during brain development. *Proc. Natl. Acad. Sci.* **79:** 4892.

Greene, L.A. and A.S. Tischler. 1976. Establishment of a nonadrenergic clonal line of rat adrenal pheochromocytoma cells which respond to nerve growth factor. *Proc. Natl. Acad. Sci.* **73:** 2424.

Greene, L.A., R.K.H. Liem, and M.L. Shelanski. 1983. Regulation of a high molecular weight microtubule-associated protein in PC12 cells by nerve growth factor. *J. Cell Biol.* **96:** 76.

Gunning, P.W., G.E. Landreth, M.A. Bothwell, and E.M. Shooter. 1981. Differential and synergistic actions of nerve growth factor and cyclic AMP in PC12 cells. *J. Cell Biol.* **89:** 240.

Izant, J.G. and H. Weintraub. 1984. Inhibition of thymidine kinase gene expression by anti-sense RNA: A molecular approach to genetic analysis. *Cell* **36:** 1007.

Keates, R.A. and R.H. Hall. 1975. Tubulin requires an accessory protein for self assembly into microtubules. *Nature* **257:** 418.

Lindwall, G. and R.D. Cole. 1984. Phosphorylation affects the ability of tau protein to promote microtubule assembly. *J. Biol. Chem.* **259:** 5301.

Luckenbill-Edds, L., C. Van Horn, and L.A. Greene. 1979. Fine structure of initial outgrowth of processes induced in a pheochomocytoma cell line (PC12) by nerve growth factor. *J. Neurocytol.* **8:** 493.

Mareck, A., A. Fellows, J. Francon, and J. Nunez. 1980. Changes in composition and activity of microtubule-associated proteins during brain development. *Nature* **284:** 353.

Murphy, D.B. and G.G. Borisy. 1975. Association of high molecular weight proteins with microtubules and their role in microtubule assembly *in vivo*. *Proc. Natl. Acad. Sci.* **72:** 2696.

Olmstead, J.B. 1981. Tubulin pools in differentiating neuroblastoma cells. *J. Cell Biol.* **89:** 418.

Pallas, D. and F. Solomon. 1982. Cytoplasmic microtubule-associated proteins: Phosphorylation at novel sites is correlated with their incorporation into assembled microtubules. *Cell* **30:** 407.

Pfeffer, S.R., D.G. Drubin, and R.J. Kelley. 1983. Identification of three coated vesicle components as α and β tubulin linked to a phosporylated 50,000 dalton polypeptide. *J. Cell Biol.* **97:** 40.

Seeds, N.W. and R.B. Maccioni. 1978. Proteins from morphologically differentiated neuroblastoma cells promote tubulin polymerization. *J. Cell Biol.* **76:** 547.

Sloboda, R.D., S.A. Rudolph, J.C. Rosenbaum, and P. Greengard. 1975. Cyclic-AMP-dependent endogenous phosphorylation of a microtubule-associated protein. *Proc. Natl. Acad. Sci.* **72:** 177.

Solomon, F., M. Magendantz, and A. Salzman. 1979. Identification with cellular microtubules of one of the co-assembling microtubule-associated proteins. *Cell* **18:** 431.

Vallee, R.B. and G.G. Borisy. 1977. Removal of the projections from cytoplasmic microtubules *in vitro* by digestion with trypsin. *J. Biol. Chem.* **252:** 377.

Weingarten, M.D., A.H. Lockwood, S.Y. Hwo, and M.W. Kirschner. 1975. A protein factor essential for microtubule assembly. *Proc. Natl. Acad. Sci.* **72:** 1858.

Expression and Cellular Regulation of Microtubule Proteins

Irith Ginzburg and Uriel Z. Littauer
Department of Neurobiology
The Weizmann Institute of Science
Rehovot 76100, Israel

The dynamic nature of microtubules and the broad spectrum of cellular processes in which they are involved suggest that the growth of microtubules is controlled both temporally and spatially in all cells. The detailed mechanisms by which microtubules regulate this process are not well understood (Littauer and Ginzburg 1984; Scherson et al. 1984). Since individual microtubules are not contractile in nature, it has been proposed that the generation of movement by microtubules depends on their treadmilling (Wegner 1976; Kirschner 1980; Margolis and Wilson 1981). However, the treadmilling model is based predominantly on in vitro reconstitution studies. Recently, fluorescently tagged microtubular proteins (Kreis et al. 1982) or MAP 2 (Scherson et al. 1984) were microinjected into living cells. This procedure, combined with fluorescence photobleaching recovery experiments, allowed us to follow in vivo the fate of a photobleached spot on fluorescent microtubules and demonstrated that this spot remained essentially immobile. These results suggest that treadmilling in microtubules does not exist or is limited in vivo to a maximal rate of about 4 μm/hour.

Another mechanism by which the cell may control microtubule assembly involves the availability and the change in composition of the diverse microtubule-associated proteins (MAPs) and their interaction with the various tubulin isotypes. It therefore appears that different regulatory signals for microtubule polymerization may be present in different cells and perhaps even within a single cell. Thus, during rat brain development, there is an increase in the number as well as a change in the distribution of the various tubulin isoforms (Gozes and Littauer 1978; Dahl and Weibel 1979) and τ factors (Mareck et al. 1980; Ginzburg et al. 1982). Moreover, τ factors isolated from adult rat brain have a better ability to promote in vitro tubulin polymerization than those isolated from newborn brain (Fellous et al. 1976; Francon et al. 1978). In addition, enhanced synthesis of two τ factors was observed in regenerating goldfish retinae, which coincides with enhanced synthesis of a β_2-isotubulin form (Ginzburg et al. 1983b; Neumann et al. 1983). Other studies show that there is a dissimilarity in the distribution of isotubulins isolated from various brain regions (Gozes et al. 1980) and that MAP 2 associates preferentially with dendritic microtubules of neurons, rather than with those of axons (Matus et al. 1981; Vallee 1982).

It is therefore suggested that the various isotubulin and MAP forms are not randomly distributed within a given microtubular system; rather, their organization is directly coupled to their function. It was of interest to define the levels of regulation of these proteins and to determine whether the observed microheterogeneity is due to the differential expression of a multigene family.

RESULTS AND DISCUSSION

Modulation of Tubulin mRNA Levels in Neuronal and Nonneuronal Cells

Previous studies have shown that during the first 10 days after birth, the percentage of cytoplasmic rat brain tubulin increases moderately or remains almost constant; thereafter, a progressive decrease in the percentage of cytoplasmic brain tubulin occurs until the rat reaches adulthood (Schmitt et al. 1977; Lennon et al. 1980). The changes in tubulin content during brain development appear in parallel for both neurons and glia and result from changes in the proportion of the corresponding mRNA (Gozes et al. 1975, 1977; Bond and Farmer 1983; Ginzburg et al. 1983a). The changes in tubulin protein levels observed in the developing brain correlate with the differentiation of neurons, axon formation, and increase in the expression of brain-specific isotubulins.

It should be noted that tubulin is not only confined to the cytoplasm, but is also found to be associated with various membranes. A striking feature is the observation that the α- and β-tubulin subunits differ in their association with brain presynaptic membranes (Gozes and Littauer 1979). It appears that the α subunit is an integral synaptic membrane protein, whereas most of the β subunit can be easily dissociated from these membranes. Whether the membrane-associated tubulin functions by a mechanism other than polymerization into microtubules remains to be determined.

The high complexity of brain mRNA is generally interpreted as reflecting the heterogeneity of cell types present in this organ. Cloned cells originating from the nervous system that retain many of the morphological and physiological properties of neurons and glia in vivo were therefore used as convenient models for the study of differentiated cell functions. Mouse neuroblastoma cells (clone N18TG-2) and rat glioma cells (clone C6BU-1), as well as the hybrid cell line (clone NG108-15) derived from these parental cell lines, differentiate morphologically and functionally in response to dibutyryl cAMP (Bt_2cAMP) treatment (Kimhi 1981; Soreq et al. 1983; Littauer et al. 1984). The morphological changes most likely involve microtubules and microfilaments, since cAMP-induced neurite formation was shown to be blocked by vinblastine and cytochalasin B (Seeds et al. 1970; Isenberg et al. 1979). One of the earliest events observed after administration of Bt_2cAMP to neuroblastoma cells is a marked flattening of the cells (within 1 hour); we therefore assumed that at least one of the primary events regulated by this agent might involve alterations in cytoskeletal protein synthesis. Our results showed that the effect of Bt_2cAMP on tubulin mRNA levels is biphasic. The induction of tubulin sequences was rapid and was evident after the first few hours of Bt_2cAMP treatment (Ginzburg et al. 1983a,b); it then leveled off, reaching maximal values at 12 hours, followed by a down-regulation of the tubulin mRNA levels to half of the value found in untreated cells. In the glioma cells, Bt_2cAMP also rapidly increased the level of tubulin mRNA sequences, reaching a maximum within 6 hours. In these cells, however, the subsequent decrease in tubulin mRNA content depends on the culture's phase of growth. Cells at the logarithmic growth phase do not down-regulate the tubulin mRNA content even after prolonged treatment with Bt_2cAMP, whereas confluent cells do. The hybrid cell line manifested intermediate characteristics in Bt_2cAMP regulation of tubulin mRNA levels. The time course of induction and down-regulation of tubulin mRNA content observed in the hybrid cells were similar to those of the parent neuroblastoma; however, the sensitivity of the hybrid cells to induction was like that of glioma cells and was eightfold higher than the initial level. Further analysis of these neuronal cells showed that modulation is more pronounced for the β-tubulin mRNA level. It also appears that in different cell lines, various inducers may differ in their effect on tubulin mRNA levels. Thus, serum deprivation of cultured brain neuroblastoma B104 cells (S. Rybak et al., unpubl.) or addition of bovine brain gangliosides to hybrid cell line SB21B1 caused a persistent induction of tubulin mRNA levels (Rybak et al. 1983) and was dependent on the presence of the inducer.

Some insight into the regulation of tubulin genes was obtained from our studies on nonneuronal cells. The effect of interferon treatment on Ramos cells, a human cell line of lymphoblastoid origin, was examined (Fellous et al. 1982). Blot hybridization with labeled tubulin cDNA showed that treatment with either α- or β-interferon induced a marked increase in the amounts of tubulin mRNA sequences, i.e., four and seven times the control level, respectively. The specific induction of tubulin mRNA in interferon-treated cells was abolished when these cells were pretreated with colchicine, suggesting two independent regulatory sites for tubulin mRNA expression.

The control of tubulin gene expression was also studied during the cell cycle of the unicellular organism *Tetrahymena pyriformis* (Zimmerman et al. 1983). The results show a periodicity of tubulin mRNA levels during the cell cycle. It was suggested that the fluctuation of tubulin mRNA levels may differ for the various isotubulin forms and may point to those involved in cytokinesis. It is interesting to note that in both of these nonneuronal systems, the modulation of the α-tubulin mRNA levels was more pronounced than for the β-tubulin mRNA.

Microheterogeneity of Tubulin

Brain α- and β-tubulin subunits display extensive microheterogeneity compared with tubulin from other tissues (Gozes and Littauer 1978). By using isoelectric focusing gels, it was possible to resolve 9–11 isoforms from rat brain tubulin

(Gozes and Littauer 1978; Dahl and Weibel 1979). Moreover, amino acid analysis of porcine tubulin indicates that the various α- and β-isotubulins differ from each other, within each group, by only a few amino acid replacements (Krauhs et al. 1981; Ponstingl et al. 1981).

Although the overall synthesis of tubulin decreases during rat brain development, there is an increase in the number of isotubulin forms and a change in their relative proportions during the maturation process (Gozes and Littauer 1978; Dahl and Weibel 1979; Gozes et al. 1980). These changes in brain tubulin microheterogeneity are regulated, at least in part, at the mRNA level (Gozes et al. 1980) and appear to be completed by the time the rat is 10–12 days of age (Dahl and Weibel 1979), which coincides with intense elaboration of neuronal processes. However, modulation of the relative proportions among α- and β-isotubulins follows different patterns during the various stages of brain development. Most of the changes in the proportion of the mouse brain α-isotubulins occur just before birth, whereas the evolution of the more acidic β-tubulin isoforms takes place throughout brain development. Thus, it appears that the α- and β-isotubulins are regulated independently (Denoulet et al. 1982; Wolff et al. 1982) and might be correlated with the changing pattern from proliferating cells to those undergoing neuronal differentiation.

Modulation of mRNA for MAPs

In addition to tubulin, various MAPs have been identified in preparations of cytoplasmic microtubules. Two major groups of MAPs have been isolated from neural tissues. These include high-molecular-mass components termed MAP 1 (350 kD) and MAP 2 (270 kD) (Borisy et al. 1975; Sloboda et al. 1976) and a number of polypeptides in the molecular-mass range of 55–62 kD, known collectively as τ proteins (Weingarten et al. 1975). It appears that different regulatory signals for microtubule polymerization may be present in different cells and perhaps within a single cell. This may result, among other factors, from changes in the distribution of various isotubulins and the composition of MAPs. The τ factors appear to be composed of four to six closely spaced bands (55–62 kD) when subjected to electrophoresis in SDS-polyacrylamide gels and a multiplicity of forms when analyzed by two-dimensional gel electrophoresis (Cleveland et al. 1977). During brain development, there is a marked increase in the diversity of τ factors and an enhancement of τ ability to induce, in vitro, microtubule assembly (Mareck et al. 1980). Newborn rat brain consists of two τ forms (τ_0), each having somewhat different molecular weights than the four to six τ components associated with microtubules from 12-day-old brain (τ_{12}) (Mareck et al. 1980; Ginzburg et al. 1982). However, the two groups of proteins are immunologically related and cross-react with antibodies to τ_{12} proteins (Ginzburg et al. 1982).

Our experiments involving cell-free translation of mRNA indicated that the major differences in the composition of τ proteins from newborn and developing brain are controlled at the mRNA level. The mRNA isolated from newborn rat brain directed the synthesis of five τ proteins, two of which are specific for newborn brain, whereas the other three forms are characteristic of the adult brain. Thus, the appearance in newborn rat brain of mRNA species specific for three mature τ forms precedes the phase of synthesis of these proteins in vivo. This observation suggests that during brain development, a translational control mechanism regulates the expression of several stage-specific τ proteins. In contrast, mRNA from 12-day-old rat brain directed the synthesis of four τ_{12} proteins, one of which is not synthesized with mRNA from newborn rat brain. The induction in the capacity of the mRNA to code for mature τ forms is accompanied by the disappearance of mRNA species that direct the synthesis of newborn rat brain τ (Ginzburg et al. 1982). These results indicate that, although the various τ forms are closely related, there are several different mechanisms for controlling the expression of these proteins. It thus appears that the basis for tubulin and τ heterogeneity and the control of expression of these proteins may be directly related to the functional properties of the microtubules.

Construction of Specific DNA Probes

A more precise insight into the involvement of the multiple tubulin genes in microtubule function required the construction of specific DNA probes for the individual gene transcripts. The α- and β-tubulin subunits are highly conserved proteins; consequently, DNA probes containing the coding sequence are unable to distinguish among the different α-isotubulin or β-isotubulin gene transcripts. From the limited data available, it appears that the various isotubulin mRNAs are distinguished by their highly divergent 3′-untranslated regions (Ginzburg et al. 1981; Lemischka et al. 1981; Cowan et al. 1983; Ginzburg et al. 1983b). Several approaches have been employed to isolate such specific DNA probes. It was also clear that only high-efficiency cloning techniques may yield some of the less-abundant cDNA clones. Several libraries were constructed in λgt10 phage. These include cDNA libraries from rat brain at different developmental stages and cDNA libraries from various neuroblastoma and glioma cells. The latter cell lines have previously been shown to possess a specific subset of tubulin isoforms, as compared with the complex pattern observed in brain (Gozes et al. 1979). Figure 1 shows Southern blot analysis of *Eco*RI fragments isolated from mouse neuroblastoma N18TG-2 and rat glioma C6BU-1 DNA, which is hybridized with a coding fragment derived from α-tubulin cDNA clone pT25 (Ginzburg et al. 1981). The hybridization pattern shows a complex tubulin gene family very similar to that obtained with the parental DNA isolated from rat and mouse brain, respectively. Thus, most of the tubulin genes are not lost during establishment of these cell lines in culture, and the reduced expression of tubulin isotypes appears to be controlled at the mRNA level. Using mRNA isolated from these cells for the construction of cDNA libraries is therefore advantageous, since it might be easier to track the cloned cDNA to its respective tubulin isotype. It is also interesting to note that results obtained from in situ hybridization of tubulin DNA probes to chromosomal spreads of these cells indicate that the tubulin genes are not linked but rather spread on several chromosomes. Figure 2 shows that there are at least six to eight hybridization sites localized on chromosomes from rat glioma C6BU-1 cells.

Recombinant cDNA libraries were initially constructed by inserting double-stranded DNA into the *Pst*I site of pBR322. More recently selected mRNA populations were converted into double-stranded cDNA according to the method of Gubler and Hoffman (1983). This method has been modified from the procedure

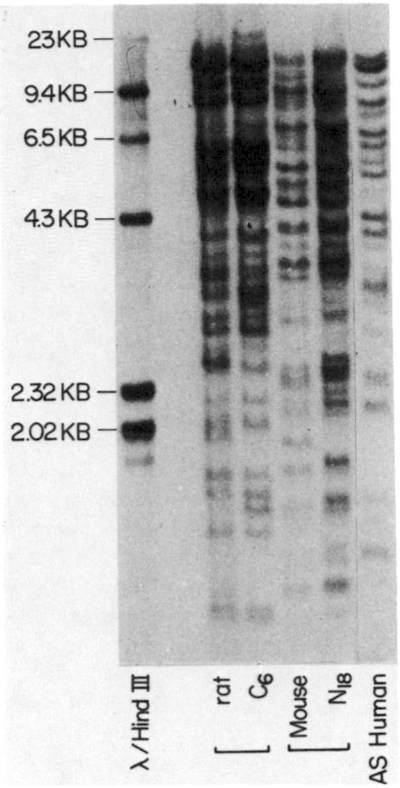

Figure 1
Genomic Southern blot of DNA isolated from rat, C6BU-1 rat glioma, mouse, N18TG-2 mouse neuroblastoma, and human neuroblastoma SK-N-AS. DNA samples were digested with *Eco*RI restriction endonuclease and resolved on a 0.8% agarose gel. The blots were hybridized with a nick-translated ^{32}P-labeled insert isolated from tubulin cDNA clone pT25.

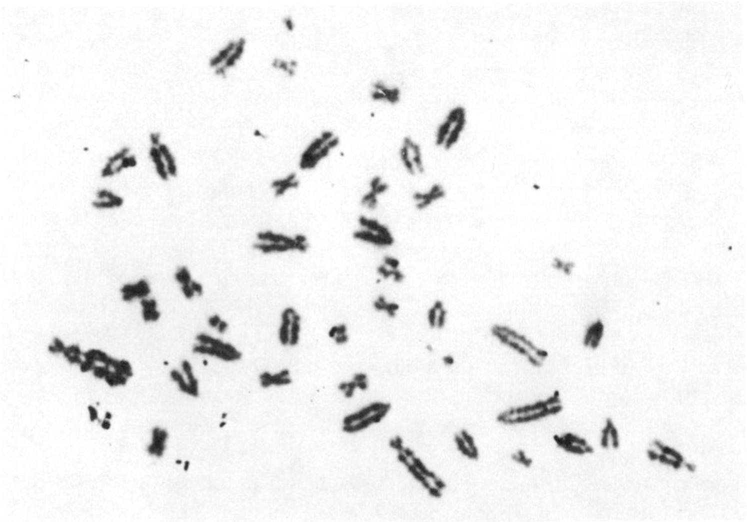

Figure 2
In situ hybridization of tubulin ^{32}P-labeled cDNA (pT25) to chromosome spread from rat glioma C6BU-1 cells.

of Okayama and Berg (1982) and renders large cDNA molecules. The double-stranded cDNA molecules thus derived were cloned in the high-efficiency phage λgt10 cloning system using *Eco*RI linkers. In contrast to genomic libraries that also contain pseudogenes, the cDNA libraries score only sequences derived from mRNA transcripts. The tubulin cDNA clones were identified by in situ colony hybridization to [^{32}P]cDNA derived from rat α-tubulin (Ginzburg et al. 1980, 1981) or chick β-tubulin (Valenzuela et al. 1981). Alternatively, we used ^{32}P-labeled synthetic oligonucleotides that are homologous to the 3′-coding sequences of α- and β-tubulin. By using these methods, we identified over 100 tubulin cDNA clones. These tubulin cDNA clones were further hybridized to short synthetic sequences (20 nucleotides long) that are homologous to 3′-untranslated regions of the several clones we have already sequenced. This allowed us to select new tubulin cDNA clones that differ in their 3′-untranslated sequences. Currently, these synthetic oligonucleotides are used as probes to analyze which of the tubulin genes are brain-specific and how they are developmentally regulated. The specific oligonucleotides are also used for in situ mRNA hybridization studies to determine the expression of specific microtubule components within individual cells in various regions of the central nervous system.

Tubulin MAP-binding Sites

We have developed a specific binding assay for the interaction of ^{125}I-labeled MAPs with tubulin or its cleavage peptides. To identify the tubulin MAP-binding domains, we have examined, in collaboration with H. Ponstingl (German Cancer Research Center, Heidelberg), the binding of rat brain ^{125}I-labeled MAP 2 or ^{125}I-labeled τ factors to 46 cleavage peptides derived from pig α- and β-tubulins. Our results show that MAP 2 will specifically interact with only two peptides located at the carboxyl terminus of β-tubulin, between positions 407 and 415 and positions 416 and 445. Strong binding sites for τ factors were also located at the carboxyl terminus of β-tubulin between positions 416 and 445. In addition, τ factors, but not MAP 2, interacted with two peptides derived from α-tubulin. The first peptide is located near the amino terminus of α-tubulin between positions 37 and 78 and a second binding site is located between positions 314 and 327.

Comparison of the amino acid sequences of human β-tubulin isotypes (Hall et al. 1983) with those of pig (Krauhs et al. 1981), chicken (Valenzuela et al. 1981), and yeast (Neff et al. 1983) reveals a variable domain within the 15 carboxyterminal amino acids. This same region was found by our studies to include, at least in part, the binding site to MAP 2 and τ factors. It is tempting to speculate that the variability in amino acid sequence in this region will determine the specificity of interaction of β-tubulin isotypes with the various MAPs, thus generating functionally different microtubules.

ACKNOWLEDGMENTS

This work was supported in part by grants from the United States–Israel Binational Science Foundation, the Forchheimer Center for Molecular Genetics, the National Council for Research and Development, Israel, and the Deutsches Krebsforschungzentrum, Heidelberg, Germany.

REFERENCES

Bond, J.F. and S.R. Farmer. 1983. Regulation of tubulin and actin mRNA production in rat brain: Expression of a new β-tubulin mRNA with development. *Mol. Cell. Biol.* **3:** 1333.

Borisy, G.G., J.M. Marcum, J.B. Olmsted, D.B. Murphy, and K.A. Johnson. 1975. Purification of tubulin and associated high molecular weight proteins from porcine brain and characterization of microtubule assembly *in vitro*. *Ann. N.Y. Acad. Sci.* **253:** 107.

Cleveland, D.W., S.Y. Hwo, and M.W. Kirschner. 1977. Physical and chemical properties of purified TAU factor and the role of TAU in microtubule assembly. *J. Mol. Biol.* **116:** 227.

Cowan, N.J., P.R. Dobner, E.V. Fuchs, and D.W. Cleveland. 1983. Expression of human α-tubulin genes: Interspecies conservation of 3′-untranslated regions. *Mol. Cell. Biol.* **10:** 1738.

Dahl, J.L. and V.J. Weibel. 1979. Changes in tubulin heterogeneity during postnatal development of rat brain. *Biochem. Biophys. Res. Commun.* **86:** 822.

Denoulet, P., B. Edde, C. Jeantet, and F. Gros. 1982. Evolution of tubulin heterogeneity during mouse brain development. *Biochimie* **64:** 165.

Dustin, P., ed. 1978. *Microtubules*. Springer-Verlag, Berlin.

Fellous, A., I. Ginzburg, and U.Z. Littauer. 1982. Modulation of tubulin mRNA levels by interferon in human lymphoblastoid cells. *EMBO J.* **1:** 835.

Fellous, A., J. Francon, A.M. Lennon, and J. Nunez. 1976. Initiation of neurotubulin polymerization and rat brain development. *FEBS Lett.* **64:** 400.

Francon, J., A. Fellous, A.M. Lennon, and J. Nunez. 1978. Requirement for "factor(s)" for tubulin assembly during brain development. *Eur. J. Biochem.* **85:** 43.

Ginzburg, I., L. Behar, D. Givol, and U.Z. Littauer. 1981. The nucleotide sequence of rat α-tubulin: 3′-end characteristics and evolutionary conservation. *Nucleic Acids Res.* **9:** 2691.

Ginzburg, I., S. Rybak, Y. Kimhi, and U.Z. Litauer. 1983a. Biphasic regulation of dibutyryl cyclic AMP: Of tubulin and actin mRNA levels in neuroblastoma cells. *Proc. Natl. Acad. Sci.* **80:** 4243.

Ginzburg, I., T. Scherson, D. Giveon, L. Behar, and U.Z. Littauer. 1982. Modulation of mRNA for microtubule-associated proteins during brain development. *Proc. Natl. Acad. Sci.* **79:** 4892.

Ginzburg, I., A. de Baetselier, M.D. Walker, L. Behar, H. Lehrach, A.M. Frischauf, and U.Z. Littauer. 1980. Brain tubulin and actin cDNA sequences: Isolation of recombinant plasmids. *Nucleic Acids Res.* **8:** 3553.

Ginzburg, I., T. Scherson, S. Rybak, Y. Kimhi, D. Neuman, M. Schwartz, and U.Z. Littauer. 1983b. Expression of mRNA for microtubule proteins in the developing nervous system. *Cold Spring Harbor Symp. Quant. Biol.* **48:** 783.

Gozes, I. and U.Z. Littauer. 1978. Tubulin microheterogeneity increases with rat brain maturation. *Nature* **276:** 411.

———. 1979. The α-subunit of tubulin is preferentially associated with brain presynaptic membrane. *FEBS Lett.* **99:** 86.

Gozes, I., A. de Baestselier, and U.Z. Littauer. 1980. Translation *in vitro* of rat brain mRNA coding for a variety of tubulin forms. *Eur. J. Biochem.* **103:** 13.

Gozes, I., D. Saya, and U.Z. Littauer. 1979. Tubulin microheterogeneity in neuroblastoma and glioma cell lines differs from that of brain. *Brain Res.* **171:** 171.

Gozes, I., H. Schmitt, and U.Z. Littauer. 1975. Translation *in vitro* of rat brain messenger RNA coding for tubulin and actin. *Proc. Natl. Acad. Sci.* **72:** 701.

Gozes, I., M.D. Walker, A.M. Kaye, and U.Z. Littauer. 1977. Synthesis of tubulin and actin by neuronal and glial nuclear preparations from developing rat brain. *J. Biol. Chem.* **252:** 1819.

Gubler, U. and B.J. Hoffman. 1983. A simple and very efficient method for generating cDNA libraries. *Gene* **25:** 263.

Hall, J.L., L. Dudley, P.R. Dobner, S.A. Lewis, and N.J. Cowan. 1983. Identification of two human α-tubulin isotypes. *Mol. Cell. Biol.* **3:** 854.

Isenberg, G., J.V. Small, and G.W. Kreutzberg. 1979. The cytoskeleton and its influence on shape, motility and receptor segregation in neuroblastoma cell. *Prog. Brain Res.* **51:** 45.

Kimhi, Y. 1981. Nerve cells in clonal system. In *Excitable cells in tissue culture* (ed. P.G. Nelson and M. Lieberman), p. 173. Plenum Press, New York.

Kirschner, M. 1980. Implications of treadmilling for the stability and polarity of actin and tubulin polymerization *in vivo. J. Cell Biol.* **86:** 330.

Krauhs, E., M. Little, T. Kempf, R. Hofer-Warbinek, W. Ade, and H. Ponstingl. 1981. Complete amino acid sequence of β-tubulin from porcine brain. *Proc. Natl. Acad. Sci.* **78:** 4156.

Kreis, T.E., T. Scherson, U.Z. Littauer, J. Schlessinger, and B. Geiger. 1982. Dynamics of microtubules in living cultured cells. *J. Cell Biol.* **95:** 348a.

Lemischka, I.R., S. Farmer, V.R. Racaniello, and P.A. Sharp. 1981. Nucleotide sequence and evolution of a mammalian α-tubulin messenger RNA. *J. Mol. Biol.* **151:** 101.

Lennon, A.M., J. Francon, A. Fellous, and J. Nunez. 1980. Rat, mouse, and guinea pig brain development and microtubule assembly. *J. Neurochem.* **35:** 804.

Littauer, U.Z. and I. Ginzburg. 1984. Expression of microtubule proteins in brain. In *Gene expression in brain* (ed. C. Zomzely-Neurath and W.A. Walker). J. Wiley and Sons, New York. (In press.)

Littauer, U.Z., A. Zutra, S. Rybak, and I. Ginzburg. 1984. The expression of tubulin and various enzyme activities during neuroblastoma differentiation. In *Advances in neuroblastoma research* (ed. A.E. Evans). Alan R. Liss, New York. (In press.)

Mareck, A., A. Fellous, J. Francon, and J. Nunez. 1980. Changes in composition and activity of microtubule-associated proteins during brain development. *Nature* **284:** 353.

Margolis, R.L. and L. Wilson. 1981. Microtubule treadmills—Possible molecular machinery. *Nature* **293:** 705.

Matus, A., R. Bernhardt, and T. Hugh-Jones. 1981. High molecular weight microtubule-associated proteins are preferentially associated with dendritic microtubules in brain. *Proc. Natl. Acad. Sci.* **78:** 3010.

Neff, N.F., J.H. Thomas, P. Grisafi, and D. Botstein. 1983. Isolation of the β-tubulin gene from yeast and demonstration of its essential function *in vivo. Cell* **33:** 211.

Neumann, D., T. Scherson, I. Ginzburg, U.Z. Littauer, and M. Schwartz. 1983. Regulation of mRNA levels for microtubule proteins during nerve regeneration. *FEBS Lett.* **162:** 270.

Okayama, H. and P. Berg. 1982. High efficiency cloning of full length cDNA. *Mol. Cell. Biol.* **2:** 161.

Ponstingl, H., E. Krauhs, M. Little, and T. Kempf. 1981. Complete amino acid sequence of α-tubulin from porcine brain. *Proc. Natl. Acad. Sci.* **78:** 2757.

Rybak, S., I. Ginzburg, and E. Yavin. 1983. Gangliosides stimulate neurite outgrowth and induce tubulin mRNA accumulation in neural cells. *Biochem. Biophys. Res. Commun.* **116:** 974.

Scherson, T., T. Kreis, J. Schlessinger, U.Z. Littauer, G.G. Borisy, and B. Geiger. 1984. Dynamic interactions of fluorescently labeled microtubule associated proteins in living cells. *J. Cell Biol.* **99:** 425.

Schmitt, H., I. Gozes, and U.Z. Littauer. 1977. Decrease in levels and rates of synthesis of tubulin and actin in developing rat brain. *Brain Res.* **121:** 327.

Seeds, N.W., A.G. Gilman, T. Amano, and M.W. Nirenberg. 1970. Regulation of axon formation by clonal lines of a neural tumor. *Proc. Natl. Acad. Sci.* **66:** 160.

Sloboda, R.D., W.L. Dentler, R.A. Bloodgood, B.R. Telzer, S. Granett, and J.L. Rosenbaum. 1976. Microtubule-associated proteins (MAPs) and the assembly of microtubules *in vitro. Cold Spring Harbor Conf. Cell Proliferation* **3:** 1171.

Soreq, H., R. Miskin, A. Zutra, and U.Z. Littauer. 1983. Modulation in the levels and localization of plasminogen activator in differentiating neuroblastoma cells. *Dev. Brain Res.* **7:** 257.

Valenzuela, P., M. Quiroga, J. Zaldivar, W.J. Rutter, M.W. Kirschner, and D.W. Cleveland. 1981. Nucleotide and corresponding amino acid sequences encoded by α and β tubulin mRNAs. *Nature* **289:** 650.

Vallee, R.B. 1982. A Taxol-dependent procedure for the isolation of microtubules and microtubule-associated proteins (MAPs). *J. Cell Biol.* **92:** 435.

Wegner, A. 1976. Head-to-tail polymerization of actin. *J. Mol. Biol.* **108:** 139.

Weingarten, M.D., A.H. Lockwood, S.Y. Hwo, and M.W. Kirschner. 1975. A protein factor essential for microtubule assembly. *Proc. Natl. Acad. Sci.* **72:** 1858.

Wolff, A., P. Denoulet, and Ĉ. Jeantet. 1982. High level of tubulin microheterogeneity in the mouse brain. *Neurosci. Lett.* **31:** 323.

Zimmerman, A.M., S. Zimmerman, J. Thomas, and I. Ginzburg. 1983. Control of tubulin and actin gene expression in *Tetrahymena pyriformis* during the cell cycle. *FEBS Lett.* **164:** 318.

Coordinate Expression of the Four Tubulin Genes in *Chlamydomonas*

Karen Brunke,*† James Anthony,*†
Fred Kalish,† Edmund Sternberg,*‡
and Donald Weeks†
**Institute for Cancer Research*
Fox Chase Cancer Center
Philadelphia, Pennsylvania 19111
†Zoecon Corporation
Palo Alto, California 94304

The removal of flagella from *Chlamydomonas reinhardtii* induces a burst of tubulin production in this unicellular alga. This increased synthesis begins soon after deflagellation, peaking quickly (about 45–60 min), after which time synthesis starts to decline as flagella approach full length (Weeks et al. 1977). Both the α- and β-tubulin subunits are produced following flagellar excision (Weeks and Collis 1976; Lefebvre et al. 1978; Minami et al. 1981; Brunke et al. 1982a). In gametes, this synthesis accounts for 10–15% of total cellular protein synthesis. Increased synthesis of these subunits begins within minutes of flagellar excision and precedes the initiation of flagellar elongation, which begins at about 8–10 minutes after deflagellation.

A simultaneous increase in production of four distinct species of tubulin mRNAs accompanies the induction of tubulin synthesis (Brunke et al. 1982b; Schloss et al. 1984). The two α-tubulin and two β-tubulin mRNAs are produced coordinately from a set of four unclustered nuclear genes (Silflow and Rosenbaum 1981; Brunke et al. 1982b). As with tubulin synthesis, production of all four mRNAs peaks and then decreases as flagella attain their full length (Minami et al. 1981; Silflow and Rosenbaum 1981; Brunke et al. 1982b). Apparently, the regulatory mechanism for the synchronous production of tubulin mRNAs operates, at least in part, at the transcriptional level. This is based on the finding that transcription studies in vitro ("runoff" experiments) using nuclei isolated from deflagellated

‡Present address: Smith-Kline and French Laboratories, 709 Swedeland, Swedeland, Pennsylvania 19479.

cells yield more tubulin transcripts than do their counterpart nuclei from nondeflagellated cells (Keller et al. 1984). Elevated tubulin production and appearance of the four mRNAs for tubulin also occur during cell division (Weeks and Collis 1979), another time when tubulin is required by the cell for microtubule assembly.

For a family of genes that are not tightly clustered to be induced coordinately in response to a given stimulus, all members of the gene family must possess a feature in common that allows recognition by a *trans*-acting regulatory signal. We wanted to know whether such a recognition element was present in the *Chlamydomonas* tubulin genes. Comparison of other eukaryotic gene systems that respond to developmental, hormonal, or environmental stimuli in a synchronous fashion has revealed one major common feature, namely, a short stretch of DNA (9–24 bp) of similar, but not identical, sequence that conforms to a particular consensus sequence (Jones and Kafatos 1980; Barta et al. 1981; Donehower et al. 1981; Grez et al. 1981; Bienz and Pelham 1982; Cochet et al. 1982; Muskavitch and Hogness 1982; Pelham 1982; Pelham and Bienz 1982; Struhl 1982; Hinnebusch and Fink 1983). Such consensus-sequence elements are usually located within 50–300 bp upstream of the transcription-initiation site (Davidson et al. 1983) and are often found repeated in this region. Davidson et al. (1983) have demonstrated by statistical analyses of a number of such coordinately regulated gene sets that the upstream location of these consensus-sequence elements is not likely to occur by chance alone.

In several systems, definitive evidence has now been obtained that the short regions of nucleotide homology discussed above are involved in modulation of inducible transcription levels. By in vitro mutagenesis, the promoter regions of particular gene sets have been altered so as to test the importance of consensus-sequence stretches (Pelham 1982; Pelham and Bienz 1982; Struhl 1982; Donahue et al. 1983). After reintroduction of these altered genes by transformation of homologous or heterologous cell systems, changes in transcriptional regulation have been correlated with specific sequence alterations. Thus, in the case of the heat-shock protein genes of *Drosophila*, induction of transcription during the heat-shock response is associated with a 15-bp consensus-sequence element (Bienz and Pelham 1982; Pelham and Bienz 1982). In another example, a number of co-regulated genes involved in general amino acid metabolism in yeast possess sequence stretches homologous to a single consensus sequence of 9 bp (Struhl 1982; Donahue et al. 1983; Hinnebusch and Fink 1983). Only one copy of a sequence element homologous to this consensus sequence need be present for induction, although multiple sequence elements usually occur in the promoter regions of these genes.

In this paper, we describe experiments designed to look for potential control sequences in the promoter regions of the four tubulin genes of *Chlamydomonas*. Our initial findings show that the *Chlamydomonas* tubulin genes, like other coordinately regulated eukaryotic genes, each contains multiple copies of sequences in their 5′-flanking regions homologous to a common consensus sequence. In addition, two of the four genes have promoterlike regions that are not apparently utilized for mRNA synthesis during tubulin induction or at any other period of the life and cell cycle studied thus far (F. Kalish and K. Brunke, unpubl.). We have termed these regions pseudopromoters.

DISCUSSION

Subcloning Fragments of the Genes and DNA Sequence Analysis

In previous studies (Brunke et al. 1982b) we described the preparation of a library of *Chlamydomonas* genomic DNA fragments in λ vector Ch 30, which was used in selection of multiple clones for each of the four tubulin genes. Prior to DNA sequence analysis of the promoter regions of these genes, subclones were constructed from the genomic clones by insertion of restriction fragments containing 5′ regions into the plasmid vector pBR322 or pUC9. Extensive additional restriction map information was then obtained for each subclone to aid in selection of fragments to be sequenced, using the chemical method of Maxam and Gilbert (1980). A portion of the 5′ region of each gene was first sequenced to locate and analyze the tubulin-coding regions. Comparison of the resulting sequence data with previously published sequence data for chicken tubulin genes (Valenzuela et al. 1981) shows that only 1 amino acid residue in the first 15 is different between chicken and *Chlamydomonas* β-tubulin, whereas 3 of 15 residues differ in the α-tubulin subunit. Thus, there is notable conservation in the amino-terminal amino acid sequence of tubulin from *C. reinhardtii* and chicken, as well as other organisms (Luduena and Woodward 1973; Krauhs et al. 1981; Valenzuela et al. 1981; Lemischka and Sharp 1982; Lee et al. 1983; Neff et al. 1983).

5′-noncoding Region Sequences

Adjacent to the coding-region sequences in eukaryotic genes are regions that are transcribed but do not encode information that is translated. The 5′-noncoding regions of the *Chlamydomonas* tubulin genes (Brunke et al. 1984) vary in their lengths (140, 141, 159, and 132 bp for the α_1-, α_2-, β_1-, and β_2-tubulin genes, respectively) and show intermittent stretches of homology within the α-tubulin gene set and to a lesser degree within the β-tubulin gene set. Two notable features of these transcribed but untranslated regions are described below.

1. The initiator ATG codon is preceded by a conserved sequence in all four genes, which forms the consensus hexanucleotide, GCAA(A/C)C. Similar sequences precede the initiator ATG in the rat α-tubulin gene (GCAACC) (Lemischka and Sharp 1982) and in the human β-tubulin gene (TTAACC) (Lee et al. 1983). These short sequences adjacent to the ATG initiator codon all share features first noted by Kozak (1984); i.e., there is a deoxyadenosine residue at −3 from the ATG codon and there is a tendency for deoxycytidine to be in positions −1, −2, and −5. One exception to the observations of Kozak is that these genes all have a deoxyadenosine residue instead of a deoxycytidine at −4.
2. The mRNA sense strand of the DNA in the 5′-noncoding region is highly CT-rich in all four genes. The region about 5–50 nucleotides upstream of the ATG initiator codon contains almost exclusively C and T residues. Stretches of 40 or more nucleotides occur in the region 100–150 nucleotides upstream of the ATG, which contain few if any Gs.

It is unknown whether either of these features in the noncoding region of the genes might influence their synchronous transcription in response to regulatory signals. Further studies are needed to determine the role(s) these sequence features may have in induction response, ribosome binding, mRNA structure, and translational efficiency.

Promoter Region Sequences

Initiation of transcription in eukaryotic cells has been characterized by the presence of an AT-rich sequence, the TATA box, found approximately 30 bp upstream of the mRNA initiation site. Sequence information from regions upstream of the ATG initiator codon in the four tubulin genes (Fig. 1) (Brunke et al. 1984) suggests a number of potential TATA box sequences in each gene that may be involved in alignment of RNA polymerase II molecules with respect to the mRNA initiation site. Determination of the authentic mRNA initiation site (cap site) was necessary to determine the corresponding TATA box sequence. Primer-extension experiments (Ghosh et al. 1978) were performed to locate these sites. As diagramed in Figure 2 for the β_2-tubulin gene, the basic strategy of these experiments involves the isolation of a small restriction fragment of DNA (25–35 bp), which is subsequently labeled at its 5′ end. This fragment is then denatured and reannealed to polyadenylated mRNA under conditions that favor RNA-DNA hybrid formation. The RNA-DNA hybrid is next extended to the 5′ end of the mRNA using reverse transcriptase. The extended DNA product is then isolated and sized as shown on the gel in Figure 2 for the β_2-tubulin gene. In this case, the primary transcription-initiation site is at the deoxyadenosine residue located 132 bp upstream of the initiator ATG codon. This type of analysis was performed for all four tubulin genes. The positions of the primary mRNA initiation sites determined in this manner are designated in Figure 1 as +1.

Potential Regulatory Sequences

Once the mRNA initiation sites had been determined, upstream sequences (as far as 1000 bp upstream of the ATG initiator codon) were compared for features of potential regulatory significance common to all four genes. On the basis of studies of regulatory elements in the 5′-flanking region of genes in other systems (Baker and Ziff 1981; Benoist and Chambon 1981; McKnight and Kingsbury 1981; Tsuda and Suzuki 1981; Darnell 1982; McGinnis et al. 1983), most such elements are located within 250 bp of the gene. For the particular case of coordinately regulated genes, these elements usually occur even closer, within 100 bp of the TATA box sequence (see above). Computer analyses of our sequence data reveal a variety of structural elements in the upstream regions of the tubulin genes, including direct and inverted repeats, palindromic sequences, dyads of symmetry, and regions of shared homology (a number of these elements are noted in Fig. 1). Most such elements are not found common to all four genes and are therefore poor candidates for *cis*-acting regulatory elements. There are, however, two features that are clearly characteristic of the promoter regions in all four tubulin genes.

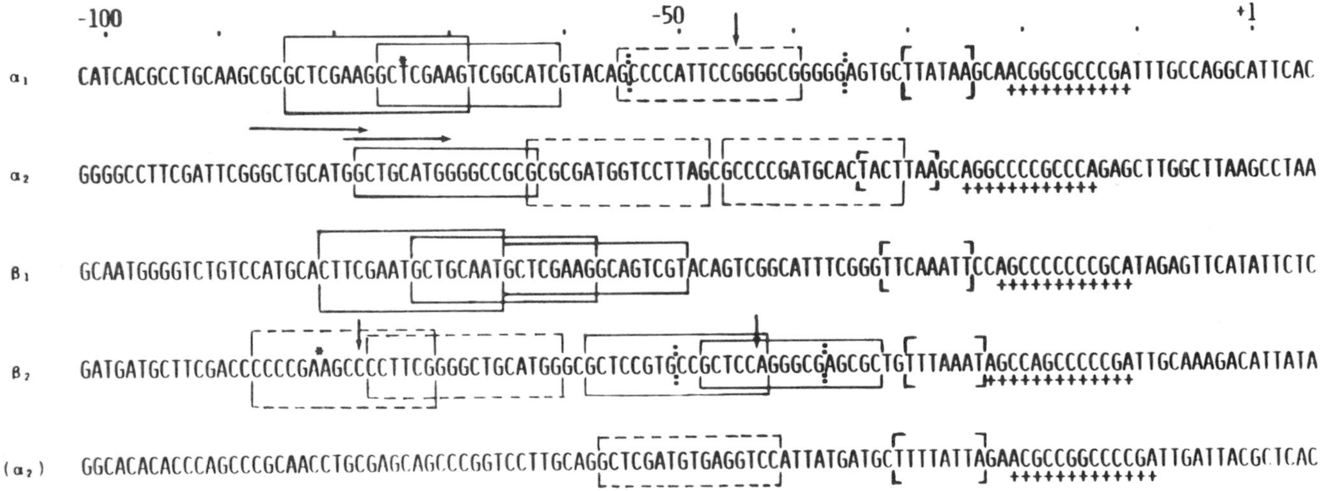

Figure 1

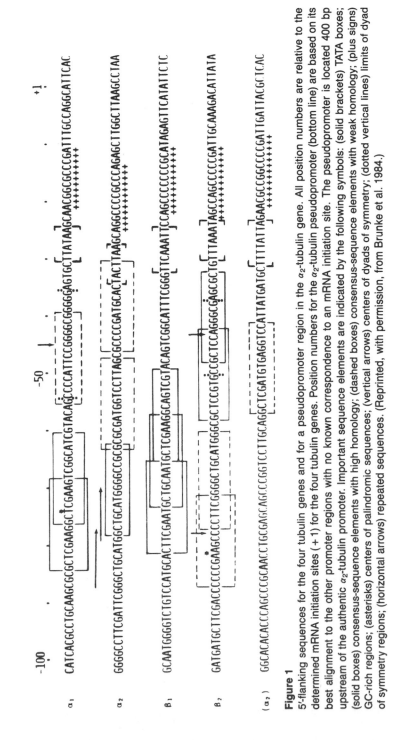

5′-flanking sequences for the four tubulin genes and for a pseudopromoter region in the α_2-tubulin gene. All position numbers are relative to the determined mRNA initiation sites (+1) for the four tubulin genes. Position numbers for the α_2-tubulin pseudopromoter (bottom line) are based on its best alignment to the other promoter regions with no known correspondence to an mRNA initiation site. The pseudopromoter is located 400 bp upstream of the authentic α_2-tubulin promoter. Important sequence elements are indicated by the following symbols: (solid brackets) TATA boxes; (solid boxes) consensus-sequence elements with high homology; (dashed boxes) consensus-sequence elements with weak homology; (plus signs) GC-rich regions; (asterisks) centers of palindromic sequences; (vertical arrows) centers of dyads of symmetry; (dotted vertical lines) limits of dyad of symmetry regions; (horizontal arrows) repeated sequences. (Reprinted, with permission, from Brunke et al. 1984.)

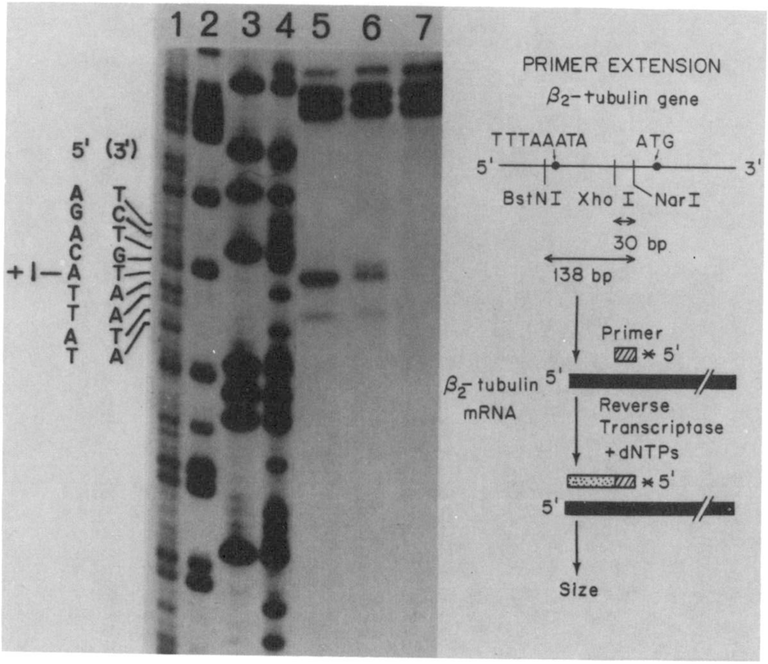

Figure 2
Determination of the transcription-initiation site for the β_2-tubulin mRNA by primer-extension analysis. The diagram outlines the protocol followed in primer-extension experiments with double-stranded primers. When single-stranded oligomers were used, the denaturation and annealing steps were omitted. The radioactive extension product was isolated and sized on 8% polyacrylamide–7 M urea gels as shown here for primer extension of the β_2-tubulin mRNA. The DNA sequence reactions in lanes *1–4* are for the 138-bp DNA fragment indicated in the diagram that extends beyond the cap site. These markers allowed accurate determination of the size of the extension products (lanes *5* and *6*). A 1.5-bp correction was taken into account when comparing sequenced marker bands with primer-extension bands. Lane *7* is a control extension done in the absence of mRNA to distinguish artifacts from extension products. (Reprinted, with permission, from Brunke et al. 1984.)

The first feature is a 10–11-bp GC-rich region located between the TATA box and the mRNA initiation site (underscored by plus symbols in Fig. 1). This sequence is flanked on either end with a deoxyadenosine residue. Although a number of viral and eukaryotic genes, including some encoding structural proteins, have a similar GC-rich stretch in this region (Lemischka and Sharp 1982; Bensimhon et al. 1983; Everett et al. 1983), this GC sequence in the tubulin genes of *Chlamydomonas* is long in comparison. The high stability of this structure makes it a good candidate for recognition by enzymatic elements involved in transcription (Bensimhon et al. 1983), although a role of enhancement of transcriptional efficiency is at this point purely speculative.

The second feature is the 16-bp consensus sequence, GCTC(G/C)AAGGC(G/T)(G/C)–(C/A)(C/A)G. This consensus sequence was derived by comparison of sequences sharing at least 67% homology with each other. Each position within the consensus sequence was determined using rules previously outlined (Davidson et al. 1983). At least three copies of sequence elements with varying degrees of homology with the consensus sequence are located immediately 5′ to each of the tubulin genes (Fig. 1, boxed regions). For every gene, one or more copies are at least 73% homologous to the consensus sequence. Also, it is interesting to consider that the absence of a given nucleotide may be as significant as homology with a single nucleotide from a functional perspective. For example, in the 13th position of the consensus sequence (denoted by a dash) there is a nearly equal probability for a G, C, or T residue. However, in no case is an A residue observed in the 15 sequences from which the consensus sequence was derived. Note that the greatest sequence conservation is in the 10-bp core, GCTC(G/C)AAGGC. Judging from the relatively short consensus sequences involved in the regulation of other inducible gene families (Bienz and Pelham 1982; Pelham and Bienz 1982; Donahue et al.1983), this 10-bp core sequence may be of sufficient length to function as a regulatory element in induction of the tubulin genes in *C. reinhardtii*. In addition, the location of this element in the promoter region of the tubulin genes is similar to that observed previously for the consensus sequences found in both the heat-shock protein genes of *Drosophila* and the amino acid metabolism genes of yeast.

Finally, the consensus sequence element is located within the first 150 bp upstream of the cap site and is not found randomly repeated further upstream. Instead, copies of this element are clustered and frequently found overlapping in the region from −35 to −90 bp relative to the cap site. Sequenced tubulin genes from other organisms do not have an even remotely similar sequence in their promoter regions (Lemischka and Sharp 1982; Lee et al. 1983). If this element is indeed involved in coordinate induction of the tubulin genes, further studies will be required to determine what portions of the sequence are required in induction. Furthermore, as tubulin induction is required at other times such as cell division, it will be interesting to see if the same or different regulatory elements are involved in its induction during these periods.

Our focus to this point has been on elements involved in the coordinate regulation of the tubulin genes. However, it is possible that at certain specific times during the life or cell cycle of *Chlamydomonas*, selective transcription of an α-β-tubulin gene set occurs. With respect to this, the α_1- and β_2-tubulin genes share a number of common features in their promoter regions that are not present in the other two tubulin genes. The most striking of these is the association of the consensus-sequence elements with sequences that contain dyads of symmetry, a feature also found in the promoter regions of certain heat-shock protein genes of *Drosophila* (Pelham 1982). Also common to the α_1-β_2-tubulin gene set is an area of close homology in the region adjacent to the cap site. We are currently investigating a possible role for these shared features.

Pseudopromoter Regions

The presence of the consensus-sequence element just upstream of the TATA box sequence and the highly characteristic GC-rich region immediately following the TATA sequence makes the tubulin promoter regions quite distinguishable from adjacent sequence regions (see Fig. 1). These features indicate that the β_1- and the α_2-tubulin genes both have two potential promoter regions upstream of the initiator ATG codon. This raises the possibility that alternative transcription-initiation sites may exist for these two genes. Because of the diverse roles tubulin plays in the cell, one way of achieving functional specificity in an organism with a limited number of tubulin genes such as *Chlamydomonas* is to have slightly different mRNAs specified by a single gene. This method of achieving developmental specificity is known to occur, for example, with the alcohol dehydrogenase gene of *Drosophila* and the α-amylase gene of the mouse (Young et al. 1981; Benyajati et al. 1983).

The β_1-tubulin gene contains two potential promoter regions separated by only 50 bp. Primer-extension experiments (Brunke et al. 1984) have demonstrated that the deoxyadenosine residue 159 bp from the ATG initiator codon is the preferred cap site for mRNA produced in nondividing gametic cells of *Chlamydomonas*. At least in this life-cycle stage, the downstream promoter region for the β_1-tubulin gene does not appear to be used.

The two potential promoter regions in the α_2-tubulin gene are separated by 400 bp. By primer-extension analysis using mRNA isolated from nondividing gametic cells, the mRNA initiation site for the α_2-tubulin gene appears to be 141 bp upstream of the ATG initiator codon, a position typical of the other three tubulin genes. The presence of a deoxycytidine residue in its TATA box sequence, however, initially made the potential effectiveness of this promoter somewhat questionable. In contrast, the upstream promoter has a TATA sequence with a stretch of eight As and Ts (see Fig. 1). This far upstream promoter region might represent the authentic promoter for the α_2-tubulin gene should the unprocessed transcript contain an intervening sequence in the nontranslated region 5′ to our selected primer. Or, as described earlier, differential transcription of the α_2-tubulin gene could result in usage of both promoter regions. To clarify which one (or potentially both) of these promoters is recognized in vivo, primer-extension experiments were performed using a synthetic primer that enabled us to obtain a highly radioactive extension product (Brunke et al. 1984). Sequence analysis of this ^{32}P-labeled extension product has demonstrated that the cap site located 141 bp upstream of the ATG codon is the one used to produce mRNA in gametic cells. Thus, no intron interrupts the noncoding region in the α_2-tubulin gene, and the authentic promoter, at least in gametic cells, is the one closest to the ATG initiator codon. The possibility of differential promoter usage at different life- or cell-cycle stages was examined next.

As shown in part in Figure 3, primer-extension analyses of different mRNA populations from various life- and cell-cycle stages (F. Kalish and K. Brunke, unpubl.) using the synthetic α_2-tubulin gene primer give the same extension product as obtained using gametic cells. The major band produced corresponds to the downstream promoter in the α_2-tubulin gene. (The two bands above the major band are artifacts. One of these bands corresponds to another mRNA species other than tubulin, which is detected with the particular 16-bp oligomeric primer used in this experiment.) Combined with recent information we have obtained using zygotic RNA in primer-extension analyses (F. Kalish and K. Brunke, unpubl.), we can conclude that the upstream promoter is not used to any measurable extent in gametic cells, zygotic cells prior to germination, vegetative cells grown asynchronously, vegetative cells grown synchronously and harvested dur-

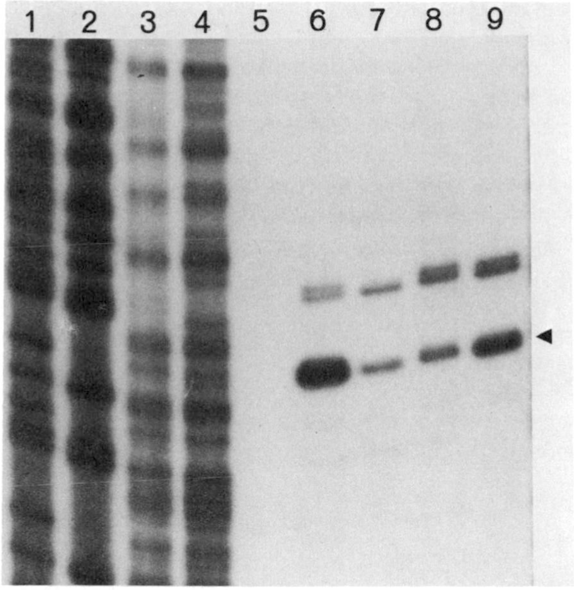

Figure 3
Comparison of transcription-initiation sites for the α_2-tubulin mRNA using poly(A)$^+$ mRNAs from different cell-cycle stages. Primer-extension reactions are with poly(A)$^+$ mRNA plus a 16-bp single-stranded oligomer as primer. (Lanes *1–4*) DNA sequencing markers (pBR322, labeled at *Hin*dIII site); (lane *5*) control extension in the absence of mRNA. For substrate in primer-extension experiments, poly(A)$^+$ mRNAs were isolated from deflagellated gametic cells (lane *6*), asynchronously growing vegetative cells (lanes *7,8*), and synchronously growing vegetative cells isolated during cell division (lane *9*). Arrowhead indicates the major extension product. The two minor bands are artifacts. (Note that one band can be eliminated from lane *7* by alteration of reverse transcriptase concentration.) Both the major and minor bands in lane *7* were sequenced. The major band corresponds to the mRNA initiation site shown in Fig. 1. The minor band is not related to the tubulin genes and appears to be an extension product of another mRNA that annealed to the 16-bp oligonucleotide primer used in this study.

ing cell division, and gametic cells harvested following deflagellation. Primer-extension experiments using a synthetic oligomer specific for an open reading frame region downstream from the α_2-tubulin pseudopromoter do not detect any mRNA produced using this promoter. Thus, this promoterlike region appears to be nonfunctional in all life- and cell-cycle stages we have tested.

We have designated the upstream promoterlike region in the α_2-tubulin gene as well as the downstream promoterlike region in the β_1-tubulin gene as pseudopromoters because of their deceptive similarity to the authentic promoter regions of the tubulin genes of *Chlamydomonas.* Why these promoter regions are not utilized in vivo is puzzling, and studies are currently under way to clarify what features are necessary for a functional tubulin gene promoter. One striking feature that may discriminate between promoter usage during tubulin gene induction is the absence of a sequence that has strong homology with the consensus sequence in the pseudopromoter region of the β_1-tubulin gene. The absence of a G or C residue in the highly conserved fifth position of its consensus-sequence element may prove important. A similar need for particular nucleotides in highly conserved positions of the sequence element may also explain the apparent lack of function of the α_2-tubulin pseudopromoter during tubulin induction. Because both the β_1- and the α_2-tubulin pseudopromoters have only a single copy of the consensus-sequence element in the 70-bp region upstream of the TATA box, alteration in this sequence may prevent productive interaction with regulatory molecules and would thus prevent the induction process. Although these are all plausible explanations for why increases in tubulin gene transcription may not involve the pseudopromoter regions, it is unclear what feature of these regions prevents the use of the promoterlike elements in basal-level expression of the tubulin genes. A further understanding of how the tubulin gene promoters interact with regulatory molecules and the site for this interaction in the DNA must await future experiments in which the promoter regions are modified in vitro and reintroduced into *Chlamydomonas* cells for in vivo testing.

SUMMARY

The 5′ promoter regions of the four coordinately regulated tubulin genes of *Chlamydomonas* have been sequenced, and these data have been analyzed. Transcription-initiation sites have been located by primer-extension analyses at 140, 141, 159, and 132 bp upstream of the translation initiator codon for the α_1-, α_2-, β_1-, and β_2-tubulin genes, respectively. In addition, the 5′-noncoding region and a portion of the coding region for each tubulin gene have been sequenced. Sequence data for the predicted aminoterminal amino acid sequences are quite similar between the *Chlamydomonas* tubulins and tubulins from other eukaryotes. The 5′-noncoding regions for the tubulin genes are strikingly G-deficient, and this characteristic feature cannot be ruled out in our search for regions of potential regulatory significance. Also of potential significance in regulation of tubulin induction is a GC-rich region that flanks the TATA box sequence in the four genes. Nonetheless, in view of the types of short consensus elements that have been found significant in other systems, which, like the tubulin genes of *Chlamydomonas*, show rapid and coordinate induction of mRNA synthesis, the presence of a 16-bp consensus sequence in a position analogous to these elements in all four tubulin genes suggests that this sequence may play a role in the coordinate expression of these spatially separated genes. Modification, rearrangement, and deletion of these sequence elements, followed by reintroduction of the altered sequences back into *Chlamydomonas* by transformation, will be necessary to test what role, if any, these sequence elements play in the regulation of tubulin gene expression.

Finally, the presence of pseudopromoter sequences in two of the four genes offers the opportunity to determine the basic elements necessary for promoter function in a eukaryotic system. From this, we may be able to obtain a better understanding of what components are necessary for expression of the authentic tubulin genes.

ACKNOWLEDGMENTS

We thank L. Gaitmaitan for teaching us DNA sequencing, R. Scott for help with primer-extension protocols, and S. Hazel, P. Dietrich, N. Malin, C. Cannon, and S. Chow for technical assistance. A special thanks to Dr. Jack Ohms at Beckman Instruments in Palo Alto for preparation of one of the synthetic oligonucleotide primers discussed in this manuscript. For part of this work, K.J.B. was supported by a postdoctoral training grant from the National Institutes of Health. This work was supported by the National Science Foundation (grant PCM-8302639), a grant from the Commonwealth of Pennsylvania, and the Zoecon Corporation.

REFERENCES

Baker, C. and E.J. Ziff. 1981. Promoter and heterogeneous 5′ termini of the messenger RNAs of adenovirus serotype 2. *J. Mol. Biol.* **149:** 189.

Barta, A., R.I. Richards, J.D. Baxter, and J. Shine. 1981. Primary structure and evolution of rat growth hormone gene. *Proc. Natl. Acad. Sci.* **78:** 4867.

Benoist, C. and P. Chambon. 1981. *In vivo* sequence requirements of the SV40 early promoter region. *Nature* **290:** 304.

Bensimhon, M., J. Gabarro-Arpa, R. Ehrlich, and C. Reiss. 1983. Physical characteristics in eukaryotic promoters. *Nucleic Acids Res.* **11:** 4521.

Benyajati, C., N. Spoerel, H. Haymerle, and M. Ashburner. 1983. The messenger RNA for alcohol dehydrogenase in *Drosophila melanogaster* differs in its 5′ end in different developmental stages. *Cell* **33:** 125.

Bienz, M. and H.R.B. Pelham. 1982. Expression of a *Drosophila* heat-shock protein in *Xenopus* oocytes: Conserved and divergent regulatory signals. *EMBO J.* **1:** 1583.

Brunke, K.J., P.S. Collis, and D.P. Weeks. 1982a. Post-translational modification of tubulin dependent on organelle assembly. *Nature* **297:** 516.

Brunke, K.J., J.G. Anthony, E.J. Sternberg, and D.P. Weeks. 1984. Repeated consensus sequence and pseudopromoters in the four coordinately regulated tubulin genes of *Chlamydomonas reinhardi. Mol. Cell. Biol.* **4:** 1115.

Brunke, K.J., E.E. Young, B.U. Buchbinder, and D.P. Weeks. 1982b. Coordinate regulation of the four tubulin genes of *Chlamydomonas reinhardi. Nucleic Acids Res.* **10:** 1295.

Cochet, M., A.C.Y. Chang, and S.N. Cohen. 1982. Characterization of the structural gene and putative 5′-regulatory sequences for human proopiomelanocortin. *Nature* **297:** 335.

Darnell, J.E., Jr. 1982. Variety in the level of gene control in eukaryotic cells. *Nature* **297:** 365.

Davidson, E.H., H.T. Jacobs, and F.J. Britten. 1983. Very short repeats and coordinate induction of genes. *Nature* **301:** 468.

Donahue, T.F., R.S. Daves, G. Lucchini, and G.R. Fink. 1983. A short nucleotide sequence required for regulation of *HIS4* by the general control system of yeast. *Cell* **32:** 89.

Donehower, L.A., A.L. Huang, and G.L. Hager. 1981. Regulatory and coding potential of the mouse mammary tumor virus long terminal redundancy. *J. Virol.* **37:** 226.

Everett, R.E., D. Baty, and P. Chambon. 1983. The repeated GC-rich motifs upstream from the TATA box are important elements of the SV40 early promoter. *Nucleic Acids Res.* **11:** 2447.

Ghosh, P.K., V.B. Reddy, J. Swinschoe, P. Lebowitz, and S.M. Weissman. 1978. Heterogeneity and 5′-terminal structures of the late RNAs of simian virus 40. *J. Mol. Biol.* **126:** 813.

Grez, M., H. Land, K. Giesecke, G. Schutz, A. Jung, and A.E. Sippel. 1981. Multiple mRNAs are generated from the chicken lysozyme gene. *Cell* **25:** 743.

Hinnebusch, A.G. and G.R. Fink. 1983. Repeated DNA sequences upstream from *HIS1* also occur at several other co-regulated genes in *Saccharomyces cerevisiae. J. Biol. Chem.* **258:** 5238.

Jones, C.W. and F.C. Kafatos. 1980. Structure, organization and evolution of developmentally regulated chorion genes in a silkmoth. *Cell* **22:** 855.

Keller, L.R., J.A. Schloss, C.D. Silflow, and J.L. Rosenbaum. 1984. Transcription of α- and β-tubulin genes *in vitro* in isolated *Chlamydomonas reinhardi* nuclei. *J. Cell Biol.* **98:** 1138.

Kozak, M. 1984. Compilation and analysis of sequences upstream from the translational start site in eukaryotic mRNAs. *Nucleic Acids Res.* **12:** 857.

Krauhs, E., M. Little, T. Kelpf, R. Hofer-Warbinek, W. Ade, and H. Ponstingl. 1981. Complete amino acid sequence of β-tubulin from porcine brain. *Proc. Natl. Acad.Sci.* **78:** 4156.

Lee, M.G.-S., S.A. Lewis, C.D. Wilds, and N.J. Cowan. 1983. Evolutionary history of a multigene family: An expressed human β-tubulin gene and three processed pseudogenes. *Cell* **33:** 477.

Lefebvre, P.A., S.A. Nordstrom, J.E. Moulder, and J.L. Rosenbaum. 1978. Flagellar elongation and shortening in *Chlamydomonas. J. Cell Biol.* **78:** 8.

Lemischka, I. and P.A. Sharp. 1982. The sequences of an expressed rat α-tubulin gene and a pseudogene with an inserted repetitive element. *Nature* **300:** 330.

Luduena, R.F. and D.O. Woodward. 1973. Isolation and partial characterization of α- and β-tubulin from outer doublets of sea-urchin sperm and microtubules of chick-embryo brain. *Proc. Natl. Acad. Sci.* **70:** 3594.

Maxam, A.M. and W. Gilbert. 1980. Sequencing end-labeled DNA with base-specific chemical cleavages. *Methods Enzymol.* **65:** 499.

McGinnis, W., A.W. Shermoen, and S.K. Beckendorf. 1983. A transposable element inserted just 5′ to a *Drosophila* glue protein gene alters gene expression and chromatin structure. *Cell* **34:** 75.

McKnight, S. and R. Kingsbury. 1982. Transcriptional control signals of a eukaryotic protein-coding gene. *Science* **217:** 316.

Minami, S.A., P.S. Collis, E.E. Young, and D.P. Weeks. 1981. Tubulin induction in *C. reinhardii*: Requirement for tubulin mRNA synthesis. *Cell* **24:** 89.

Muskavitch, M.A.T. and D.S. Hogness. 1982. An expandable gene that encodes a *Drosophila* glue protein is not expressed in variants lacking remote upstream sequences. *Cell* **29:** 1041.

Neff, N.F., J.H. Thomas, P. Grisati, and D. Botstein. 1983. Isolation of the β-tubulin gene from yeast and demonstration of its essential function *in vivo. Cell* **33:** 211.

Pelham, H.R.B. 1982. A regulatory upstream promoter element in the *Drosophila* hsp70 heat-shock gene. *Cell* **30:** 517.

Pelham, H.R.B. and M. Bienz. 1982. A synthetic heat-shock promoter element confers heat-inducibility on the herpes simplex virus thymidine kinase gene. *EMBO J.* **1:**1473.

Schloss, J.A., C.D. Silflow, and J.L. Rosenbaum. 1984. mRNA abundance changes during flagellar regeneration in *Chlamydomonas reinhardii. Mol. Cell. Biol.* **4:** 424.

Silflow, C.D. and J.L. Rosenbaum. 1981. Multiple α- and β-tubulin genes in *Chlamydomonas* and regulation of tubulin mRNA levels after deflagellation. *Cell* **24:** 81.

Struhl, K. 1982. Regulatory sites for *his3* gene expression in yeast. *Nature* **300:** 284.

Tsuda, M. and Y. Suzuki. 1981. Faithful transcription initiation of fibroin gene in a homologous cell-free system reveals an enhancing effect of 5′ flanking sequence far upstream. *Cell* **27:** 175.

Valenzuela, P., M. Quiroga, J. Zaldivar, W.J. Rutter, M.W. Kirschner, and D.W. Cleveland. 1981. Nucleotide and corresponding amino acid sequences encoded by α- and β-tubulin mRNAs. *Nature* **289:** 650.

Weeks, D.P. and P.S. Collis. 1976. Induction of microtubule protein synthesis in *Chlamydomonas reinhardi* during flagellar regeneration. *Cell* **9:** 15.

———. 1979. Induction and synthesis of tubulin during the cell cycle and life cycle of *Chlamydomonas reinhardi. Dev. Biol.* **69:** 400.

Weeks, D.P., P.S. Collis, and M.A. Gealt. 1977. Control of induction of tubulin synthesis in *Chlamydomonas reinhardi. Nature* **268:** 667.

Young, R.A., O. Hagenbuchle, and U. Schibler. 1981. A single mouse α-amylase gene specifies two different tissue-specific mRNAs. *Cell* **23:** 451.

Influence of Differential Keratin Gene Expression on the Structure and Properties of the Resulting 8-nm Filaments

Elaine Fuchs, Kwan Hee Kim,
Israel Hanukoglu,* Douglas Marchuk,
Angela Tyner, and Sean McCrohon
Department of Biochemistry
The University of Chicago
Chicago, Illinois 60637

Epithelial tissues represent the major protective covering and lining of all vertebrate organisms. Despite the diverse morphologies of these tissues, the predominant and unifying features in all of them are bundles of 8-nm keratin filaments. These filaments are unique to epithelia and are especially dense in the epidermal layers of the skin and epidermal appendages, including hair, wool, horn, hoof, and quill. They are most likely responsible for the strength and special structural characteristics of these tissues. The filaments are composed of subunits that constitute a family of related proteins known as α-keratins. In epidermal cells, a subset of this family is assembled into a cytoplasmic network of 8-nm filaments that loop through the desmosomal plaques in the plasma membrane. In contrast, in epidermal appendages such as hair and wool, a different subset of keratins forms very orderly arrays of microfibrils embedded in a rigid matrix of other proteins. In this paper, we review recent studies on the sequences and structural characteristics of the different keratin subunits that constitute the backbone of the 8-nm filaments in these two distinct tissues. These investigations have provided a possible explanation for the differential expression of the multiple keratins and for the morphological variations in the keratin filament organization in different cell types.

Keratins Constitute a Family of Related Proteins That Can be Further Dissected into Two Distinct Types

For any particular species (e.g., the human), there are about 20 different keratin polypeptides (M_r = 40,000–70,000) according to analysis by two-dimensional gel

*Present address: Department of Biology, Technion-Israel Institute of Technology, Technion City, Haifa, Israel 32000.

electrophoresis (for review, see Moll et al. 1982). These keratins are not all expressed in any tissue at any one time. Rather, the expression of keratins is perhaps one of the most variable of all protein families that are known: Keratins are differentially expressed in different tissues, at different developmental stages, upon malignant transformation of an epithelial tissue, and, in certain instances, when the environment of a cell is changed. In the epidermis, the expression of keratins changes as a basal cell undergoes a commitment to terminally differentiate. Typically, a subset of three to six keratins are expressed in an epithelial cell at a given time.

As judged from their amino acid compositions, one-dimensional peptide maps, and immunological cross-reactivity, the keratins from diverse cell types have been shown to share at least some similarities (Jones 1975, 1976; Steinert and Idler 1975; Dale et al. 1976; Fuchs and Green 1978; Weber et al. 1980; Bowden and Cunliffe 1981; Skerrow and Skerrow 1983). However, these biochemical studies have also revealed differences among the individual members of the keratin family (Fuchs and Green 1978; Bowden and Cunliffe 1981; Moll et al. 1982; Woodcock-Mitchell et al. 1982; Nelson and Sun 1983; Skerrow and Skerrow 1983). It became apparent that different sequences must exist for these proteins when it was discovered that there are different mRNAs for many of the different epithelial keratins (Fuchs and Green 1979, 1980; Schweizer and Goerttler 1980; Bladon et al. 1982; Kim et al. 1983; Magin et al. 1983; Roop et al. 1983; Schweizer and Winter 1983).

When mRNAs coding for the epidermal keratins were isolated and cloned as cDNAs (Fuchs et al. 1981; Roop et al. 1983), it was discovered that, in fact, most if not all of the human epithelial keratins could be dissected into two distinct groups of sequences according to the ability of their corresponding mRNAs to cross-hybridize with one another (Fuchs et al. 1981; Kim et al. 1983, 1984). At the nucleic acid level, the two groups of keratin sequences share a very low homology (<30%), whereas the individual members within each group are very similar (50–90% homologous). Recent amino acid sequence data have confirmed these relations (Crewther et al. 1980; Hanukoglu and Fuchs 1982, 1983; Dowling et al. 1983; Steinert et al. 1983). The relatively acidic and usually small (40K–56K) keratins have been designated as type I, whereas the basic and frequently larger (53K–67K) keratins have been classified as type II (Fuchs and Marchuk 1983; Hanukoglu and Fuchs 1983).

Each of the two types of keratins is encoded by a small multigene family (four to ten genes each), the members of which are differentially expressed in different tissues (Fuchs et al. 1981; Fuchs and Marchuk 1983). At least one member of each group appears to be expressed in all epithelial cells at all times, suggesting that the two different types of keratin polypeptides play an intimate and important role in the assembly of the 8-nm filament (Fuchs and Marchuk 1983; Kim et al. 1983; Nelson and Sun 1983).

Like other intermediate filament protein subunits, the keratins have the capacity to assemble in vitro in the absence of any auxiliary proteins or factors. However, whereas the other intermediate filament subunits can form homopolymers, no single keratin polypeptide by itself is sufficient for filament assembly to occur (Steinert et al. 1976; Lee and Baden 1976; Milstone 1981). Recent in vitro assembly studies have demonstrated that as conditions for filament formation are adjusted to decreasing stringency, pairs of keratin polypeptides—one of each type—begin to sort out from a mixture of subunits and polymerize into filaments (Franke et al. 1983). This discovery strongly supports the notion that both the type I and type II keratins are essential for filament formation and that the 8-nm filaments in different epithelia are composed of different combinations of type I and type II keratins.

Assembly of the Keratin Filament: The Coiled-coil of the Protofilament

Regardless of the subunit composition of the keratin filament, the resulting filament has a diameter of 8–10 nm and an axial periodicity of 21 nm. This is true for virtually all intermediate filaments (Henderson et al. 1982; Milam and Erikson 1982). Early X-ray diffraction studies on wool keratin microfibrils indicated that this structure contains an undetermined number of closely packed protofilaments (2 nm dia) composed of two or three largely α-helical polypeptide subunits intertwined in a coiled-coil configuration (Crick 1953; Pauling and Corey 1953). Higher-order supercoiling of protofilaments seems to take place to give rise to protofibrils (4.5 nm dia), which then interact to form the 8-nm keratin filament (Aebi et al. 1983). A similar X-ray pattern has been shown for all intermediate filaments, and it has therefore been assumed that once determined, the details of the structure of one filament will apply to all 8-nm filaments.

The number of polypeptide strands that intertwine to form the protofilament has been a question of considerable interest. Initially, physicochemical studies conducted by Skerrow et al. (1973) and later by Steinert (1978) suggested that the correct number of α-helical chains in the coiled-coil structure was three. In contrast, model building using partial sequence data for the wool keratins demonstrated that these helical segments are compatible with two-stranded, but not three-stranded, protofilaments (McLachlan 1978). Nearest-neighbor cross-linking analysis of the polypeptide chains within an intermediate filament supports this two-stranded model for the structure of the protofilament (Ahmadi and Speakman 1978; Geisler and Weber 1982).

To explore the possibility that a type I keratin and a type II keratin might interact to form the coiled-coil backbone of the keratin protofilament, we have obtained the predicted amino acid sequences from near-full-length cloned cDNAs that are complementary to the mRNAs of a type I (50K) and a type II (56K) keratin from cultured human epidermal cells (Hanukoglu and Fuchs 1982, 1983). To complete the protein sequences, we have isolated and partially characterized the genes encoding these two keratins.

Since the two keratin cDNAs did not cross-hybridize with each other, it was not surprising that the two keratins showed only about 25% sequence identity at the amino acid sequence level (Hanukoglu and Fuchs 1983). The remarkable feature of these two distantly related sequences was that they showed nearly identical secondary structures (Hanukoglu and Fuchs 1983). The general features of this structure are diagramed in Figure 1. Computer-assisted analyses of these sequences showed that for both keratins, a central, approximately 300-amino-acid-residue long region exists consisting of four richly α-helical domains. The helical region is not continuous, but rather is demarcated by three regions in which β-

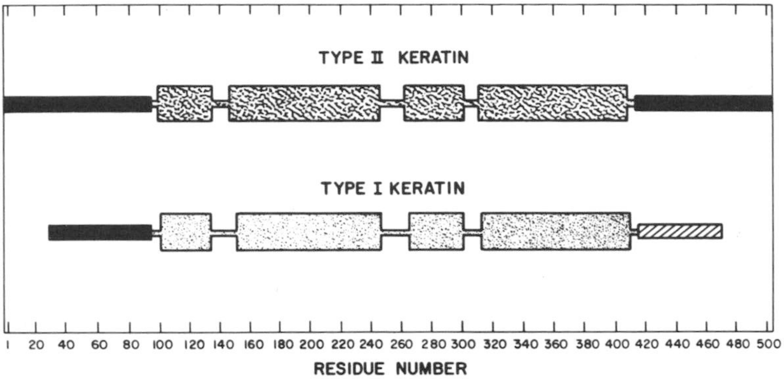

Figure 1
Comparison of the predicted secondary structures of the 56K type II and 50K type I keratins that are coordinately expressed in cultured human epidermal cells. The predicted amino acid sequences of the two keratins and their predicted secondary structures were determined previously (Hanukoglu and Fuchs 1982, 1983). The values on the abscissa indicate amino acid residue number. The sequences have been aligned according to their predicted secondary structures. The Chou and Fasman (1978, 1979) and Garnier et al. (1978) methods used in our analyses make predictions for four conformational states: α-helix, β-sheet, β-turn, and random coil. For each keratin, central α-helical domains are represented by the thick bars. The thin internal bars mark the positions of the β-turns, or helix breakers, that are predicted by the Chou and Fasman (1978, 1979) method for all intermediate filament protein sequences. The bars flanking the central helical domains represent the nonhelical termini. Largely different sequence homologies are indicated by different shades of grey.

turns, or helix breaking, are predicted with high probability. Despite the difference in the molecular weights of the two keratins, the size of their helical domains seems to be nearly identical. The difference in the sizes of the two keratins resides solely in their nonhelical termini, which are larger for the type II (56K) keratin. Recent analysis of an unusually large type I (59K) keratin from differentiating mouse epidermis reveals a sequence that is compatible with the structure predicted for the human epidermal keratins (Steinert et al. 1983). It is interesting to note that these structural features also apply to the sequences of other intermediate filament subunits (Hanukoglu and Fuchs 1982, 1983; Geisler and Weber 1982; Geisler et al. 1982; Dowling et al. 1983; Fuchs and Hanukoglu 1983). These studies confirm earlier suggestions that the intermediate filaments share a common structure (Steinert et al. 1978; Geisler and Weber 1981, 1982; Hong and Davison 1981; Pruss et al. 1981).

The central α-helical domains of the two human epidermal keratin subunits are likely to be involved in the coiled-coil of the protofilament. The sequence characteristic of a periodic interchain interaction, i.e., a heptade repeat of **a** b c **d** e f g, where **a** and **d** are hydrophobic residues (McLachlan and Stewart 1975), has been identified in the helical portions of all sequenced intermediate filament proteins, including the keratins. In addition, an interesting periodicity in positively and negatively charged residues has been observed for different subunits, indicating that electrostatic interactions may also play an important role in stabilizing the coiled-coil structure (Parry et al. 1977; McLachlan and Stewart 1982). In summary, the similar structural aspects of the two different types of keratins and the presence of a common heptade repeat of hydrophobic residues make it feasible that a type I keratin and a type II keratin could form a coiled-coil dimer as the basic structural unit of the keratin protofilament.

Differences between Individual Members of the Same Keratin Type Reside in Their Nonhelical Termini

All keratin filaments in all epithelial tissues seem to have very similar structures, even though they are composed of different pairs of keratin subunits. It is important to know whether the differential expression of keratins has some underlying functional significance or whether it is merely a consequence of evolution. At present, sequence data limit our comparison to the keratins from only two different tissues: epidermis and wool. Nonetheless, this comparison has revealed an interesting pattern. For any two differentially expressed keratins of the same type, the α-helical domains seem to be highly homologous, whereas the nonhelical termini are quite different (Fig. 2). This is in striking contrast to a comparison of two coordinately expressed keratins of opposite types, where the termini are very similar, but the helical domains are different (Fig. 1).

The exciting picture that is emerging from amino acid sequence data is that the nonhelical terminal domains of pairs of coexpressed subunits contain similar, but highly unusual, sequences that not only distinguish them from all other proteins, but also lend uniqueness to each pair of type I and type II keratins. For the human epidermal keratins, the aminoterminal segment of the type I keratin and the amino- and carboxyterminal segments of the type II keratins are rich in gly-

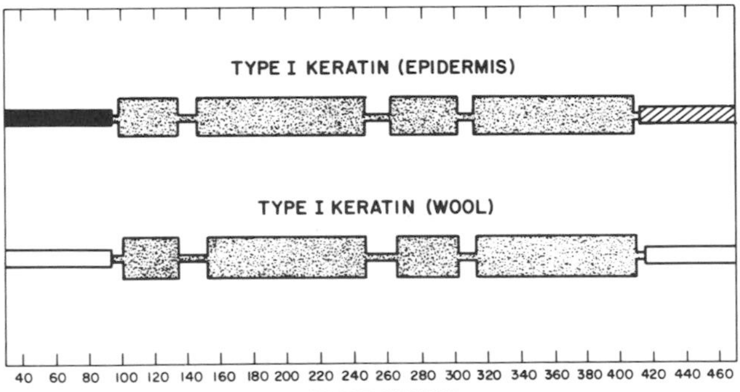

Figure 2
Comparison of the predicted secondary structures of the 50K epidermal type I keratin and the wool type I keratin that are differentially expressed in different tissues. The predicted secondary structures were determined previously (Hanukoglu and Fuchs 1982, 1983; Dowling et al. 1983). The values on the abscissa indicate amino acid residue number. The sequences have been aligned according to their predicted secondary structures, which were determined as indicated in Fig. 1. The thick bars represent α-helical domains, the thin internal bars mark positions where the helix is broken, and the bars flanking the central helical domains represent the nonhelical termini. Largely different sequence homologies are indicated by different shades of gray.

cine and serine residues and frequently contain distinct inexact repeat sequences of Gly-Gly-Gly-*X*, where *X* is either Phe, Tyr, or Leu. These repeats are frequently interrupted with stretches of 4–6 serine residues. The amino- and carboxyterminal domains of the mouse epidermal type I (59K) keratin also contain a similar inexact repeat pattern of Gly-Gly-Gly-Ser-*Y*, where *Y* is either Phe or Tyr. To our knowledge, these interesting sequences have not been identified in any other proteins, with the possible exception of some of the wool matrix proteins, and they appear to be curiously characteristic of epidermal keratins.

In striking contrast to the epidermal keratins, the terminal regions of the closely related wool microfibrillar keratin sequences show a conspicuous absence of the glycine repeat sequences described above, but rather they contain a preponderance of cysteine residues. Table 1 illustrates by amino acid compositions the remarkable variation in the terminal sequences of different keratins of the same type and, conversely, the high degree of similarity that exists within their central helical domains.

Nonhelical Amino and Carboxyl Termini: Extending the Protofibril and Assembling the Filament

A clue as to the function of the variable nonhelical terminal segments of the keratins stems from physicochemical studies of Skerrow et al. (1973) and Steinert et al. (1983). When keratin filaments are subjected to limited proteolysis, the nonhelical termini of the subunits are selectively digested and the filament is partially disassembled. Short rods form that resemble a single coiled-coil segment of an individual protofilament or protofibril. Since these rods cannot by themselves assemble into either protofibrils or intermediate filaments, it seems that at least a part of the terminal segments of the keratin subunits must be necessary not only for end-to-end linkage of the subunits, but also for the interprotofibrillar interactions required to form the 8-nm diameter of the intermediate filament.

It is readily apparent that if the nonhelical terminal domains of the intermediate filament subunits interact end-to-end to extend the length of the filament and side-by-side to produce at least some of the interprotofibrillar interactions, then the resulting properties of the 8-nm filaments will differ widely depending on the subunit selected (Geisler and Weber 1982; Fuchs and Hanukoglu 1983). Thus, for example, the ends of the microfibrillar keratins can be cross-linked through disulfide bonding to form the highly stable and insoluble filaments characteristic of wool fibers. The glycine-serine-rich termini of the epidermal keratin subunits, on the other hand, would be expected to give rise to filaments with the same overall structure but with very different solubility properties.

It is interesting that the type I and type II keratin subunits that are coordinately expressed in different tissues have similar, if not homologous, amino acid sequences in their terminal nonhelical segments. We do not yet know whether an 8-nm filament can form in vitro between a type I epidermal keratin with glycine-rich termini and a type II wool keratin with cysteine-rich termini. However, it seems that the choice of which keratins will pair with each other in forming the 8-nm filament is tightly regulated at the level of transcription, and this may be functionally and structurally important to the overall characteristics of the resulting 8-nm filament network.

Table 1
Amino Acid Compositions of the Different Structural Domains of Epidermal (Cytoskeletal) and Wool (Microfibrillar) Keratins

	Amino terminus				Central region					Carboxyl terminus				
	WT-I	*MT-I*	*HT-I*	*WT-II*[a]	*WT-I*	*MT-I*	*HT-I*	*HT-II*[a]	*WT-II*[a]	*WT-I*[a]	*MT-I*	*HT-I*	*HT-II*	*WT-II*[a]
Ala	3.7	0.0	5.2	5.5	4.8	6.3	8.0	10.0	8.4	12.1	0.9	0.0	2.3	6.3
Arg(+)	5.6	2.8	3.5	5.5	9.0	7.0	7.7	7.1	9.7	6.1	6.2	8.7	2.3	4.8
Asn	11.1	0.7	0.9		7.1	6.7	6.1	5.9	4.6	0.0	0.0	2.2	2.3	
Asp(−)	1.9	0.7	0.9	(7.3)	4.8	4.8	6.4	8.2	5.1	0.0	0.9	6.5	0.0	(6.3)
Cys	14.8	2.1	2.6	6.4	2.6	0.6	0.6	0.7	3.0	18.2	0.9	0.0	0.0	17.5
Gln	0.0	0.7	0.9		8.7	8.6	6.4	7.1	3.8	0.0	0.9	8.7	2.3	
Glu(−)	1.9	0.0	0.9	(12.8)	15.4	13.3	14.1	11.9	16.5	0.0	0.0	2.2	1.1	(3.2)
Gly	9.3	49.6	35.0	9.2	1.0	2.9	2.9	3.0	2.5	6.1	58.4	4.3	29.5	7.9
His	0.0	0.0	0.0	0.0	0.6	0.3	0.6	1.1	0.8	0.0	0.0	4.3	1.1	0.0
Ile	1.9	0.0	2.6	4.6	4.2	4.8	4.5	4.8	6.3	3.0	0.0	2.2	3.4	0.0
Leu	5.6	3.5	6.0	11.9	14.7	14.6	12.8	12.3	9.7	3.0	0.0	2.2	4.5	7.9
Lys(+)	0.0	0.7	1.7	5.5	3.5	5.4	6.1	7.4	5.9	3.0	0.9	6.5	3.4	0.0
Met	0.0	0.0	1.7	0.9	0.3	1.9	3.8	2.6	1.7	0.0	0.0	2.2	0.0	0.0
Phe	7.4	9.9	6.0	4.6	1.0	0.6	1.3	1.5	1.3	0.0	0.9	2.2	1.1	0.0
Pro	9.3	0.0	0.9	5.5	1.0	1.0	1.0	0.4	0.4	18.2	0.9	0.0	0.0	4.8
Ser	18.5	22.7	21.6	7.3	7.4	7.6	4.5	4.1	5.9	12.1	22.1	26.1	26.1	17.5
Thr	3.7	0.7	3.5	4.6	4.5	5.4	4.5	3.7	3.4	9.1	0.9	8.7	5.7	6.3
Trp	1.9	0.0	0.0	0.0	0.3	0.3	0.6	0.4	0.4	0.0	0.0	0.0	0.0	0.0
Tyr	0.0	3.5	3.5	3.7	2.9	3.2	3.2	3.3	3.8	3.0	6.2	0.0	5.7	0.0
Val	3.7	2.1	2.6	4.6	6.4	4.8	4.8	4.5	6.8	6.1	0.0	13.0	9.1	17.5

The values are presented as percentages and are based on the sequences from the following sources: (1) WT-I: type I microfibrillar keratin from sheep wool (Crewther et al. 1978, 1980; Gough et al. 1978; Dowling et al. 1983); (2) MT-I: type I cytoskeletal keratin from mouse epidermis (59K) (Steinert et al. 1983); (3) HT-I: type I cytoskeletal keratin from human epidermal cells (51K) (Hanukoglu and Fuchs 1982); (4) HT-II: type II cytoskeletal keratin from human epidermal cells (56K) (Hanukoglu and Fuchs 1983); (5) WT-II: type II microfibrillar keratin from sheep wool (Crewther et al. 1978, 1980; Gough et al. 1978; Dowling et al. 1983). The "Central region" represents the segment extending from the first amino acid of helical region I to the last amino acid of helical region IV.

[a]The composition is based on incomplete sequence for this region.

It is likely that filaments composed of different subunits will be different in ways other than merely their solubility properties. Recent physicochemical studies by Steinert et al. (1983) have indicated that a part of the nonhelical termini can be removed by very mild proteolysis without influencing the structure of the 8-nm filament. This would suggest that the extreme ends of the keratin subunits may reside on the surface of the 8-nm filament, an idea which is supported by electron microscopy studies (Steven et al. 1982; Aebi et al. 1983). If this is indeed correct, then the surface of different keratin filaments may be coated with different sequences, depending on the particular subunits chosen for assembly. This could influence the ability of the resulting filaments to interact with different cytoskeletal components, including the known desmosome-keratin filament associations (Drochmans et al. 1978) and other as yet unidentified filament-protein, filament-factor, or filament-membrane interactions.

Keratins Expressed in Different Tissues Seem to be Transcribed from Different Genes

If different pairs of type I and type II keratins are coexpressed in different tissues giving rise to 8-nm filaments with different properties, then the differential expression of these proteins must be precisely regulated. At least in some instances, this regulation seems to be at the level of the gene, since there are multiple genes (four to ten) for each of the two keratin classes and these genes seem to be differentially expressed in different tissues (Fuchs et al. 1981; Kim et al. 1984a).

Changes in Keratin Subunits Associated with a Commitment of the Epidermal Cell to Terminally Differentiate

In the epidermis, where keratin filaments are especially abundant, a number of changes have been documented in the pattern of keratins produced by a cell as it undergoes terminal differentiation (Dale et al. 1976; Baden and Lee 1978; Skerrow and Hunter 1978; Fuchs and Green 1980; Schlegel et al. 1980; Viac et al. 1980; Woodcock-Mitchell et al. 1982; Skerrow and Skerrow 1983; Warhol et al. 1983). Cells of the inner basal layer produce two major type I keratins (46K and 50K) and two major type II keratins (56K and 58K). Once a cell has been triggered to terminally differentiate and begins its subsequent migration to the cell surface, it produces new keratins of 56.5K and 67K in addition to the preexisting keratins. These changes arise from the synthesis of new keratin mRNAs in the terminally differentiating spinous cells (Fuchs and Green 1980). The newly synthesized keratins have been a curiosity, since their expression appears to be unique to terminally differentiating keratinocytes. Their expression has been shown to be negatively regulated by the fat-soluble factor, vitamin A (Fuchs and Green 1981; Kim et al. 1984b). Although other keratins have been shown to be positively regulated by the vitamin, these two keratins are the only ones that are under such negative control.

To explore the possible functional or structural significance of these differentiation-specific keratins in greater detail, we have determined their relation to the two major types of keratin sequences. Using immunoblot analysis with antisera

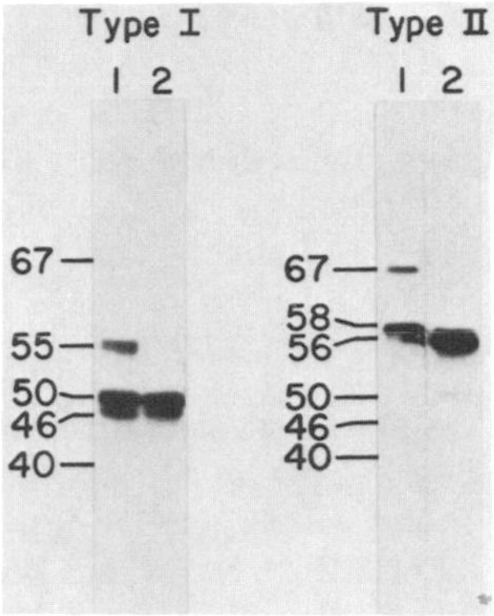

Figure 3
Shift to large keratins during terminal differentiation does not disrupt the balance of type I (*left*) and type II (*right*) keratins. Antibodies were raised against electrophoretically purified 50K (type I) and 56K (type II) human keratins. These antisera were used in immunoblot analysis to compare the presence of immunoreactive forms of both classes of keratins in differentiating and basal-type epidermal cells. (*Left*) Immunoblot with anti-type I keratin antisera; (*right*) immunoblot with anti-type II keratin antisera. Proteins were extracted from (lane *1*) whole human epidermis; (lane *2*) cultured human basal epidermal cells. M_r is shown 10^{-3}. (Data from Fuchs and Marchuk 1983; Kim et al. 1984a.)

specific for the type I and type II keratins, we have demonstrated that the 56.5K keratin belongs to the type I class, whereas the 67K keratin is a member of the type II class (Fig. 3). Two-dimensional gel electrophoresis has shown that the 67K keratin migrates with a basic isoelectric focusing point (7.4), whereas the 56.5K keratin has an acidic isoelectric focusing point (5.0) (Kim et al. 1984a). When mRNA from human epidermis was resolved by formaldehyde gel electrophoresis and subsequently hybridized with ^{32}P-labeled type I and type II keratin cDNA probes, it was confirmed that the large 67K keratin mRNA belongs to the type II class, whereas the 56.5K keratin mRNA is a member of the type I class (Fig. 4).

We do not yet know whether this shift to the synthesis of unusually large type I and type II keratins arises from transcription of new keratin genes or alternatively from a change in the transcription of keratin genes that are already expressed in the basal epidermal cells. However, it is clear from our results that the commitment to terminal differentiation results in a shift to the synthesis of unusually large keratins of both types, without necessarily causing a disruption in the ratio of the two types within the cell. In the future, as we learn more about the genetic complexities of the two keratin multigene families, we hope to begin

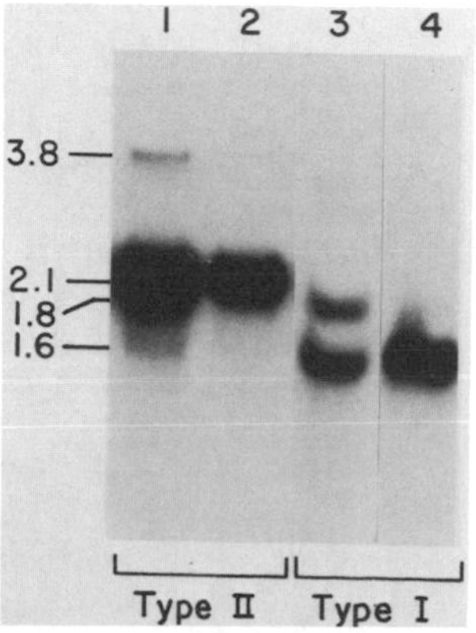

Figure 4
Shift to large type I and type II keratins during terminal differentiation occurs at the level of mRNA synthesis. Cloned cDNAs complementary to the mRNAs for the 56K type II (*left*) and the 50K type I (*right*) epidermal keratins were radiolabeled and used as probes to detect the presence of keratin mRNAs in basal and differentiating epidermal cells. mRNAs were extracted from these cells as described previously (Fuchs and Green 1980) and resolved by formaldehyde agarose gel electrophoresis (Fuchs et al. 1981). (Lanes *1,3*) mRNA from whole human epidermis; (lanes *2,4*) mRNA from cultured human basal epidermal cells. Sizes (in kilobase pairs) of hybridizing mRNAs were determined from the electrophoretic mobilities of eukaryotic and prokaryotic rRNA markers. (Data from Kim et al. 1984a.)

to elucidate the details of the closely linked coordinate regulation of different subsets of keratins to assemble into 8–10-nm filaments with common structures but, in some cases, dramatically different properties. This knowledge should be useful in determining the significance of the shift to the synthesis of new subsets of keratins in terminally differentiating epidermis and in other epithelia.

ACKNOWLEDGMENTS

We thank Ms. Claire Doyle for her expeditious typing of the manuscript. E.F. is a Searle Scholar and the recipient of a National Science Foundation Presidential Young Investigator Award and a National Institutes of Health Career Development Award. K.H.K. is a postdoctoral fellow funded by the National Cancer Institute. D.M. and A.T. are National Institutes of Health predoctoral trainees. This work was supported by a grant from the National Institutes of Health.

REFERENCES

Aebi, U., W.E. Fowler, P. Rew, and T.-T. Sun. 1983. The fibrillar substructure of keratin filaments unraveled. *J. Cell Biol.* **97:** 131.

Ahmadi, B. and P.T. Speakman. 1978. Superimidate crosslinking shows that a rod-shaped, low cysteine, high helix protein prepared by limited proteolysis of reduced wool has four protein chains. *FEBS Lett.* **94:** 365.

Baden, H.P. and L.D. Lee. 1978. Fibrous protein of human epidermis. *J. Invest. Dermatol.* **71:** 148.

Bladon, P.T., P.E. Bowden, W.J. Cunliffe, and E.J. Wood. 1982. Prekeratin biosynthesis in human scalp epidermis. *Biochem. J.* **208:** 179.

Bowden, P.E. and W.J. Cunliffe. 1981. Modification of human prekeratin during epidermal differentiation. *Biochem. J.* **199:** 145.

Chou, P.Y. and G.D. Fasman. 1978. Prediction of the secondary structure of proteins from the amino acid sequence. *Adv. Enzymol.* **47:** 45.

———. 1979. Prediction of β-turns. *Biophys. J.* **26:** 367.

Crewther, W.G., L.M. Dowling, and A.S. Inglis. 1980. Amino acid sequence data from a microfibrillar protein of α-keratin. In *Proceedings of the 6th Quinquennial International Wool Textile Conference.* vol. 2, p. 79.

Crewther, W.G., A.S. Inglis, and N.M. McKern. 1978. Amino acid sequences of α-helical segments of *S*-carboxymethylkerateine-A. *Biochem. J.* **173:** 365.

Crick, F.H.C. 1953. The packing of α-helices: Simple coiled-coils. *Acta Crystallogr.* **6:** 689.

Dale, B.A., L.B. Stern, M. Rabin, and L.-Y. Huang. 1976. The identification of fibrous proteins in fetal rat epidermis by electrophoretic and immunological techniques. *J. Invest. Dermatol.* **66:** 230.

Dowling, L.M., D.A.D. Parry, and L.G. Sparrow. 1983. Structural homology between α-keratin and the intermediate filament proteins desmin and vimentin. *Biosci. Rep.* **3:** 73.

Drochmans, P., C. Freudenstein, J.-C. Wanson, L. Laurent, T.W. Keenan, L. Stadler, R. LeLoup, and W.W. Franke. 1978. Structure and biochemical composition of desmosomes and tonofilaments isolated from calf muzzle epidermis. *J. Cell Biol.* **79:** 427.

Franke, W.W., D.L. Schiller, R. Hatzfeld, and S. Winter. 1983. Protein complexes of intermediate-sized filaments: Melting of cytokeratin complexes in urea reveals different polypeptide separation characteristics. *Proc. Natl. Acad. Sci.* **80:** 7113.

Fuchs, E. and H. Green. 1978. The expression of keratin in genes in epidermis and cultured epidermal cells. *Cell* **15:** 887.

———. 1979. Multiple keratins of cultured human epidermal cells are translated from different mRNA molecules. *Cell* **17:** 573.

———. 1980. Changes in keratin gene expression during terminal differentiation of the keratinocyte. *Cell* **19:** 1033.

———. 1981. Regulation of terminal differentiation of cultured human keratinocytes by vitamin A. *Cell* **25:** 617.

Fuchs, E. and I. Hanukoglu. 1983. Unravelling the structure of the intermediate filament. *Cell* **34:** 332.

Fuchs, E. and D. Marchuk. 1983. Type I and type II keratins have evolved from lower eukaryotes to form the epidermal intermediate filaments in mammalian skin. *Proc. Natl. Acad. Sci.* **80:** 5857.

Fuchs, E., S. Coppock, H. Green, and D. Cleveland. 1981. Two distinct classes of epidermal keratin genes and their evolutionary significance. *Cell* **27:** 75.

Garnier, J., D.J. Osguthorpe, and B. Robson. 1978. Analysis of the accuracy and implications of simple methods for predicting the secondary structure of globular proteins. *J. Mol. Biol.* **120:** 97.

Geisler, N. and K. Weber. 1981. Self-assembly *in vitro* of the 68,000 molecular weight component of the mammalian neurofilament triplet proteins into intermediate-sized filaments. *J. Mol. Biol.* **151:** 565.

———. 1982. The structural relation between intermediate filament proteins in living cells and the α-keratins of sheep wool. *EMBO J.* **1:** 1155.

Geisler, N., E. Kaufmann, and K. Weber. 1982. Protein chemical characterization of three structurally distinct domains along the protofilament unit of desmin 10 nm filaments. *Cell* **30:** 277.

Gough, K.H., A.S. Inglis, and W.G. Crewther. 1978. Amino acid sequences of α-helical segments from *S*-carboxymethylkerateine-A. *Biochem. J.* **173:** 373.

Hanukoglu, I. and E. Fuchs. 1982. The cDNA sequence of a human epidermal keratin: Divergence of sequence but conservation of structure among intermediate filament proteins. *Cell* **31:** 243.

———. 1983. The cDNA sequence of a type II cytoskeletal keratin reveals constant and variable structural domains among keratins. *Cell* **33:** 915.

Henderson, D., N. Geisler, and K. Weber. 1982. A periodic ultrastructure in intermediate filaments. *J. Mol. Biol.* **155:** 173.

Hong, B.-S. and P.F. Davison. 1981. Isolation and characterization of a soluble, immunoactive peptide of glial fibrillary acidic protein. *Biochim. Biophys. Acta* **670:** 139.

Jones, L.N. 1975. The isolation and characterization of α-keratin microfibrils. *Biochim. Biophys. Acta* **412:** 91.

———. 1976. Studies on microfibrils from α-keratin. *Biochim. Biophys. Acta* **446:** 515.

Kim, K.H., D. Marchuk, and E. Fuchs. 1984a. Expression of unusually large keratins during terminal differentiation: The balance of type I and type II keratins is not disrupted. *J. Cell Biol.* (in press).

Kim, K.H., J.G. Rheinwald, and E. Fuchs. 1983. Tissue-specificity of epithelial keratin: Differentiation expression of mRNAs from two multigene families. *Mol. Cell. Biol.* **3:** 495.

Kim, K.H., F. Schwartz, and E. Fuchs. 1984. Differences in keratin synthesis between normal and epithelial cells and squamous cell carcinomas are mediated by vitamin A. *Proc. Natl. Acad. Sci.* **81:** 4280.

Lee, L.D. and H.P. Baden. 1976. Organisation of the polypeptide chains in mammalian keratin. *Nature* **264:** 377.

Magin, T.M., J.L. Lorcano, and W.W. Franke. 1983. Translational products of mRNAs coding for non-epidermal cytokeratin. *EMBO J.* **2:** 1387.

McLachlan, A.D. 1978. Coiled-coil formation and sequence regularities in the helical regions of α-keratin. *J. Mol. Biol.* **124:** 297.

McLachlan, A.D. and M. Stewart. 1975. Tropomyosin coiled-coil interactions: Evidence for an unstaggered structure. *J. Mol. Biol.* **98:** 293.

———. 1982. Periodic charge distribution in the intermediate filament proteins desmin and vimentin. *J. Mol. Biol.* **162:** 693.

Milam, L. and H.P. Erickson. 1982. Visualization of a 21-nm axial periodicity in shadowed keratin filaments and neurofilaments. *J. Cell Biol.* **94:** 592.

Milstone, L.M. 1981. Isolation and characterization of two polypeptides that form intermediate filaments in bovine esophageal epithelium. *J. Cell Biol.* **88:** 317.

Moll, R., W.W. Franke, D. Schiller, B. Geiger, and R. Krepler. 1982. The catalog of human cytokeratins: Patterns of expression in normal epithelia, tumors, and cultured cells. *Cell* **31:** 11.

Nelson, W.G. and T.-T. Sun. 1983. The 50-k dalton keratin classes as molecular markers for stratified squamous epithelia: Cell culture studies. *J. Cell Biol.* **97:** 244.

Parry, D.A.D., W.G. Crewther, R.D. Fraser, and T.P. MacRae. 1977. Structure of the α-keratin: Structural implications of the amino acid sequence of the type I and type II chain segments. *J. Mol. Biol.* **113:** 449.

Pauling, L. and R.B. Corey. 1953. Compound helical configurations of polypeptide chains: Structure of proteins of the α-keratin type. *Nature* **171:** 59.

Pruss, R.M., R. Mirsky, M.C. Raff, R. Thorpe, A.J. Dowling, and B.H. Anderton. 1981. All classes of intermediate filaments share a common antigenic determinant defined by a monoclonal antibody. *Cell* **27:** 419.

Roop, D.R., P. Hawley-Nelson, C.K. Cheng, and S.H. Yuspa. 1983. Keratin gene expression in mouse epidermis and cultured epidermal cells. *Proc. Natl. Acad. Sci.* **80:** 716.

Schlegel, R., S. Banks-Schlegel, and G.S. Pinkus. 1980. Immunohistochemical localization of keratin in normal human tissues. *Lab. Invest.* **42:** 91.

Schweizer, J. and K. Goerttler. 1980. Synthesis *in vitro* of keratin polypeptides directed by mRNA isolated from newborn and adult mouse epidermis. *Eur. J. Biochem.* **112:** 243.

Schweizer, J. and H.J. Winter. 1983. Keratin biosynthesis in normal mouse epithelia and in squamous cell carcinomas. *J. Biol. Chem.* **258:** 13268.

Skerrow, D. and L. Hunter. 1978. Protein modifications during the keratinization of normal and psoriatic human epidermis. *Biochim. Biophys. Acta* **537:** 474.

Skerrow, D. and C.J. Skerrow. 1983. Tonofilament differentiation in human epidermis: Isolation and polypeptide chain composition of keratinocyte subpopulation. *Exp. Cell Res.* **43:** 27.

Skerrow, D., G. Matoltsy, and M. Matoltsy. 1973. Isolation and characterization of the helical regions of epidermal prekeratin. *J. Biol. Chem.* **248:** 4820.

Steinert, P.M. 1978. Structure of the three-chain unit of the bovine epidermal keratin filament. *J. Mol. Biol.* **123:** 49.

Steinert, P.M. and W.W. Idler. 1975. The polypeptide composition of bovine epidermal α-keratin. *Biochem. J.* **151:** 603.

Steinert, P.M., W.W. Idler, and S.B. Zimmerman. 1976. Self-assembly of bovine epidermal keratin filaments *in vitro*. *J. Mol. Biol.* **108:** 547.

Steinert, P.M., S.B. Zimmerman, J.M. Starger, and R.D. Goldman. 1978. Ten-nanometer filaments of hamster BHK-21 cells and epidermal keratin filaments have similar structures. *Proc. Natl. Acad. Sci.* **75:** 6098.

Steinert, P.M., R.H. Rice, D.R. Roop, B.L. Trus, and A.C. Steven. 1983. Complete amino acid sequence of a mouse epidermal keratin subunit and implications for the structure of intermediate filaments. *Nature* **302:** 794.

Steven, A.C., J. Wall, J. Hainfeld, and P.M. Steinert. 1982. Structure of fibroblastic intermediate filaments: Analysis by scanning transmission electron microscopy. *Proc. Natl. Acad. Sci.* **79:** 3101.

Viac, J., M.J. Staquet, J. Thivolet, and C. Goujon. 1980. Experimental production of antibodies against stratum corneum keratin polypeptides. *Arch. Dermatol. Res.* **267:** 179.

Warhol, M.J., G.S. Pinkus, S.P. Banks-Schlegel. 1983. Localization of keratin proteins in the human epidermis by a postembedding immunoperoxidase technique. *J. Histochem. Cytochem.* **31:** 716.

Weber, K., M. Osborn, and W.W. Franke. 1980. Antibodies against merokeratin from sheep wool decorate cytokeratin filaments in non-keratinizing epithelial cells. *Eur. J. Cell Biol.* **23:** 110.

Woodcock-Mitchell, J., R. Eichner, W.G. Nelson, T.-T. Sun. 1982. Immunolocalization of keratin polypeptides in human epidermis using monoclonal antibodies. *J. Cell Biol.* **95:** 580.

Identification of Different Types of Keratin Polypeptides Distinguished within the Acidic Cytokeratin Subfamily: Corresponding Cytokeratins in Diverse Species

José L. Jorcano,*† Jürgen K. Franz,* Michael Rieger,* Thomas M. Magin,* and Werner W. Franke*

*Division of Membrane Biology and Biochemistry Institute of Cell and Tumor Biology, German Cancer Research Center; †Center of Molecular Biology, University of Heidelberg, D-6900 Heidelberg, Federal Republic of Germany

Cytokeratins are a family of polypeptides that polymerize to form the intermediate-sized filament cytoskeleton characteristic of vertebrate epithelial cells (for review, see Franke et al. 1982; Lazarides 1982; Fuchs and Hanukoglu 1983). In mammals, this family is composed of approximately 20 members, ranging in molecular weight from 40,000 to 68,000 and in isoelectric pH values from 5 to 8 (Franke et al. 1981a,b,c; Wu and Rheinwald 1981; Moll et al. 1982; Schiller et al. 1982; Tseng et al. 1982). In humans, the most intensely studied species, 19 different polypeptides have been distinguished so far (Fig. 1A) (Moll et al. 1982) and 22 different polypeptides have been distinguished in bovine epithelia (Fig. 1B) (Franke et al. 1982; Schiller et al. 1982). Like other intermediate filament proteins, cytokeratins possess a common general structure, consisting of a central α-helical core of approximately 310 amino acids flanked by two non-α-helical tails of variable sizes (Geisler and Weber 1982a, 1983; Hanukoglu and Fuchs, 1982, 1983; Quax-Jeuken et al. 1983; Steinert et al. 1983). However, systematic differences between cytokeratin polypeptides have also been noted. For example, nucleic acid hybridization data (Fuchs et al. 1981; Kim et al. 1983), peptide map analysis (Schiller et al. 1982), and reactivity with certain monoclonal antibodies (Gigi et al. 1982; Tseng et al. 1982; Eichner et al. 1984) clearly divide the cytokeratin family into two subfamilies. The acidic subfamily (type I keratins, according to Hanukoglu and Fuchs 1982, 1983) consists of the group of relatively small (M_r = 40,000–60,000) and more acidic (isoelectric pH between 4.5 and 6.0 on two-dimensional gel electrophoresis in solutions containing 9.5 M urea) poly-

  0-87969-174-3/84 $1.00

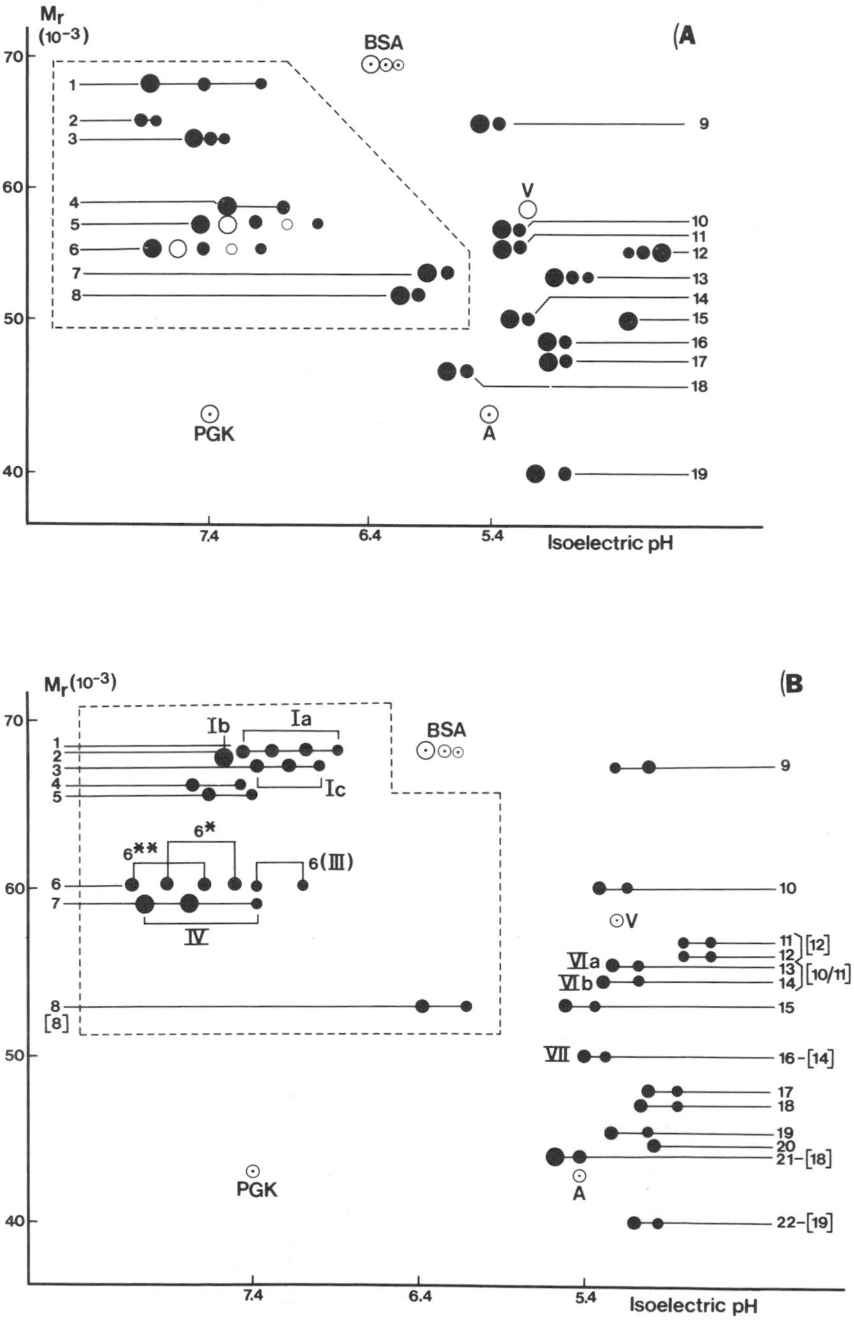

Figure 1 (*See facing page for legend.*)

peptides, which are related to the type I keratins of sheep wool (Crewther and Dowling 1971; Crewther et al. 1980, 1983).

The basic subfamily (also referred to as type II keratins because of their similarity to the type II sheep wool keratins [see same references]) includes polypeptides that are generally larger (M_r = 52,000–68,000) and more positively charged (isoelectric pH between 6 and 8) than those of the acidic family. The difference in characteristics of these subfamilies is probably of functional importance since the structural subunits of the cytokeratin intermediate filament are keratin tetramers containing specific combinations of two basic and two acidic polypeptides (Franke et al. 1983a, 1984; Quinlan et al. 1984) arranged in coiled coils (for wool keratins, see also Crewther et al. 1983; Gruen and Woods 1983). The same characteristics that divided the keratins into two clear subfamilies (nucleic acid cross-hybridization, peptide map comparison, and immunological cross-reaction; see references cited above) also led to the belief that the homologies among the members of each subfamily were very high; indeed, several reports demonstrated a high degree of cross-hybridization between mRNAs coding for various members of the same subfamily (Fuchs et al. 1981; Kim et al. 1983; Roop et al. 1983). However, using more stringent hybridization conditions, we have recently shown that the degree of nucleotide sequence divergence between members of the same subfamily is sufficient to allow the specific hybridization of the mRNAs coding for the diverse polypeptides of the same subfamily (cf. Kim et al. 1983; Jorcano et al. 1984a). Here, we present a comparison of amino acid sequences of two different polypeptides of the same subfamily of the same species (i.e., the two acidic keratin polypeptides coexpressed in bovine epidermis) and demonstrate that profound sequence differences can exist between members of the same keratin subfamily. We have also compared the acidic bovine keratins with those expressed in the epidermis of man, mouse, and frog and have found that the corresponding keratins of different species share characteristic sequence homologies not only in the α-helical region, but also in the carboxyterminal non-α-helical region.

Figure 1
Schematic diagram of human (*A*) and bovine (*B*) cytokeratin polypeptides as separated by two-dimensional gel electrophoresis. Abscissa: isoelectric pH values in gels containing 9.5 M urea; ordinate: apparent relative molecular weights (M_r) as determined by SDS-polyacrylamide gel electrophoresis. Horizontal series of closed circles represent isoelectric variants of the same polypeptide, the diameter of the circles being proportional to the intensity of the Coomassie-blue staining. (○) Minor variants of the specific polypeptide not always recognized in the various cell types analyzed. 3-Phosphoglycerokinase (PGK), bovine serum albumin (BSA), α-actin (A), and vimentin (V) were used as reference proteins in coelectrophoresis. Human keratins in *A* are numbered according to the catalog of Moll et al. (1982). Roman numerals in *B* designate bovine epidermal keratins (cf. Jorcano et al. 1984a); arabic numerals in *B* give the numbers of the bovine catalog of Schiller et al. (1982); bracketed numbers in *B* give the catalog no. of the human cytokeratins (Moll et al. 1982) that most likely correspond to the specific bovine cytokeratins, as judged from immunological data, tryptic peptide maps, and similar patterns of cell-type-specificity of expression. Dashed line encloses the members of the basic subfamily.

Carboxyterminal Amino Acid Sequences of Bovine Keratins VIb and VII

We have recently isolated and characterized cDNA clones coding for all major bovine epidermal keratins, i.e., components VII (M_r = 50,000), VIb (M_r = 54,000), IV (M_r = 59,000), III (M_r = 60,000), and Ia-c (M_r = 68,000) (for details, see Jorcano et al. 1984a; for nomenclature, see Franke et al. 1981a; Schiller et al. 1982). Components VIb and VII belong to the acidic subfamily, whereas components Ia-c, III, and IV are representatives of the basic subfamily expressed in this tissue.

We have sequenced the inserts of the clones coding for the two acidic keratins (Jorcano et al. 1984b). Figure 2 presents the nucleotide sequence of clone pKBVIb[1], coding for component VIb, that contains 52% of the estimated length of the mRNA (2100 nucleotides) (Jorcano et al. 1984a) and most likely includes the 3′ end of the mRNA, since it carries an AATAAA polyadenylation signal 24 nucleotides before its end. The sequence contains an open reading frame coding for 202 amino acids, leaving a relatively long 3′-untranslated end of approximately 500 bases, in agreement with our previous estimations (Franke et al. 1983b). After this first stop codon, the reading frame again remains open for a stretch of 126 bases, ending eventually at position 732 with a UAA stop codon. The first 157 amino acids define a region highly homologous to the carboxyl end of the α-helix of other keratins (see below), which ends with the sequence TYRSLLEGE (Fig. 2, large arrowhead), very similar to that found at the ends of the α-helical regions of other intermediate filament proteins (see below) (Geisler and Weber 1982a, 1983; Hanukoglu and Fuchs 1982, 1983; Crewther et al. 1983; Quax-Jeuken et al. 1983). This α-helical region contains a tryptophan residue 118 amino acids upstream of the end of the helix (Fig. 2, small arrowhead), also typically found at this position in all intermediate filament polypeptides sequenced so far (same references as above). The sequence following the α-helical portion is very rich in glycine, a feature also found in the non-α-helical carboxy-terminal "tail" regions of two epidermal keratin polypeptides of mouse and man (Hanukoglu and Fuchs 1983; Steinert et al. 1983).

Figure 3 presents the sequence of clone pKBVII[1], coding for bovine epidermal component VII (Jorcano et al. 1984a), that includes the 3′ end of the mRNA, since it ends with a poly(A) tail that is preceded 28 nucleotides upstream by the putative polyadenylation signal AATAAA. The sequence contains a long open reading frame, coding for 93 amino acids, that ends at position 280 with a UAA stop codon. The 3′-untranslated region is 165 nucleotides long, which is again in agreement with our previous calculations based on electrophoretic estimations of the molecular weight of this mRNA (Franke et al. 1983b). The region comprising amino acids 1–44 shows a high degree of homology with the end of the α-helical domain of other keratins (see below) and, like keratin VIb, ends with a sequence TYRRLLEGE (Fig. 3, arrowhead). This is very similar to the sequence marking the ends of the α-helical core region of cytokeratin VIb and other intermediate filament proteins (for references, see above). The remaining amino acid sequence has no potential for forming an α-helix and has an amino acid composition relatively rich in serine and valine, a characteristic also found in the extra-helical carboxyl end of human epidermal keratin no. 14 (Hanukoglu and Fuchs 1982).

```
   1 GGTGATGTGAATGTGGAAATGAATGCTGCCCCAGGTGTTGATCTCACTGAACTTCTGAAT
     GlyAspValAsnValGluMetAsnAlaAlaProGlyValAspLeuThrGluLeuLeuAsn

  61 AACATGAGAAGCCAATATGAACAACTTGCCGAGAAAAATCGCAGAGACGCCGAAGCCTGG
     AsnMetArgSerGlnTyrGluGlnLeuAlaGluLysAsnArgArgAspAlaGluAlaTrp

 121 TTCAATGAAAAGAGCAAGGAACTCACTACGGAAATCAACAGTAACCTTGAACAAGTATCG
     PheAsnGluLysSerLysGluLeuThrThrGluIleAsnSerAsnLeuGluGlnValSer

 181 AGCCATAAATCTGAGATTACTGAGTTGAGACGCACTATTCAAGGCCTGGAGATTGAGCTA
     SerHisLysSerGluIleThrGluLeuArgArgThrIleGlnGlyLeuGluIleGluLeu

 241 CAGTCCCAACTAGCCCTAAAACAATCCCTGGAAGCCTCCTTGGCAGAGACAGAAGGTCGC
     GlnSerGlnLeuAlaLeuLysGlnSerLeuGluAlaSerLeuAlaGluThrGluGlyArg

 301 TATTGTGTGCAGCTCTCTCAGATTCAGTCCCAGATCTCCTCTCTGGAAGAACAACTGCAA
     TyrCysValGlnLeuSerGlnIleGlnSerGlnIleSerSerLeuGluGluGlnLeuGln

 361 CAGATTCGAGCTGAGACTGAGTGCCAGAATGCCGAGTACCAACAACTCCTGGATATTAAG
     GlnIleArgAlaGluThrGluCysGlnAsnAlaGluTyrGlnGlnLeuLeuAspIleLys

 421 ATACGACTGGAGAATGAAATTCAAACCTACCGCAGCCTGCTAGAAGGAGAGGGAAGTTCC
     IleArgLeuGluAsnGluIleGlnThrTyrArgSerLeuLeuGluGlyGluGlySerSer

 481 GGCGGCGGCTCCTACGGCGGCGGGCGCGGCTATGGCGGAAGTTCCGGCGGCGGCGGCGGC
     GlyGlyGlySerTyrGlyGlyGlyArgGlyTyrGlyGlySerSerGlyGlyGlyGlyGly

 541 GGCTACGGCGGCGGAAGTCCAGCGGCGGCTACGGCGGCGGAAGTCCAGCGGTGGCGGCCA
     GlyTyrGlyGlyGlySerProAlaAlaAlaThrAlaAlaGluValGlnArgTrpArgPro

 601 CGGCGGTAGTTCCGGCGGCAGTACGGCGGCGGAAGTCCAGTGGCGGTGGCCAGGCGGCGG
     ArgArg***PheArgArgGlnTyrGlyGlyGlySerProValAlaValAlaArgArgArg

 661 AAGTCCAGCGGAGGCCACAAATCCACCACTACAGGGTCCGTTGGAGAGTCCTCATCTAAG
     LysSerSerGlyGlyHisLysSerThrThrThrGlySerValGlyGluSerSerSerLys

 721 GGACCAAGATACTAACAAAACCAGAGTGATCAAGACAATTATTGAGGAGTTGGCACCTGA
     GlyProArgTyr***

 781 TGGTAGAGTCCTTTCATCTATGGTTGAATCGGAAACCAAGAAACACTACTATTAAGCCGC

 841 ATTAAGAGGAAAGAGTCTCCCCTCACACAGGCCATTATGCATAGATGCATGAAAAAAAAT

 901 CTCCAAGAAAACACTTCAGTCTTAATGGCCTATGGAAATAGGTCCACTCCTTGATCATAA

 961 GGTGTCTAAACGGTGTTTTGTGGTTTCTGTATTTCTCCTTTTCACTTTACCAAAAAGTAT

1021 TCTTTACTTTTTGGAAAAAAAGAAAAAACTTTCTATTCTCATTTACTAATGAATTTCAAT

1081 AAACTTTCTTACTGATGCACT
```

Figure 2

Nucleotide sequence of the insert of clone pKBVI[1] and the predicted amino acid sequence (three-letter code) of the carboxyterminal half of bovine epidermal keratin VIb. Nucleotides in this and subsequent figures are numbered in the 5′ to 3′ direction. The 3′ end does not extend to the poly(A) tail, but at position 1078, it contains a putative polyadenylation signal (AATAAA). The stop codon at position 607 (UAG; marked by three asterisks) is followed by a continuation of the reading frame and a stop codon at position 733 (UAA). Large arrowhead denotes the end of the α-helical core domain at the protein. Small arrowhead points to a tryptophan residue found at this position in all sequenced intermediate filament polypeptides (see text).

A comparison of the amino acid sequences derived for the bovine epidermal components VIb and VII of the acidic subfamily (Fig. 4) shows that these polypeptides can be divided on the basis of their homologies into two clearly differ-

```
  1 AGCGTGGAAGAGCAGCTCGCCCAGCTGCGCTGCGAGATGGAGCAGCAGAACCAGGAGTAC
    SerValGluGluGlnLeuAlaGlnLeuArgCysGluMetGluGlnGlnAsnGlnGluTyr

 61 AAGATCCTGCTGGACGTGAAGACCCGGCTGGAGCAGGAGATCGCCACCTACCGCCGCCTG
    LysIleLeuLeuAspValLysThrArgLeuGluGlnGluIleAlaThrTyrArgArgLeu
            ▼
121 CTGGAGGGCGAGGACGCCCACCTCTCCTCTTCCCAGTTCTCCTCTGGCTCTCAGTCATCC
    LeuGluGlyGluAspAlaHisLeuSerSerSerGlnPheSerSerGlySerGlnSerSer

181 AGAGATGTATCCTCCAGTCGCCAGGTTCGCACCAAAGTCGTGGATGTACACGATGGCAAG
    ArgAspValSerSerSerArgGlnValArgThrLysValValAspValHisAspGlyLys

241 GTGGTGTTCACACACGAGCAGATCGTTCGCACCAAGAACTAAGGCTTCTCCTGCTCATGC
    ValValPheThrHisGluGlnIleValArgThrLysAsn***

301 CTAGAAAGCACCCCCCAAGTGGACACAGATTTCCCTAGGGCATCCCTTCTCCTGTCCCCG

361 CCTCCCCCAAGCACTTGACTCGTGAGCCCTGTTTCACCCTTGCCCCTTCCAGCAATCAAT

421 AAAGCTTCATTAGCTGAGTTGCCCAA
```

Figure 3
Nucleotide sequence of the insert of clone pKBVII[1] and the predicted amino acid sequence (three-letter code) of the carboxyterminal portion of the bovine epidermal keratin VII. The nucleotide sequence ends with a poly(A) track (not shown) and contains a putative polyadenylation signal 28 nucleotides before the end. Asterisks mark the position of the stop codon, and the arrowhead denotes the end of the region coding for the protein α-helical core.

entiated regions. Taking as a reference point for the alignment of the end of the α-helix (position 0; this position is denoted by a large arrowhead in Figs. 4–6), the α-helical portions of both polypeptides can easily be matched, exhibiting 68% sequence identity (identical residues at same positions) and 84% sequence homology, i.e., including conservative amino acid changes that most likely will not influence the α-helical structure. The degree of homology in this region is similar to that found between the three other sequenced members of the acidic subfamily of keratins, i.e., the mouse epidermal keratin of M_r 59,000 (Steinert et al. 1983), the human epidermal keratin no. 14 of M_r 50,000 (Hanukoglu and Fuchs 1983), and the *Xenopus laevis* epidermal keratin of M_r 51,000 (Hoffmann and Franz 1984), belonging to taxonomically diverse species. On the other hand, this homology is significantly higher than the 25–30% homology found between the α-helical regions of members of the basic and acidic keratin subfamilies (Crewther et al. 1983; Hanukoglu and Fuchs 1983). Unlike the α-helical core regions, the carboxyterminal non-α-helical tails of the bovine components VIb and VII show no significant homology, indicating that this region varies considerably between members of the acidic keratin subfamily, even within the same species.

This degree of sequence divergence, which is very prominent in the extrahelical carboxyl tails but is also appreciable in the α-helical core region, may provide an explanation for the various differences observed between these two members of the bovine acidic keratin subfamily, i.e., differences in size and charge, in tryptic peptide maps, in antigenic determinants, and in the stability of the complexes they form with members of the basic subfamilies (Jorcano et al. 1984b; see also Franke et al. 1983a, 1984). The corresponding nucleotide sequences also explain why, under stringent hybridization conditions, clones pKBVIb[1] and pKBVII[1] hybridize specifically with their corresponding mRNAs in Northern blots

```
        -40       -30       -20       -10         0        10
          .         .         .         .         .         .
B VI   SLEEQLQQIRAETECQNAEYQQLLDIKIRLENEIQTYRSLLEGEGSSGGGSYGGRGYGG
       * **** * * * * ** **  *** * *** ** *** *****      *
B VII  SVEEQLAQLRCEMEQQNQEYKILLDVKTRLEQEIATYRRLLEGEDAHLSSSQFSSGSQSS
                                                  ▲
         20        30        40        50        60        70
          .         .         .         .         .         .
       SSGGGGGGYGGGSPAAATAAEVQRWRPRR*FRRQYGGGSPVAVARRRKSSGGHKSTTTGS
                            *
       RDVSSSRQVRTKVVDVHDGKVVFTHEQIVRTKN*..........................

         80                  .
          .
       VGESSSKGPRY*

       .............
```

Figure 4
Comparison of the amino acid sequences (one-letter code) of bovine epidermal keratins VIb (B VI) and VII (B VII). The sequences are aligned in such a way that the end of the α-helical regions matches (arrowhead, position 0 in this and all subsequent figures). Amino acids with a negative number belong to the α-helical core and those with a positive number are in the non-α-helical carboxyterminal part. Asterisks interrupting the coding sequences represent stop codons. Asterisks between the two sequences denote conserved amino acids. Note the high degree of homology in the α-helical region vis-à-vis the absence of significant homology in the extrahelical portion.

or in hybridization-selection-translation experiments without appreciable cross-hybridization (Jorcano et al. 1984a).

Comparison of the Bovine Epidermal Keratins VIb and VII with Epidermal Keratins of the Same Subfamily Expressed in Other Species

To determine which features of the keratin sequences are evolutionarily conserved and may be functionally important, we have compared the sequences of the bovine epidermal components VIb and VII with those of two other epidermal keratins of the same subfamily of distant species, i.e., man and mouse. Figure 5 shows a comparison of the amino acid sequences of bovine component VIb and the acidic keratin (M_r = 59,000) from mouse epidermis reported by Steinert et al. (1983). When these sequences are aligned, two regions with different degrees of homology become apparent. The α-helical core regions (amino acids in positions 0 to −156) have 88% sequence identity and 96% sequence homology, which is clearly higher than the homology between components VIb and VII from the same (bovine) species discussed above (see Fig. 4). There is only one stretch of six amino acids (positions −54 to −59) in which the two sequences diverge significantly. Interestingly, this stretch is located in a region where in the mouse keratin (M_r = 59,000), an unstability of the coiled-coil structure due to reversal in the polarity of the heptades has been suggested (Steinert et al. 1983).

Outside of the α-helical region (Fig. 5, residues 1–122 of the murine sequence), bovine component VIb and the mouse keratin (M_r = 59,000) are much more divergent. However, striking similarities not observed between the bovine components VIb and VII are also evident. Both keratins are glycine-rich and con-

```
           -150      -140      -130      -120      -110      -100
              .         .         .         .     ▼   .         .
B VI  GDVNVEMNAAPGVDLTELLNNMRSQYEQLAEKNRRDAEAWFNEKSKELTTEINSNLEQVS
      ********** ***** ****** ********** *** *** ********* **  * *
M 59  GDVNVEMNAAQGVDLTQLLNNMRNQYEQLAEKNRKDAEEWFNQKSKELTTEIDSNIAQMS

            -90       -80       -70       -60       -50       -40
              .         .         .         .         .         .
      SHKSEITELRRTIQGLEIELQSQLALKQSLEASLAETEGRYCVQLSQIQSQISSLEEQLQ
      ************ ************************      ********** ******
      SHKSEITELRRTVQGLEIELQSQLALKQSLEASLAETVESLLRQLSQIQSQISALEEQLQ

            -30       -20       -10         0        10        20
              .         .         .         .         .         .
      QIRAETECQNAEYQQLLDIKIRLENEIQTYRSLLEGEGSSGGGSYGGRGYGGSSGGGGG
      ******************** ******************* **  ****  *** ***
      QIRAETECQNAEYQQLLDIKTRLENEIQTYRSLLEGEGSSSGG--GGGRR-GGSGGGSYG
                                              ▲
             30        40        50        60        70        80
              .         .         .         .         .         .
      GYGGGSPAAATAAEVQRWRPRR*FRRQYGGGSPVAVARRRKSSGGHKSTTTGSVGESSSK
      *  **                       *****     *   * **       ** *
      GSSGGGSYGGSSGGGGSYGGSSGGGGSYGGGSSGCGGRGGGSGGG--YGGGGSSGGAGGR

             90       100       110       120
              .         .         .         .
      GPRY*

      GGGSGGGYGGGGSSGRRGGSGGFSGTSGGGDQSSKGPRY*
```

Figure 5
Comparison of the amino acid sequences of bovine epidermal keratin VIb (B VI) and mouse epidermal keratin (M_r = 59,000: M 59) (data from Steinert et al. 1983). The two sequences are aligned according to the principle introduced in Fig. 4. Solid underscoring refers to the characteristic heptapeptide GGG^S_GYGG present in the non-α-helical carboxyterminal regions of both polypeptides. Dashed underscoring denotes a conserved heptapeptide SSKGPRY sequence at the ends, although the actual stop codon of the bovine keratin (see text) very probably is located 35 amino acids upstream (asterisk at position +46). Asterisks between the sequences mark conserved amino acids. Small arrowhead points to the conserved tryptophan residue (see Fig. 2). Note pronounced homology in the α-helical region as opposed to weak matching in the non-α-helical part.

tain a consensus sequence GGG^S_GYGG that is repeated twice in the bovine sequence and four times in the murine sequence, and in the latter it occurs for another two times in slightly modulated form (SGGGYGG).

These results demonstrate that the bovine keratin VIb is much more related to an epidermal keratin expressed in an evolutionarily distant species (i.e, mouse) than to another member of the acidic bovine subfamily coexpressed in the same tissue (component VII).

A comparison of the nucleotide sequences of the bovine and mouse clones (not shown) reveals, as expected, that the parts of the mRNAs coding for the α-helical regions of the two polypeptides (nucleotides 0–471 in Fig. 2) are very conserved (85% sequence identity), most of the changes being silent A→G or T→C transitions in the third position of the codons. In contrast, the mRNA regions coding for the non-α-helical tail regions show little nucleotide sequence identity, except for very short stretches of homologous nucleotides, reflecting the high glycine content of these regions. However, the 3′-untranslated ends of both mRNAs are unexpectedly well conserved (81% sequence identity and 89% sequence homology). In the mouse mRNA, this stretch of homology starts 21 bases before the UAA stop codon and includes 7 amino acids of the carboxyterminal end (SSKGPRY; Fig. 5). Remarkably, the corresponding 21 homologous nucleotides of the bovine clone also precede a UAA signal (position 733 in Fig. 2), but in this clone there is another stop codon (UAG) located 126 bp upstream (position 607 in Fig. 1). The existence of an open reading frame between these two stop codons that codes for a sequence rich in glycine and hydroxy-amino acids and ends with the same heptapeptide sequence (SSKGPRY), together with the extreme conservation of the nucleotides following the second stop codon (UAA), suggests that the stop codon at position 607 might be the result of a cloning artifact. However, we have found the same codon in the same position in three independent cDNA clones coding for keratin component VIb (pKBVI[1], pKBVI[2], pKBVI[3]) (see Jorcano et al. 1984a), indicating that the first stop codon is real and has been acquired relatively recently during evolution. Amino acid sequence analysis of the carboxyterminal end of keratin VIb, currently under way in our laboratory, will be necessary to determine whether the UAG stop codon found at position 607 in these cDNA sequences really defines the carboxyl terminus of this polypeptide.

Figure 6 presents a comparison of bovine epidermal keratin component VII and the human epidermal keratin no. 14 ($M_r = 50{,}000$ [Hanukoglu and Fuchs 1982]). Both polypeptides not only are identical in the α-helical core regions (amino acid positions 0 to −43), but also are extremely well conserved in the non-α-helical carboxyterminal tails (amino acid positions 1–50). From this, we conclude that they represent the same keratin polypeptide in the two species. At the nucleic acid level (not shown), the comparable regions of the two clones show 82% sequence identity and 91% homology. The parts of the mRNAs coding for the α-helical polypeptide cores (nucleotides 1–132 in Fig. 3) are nearly identical (97% sequence identity), whereas the regions coding for the non-α-helical carboxyl end (nucleotides 133–280 in Fig. 3) share 88% sequence identity and 95% sequence homology, with most of the changes being transitions in the

```
        -40       -30       -20       -10         0        10
           .         .         .         .         .         .
B VII   SVEEQLAQLRCEMEQQNQEYKILLDVKTRLEQEIATYRRLLEGEDAHLSSSQFSSGSQSS
        ************************************************************
H 50    SVEEQLAQLRCEMEQQNQEYKILLDVKTRLEQEIATYRRLLEGEDAHLSSSQFSSGSQSS
                                                   ▲
         20        30        40        50
          .         .         .         .
        RDV-SSSRQVRTKVVDVHDGKVVFTHEQIVRTKN*
        *** ******************* **********
        RDVTSSSRQVRTKVVDVHDGKVVSTHEQIVRTKN*
```

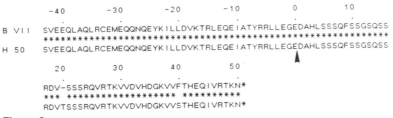

Figure 6
Comparison of the amino acid sequences of bovine epidermal keratin VII (B VII) and human keratin no. 14 ($M_r = 50{,}000$: H 50) (Hanukoglu and Fuchs 1982). The sequences have been aligned according to the principle described in Fig. 4. Asterisks between sequences denote conserved amino acids. The horizontal bar interrupting the bovine sequences has been introduced in order to maximize homology. Note exceptionally high homology in both the α-helical and the extra helical portions.

third codon position which do not result in changes of amino acid sequence. Again, in this case, the 3′-untranslated regions show a surprisingly high degree of sequence conservation (71% identity, 83% homology), which is particularly conspicuous in a stretch of 30 nucleotides surrounding the polyadenylation signals, where only two changes can be detected.

These results demonstrate, as already suggested by tryptic peptide maps and immunological experiments (Franke et al. 1981b; Debus et al. 1982; Schiller et al. 1982), that the corresponding keratins from different species are much more related to each other than different keratins from the same subfamily of the same species. This interspecies homology is also recognized in the 3′-untranslated part of the keratin mRNAs, an observation that has also been made in other cytoskeletal multigene families such as tubulins (Cowan et al. 1983) and actins (Ponte et al. 1984), indicating the importance of this noncoding mRNA region in some, yet unknown, functions.

Figure 7 presents a comparison of the acidic keratins that have been sequenced so far. It emphasizes the sequence conservation in the α-helical region and the variability of the non-α-helical carboxyl termini exhibited by the various members of this subfamily. Preliminary sequence data indicate that members of the basic subfamily display similar principles (J.L. Jorcano et al., unpubl.). The sequences of the carboxyterminal tails distinguish at least two different keratin types within the acidic subfamily. One type, which is represented by the bovine epidermal keratin VIb (M_r = 54,000) and the murine epidermal keratin (M_r = 59,000), is characterized by a wealth of glycine residues and repeats of a canonic heptapeptide GGG$^{S}_{G}$YGG. Similar oligopeptide stretches rich in glycine residues, mostly in combination with serine residues and aromatic amino acids, have also been reported to occur in the aminoterminal head regions of keratins of both subfamilies I and II (Hanukoglu and Fuchs 1982, 1983; Steinert et al. 1983). This glycine-rich sequence may be related to special differentiation processes in certain multistratified epithelia such as epidermis, as this type of keratin is only expressed in certain epithelia that are characterized by their potential for keratinization (cf. Jorcano et al. 1984a). From immunological experiments, we would predict that human epidermal keratins nos. 10 and 11 (M_r = 56,000 and 56,500) also belong to the same type of acidic keratins as bovine keratins VIa and VIb (Jorcano et al. 1984b).

The carboxyl end of the other type of acidic keratin expressed in epidermis is relatively rich in serine and valine residues and reveals a subterminal consensus heptapeptide DGKV$^{S}_{F}$T that, remarkably, is also found in the sequence of an amphibian epidermal type I keratin, i.e., the keratin with a relative molecular weight of 51,000 (Hoffmann and Franz 1984). At the moment, no specific functions can be ascribed to the carboxyterminal tail domains of keratin polypeptides, although the tail regions have been reported to be required for the in vitro assembly into intermediate filaments of keratins and other intermediate filament proteins (Steinert 1978; Geisler et al. 1982b). On the other hand, it also seems plausible that this terminal domain, which in assembled intermediate filaments might somewhat protrude from the α-helical backbone, is involved in the specific binding of other cellular proteins ("intermediate-filament-associated proteins"). That the different sequences in the carboxyterminal tail regions of different keratin types are

```
                -90           -80           -70           -60           -50
                 .             .             .             .             .
M  59   QMSSHKSEITELRRTVQGLEIELQSQLALKQSLEASLAETVESLLRQLSQ
B  VI   QVSSHKSEITELRRTIQGLEIELQSQLALKQSLEASLAETEGRYCVQLSQ
B  VII  ——————————————————————————————————————————————————
H  50   LVQSGKSEISELRRTMQNLEIELQSQLSMKASLENSLEETKGRYCMQLAQ
X1 51   QQKESKSELTELKTTLQSLEIELQSQLAMKKSLEMTLAEVEGSFCMKLSR
SW  I   QLQCNQEEIIELRRTVNALQVELQAQHNLRDSLENTLTETEARYSCQLNQ

                -40           -30           -20           -10            0
                 .             .             .             .             .▼
M  59   IQSQISALEEQLQQIRAETECQNAEYQQLLDIKTRLENEIQTYRSLLEGE
B  VI   IQSQISSLEEQLQQIRAETECQNAEYQQLLDIKIRLENEIQTYRSLLEGE
B  VII  ————————SVEEQLAQLRCEMEQQNQEYKILLDVKTRLEQEIATYRRLLEGE
H  50   IQEMIGSVEEQLAQLRCEMEQQNQEYKILLDVKTRLEQEIATYRRLLEGE
X1 51   LQEMIVNVEEQIARLKGESECQTAEYQQLLDIKTRLENEIETYRRLLDGD
SW  I   VQSLISNVESQLAEIRGDLERQNQEYQVLLDVRARLECEINTYRGLLDSE

                +10           +20           +30           -40           -50
                 .             .             .             .             .
M  59   GSSSGG——GGFRR—GGSGGGSYGGSSGGGSYGGSSGGGGSYGGSSGGGGS
B  VI   GSSGGGSYGGGRGYGGSSGGGGGGYGGGSPAAATAAEVQRWRPRR*FRRQ
B  VII  DAHLSSSQFSSGSQSSRDVSSS—RQVRTKVVDVH——DGKVVFTHEQIVRT
H  50   DAHLSSSQFSSGSQSSRDVTSSSRQIRTKVMDVH——DGKVVSTHEQVLRT
X1 51   LSKPKSGGGTSTNTGSTSSKGSTRTVKRREIIEEVVDGKVVSTKVVDM*—
SW  I   DCKLACGKPLTPCISSPCAPAAPCTTCVVPSSCGRRY*

                +60           +70           +80           +90          +100
                 .             .             .             .             .
M  59   YGGGSSGCGGRGGGSGGGYGGGGSSGGAGGRGGGSGGGYGGGGSSGRRGG
B  VI   YGGGSPVAVARRRKSSGGHKSTTTGSVGESSSKGPRY*
B  VII  KN*
H  50   KN*

               +110          +120
                 .             .
M  59   SGGFSGTSGGGDQSSKGPRY*
```

Figure 7
Sequence relationships of the carboxyterminal regions of diverse keratin polypeptides of the acidic subfamily in different species. The alignment has been made according to the principle described in Fig. 4. Abbreviations for the individual keratin polypeptides are M 59 (mouse epidermal keratin, $M_r = 59{,}000$; Steinert et al. 1983), B VI (bovine epidermal keratin VIb), B VII (bovine epidermal keratin VII), H 50 (human epidermal keratin no. 14, $M_r = 50{,}000$; Hanukoglu and Fuchs 1982), XI 51 (*Xenopus laevis* epidermal keratin, $M_r = 51{,}000$; Hoffmann and Franz 1984), and SW I (sheep wool type I keratin polypeptide 8c-1; Crewther et al. 1980, 1983). Arrowhead points to the end of the α-helical domain. Asterisks represent stop codons. Stretches underscoring individual polypeptides denote sequences characteristic of one of the keratin types identified within the acidic subfamily I (see text). In the M 59 sequence, the canonic heptapeptide GGG$^{S}_{G}$YGG occurs four times, and another two times in the modulated form SGGGYGG (denoted by broken underscoring). Bold letters mark amino acids that are highly conserved.

functionally important is also suggested from the coexistence of different types of the same subfamily in the same cell and from the tissue specificity of their expression. For example, bovine keratins VIa and VIb are expressed together with keratin VII only in keratinocytes of epidermis and a few related mucosae (J.L. Jorcano et al., unpubl.), whereas keratin VII is also expressed in various other tissues, showing that the expression of these two types of acidic keratin polypeptides is independently regulated by different signals and/or mechanisms.

We have recently isolated genomic clones coding for all the keratins expressed in the bovine epidermis. These clones should allow us to study the nature of the regulatory mechanisms involved in the cell-type-specific expression as well as the chromosomal structure and localization of these genes.

ACKNOWLEDGMENTS

We are indebted to Dr. P. Cowin (this institute) for reading and correcting the manuscript and to Dr. A. Suhai (this center) for his advice on using the computer. The work was supported in part by the German Ministry for Research and Technology (ZMBH-Project) and by the Deutsche Forschungsgemeinschaft (grant to W.W.F.).

REFERENCES

Cowan, N.J., P.R. Dobner, E.V. Fuchs, and D.W. Cleveland. 1983. Expression of human α-tubulin genes: Interspecies conservation of 3′ untranslated regions. *Mol. Cell. Biol.* **3:** 1738.

Crewther, W.G. and L.M. Dowling. 1971. The preparation and properties of large peptides from the helical regions of the low-sulfur proteins of wool. *Appl. Polym. Symp.* **18:** 1.

Crewther, W.G., L.M. Dowling, and A.S. Inglis. 1980. Amino acid sequence data from a microfibrillar protein of α-keratin. In *Proceedings of the 6th International Wool Textile Research Conference,* p. 79. Pretoria, South Africa.

Crewther, W.G., L.M. Dowling, P.M. Steinert, and D.A.D. Parry. 1983. Structure of intermediate filaments. *Int. J. Biol. Macromol.* **5:** 267.

Debus, E., K. Weber, and M. Osborn. 1982. Monoclonal cytokeratin antibodies that distinguish simple from stratified squamous epithelia: Characterization on human tissues. *EMBO J.* **1:** 1641.

Eichner, R., P. Bonitz, and T.-T. Sun. 1984. Classification of epidermal keratins according to their immunoreactivity, isoelectric point, and mode of expression. *J. Cell Biol.* **98:** 1388.

Franke, W.W., H. Denk, R. Kalt, and E. Schmid. 1981a. Biochemical and immunological identification of cytokeratin proteins in hepatocytes of mammalian liver tissue. *Exp. Cell Res.* **131:** 299.

Frank, W.W., D.L. Schiller, M. Hatzfeld, and S. Winter. 1983a. Protein complexes of intermediate-sized filaments: Melting of cytokeratin complexes in urea reveals different polypeptide separation characteristics. *Proc. Natl. Acad. Sci.* **80:** 7113.

Franke, W.W., H. Müller, S. Mittnacht, H.-P. Kapprell, and J.L. Jorcano. 1983b. Significance of two desmosome plaque-associated polypeptides of molecular weights 75,000 and 83,000. *EMBO J.* **2:** 2211.

Franke, W.W., S. Winter, C. Grund, E. Schmid, D.L. Schiller, and E.-D. Jarasch. 1981c. Isolation and characterization of desmosome-associated tonofilaments from rat intestinal brush border. *J. Cell Biol.* **90:** 116.

Franke, W.W., D.L. Schiller, M. Hatzfeld, T.M. Magin, J.L. Jorcano, S. Mittnacht, E. Schmid, J.A. Cohlberg, and R.A. Quinlan. 1984. Cytokeratins: Complex formation, biosynthesis and interactions with desmosomes. *Cancer Cells* **1:** 177.

Franke, W.W., D.L. Schiller, R. Moll, S. Winter, E. Schmid, I. Engelbrecht, H. Denk, R. Krepler, and B. Platzer. 1981b. Diversity of cytokeratins: Differentiation specific expression of cytokeratin polypeptides in epithelial cells and tissues. *J. Mol. Biol.* **153:** 933.

Franke, W.W., E. Schmid, D.L. Schiller, S. Winter, E.-D. Jarasch, R. Moll, H. Denk, B.W. Jackson, and K. Illmensee. 1982. Differentiation-related expression of proteins of intermediate-sized filaments in tissues and cultured cells. *Cold Spring Harbor Symp. Quant. Biol.* **46:** 431.

Fuchs, E.V., S.M. Coppock, H. Green, and D.W. Cleveland. 1981. Two distinct classes of keratin genes and their evolutionary significance. *Cell* **27:** 75.

Fuchs, E. and I. Hanukoglu. 1983. Unraveling the structure of the intermediate filaments. *Cell* **34:**332.

Geisler, N. and K. Weber. 1982a. The amino acid sequence of chicken muscle desmin provides a common structural model for intermediate filament proteins. *EMBO J.* **1:** 1649.

———. 1983. Amino acid sequence data on glial fibrillary acidic protein (GFA), implications for the subdivision of intermediate filaments into epithelial and non-epithelial members. *EMBO J.* **2:** 2059.

Geisler, N., E. Kaufmann, and K. Weber. 1982b. Protein chemical characterization of three structurally distinct domains along the protofilament unit of desmin 10 nm filaments. *Cell* **30:** 277.

Gigi, O., B. Geiger, Z. Eshhar, R. Moll, E. Schmid, S. Winter, D.L. Schiller, and W.W. Franke. 1982. Detection of a cytokeratin determinant common to diverse epithelial cells by a broadly cross-reacting monoclonal antibody. *EMBO J.* **1:** 1429.

Gruen, L.C. and E.F. Woods. 1983. Structural studies on the microfibrillar proteins of wool. Interaction between α-helical segments and reassembly of a four-chain structure. *Biochem. J.* **209:** 587.

Hanukoglu, I. and E. Fuchs. 1982. The cDNA sequence of a human epidermal keratin: Divergence of sequence but conservation of structure among intermediate filament proteins. *Cell* **31:** 243.

———. 1983. The cDNA sequence of a type II cytoskeletal keratin reveals constant and variable structural domains among keratins. *Cell* **33:** 915.

Hoffmann, W. and J.K. Franz. 1984. Amino acid sequence of the carboxy-terminal part of an acidic type I cytokeratin of molecular weight 51,000 from *Xenopus laevis* epidermis as predicted from the cDNA sequence. *EMBO J.* **3:** 1301.

Jorcano, J.L., T.M. Magin, and W.W. Franke. 1984a. Cell type-specific expression of bovine keratin genes as demonstrated by the use of complementary cDNA clones. *J. Mol. Biol.* **176:** 21.

Jorcano, J.L., M. Rieger, J.K. Franz, D.L. Schiller, R. Moll, and W.W. Franke. 1984b. Identification of two types of keratin polypeptides within the acidic cytokeratin subfamily I. *J. Mol. Biol.* (in press).

Kim, K.H., J.G. Rheinwald, and E.V. Fuchs. 1983. Tissue specificity of epithelial keratins: Differential expression of mRNAs from two multigene families. *Mol. Cell. Biol.* **3:** 495.

Lazarides, E. 1982. Intermediate filaments: A chemically heterogeneous, developmentally regulated class of proteins. *Annu. Rev. Biochem.* **51:** 219.

Moll, R., W.W. Franke, D.L. Schiller, B. Geiger, and R. Krepler. 1982. The catalog of human cytokeratins: Patterns of expression in normal epithelia, tumors and cultured cells. *Cell* **31:** 11.

Ponte, P., S.-Y. Ng, J. Engel, J. Gunning, and L. Kedes. 1984. Evolutionary conservation in the untranslated regions of actin mRNAs: DNA sequence of a human beta-actin cDNA. *Nucleic Acids Res.* **12:** 1687.

Quax-Jeuken, Y.E.F., W.J. Quax, and H. Bloemendal. 1983. Primary and secondary structure of hamster vimentin predicted from the nucleotide sequence. *Proc. Natl. Acad. Sci.* **80:** 3548.

Quinlan, R.A., J.A. Cohlberg, D.L. Schiller, M. Hatzfeld, and W.W. Franke. 1984. Heterotypic tetramer (A_2D_2) complexes of non-epidermal keratins isolated from cytoskeletons of rat hepatocytes and hepatoma cells. *J. Mol. Biol.* **2:** 365.

Roop, D.R., P. Hawley-Nelson, C.K. Cheng, and S.H. Yuspa. 1983. Keratin gene expression in mouse epidermis and cultured epidermal cells. *Proc. Natl. Acad. Sci.* **80:** 716.

Schiller, D.L., W.W. Franke, and B. Geiger. 1982. A subfamily of relatively large and basic cytokeratin polypeptides as defined by peptide mapping is represented by one or several polypeptides in epithelial cells. *EMBO J.* **1:** 761.

Steinert, P.M. 1978. Structure of the three-chain unit of the bovine epidermal keratin filament. *J. Mol. Biol.* **123:** 49.

Steinert, P.M., R.H. Rice, D.R. Roop, B.L. Trus, and A.C. Steven. 1983. Complete amino acid sequence of a mouse epidermal keratin subunit and implications for the structure of intermediate filaments. *Nature* **302:** 794.

Tseng, S.C.G., M.J. Jarvinen, W.G. Nelson, J.-W. Huang, J. Woodcock-Mitchell, and T.-T. Sun. 1982. Correlation of specific keratins with different types of epithelial differentiation. Monoclonal antibody studies. *Cell* **30:** 361.

Wu, Y.-J. and J.G. Rheinwald. 1981. A new small (40 kd) keratin filament protein made by some cultured human squamous cell carcinomas. *Cell* **25:** 627.

Changes in Keratin Gene Expression during Differentiation

Dennis R. Roop, Rune Toftgard, and Stuart H. Yuspa
Laboratory of Cellular Carcinogenesis and Tumor Promotion, National Cancer Institute National Institutes of Health Bethesda, Maryland 20205

Mark S. Kronenberg and James H. Clark
Department of Cell Biology, Baylor College of Medicine, Houston, Texas 77030

The intermediate filaments found in keratinocytes and other epithelial cells can be formed from approximately 20 different keratin subunits (Lazarides 1980; Steinert 1982). Only a limited number of these subunits are expressed in an individual epithelial cell at any given time. Various factors influence which subunits are expressed, such as cell type (Moll et al. 1982), period of embryonic development (Banks-Schlegel 1982), degree of differentiation (Dale and Stern 1975; Baden and Lee 1978; Skerrow and Hunter 1978; Fuchs and Green 1980), and growth environment of the cell (Steinert and Yuspa 1978; Sun and Green 1978; Doran et al. 1980; Roop et al. 1983). The mechanism that regulates the expression of this family of genes is not known. Recently, we have isolated and characterized cDNA clones that correspond to keratin genes expressed at different differentiation states in epidermal cells (Roop et al. 1983 and in prep.). In an attempt to define the molecular mechanism by which this family of genes is regulated, we have studied the expression of these genes during normal differentiation induced by a steroid hormone and during altered differentiation induced by the tumor promoter, TPA.

Isolation of cDNA Clones for Keratin mRNAs Expressed at Different States of Differentiation

We have isolated and characterized cDNA clones that correspond to the major keratins synthesized in mouse epidermis, the 50-, 55-, 55_b-, 59-, 60-, and 67-kD keratins (Roop et al. 1983 and in prep.). The keratin genes represented by these cDNA clones can, in general, be divided into two subsets on the basis of their expression with respect to the state of differentiation. The 50-, 55_b, and 60-kD keratin genes appear to be expressed mainly in proliferating basal cells within the epidermis, as judged from the transcript levels shown in Figure 1. These genes are expressed at high levels in primary cultures of epidermal cells that proliferate rapidly and grow as a monolayer in medium containing low Ca^{++} (0.02 mM). On the basis of several characteristics, we believe these cells to be analogous to basal cells found in intact epidermis (Yuspa et al. 1980). The reduced concentration of transcripts of these keratin genes in RNA isolated from adult and newborn epidermis (Table 1) presumably reflects the relative contribution of RNA in basal cells (which consist of a single layer) to that of total epidermis (which consists of several cell layers at different stages of differentiation). These results suggest that synthesis of these keratin mRNAs ceases as cells differentiate and leave the basal layer. It is also possible that the stability of these mRNAs is altered in cells committed to differentiate.

The 55-, 59-, and 67-kD keratin genes appear to be preferentially expressed in differentiated epidermal cells due to the absence of transcripts for these genes in proliferating primary cell cultures (Fig. 1). The higher concentration of transcripts for these genes in newborn epidermis as compared with adult epidermis suggests that newborn epidermis contains more differentiated cell layers (Table 1). The hyperplastic nature of newborn mouse epidermis has been documented morphologically and by cell kinetic studies.

Additional data supporting the conclusions based on keratin mRNA levels have been obtained with antisera that are monospecific for individual keratin subunits. Using synthetic peptides corresponding to the carboxyterminal residues of the 55_b-, 59-, 60-, and 67-kD keratins, we have been able to produce antisera that are monospecific for these subunits (Roop et al. 1984a,b). Antisera against the 55_b- and 60-kD keratins stain the basal and suprabasal layers, indicating that these genes are initially expressed in basal cells. In contrast, antisera against the 59- and 67-kD keratins only stain the suprabasal layers, which suggests that these keratin genes are only expressed in differentiated cells.

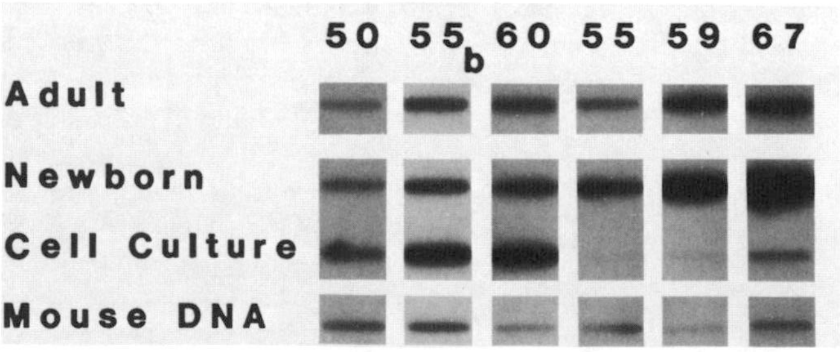

Figure 1
Quantitation of keratin mRNAs synthesized by mouse epidermal cells in vivo and in vitro. Equivalent amounts (as determined by poly[A] content) of total RNA from adult mouse epidermis, newborn mouse epidermis, or primary cultures of mouse epidermal cells were analyzed by the slot-blot technique (D.R. Roop et al., in prep.). The keratin cDNA probes used for this analysis are indicated. Mouse DNA (10 μg) was included on each blot to indicate the relative degree of hybridization of each probe.

Table 1
Relative Concentration of Keratin mRNAs Synthesized by Mouse Epidermal Cells In Vivo and In Vitro

	Peak area (% of control)[a]					
Sample	*50*	*55_b*	*60*	*55*	*59*	*67*
Newborn	142	65	111	1175	690	702
Cell culture[b]	435	244	255	18	6	20

Autoradiogram shown in Fig. 1 was scanned densitometrically, and the area under each peak was calculated.

[a]Data are expressed relative to the keratin mRNA levels observed for control RNA isolated from adult mouse epidermis.

[b]Primary epidermal cells were derived from newborn BALB/c mice and were grown under low Ca^{++} conditions (0.02 mM). These cells are considered to represent proliferating basal cells.

Hormone Induction of Keratin Gene Expression

To study the mechanism by which keratin genes are expressed, it is necessary to have a system in which their expression is easily manipulatable. Rat vaginal epithelium is well suited for this purpose, since the state of differentiation, and thus the degree of keratinization, of this epithelium is hormone-dependent. The expression of both the proliferation-associated and the differentiation-associated keratin genes can be induced in ovariectomized animals after exposure to estradiol. Vaginal epithelium isolated from ovariectomized animals prior to exposure to estradiol is thin and shows no evidence of specialization. The very low level of expression of both subsets of keratin genes in this epithelium is consistent with the low degree of keratinization observed morphologically (D.R. Roop et al., in prep.). There is a dramatic induction of the proliferation-associated genes (the 50-, 55_b-, and 60-kD keratin genes) within 24 hours after exposure to estradiol, and this correlates with the onset of proliferation (Table 2). The decrease in the level of transcripts of these genes observed at 48 hours may result from a dilution effect due to the concentration of differentiated cells present at this time or may indicate that synthesis of these mRNAs ceases in differentiated cells. A decrease in the stability of these mRNAs may also occur in differentiated cells.

Table 2
Hormone Induction of Keratin mRNAs in Rat Vaginal Epithelium

	Peak area (% of control)[a]					
Sample	*50*	*55_b*	*60*	*55*	*59*	*67*
Estradiol (24 hr)	484	647	1573	1460	918	374
Estradiol (48 hr)	158	200	604	8522	1369	1366

Total RNA was isolated from vaginal epithelium obtained from ovariectomized rats and at 24 and 48 hr after exposure to estradiol. The RNA was subjected to slot-blot analysis as described in Fig.1.

[a]Data are expressed relative to the keratin mRNA levels observed for control RNA isolated from vaginal epithelium obtained from ovariectomized rats.

Induction of expression of the differentiation-associated genes (the 55-, 59-, and 67-kD keratin genes) is also observed 24 hours after exposure to estradiol; however, transcripts of these genes continue to accumulate at 48 hours. The high concentration of transcripts of these genes observed at this time correlated with the pronounced stratification and keratinization observed morphologically. The induction of these genes also appears to occur sequentially, in that induction of the 55-kD keratin gene appears to occur first, followed by the 59-kD keratin gene, and finally the 67-kD keratin gene.

These results establish firmly that induction of expression of these keratin genes occurs at the level of transcription, since only very low levels of transcripts were found in the uninduced epithelium. There is also a suggestion that the stability of proliferation-associated keratin mRNAs may be regulated posttranscriptionally at the level of mRNA stability. These results also indicate that expression of these subsets of keratin genes is correlated with the state of differentiation of vaginal epithelial cells, as was observed for epidermal cells. Finally, the induction of the 55-, 59-, an 67-kD keratin genes in terminally differentiating epithelial cells appears to be highly regulated in a coordinate fashion.

Perturbation of Normal Differentiation Affects Keratin Gene Expression

It was of interest to determine whether cells altered in their normal differentiation program would exhibit changes in keratin gene expression. The tumor promoter, TPA, was chosen for this study, since its effects on epidermal differentiation are well documented (Yuspa et al. 1983). A dramatic decrease in the level of the differentiation-associated keratin mRNAs to less than 13% of control epidermis was observed within 12 hours after exposure to TPA (Table 3). The mRNA level of the 67-kD keratin returned to control values within 48 hours. The mRNA level of the 59-kD keratin did not return to control values until 7 days after treatment, whereas the level of the 55-kD keratin mRNA was still only half of the control values at this time. The mRNA level of the 55_b-kD keratin was elevated above control values 12, 24, and 48 hours after treatment. A slight increase in the level of the 50-kD keratin mRNA was also observed at these times.

Table 3
Alteration of Keratin mRNA Levels by the Tumor Promoter, TPA

	Peak area (% of control)[a]					
Sample	*50*	*55_b*	*60*	*55*	*59*	*67*
TPA						
12 hr	140	136	51	13	6	6
24 hr	137	280	99	11	3	13
48 hr	136	295	136	25	61	160
72 hr	138	105	85	28	56	146
7 days	91	70	77	47	91	131

Total RNA was isolated from control adult mouse epidermis and at different time points after treatment with TPA (5 μg at 0 hr). The RNA was subjected to slot-blot analysis as described in Fig. 1.

[a]Data are expressed relative to the keratin mRNA levels observed for control RNA isolated from adult mouse epidermis.

The decrease in mRNAs corresponding to the differentiation-associated keratins probably reflects an acceleration of terminal cell maturation, which is mediated by TPA. Recently, we have been able to show that this decrease occurs within 4 hours after exposure to TPA. This rapid decrease in mRNA levels suggests that a decrease in mRNA synthesis is accompanied by an increase in the rate of degradation of these mRNAs. The recovery of the 67-kD keratin mRNA to control levels prior to that of the 59- and 55-kD keratins represents a reversal in the sequence of appearance that was observed for the induction of these mRNAs in vaginal epithelium. This may indicate that the coordinated program of keratin gene expression observed during normal differentiation is uncoupled when the transit time from the basal to upper differentiated layers is decreased. This may also suggest a prerequisite role for the 67-kD keratin in the maturation process.

The limited effect on mRNA levels of the proliferation-associated keratins, with the exception of the increases noted for the 50- and 55_b-kD keratins, indicates that the initial increase in keratin mRNA instability is restricted to the differentiated cell layers and is consistent with an increase in the rate of cell maturation. The increase in the 50- and 55_b-kD keratin mRNAs may result from expansion of the basal cell compartment due to the hyperplasia that follows TPA treatment.

CONCLUSION

The data presented here indicate that expression of different subsets of keratin genes is highly correlated with the state of differentiation. Changes in the expression of these keratin genes during differentiation appear to occur not only at the level of transcription, but also quite likely at the level of mRNA stability. The requirement for the induction of a new subset of keratin genes in cells committed to terminal differentiation is not understood. However, an explanation may reside in the sequence of the subunits encoded by these genes. The amino acid sequences of keratins expressed in proliferating and terminally differentiating epidermal cells have been deduced from the nucleotide sequence of their corresponding cDNA clones. All of these keratin subunits exhibit common structural features within the central domain that participates in the formation of interchain coiled coils. However, keratins that are expressed at different states of differentiation contain variations in sequences located in the nonhelical terminal domains, which may produce changes in keratin filament properties and function during differentiation (Steinert et al. 1984). We are currently analyzing genomic sequences of these keratin genes to determine whether common regulatory sequences are shared by the genes expressed coordinately during differentiation.

ACKNOWLEDGMENT

We thank Christina K. Cheng for excellent technical assistance.

REFERENCES

Baden, H.P. and L.D. Lee. 1978. Fibrous proteins of human epidermis. *J. Invest. Dermatol.* **71:** 148.

Banks-Schlegel, S.P. 1982. Keratin alterations during embryonic epidermal differentiation: A presage of adult epidermal maturation. *J. Cell Biol.* **93:** 551.

Dale, B.A. and I.B. Stern. 1975. Sodium dodecyl sulfate-polyacrylamide gel electrophoresis of proteins of newborn rat skin. II. Keratohyalin and stratum corneum proteins. *J. Invest. Dermatol.* **65:** 223.

Doran, T.I., A. Vidrich, and T.-T. Sun. 1980. Intrinsic and extrinsic regulation of the differentiation of skin, corneal and esophageal epithelial cells. *Cell* **22:** 17.

Fuchs E. and H. Green. 1980. Changes in keratin gene expression during terminal differentiation of the keratinocyte. *Cell* **19:** 1033.

Lazarides, E. 1980. Intermediate filaments as mechanical integrators of cellular space. *Nature* **283:** 249.

Moll, R., W.W. Franke, D.L. Schiller, B. Geiger, and R. Krepler. 1982. The catalog of human cytokeratins: Patterns of expression in normal epithelia, tumors and cultured cells. *Cell* **31:** 11.

Roop, D.R., P. Hawley-Nelson, C.K. Cheng, and S.H. Yuspa. 1983. Keratin gene expression in mouse epidermis and cultured epidermal cells. *Proc. Natl. Acad. Sci.* **80:** 716.

Roop, D.R., C.K. Cheng, R. Toftgard, J.R. Stanley, P.M. Steinert, and S.H. Yuspa. 1984a. The use of cDNA clones and monospecific antibodies as probes to monitor keratin gene expression. *Ann. N.Y. Acad. Sci.* (in press).

Roop, D.R., C.K. Cheng, L. Titterington, C.A. Meyers, J.R. Stanley, P.M. Steinert, and S.H. Yuspa. 1984b. Synthetic peptides corresponding to keratin subunits elicit highly specific antibodies. *J. Biol. Chem.* **259:** 8037.

Skerrow, D. and I. Hunter. 1978. Protein modifications during keratinizaton of normal and psoriatic human epidermis. *Biochim. Biophys. Acta* **537:** 474.

Steinert, P.M. 1982. Intermediate filaments. In *Electron microscopy of proteins* (ed. J.R. Harris), vol. 1, p. 125. Academic Press, London.

Steinert, P. and S.H. Yuspa. 1978. Biochemical evidence for kertinization by mouse epidermal cells in culture. *Science* **200:** 1491.

Steinert, P.M., W.W. Idler, X.-M. Zhou, L.D. Johnson, D.A.D. Parry, A.C. Steven, and D.R. Roop. 1984. Structural and functional implications of amino acid sequences of keratin intermediate filament subunits. *Ann. N.Y. Acad. Sci.* (in press).

Sun, T.-T. and H. Green. 1978. Keratin filaments of cultured human epidermal cells: Formation of intermolecular disulfide bonds during terminal differentiation. *J. Biol. Chem.* **253:** 2053.

Yuspa, S.H., T. Ben, and H. Hennings. 1983. The induction of epidermal transglutaminase and terminal differentiation by tumor promoters in cultured epidermal cells. *Carcinogenesis* **4:** 1413.

Yuspa, S.H., P. Hawley-Nelson, J.R. Stanley, and H. Hennings. 1980. Epidermal cell culture. *Transplant. Proc.* **12** Suppl. 1: 114.

Regulation of the Expression of Genes Coding for the Intermediate Filament Subunits Vimentin, Desmin, and Glial Fibrillary Acidic Protein

Yassemi G. Capetanaki, John Ngai, and Elias Lazarides
Division of Biology
California Institute of Technology
Pasadena, California 91125

The cytoskeletal architecture of higher eukaryotic cells is composed of three different filamentous networks: microtubules, actin microfilaments, and intermediate filaments. The intermediate filaments comprise a heterogeneous class of filaments that is composed of biochemically and immunologically distinct subunit proteins. These subunits are encoded by a large multigene family subdivided into five major classes, the members of which exhibit, in general, tissue specificity in their expression: Keratins are expressed in epithelial cells, neurofilaments predominantly in neurons, glial filaments (glial fibrillary acidic protein [GFAP]) in astrocytes, desmin in muscle cells, and vimentin mainly in cells of mesenchymal origin, but also in cells of nonmesenchymal origin and in many undifferentiated cells and cultured cells (for review, see Lazarides 1980, 1982).

X-ray diffraction studies have shown that the different intermediate filament subunits are structurally related, since they all contain a considerable amount of α-helix and have a common coiled-coil structure (Steinert et al. 1980). In addition, certain monoclonal antibodies recognize a determinant common to all classes of intermediate filament subunits (Pruss et al. 1981; Dellagi et al. 1982). These observations support the hypothesis that all classes of intermediate filament subunits share conserved areas of amino acid sequence responsible for their conserved structure and areas of divergent structure responsible for functional specializations in different cell types (Lazarides 1981, 1982). Amino-acid-sequence data indeed have revealed high homology between chicken muscle desmin and porcine vimentin of up to 80% in some regions (Geisler and Weber 1981, 1982). Additionally, further sequence data for the porcine 68-kD neurofilament subunit (Geisler et al. 1982), as well as the bovine glial filament subunit (Hong and Dav-

Molecular Biology of the Cytoskeleton. © Cold Spring Harbor Laboratory. 0-87969-174-3/84 $1.00

ison 1981; Geisler and Weber 1983), have shown substantial, albeit lower, homologies with chicken desmin. Much lower homology (53%) has been shown between some keratins and the other intermediate filament subunits (Hanukoglu and Fuchs 1982, 1983).

One of the issues concerning the existence and expression of intermediate filaments is their function. Two fully differentiated and structurally well-defined cell types, chicken skeletal muscle, which expresses both desmin and vimentin (Granger and Lazarides 1978, 1979), and chicken erythrocytes, which express only vimentin (Granger et al. 1982), have been studied in this laboratory to address this issue. Structural and biochemical studies have shown that a potential role of these two intermediate filaments, at least in these two cell types, is to form a transcytoplasmic integrating matrix by linking individual myofibrils laterally at their Z disks and to the sarcolemma in skeletal muscle (Granger and Lazarides 1978, 1979; Lazarides 1980, 1982) and by interconnecting the nucleus with the plasma membrane in erythrocytes (Granger and Lazarides 1982). However, irrespective of the functional attributes of the intermediate filaments, the above studies have clearly established that these filaments are an essential component of the developmental programs of different cell lineages.

To investigate the regulation of the expression, assembly, and morphogenesis of intermediate filaments, we have constructed recombinant clones for different intermediate filament proteins (specifically for vimentin, desmin, and GFAP) and have used them to approach these issues. Here, we summarize our recent studies concerning the expression of vimentin and desmin in different tissues, during avian and mammalian erythropoiesis, and during myogenesis in vitro (Capetanaki et al. 1983, 1984; Ngai et al. 1984). These studies have revealed that expression of intermediate filaments is linked to the terminal differentiation program of certain cell lineages and have allowed us to propose a generalized model for the regulation of intermediate filament expression during terminal differentiation.

Tissue Specificity in the Expression of the Vimentin and Desmin Genes

In general, fully differentiated cells express one class of intermediate filament subunits that typifies that cell type. However, vimentin shows an interesting type of coexpression with some cell-type-specific subunits. It has been found to coexist with desmin in various ratios in different muscle types (Granger and Lazarides 1979; Berner et al. 1981; Mikawa et al. 1981; Osborn et al. 1981; Schmid et al. 1982), with GFAP in certain adult glial cells (Chiu et al. 1981; Schnitzer et al. 1981; Tapscott et al. 1981b; Yen and Fields 1981), neurofilament polypeptides during neuronal differentiation (Tapscott et al. 1981a,b; Dräger 1983) and erythroid differentiation (Granger and Lazarides 1983), and keratins during epithelial differentiation (Franke et al. 1979). With the exception of vimentin, there is no other instance known of coexpression of two or more of the other cell-type-specific subunits. For instance, none of the muscle cells studied to date express keratins, neurofilament subunits, or GFAP with desmin. Similarly, no nerve cell to date has been found to express desmin or keratins together with its cell-type-specific neurofilament subunits. Is the tissue-specific expression of the various intermediate filament subunits regulated at the transcriptional or translational level, and what are the exact mechanisms governing this control? For instance, are the mechanisms utilized in the regulation of vimentin, which shows such generalized expression, the same as those of desmin, whose expression is restricted to muscle cells? Do vimentin and desmin derive from the same gene by differential splicing or are they encoded by different genes? Is the vimentin expressed in muscle cells, nonmuscle cells, and erythrocytes the product of one or more than one gene? Is the smooth- and skeletal-muscle desmin encoded by the same gene and, furthermore, the same mRNA species?

We have approached these issues by using cDNA and genomic clones for the chicken intermediate filament subunits, vimentin and desmin (Capetanaki et al. 1983, 1984). Southern blot analyses using genomic probes for vimentin and cDNA probes for desmin have shown that both these proteins are encoded by single-copy genes in the haploid chicken genome (Fig. 1) (Zehner and Paterson 1983a; Capetanaki et al. 1983, 1984). Under standard conditions of hybridization, the chicken vimentin cDNA and genomic clones do not cross-hybridize with desmin sequences and vice versa. However, a desmin cDNA probe that corresponds to the carboxyterminal half of the second rod domain of the desmin protein exhibited cross-hybridization with sequences specific for other intermediate filament subunits, most notably with GFAP, and less with keratin, even though it did not exhibit any cross-hybridization with chicken vimentin sequences. This finding was somewhat unexpected since amino-acid-sequence comparison between chicken desmin and mammalian vimentin has revealed that these proteins are the most closely related intermediate filament subunits (see below) (Geisler and Weber 1983). Nevertheless, this observation allowed us to isolate GFAP-specific cDNA clones from a chicken spinal-cord library and to show that this intermediate filament subunit is also represented as a single copy in the haploid chicken genome (Fig. 1). These observations directly answered some of the above stated questions: (1) Vimentin and desmin are not products of the same gene; (2) the vimentin expressed in muscle, nonmuscle cells, and erythrocytes derives from one and the same gene; and (3) smooth-muscle and skeletal-muscle desmin are the products of the same gene. However, the above studies could not show whether each of these single genes give rise to more than one protein.

The modes of expression of the single vimentin and desmin genes in the different cell types have been studied by RNA blot analyses (Capetanaki et al. 1983, 1984), and the similarities and differences in the expression of these two genes are demonstrated in Figure 2. First, as shown in Figure 2A, the single vimentin gene is transcribed into two mature mRNA species with approximate lengths of 2.0 kb and 2.3 kb (Capetanaki et al. 1983; Zehner and Paterson 1983a). These two mRNAs exhibit an interesting differential expression; whereas both mRNA species are present in fibroblasts, muscle cells, spinal cord, and lens, erythroid cells from 10- and 15-day-old chicken embryos express predominantly the lower-molecular-weight RNA (Fig. 2A) (Capetanaki et al. 1983). By hybridizing RNA blots to different fragments of the vimentin gene, we have shown that the difference between these two vimentin mRNAs is due to different lengths of their 3′-noncoding regions (Fig. 3) (Capetanaki et al. 1983). Furthermore, sequencing of the 3′-untranslated region of the chicken vimentin gene has revealed two sets of tandemly repeated putative polyadenylation sites, 250 nucleotides apart (Zehner and Paterson 1983a). S1-nuclease mapping experiments have confirmed the uti-

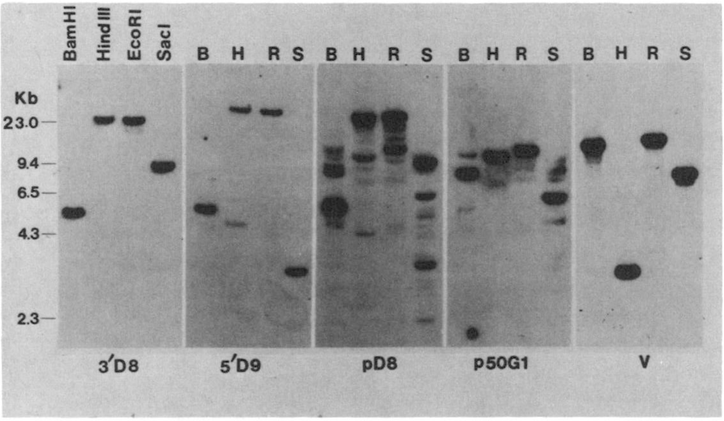

Figure 1
Single gene representation of the nonepithelial intermediate filament subunits vimentin, desmin, and GFAP in the chicken genome. High-molecular-weight chicken DNA, digested with *Bam*HI (B), *Hin*dIII (H), *Eco*RI (R), and *Sac*I (S), was separated on 0.9% agarose gels and transferred to nitrocellulose. The filters were then hybridized with the chicken desmin cDNA pD8 containing conserved sequences, the chicken desmin 5′D8 containing coding sequences specific to desmin and missing the conserved sequences, the chicken desmin 3′D8 containing 3′-noncoding sequences, the chicken GFAP cDNA p50G1 (Capetanaki et al. 1984), and the chicken vimentin genomic clone V (Capetanaki et al. 1984). The desmin probe pD8 cross-hybridizes strongly with GFAP sequences, weaker with keratin sequences, and not visibly with vimentin sequences.

lization of three out of four possible poly(A)-addition sites in chicken embryonic muscle vimentin mRNA (Zehner and Paterson 1983b). On the other hand, preliminary S1-nuclease mapping at the 5′ end of the gene has not revealed any obvious difference between these two RNAs (Y.G. Capetanaki et al., unpubl.), thus suggesting that the difference between the two vimentin mRNAs is due to the use of different polyadenylation sites and that the two mRNAs derive neither by differential splicing nor by usage of different transcription-initiation sites. This implies that muscle, nonmuscle, and erythrocyte vimentin derive from the same coding region of the single vimentin gene and hence represents a single polypeptide species.

It is clear that the differential expression of these two vimentin mRNAs is regulated either by differential termination of transcription using the different polyadenylation sites or by differential posttranscriptional processing of a larger precursor transcript to generate the correct 3′ ends. At present, there is no evidence supporting either of these two possibilities. Continuous transcription beyond the poly(A)-addition sites with subsequent endonucleolytic cleavage of the primary transcript, followed by polyadenylation, has been demonstrated for the mouse β-major globin gene (Hofer and Darnell 1981; Hofer et al. 1982; Salditt-Georgieff and Darnell 1983), as well as for the adenovirus (Nevins and Darnell 1978; Fraser et al. 1979; Nevins et al. 1980) and SV40 late transcription units (Ford and Hsu 1978; Lai et al. 1978). It is therefore possible that for vimentin, a precursor-prod-

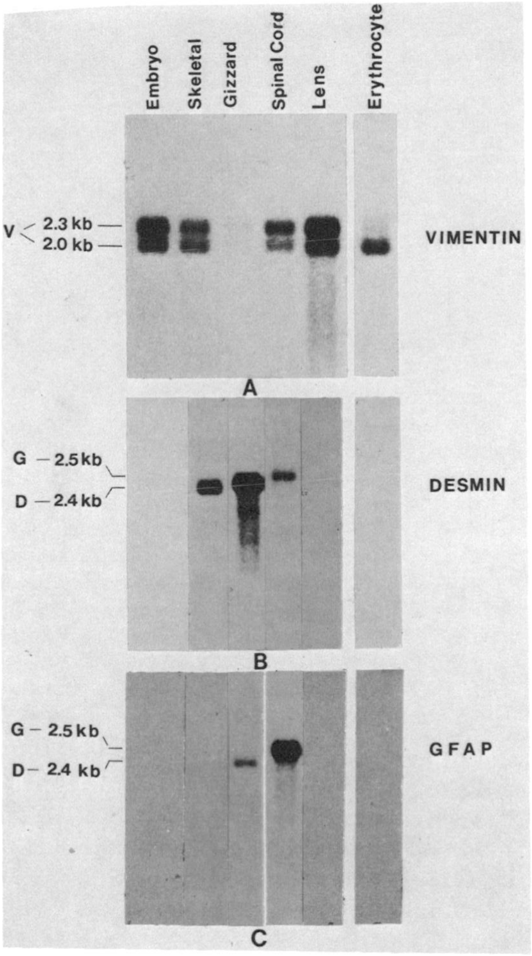

Figure 2
Comparison in the specificity of expression of the vimentin, desmin, and GFAP genes in different tissues. RNA from different chicken tissues or cells was fractionated, transferred to nitrocellulose, and hybridized with vimentin clones (*A*) (Capetanaki et al. 1983), desmin clones (*B*) (Capetanaki et al. 1984), or GFAP clones (*C*) (Capetanaki et al. 1984). The mRNAs used were isolated from 10-day-old embryos, 1-week-old skeletal muscle, 1-week-old gizzard smooth muscle, 2-week-old spinal cord, 1-week-old lens, and 15-day-old embryonic erythrocytes. Note in *A* the differential expression of the 2.0-kb and 2.3-kb vimentin mRNAs in the different tissues; chicken erythrocytes express predominantly the 2.0-kb mRNA species. (D) 2.4-kb desmin mRNA; (G) 2.5-kb GFAP mRNA; (V) 2.0-kb and 2.3-kb vimentin mRNAs.

uct relationship exists in which a primary transcript is either simultaneously cleaved and polyadenylated, giving rise to the 2.3 kb and 2.0 kb mRNAs, or first cleaved to give a 2.3-kb product that is further processed to generate the 2.0-kb mRNA.

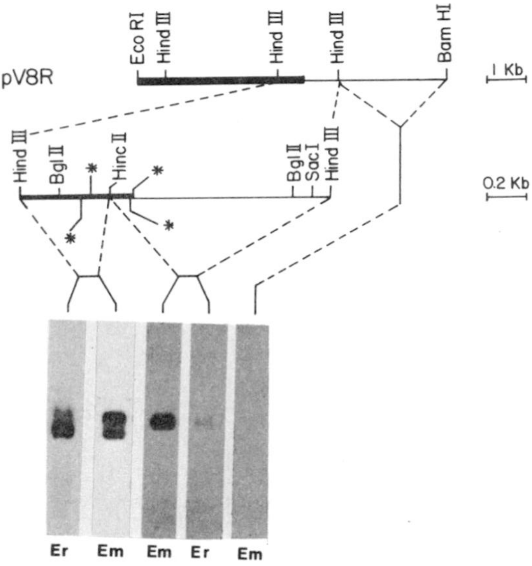

Figure 3
Size difference between the two chicken vimentin mRNAs is due to different lengths of their 3′-untranslated regions. Localization of the 3′-untranslated region responsible for the size difference of the two vimentin mRNAs was achieved by hybridizing blots of mRNA from 10-day-old embryos (Em) and 15-day-old embryonic erythrocytes (Er) with different fragments of the 3′ end of the chicken vimentin gene, including downstream flanking sequences (Capetanaki et al. 1983). The 450-bp *Hin*dIII-*Hin*cII fragment, in the case of the embryonic RNA, hybridizes to both vimentin mRNAs; the 1150-bp *Hin*cII-*Hin*dIII fragment hybridizes only to the 2.3-kb mRNA (Capetanaki et al. 1983). The preferential expression of the 2.0-kb vimentin mRNA in 15-day erythrocytes (Fig. 1) is shown once more here. Heavy line designates the vimentin gene, and the thin line designates the downstream flanking sequences. Asterisks show the four polyadenylation sites derived from the sequence data of the 3′ end of the gene (Zehner and Paterson 1983a).

Since the biological function of 3′-untranslated regions is not known, the significance of variations in their lengths is not clear. Size heterogeneity in the 3′-untranslated regions of mRNAs derived from a single gene has been described for the mouse dihydrofolate reductase gene, which codes for four mRNAs (Setzer et al. 1980), and the genes for mouse β_2-microglobulin (Parnes and Robinson 1983), α-amylase (Tosi et al. 1981), and eel calmodulin (Lagacé et al. 1983), each of which codes for at least two mRNA species. No major tissue-specific difference in the relative distribution of these different mRNAs has been reported to date; the chicken vimentin gene is the first example of its kind. The differential expression of the two size classes of vimentin mRNAs implies that the presence and usage of multiple poly(A)-addition sites in the process of creating different-size 3′-untranslated regions are biologically significant. It is possible that the differences in the 3′-untranslated region somehow affect the subcellular compartmentation of the mRNA, the relative stabilities of the two mRNAs, or both. However, it should be noted here that the vimentin gene in mammals is apparently transcribed into one mRNA species (Dodemont et al. 1982) whose 3′ terminus coincides with a position around the second poly(A)-addition site of the chicken molecule (Quax et al. 1983). Thus, the significance, if any, of the presence and differential usage of two vimentin mRNAs differing only at the 3′-noncoding region must be related to species-specific, as well as tissue-specific, requirements.

Unlike vimentin, the single desmin gene is transcribed into only one mature mRNA species of 2.4 kb in length (Fig. 2B) (Capetanaki et al. 1984). The expression of this mRNA is restricted to muscle cells and is absent from erythrocytes, spinal cord, and lens (the weaker band in spinal cord derives from cross-reaction of the desmin sequences [the conserved region] with GFAP mRNA [Capetanaki et al. 1984], as was the case in the Southern blots; for further details, see below). Comparison of A and B in Figure 2 clearly reveals the large differences in the modes of expression of these two genes. Gizzard, for instance, is the tissue with the highest desmin mRNA and the lowest vimentin mRNA expression. In contrast, erythrocytes, spinal cord, and lens express vimentin, but no desmin, mRNA. Since these patterns of vimentin and desmin mRNA expression underlie similar patterns observed at the protein level, we conclude that both the muscle-specific expression of desmin and the more generalized expression of vimentin are regulated primarily at the transcriptional level or by RNA stabilization and not at the translational level. As in the case of vimentin, it is similarly suggested that the smooth- and skeletal-muscle desmin most probably comprise one and the same protein, since the single gene is transcribed into one mRNA. However, we cannot rule out the possibility that the desmin transcripts are differentially processed in smooth and skeletal muscles to yield distinct mRNAs with indistinguishable electrophoretic mobilities. Structural analysis of the desmin gene and its transcripts will give a definitive answer to this question.

Developmental Expression of the Vimentin and Desmin Genes

Two systems have been used to study the developmental regulation in the expression of vimentin and desmin genes: erythropoiesis (Capetanaki et al. 1983; Ngai et al. 1984) and in vitro myogenesis (Capetanaki et al. 1983, 1984).

Expression of the Vimentin Gene during Avian and Mammalian Erythropoiesis

Chicken erythrocytes have a cytoplasmic network of intermediate filaments (Virtanen et al. 1979; Woodcock 1980) composed predominantly of vimentin (Granger et al. 1982). As mentioned above, the desmin gene is not expressed in these cells. However, these cells express low amounts of the 70K neurofilament subunit early in development, and its expression gradually declines by the time adulthood is reached (Granger and Lazarides 1983). We have studied the expression of the vimentin gene during erythroid development, and, as described above, we have shown first that erythroid cells from 10- and 15-day-old embryos express predominantly the 2.0-kb vimentin mRNA, whereas erythroid cells from 4-day-old embryos express both mRNA species (Fig. 4A) (Capetanaki et al. 1983). Furthermore, we have shown that there is a remarkable induction

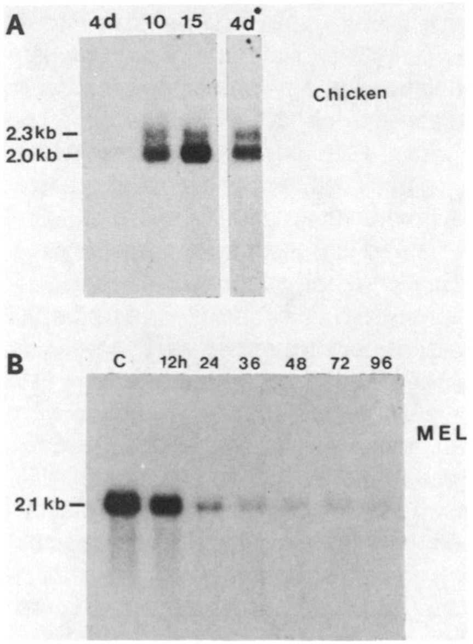

Figure 4
Developmental regulation of vimentin during erythropoiesis. (*A*) Induction in the expression of the chicken vimentin gene during erythroid development (Capetanaki et al. 1983). RNA blots were performed as described in Fig. 2, using 4-day (4d), 10-day (10d), and 15-day (15d) embryonic erythrocytes, as designated. 4d* is a sevenfold long exposure of lane 4d. (*B*) Repression of the mouse vimentin gene during MEL cell differentiation. RNA blots from uninduced cells (c) as well as cells cultured in 1.8% DMSO for 12, 24, 36, 48, 72, and 96 hr, as designated, were hybridized with chicken vimentin cDNA probe (Ngai et al. 1984).

(40–50-fold) in the accumulation of the 2.0-kb vimentin mRNA between the 4th and 15th day of embryogenesis (Fig. 4A) (Capetanaki et al. 1983). Circulating blood from 4-day-old embryos consists of 100% primitive erythroblasts; on the other hand, blood from 10-day-old embryos contains ~25% primitive cells, 35% definitive erythroblasts, and 35% mature erythrocytes, whereas in 15-day-old embryos, the mature erythrocytes comprise ~70% of the population and the remainder (~30%) are definitive erythroblasts (Bruns and Ingram 1973). Consequently, the accumulation of the 2.0-kb vimentin mRNA between the 4th and 15th day of erythroid development should be considered as lineage-specific. The tremendous induction in the expression of the chicken vimentin gene during erythropoiesis underlies similar expression changes at the protein level (I. Blikstad and E. Lazarides, unpubl.). Therefore, the accumulation of vimentin filaments during the switch of the organism from the production of the primitive series of erythroid cells to the production of the definitive series of cells is regulated by the abundance of the vimentin mRNA. This large induction in the synthesis of vimentin in the definitive series of cells suggests that vimentin filaments are an essential part of the terminal differentiation program of these cells.

One main difference between mammalian and chicken erythrocytes is that the former become anucleate during terminal differentiation, whereas the latter retain their nuclei. If the function of vimentin filaments in erythrocytes is related to the formation of a transcytoplasmic network that anchors the nucleus (Granger and Lazarides 1982), the expression of this subunit should show different developmental patterns in mammals and in chickens. Indeed, it has been shown (Dellagi et al. 1983) that vimentin is lost during human erythropoiesis in vivo, in contrast to what is observed in chickens (Granger et al. 1982). To study further the regulation of vimentin expression during mammalian erythropoiesis, we have used differentiating Friend murine erythroleukemia (MEL) cells, which provide a well-defined culture system to study the early events in erythroid differentiation (for review, see Marks and Rifkind 1978; Friend 1980). In vitro differentiation of MEL cells after chemical induction is characterized by a series of events similar to those occurring in mammalian erythropoiesis in vivo (Friend et al. 1971; Ross et al. 1972; Ebert and Ikawa 1974; Orkin et al. 1975; Arndt-Jovin et al. 1976; Gusella et al. 1976; Sassa 1976; Eisen et al. 1977a,b; Friedman and Schildkraut 1977; Hu et al. 1977; Nudel et al. 1977; MacDonald et al. 1978; Mager and Bernstein 1978; Sabban et al. 1980; Smith et al. 1982), and therefore has provided a useful system to study different events in mammalian erythropoiesis. Immunofluorescence microscopy has shown that, in contrast to what has been demonstrated during chicken erythroid development, vimentin filaments disappear from dimethylsulfoxide (DMSO)-treated MEL cells due to a rapid decrease in vimentin synthesis (Ngai et al. 1984). This reduction in vimentin synthesis is caused by a rapid and striking decrease in the abundance of vimentin mRNA (Fig. 4B) (Ngai et al. 1984). During DMSO-induced differentiation, the level of the vimentin mRNA decreases 30% by 12 hours, 70% by 24 hours, 90% by 48 hours, and ~96% by 96 hours. These data indicate that vimentin expression in differentiating MEL cells is also regulated at the mRNA level. We have further shown that hexamethylene-bis-acetamide (HMBA), another potent inducer of MEL cell terminal differentiation (Reuben et al. 1976), like DMSO, also causes a rapid reduction in vimentin mRNA, whereas hemin, which causes an increase in the globin mRNA abundance (Nudel et al. 1977; Ross and Sautner 1976; Gusella et al. 1980) but does not induce terminal differentiation (Gusella et al. 1980), causes only a twofold reduction in vimentin mRNA (Ngai et al. 1984). All of these data suggest that the repression of vimentin gene expression is associated with MEL cell terminal differentiation. However, the decrease in vimentin mRNA levels during MEL differentiation precedes the major accumulation of β-globin mRNA (Ngai et al. 1984). Since globin accumulation generally either coincides with or slightly precedes the onset of commitment (Gusella et al. 1976), this repression of the vimentin gene appears to precede commitment to terminal differentiation, although it may be associated with this event.

The opposite modes of vimentin expression in chicken and mammalian erythrocytes strongly suggest that vimentin filaments indeed may play an essential role in the terminal differentiation of erythrocytes, either by functioning to maintain the position of the nucleus in the chicken erythrocytes or, in contrast, by facilitating, with their disappearance, the enucleation during mammalian erythropoiesis. It has been found that the disappearance of vimentin from mammalian erythroblasts in vivo precedes enucleation and does not appear to be a consequence of the expulsion of the nucleus (Dellagi et al. 1983), thus supporting the idea of facilitation of enucleation mentioned above.

Developmental Regulation of the Desmin Gene during Myogenesis In Vitro

Studies using primary myogenic cultures have shown that vimentin is expressed throughout myogenesis (Gard and Lazarides 1980), whereas the synthesis of desmin is initiated upon the onset of fusion (Bennett et al. 1979; Gard and Lazarides 1980). We therefore wished to determine how the constitutive expression of vimentin and the developmental expression of desmin are related and how they are regulated. Is this noncoordinate expression the result of transcriptional or translational control?

Early in differentiation, both vimentin and desmin exhibit indistinguishable cytoplasmic distributions in the form of a filamentous network. Later in differentiation, however, after the assembly of contractile sarcomeres, both proteins show a dramatic redistribution and association with the Z disks (Bennett et al. 1979; Gard and Lazarides 1980), where they surround and link them laterally to each other and to other cytoplasmic organelles (Granger and Lazarides 1978, 1979; Gard and Lazarides 1980). Until recently, it was not known if the cytoplasmic and sarcomeric forms of vimentin or desmin are the products of the same or different genes. The regulation of this redistribution of vimentin and desmin was also unknown. It could be regulated by the expression of Z-disk-specific vimentin or desmin genes, or posttranslationally by specific chemical modifications, and/or by the appearance of specific Z-disk receptors. To probe all of these aspects, we have studied the patterns of expressions of vimentin and desmin genes during myogenesis, and we have shown first that both vimentin mRNAs are constitutively expressed throughout myogenesis (Fig. 5A) (Capetanaki et al. 1983), whereas desmin mRNA appears upon the onset of fusion (by 24 hours after plating) (Capetanaki et al. 1984). Maximal levels of desmin mRNA are observed within 3–4 days after plating, and they are maintained at fairly constant levels through 8 days of culture, when redistribution of desmin to the peripheries of the Z disks takes place (Gard and Lazarides 1980). Thus, the constitutive expression of vimentin filaments and the developmentally regulated expression of desmin filaments reflect similar expression patterns of the respective mRNAs, thus indicating that they are both regulated predominantly by mRNA abundance and not at the translational level. The induction in the accumulation of desmin mRNA during myogenic differentiation is analogous to and concurrent with the accumulation of a number of other mRNAs coding for muscle-specific proteins (Devlin and Emerson 1979; Moss and Schwartz 1981; Shani et al. 1981). Furthermore, the vimentin and desmin genes are expressed noncoordinately during myogenesis, although the levels of expression of the respective proteins appear to be regulated at the mRNA level.

Although the presence of vimentin and desmin appears to be determined by mRNA abundance, the morphogenetic redistribution of the vimentin and desmin filaments from the cytoplasmic to the sarcomeric form during myogenesis in tissue culture does not seem to be regulated either at the DNA level or at the RNA level. Both vimentin (Capetanaki et al. 1983; Zehner and Paterson 1983a) and desmin (Capetanaki et al. 1984) are represented as single copies in the haploid chicken genome, thus excluding the possibility of any Z-disk-specific vimentin or desmin gene. Furthermore, in the case of vimentin, the two mRNA species differ only in their 3′-untranslated regions, and during myogenesis, both RNAs are expressed constitutively. Similarly, the levels and size of the single desmin mRNA do not change shortly before, during, or after the transition of desmin filaments to the peripheries of the Z disks (Fig. 5). On the other hand, both vimentin and desmin show susceptibility to changes in phosphorylation by cAMP-dependent protein kinases at a time that coincides with the onset of their redistribution to the Z line (Gard and Lazarides 1982a,b). In addition, the association of desmin and vimentin with Z disks at the time of this redistribution can be inhibited re-

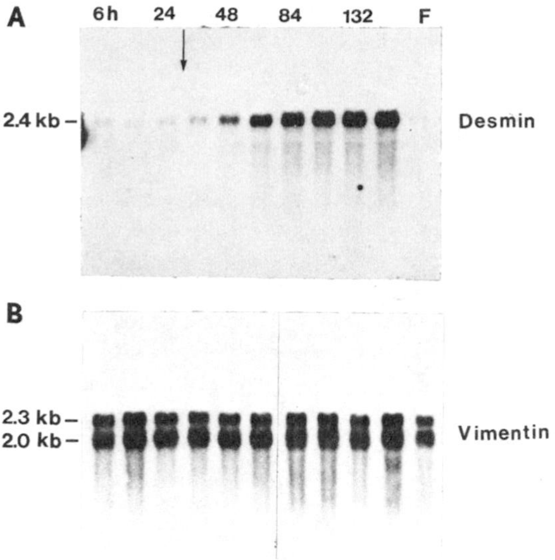

Figure 5
Developmentally regulated expression of the desmin gene versus constitutive expression of the vimentin gene during myogenesis in vitro. RNA blots prepared using total RNA from cultured myogenic cells 6, 12, 24, 36, 48, 60, 84, 108, 132, and 192 hr after plating were hybridized to either vimentin (*B*) (Capetanaki et al. 1983) or desmin-specific probes (*A*) (Capetanaki et al. 1984). (F) RNA from fibroblasts.

versibly by exposing myogenic cultures to cAMP analogs (Lazarides and Gard 1982). All of these observations suggest that the redistribution of desmin and vimentin filaments during the terminal phases of myofibril assembly is regulated primarily posttranslationally.

Expession of the GFAP Gene: Comparison with the Desmin and Vimentin Genes

Genomic DNA blot analysis using a GFAP cDNA probe has also shown that chicken GFAP is encoded by a single gene (Fig. 1) that is transcribed into one mature mRNA with a length of 2.5 kb (Fig. 2C) (Capetanaki et al. 1984). The noncoding region of this mRNA is the longest one of the different intermediate filament protein genes reported to date, since this polypeptide has the lowest molecular weight among the nonkeratin intermediate filament subunits. In particular, when vimentin (2.0 kb and 2.3 kb), desmin (2.4 kb), and GFAP (2.5 kb) mRNAs are compared, the lengths of the noncoding regions appear to be inversely proportional to the lengths of the coding regions. Lewis et al. (1984) have reported the possible existence of two GFAP genes in mice and chickens. Given our results and the present sequence comparison data (see below; Fig. 6), it is possible that GFAP is represented as a single gene in the mouse genome as well and that the extra band observed is due to cross-hybridization with another subunit, most probably desmin. RNA blots similar to those performed with vimentin

```
D   CGACACCTGAAGGATGAGATGGCCCGGCACCTGCGCGAGTACCAGGACCTGCTCAATGTCAAGATGGCCTTGGACGTGGAGATCGCCACCTACCGCAAGCTGCTGGAGGGCGAGGAGAAC 120
V   CAGAACATGAAGGAAGAGATGGCTCGTCACCTTCGTGAATACCAAGACCTGCTCAATGTCAAGATGGCTCTTGACATTGAGATAGCCACCTACAGGAAGCTACTGGAAGGCGAGGAGAGC
G   CAAAGCCTCAAGGAGGAGATGGCCCGCCACCTGCAGGAGTACCAGGATCTACTCAACGTTAAGCTAGCCCTGGACATCGAGATCGCCACCTACAGGAAATTGCTGGAGGGCGAAGAAAAC
K   CAGAAGGCCAAGCAGGACCTGGCCCGGCTGCTGAAGGAGTACCAGGAGCTGATGAATGTCAAGCTGGCCCTGGACGTGGAGATCGCCACCTACCGCAAGCTGCTGGAGGGTGAGGAGTGC

D   CGGATCAGCATCCCCATGCACCAGACCTTTGCCTCTGCTCTCAATTTCCGAGAGACCAGCCCAGA-CCAGCGTGGCTCCGAGGTGCACAC-CAAGAAG-ACAGTGATGATCAAAACCATC 240
V   AGGATTTCTCTGCCTCTTC-CCA-A-CTTT---TCTTCCCTGAACCTGAGAGAAACTAATCTGGAGTCACTCCCTCTGGTTGACACCCACTCAAAAAGAACACTCCTGATTAAGACAGTG
G   CGCATCACCAT-TCC-TGTA-CAGA-CTTT--CTCCAACCTCGAGATCCGAGAAACCAGCCTGGACACCAAATCCGTGTCAGAAGGCCACCTCAAGAGGAACATCGTGGTAAAGACTGTG
K   AGGCTGAATGGCGAAGGCGTTGGACAAGTCAACATCTCTGTGGTGCAGTCCACCGTCTCCAGTGGCTATGGCGGTGCCAGTGGTGTCGGCAGTGGCTTAGGCCTGGGTGGAGGAAGCAGC

                                                          |->
D   GAAACTCGTCATGGAGAGGTGGTGAGCGAGGCGACCCAGCAGCAGCACGAGGTGCTGTAGAAGACTCTGTCCCGGTGCCCAGCCGCCCTTCTGCCCCCCCCAAGCCCGTACTCCCCCCGA 360
V   GAAACTAGGGATGGACAGGTGATCAATGAAACCTCTCAGCATCATGATCACCTTGAATAAAAATTGCACATACTCTGTGCAACAACGCAGTCCAGCAAGAAGAAAAAAAAGAAATTCGTA
G   GAGATGCGGGATGGTGAGGTCATTAAGGACTCGAAGCAGGAGCACAAGCACGTGGTGATGTGAGGTGTGCCACCTGGTGGCCCTTGCCATGCAGTGTGAGGGCCCAAAGCTTAGCCTCAA
K   TACTCCTATGGCAGTGGTCTTGGCGTTGGAGGTGGCTTCAGTTCCAGCAGTGGCAGAGCCATTGGGGGTGGCCTCAGCTCTGTTGGAGGCGGCAGTTCCACCATCAAGTACACCACCACC

D   CACGGCCCCCTCCGGGCACGGCCCGCTCCCCCCGCACCCCCGGCTCTCTCCTTCGGCTTCGCAGGCTTCGCGCTCAGCGCAAGACCAGGCCGGGCCGGGGGCGGCAGAGAGGACGGC 480
V   TCTTAAGGAAACAGCTTTCAAGTGCCTTTACTGCAGTTTTTCAGGAGCGCAAGTAAGATTTGGATAGAAATAAGCTCTAGTTTCTAACAACTGACACCCTAAAAGATTTAGAAAAGCTT
G   ATAGGCCTGTTTGCCAGGCTCAGTTCCCACCCACAGCAGCACTTCCCTTCCTTCCCTGGATTTCTGCCTGTGTGCTGCCCAAGGCGTCAATCAGGTCATAAGCTTCATATATGGCATAGA
K   TCCTCCTCCAGCAGGAAGAGCTATAAGCACTAAAGTGCGTCTGCTAGCTCTCGGTCCCACAGTCCTCAGGCCCCTCTCTGGCTGCAGAGCCCTCTCCTCAGGTTGCCTGTCCTCTCCTGG

D   GGGGACAGGCGGCAAGG
V   TACAACACAATCTAGTT
G   TGGCATATACCCTTCAC
K   CCTCCAGTCTCCCC
```

Figure 6 (*See facing page for legend.*)

and desmin reveal that the expression of the GFAP mRNA is restricted to nervous tissues (Fig. 2C), thus indicating that its tissue-specific expression is also regulated at the mRNA level. By analogy to the conclusions regarding vimentin, desmin, and GFAP expression, the induction in the expression of the different neurofilament subunits during neuronal differentiation may also be regulated primarily at the same level, thus indicating that transcriptional regulation and/or RNA stabilization are the predominant levels of control in the expression of the different intermediate filament subunits.

Highly Conserved and Variable Nucleotide Sequences between Desmin and Other Intermediate Filament Proteins

Amino-acid-sequence comparisons among the different intermediate filament subunits have shown that these proteins possess a conserved rod domain, with head and tail pieces exhibiting striking divergence (Geisler and Weber 1982; Hanukoglu and Fuchs 1983). Desmin and vimentin are the most closely related pair of the nonepithelial (non-keratin) subunits, exhibiting a 64% homology over the last 141 amino acids of the carboxyl terminus (Geisler and Weber 1981). It is important to note that these studies compared chicken or mammalian desmin with mammalian vimentin, but not chicken vimentin. The exact amino acid sequence of chicken vimentin is not known; however, hybridization studies with chicken vimentin probes have not revealed any cross-reaction with other intermediate filament nucleotide sequences (Fig. 1) (Capetanaki et al. 1983, 1984). On the other hand, similar DNA blot analyses using desmin cDNA probes that contain regions coding for a conserved amino acid stretch corresponding to the last 140 amino acids of the rod domain of the protein (Geisler and Weber 1983) revealed multiple band patterns (Fig. 1). We ascribe this to cross-hybridization with other intermediate filament genes, particularly with GFAP and to a lesser extent with type II keratin genes, but not with vimentin. Furthermore, similar data were obtained by RNA blot analyses (Fig. 2) which showed that vimentin probes do not cross-react with other intermediate filament mRNAs, whereas desmin probes cross-react only with GFAP mRNA from spinal cord and GFAP probes cross-react with desmin mRNA from gizzard (Fig. 2). To explain and confirm these observations further, we compared the chicken desmin nucleotide sequence coding for the conserved region described above (Capetanaki et al. 1984) with that available for mouse GFAP (Lewis et al. 1984), hamster vimentin (Quax-Jeuken et al. 1983), and human type II keratin (Hanukoglu and Fuchs 1983) cDNAs. Figure 6 shows a striking 80% homology in the first 120 nucleotides between

Figure 6
Highly conserved and variable nucleotide sequences between different intermediate filament genes. A part of the chicken desmin cDNA sequence (D) (Capetanaki et al. 1984) is compared with the corresponding sequences from hamster vimentin (V) (Quax-Jeuken et al. 1983), mouse GFAP (G) (Lewis et al. 1984), and the human epidermal type II keratin KA-1 (K) (Hanukoglu and Fuchs 1983). DNA sequences homologous to desmin are underscored. The highly conserved sequences (80%) in the first 120 nucleotides correspond to the end of the rod domain of the protein. The variable region starts then with the sequence coding for the carboxyterminal tail and continues to the completely heterologous 3′-untranslated region. Arrow indicates the beginning of the noncoding region.

desmin and the other three sequences. This sequence corresponds to the carboxyl terminus of coil II of the rod domain of each protein (Geisler and Weber 1983), and, together with the start of coil Ia, the end of coil Ib and the amino terminus of coil II comprise the most conserved sequences among all intermediate filament proteins (Geisler and Weber 1983), suggesting that these sequences probably are of great structural importance. Downstream from this region, the homology decreases and there is no homology between the different subunits at the 3′-noncoding regions of their mRNAs. RNA (Fig. 2) and DNA (Fig. 1) blot analyses revealed that at least the 120-nucleotide sequence of the chicken desmin gene shown in Figure 5 shows a high nucleotide sequence homology with chicken GFAP and lesser homology with keratins, in contrast to the apparent divergence of the chicken vimentin gene. This finding is of great interest because, as previously mentioned, amino-acid-sequence comparison between chicken desmin and mammalian vimentin has revealed that among all intermediate filaments, these two subunits show the highest homology. These observations, together with the high nucleotide sequence homology observed between chicken desmin and mammalian vimentin, GFAP, and type II keratin (Fig. 6), suggest that this particular region of the vimentin gene is much less homologous to the other chicken intermediate filament subunits than to the other mammalian subunits studied. Apparently, a selective pressure(s) has operated through evolution either to cause within this region a greater divergence of vimentin sequences in chickens from other subunits or to conserve these sequences in mammals. Finally, it is interesting that the human type II keratin nucleotide sequence at the region examined is more homologous to the other intermediate filament subunits (Fig. 6) than to the human type I keratin (Hanukoglu and Fuchs 1982).

CONCLUSIONS

From the data reviewed here, we conclude that the genes for the intermediate filament subunits, vimentin, desmin, and GFAP, are represented as single copies in the haploid chicken genome (Capetanaki et al. 1983, 1984; Zehner and Paterson 1983a). Surprisingly, the chicken desmin gene shows much higher homology with the GFAP gene than with the vimentin gene (Capetanaki et al. 1984), contrary to what was expected from the high amino-acid-sequence homology observed between the chicken desmin and mammalian vimentin (Geisler and Weber 1981, 1983). The tissue-specific expression of desmin (muscle cells) and GFAP (nervous system) and the more generalized expression of vimentin reflect similar qualitative and quantitative patterns at the mRNA level (Capetanaki et al. 1983, 1984). The dramatic induction of the vimentin mRNA during chicken erythroid development (Capetanaki et al. 1983) parallels similar changes at the protein level (I. Blikstad and E. Lazarides, unpubl.). Furthermore, the disappearance of vimentin filaments during in vitro DMSO-induced MEL cell differentiation correlates with the dramatic decrease in the vimentin mRNA (Ngai et al. 1984). Similarly, the constitutive expression of vimentin and the developmental expression of desmin mRNA during myogenesis in vitro (Capetanaki et al. 1983, 1984) underlie similar patterns in the accumulation of vimentin and desmin filaments (Gard and Lazarides 1980). Hence, the expression of vimentin and desmin as well as of GFAP filaments is regulated transcriptionally or posttranscriptionally, and not at the level of translation. The differential expression of keratins during terminal differentiation of epithelial cells also appears to be regulated at the mRNA level (Fuchs and Green 1980; Jorcano et al. 1984). By analogy, the induction in the expression of neurofilament subunits during neuronal differentiation may also be regulated primarily at this level; we thus conclude that, in general, the expression of the intermediate filament proteins is regulated mainly at the transcriptional level, assuming that differential RNA stabilization (e.g., through differentiation of a lineage) does not play a significant role in changing intermediate filament mRNA abundances. In this regard, it should be mentioned that a single exception involving translational regulation has been reported in which a 67K keratin-coding mRNA is masked and not translated in mouse squamous cell carcinomas (Winter and Schweizer 1983).

Reconstitution experiments have shown that denatured vimentin and desmin molecules can copolymerize in vitro to form heteropolymers (Steinert et al. 1981). Furthermore, chemical cross-linking studies using vascular smooth-muscle tissue and BHK cells have shown that vimentin and desmin can be present in heteropolymer filaments in vivo (Quinlan and Franke 1982). Similar studies have demonstrated heteropolymers of vimentin and GFAP in cultured human glioma cells (Quinlan and Franke 1983). In chicken erythroid cells, the assembly of vimentin filaments takes place rapidly and posttranslationally from a saturable pool of soluble vimentin (Blikstad and Lazarides 1983; Moon and Lazarides 1983). This assembly is regulated by the availability of the soluble subunit pool and appears not to be limited or regulated posttranslationally (Blikstad and Lazarides 1983; Moon and Lazarides 1983). Antibody decoration of the chicken erythrocyte intermediate filament has suggested that vimentin and the 70K neurofilament subunit randomly polymerize into intermediate filaments; the modulation of the ratio of these two subunits is a function of their synthesis (Granger and Lazarides 1983). Similarly, a punctate distribution of desmin in intermediate filaments has been demonstrated in fibroblasts (Ip et al. 1983), which contain mainly vimentin and very low amounts of desmin (Gard et al. 1979). All of the above results suggest that the formation of intermediate filaments is determined by the availability of soluble subunits and takes place randomly by rapid self-assembly from small soluble pools of each subunit. Furthermore, the subunit composition of the intermediate filaments must be regulated by the mRNA abundance of the corresponding intermediate filament subunits, since we concluded above that the expression of the intermediate filament subunits is regulated at the transcriptional or posttranscriptional level, and not at the translational level. However, the morphogenetic redistribution of vimentin and desmin filaments during myogenesis, and perhaps the subcellular localization of intermediate filaments in other systems, appears to be facilitated posttranslationally by chemical modifications (e.g., phosphorylation; Gard and Lazarides 1982a,b) or by de novo availability of specific binding sites. These conclusions are presented in the form of a generalized model for the regulation of intermediate filament expression during differentiation in Figure 7.

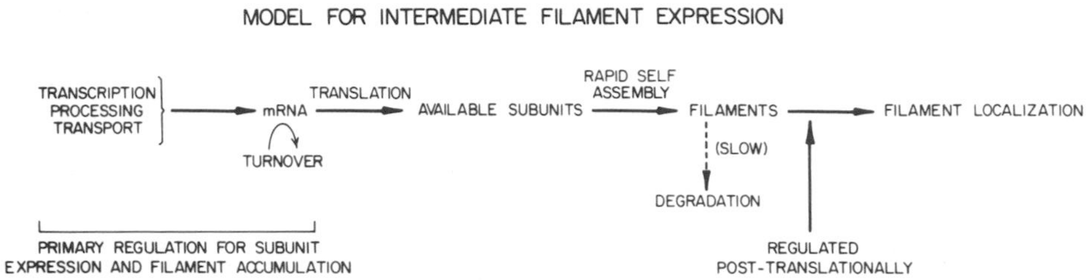

Figure 7
General model for the regulation of intermediate filament expression. The availability of the intermediate filament subunits is regulated by the abundance of their mRNA. The levels of soluble subunits determine their rapid self-assembly. The intracellular localization of filaments may be facilitated by posttranslational modification of the subunits themselves or by the de novo availability of specific protein-binding sites (see Conclusions).

ACKNOWLEDGMENTS

We thank Dr. John Cox for his comments on the manuscript. This work was supported by grants from the National Institutes of Health, the National Science Foundation, the Muscular Dystrophy Association of America, and a grant-in-aid from the American Heart Association (Greater Los Angeles Affiliate). Y.G.C. is a Lievre fellow of the American Cancer Society (California Division) and was also supported by a Muscular Dystrophy Association postdoctoral fellowship. J.N. was supported by a Gordon Ross Foundation predoctoral fellowship. E.L. is the recipient of a Research Career Development Award from the National Institutes of Health.

REFERENCES

Arndt-Jovin, D.J., W. Ostertag, H. Eisen, F. Klimek, and T.M. Jovin. 1976. Studies of cellular differentiation by automated cell separation. Two model systems: Friend virus-transformed cells and *Hydra attenuata. J. Histochem. Cytochem.* **24:** 332.

Bennett, G.S., S.A. Fellini, Y. Toyama, and H. Holtzer. 1979. Redistribution of intermediate filament subunits during skeletal myogenesis and maturation *in vitro. J. Cell Biol.* **82:** 577.

Berner, P.F., E. Frank, H. Holtzer, and A.P. Somlyo. 1981. The intermediate filament proteins of rabbit vascular smooth muscle: Immunofluorescent studies of desmin and vimentin. *J. Muscle Res. Cell Motil.* **2:** 439.

Blikstad, I. and E. Lazarides. 1983. Vimentin filaments are assembled from a soluble precursor in avian erythroid cells. *J. Cell Biol.* **96:** 1803.

Bruns, G.A.P. and V.M. Ingram. 1973. The erythroid cells and haemoglobins of the chick embryo. *Philos. Trans. R. Soc. Lond. B* **266:** 225.

Capetanaki, Y.G., J. Ngai, and E. Lazarides. 1984. Characterization and regulation in the expression of a single gene coding for the intermediate filament protein desmin. *Proc. Natl. Acad. Sci.* **81:** (in press).

Capetanaki, Y.G., J. Ngai, C.N. Flytzanis, and E. Lazarides. 1983. Tissue-specific expression of two mRNA species transcribed from a single vimentin gene. *Cell* **35:** 411.

Chiu, F.-C., W.T. Norton, and K.L. Fields. 1981. The cytoskeleton of primary astrocytes in culture contains actin, glial fibrillary acidic protein, and the fibroblast-type filament protein, vimentin. *J. Neurochem.* **37:** 147.

Dellagi, K., J.C. Brouet, J. Perreau, and D. Paulin. 1982. Human monoclonal IgM with autoantibody activity against intermediate filaments. *Proc. Natl. Acad. Sci.* **79:** 446.

Dellagi, K., W. Vainchenker, G. Vinci, D. Paulin, and J.C. Brouet. 1983. Alteration of vimentin intermediate filament expression during differentiation of human hemopoietic cells. *EMBO J.* **2:** 1509.

Devlin, R.B. and C.P. Emerson. 1979. Coordinate accumulation of contractile protein mRNAs during myoblast differentiation. *Dev. Biol.* **69:** 202.

Dodemont, H.J., P. Soriano, W.J. Quax, F. Ramaekers, J.A. Lenstra, M.A.M. Groenen, G. Bernardi, and H. Bloemendal. 1982. The genes coding for the cytoskeletal proteins actin and vimentin in warm-blooded vertebrates. *EMBO J.* **1:** 167.

Dräger, U.C. 1983. Coexistence of neurofilaments and vimentin in a neurone of adult mouse retina. *Nature* **303:** 169.

Ebert, P.S. and Y. Ikawa. 1974. Induction of δ-aminoleuolinic acid synthetase during erythroid differentiation in cultured leukemia cells. *Proc. Soc. Exp. Biol. Med.* **146:** 601.

Eisen, H., R. Bach, and R. Emery. 1977a. Induction of spectrin in erythroleukemic cells transformed by Friend virus. *Proc. Natl. Acad. Sci.* **74:** 3898.

Eisen, H., S. Nasi, C.P. Georgopoulos, D. Arndt-Jovin, and W. Ostertag. 1977b. Surface changes in differentiating Friend erythroleukemic cells in culture. *Cell* **10:** 689.

Ford, J. and M.-T. Hsu. 1978. Transcription pattern of *in vivo* labeled later simian virus 40 RNA: Equimolar transcription beyond the mRNA 3′ terminus. *J. Virol.* **28:** 795.

Franke, W.W., E. Schmid, D. Breitkreutz, M. Luder, P. Boukamp, N.E. Fusenig, M. Osborn, and K. Weber. 1979. Simultaneous expression of two different types of intermediate sized filaments in mouse keratinocytes proliferating *in vitro. Differentiation* **14:** 35.

Fraser, N.W., J.R. Nevins, E. Ziff, and J.E. Darnell, Jr. 1979. The major late adenovirus type-Z transcription unit: Termination is downstream from the last poly(A) site. *J. Mol. Biol.* **129:** 643.

Friedman, E.A. and C.L. Schildkraut. 1977. Terminal differentiation in cultured Friend erythroleukemia cells. *Cell* **12:** 901.

Friend, C.H. 1980. The regulation of differentiation in murine virus-induced erythroleukemic cells. In *Results and problems in cell differentiation* (ed. R.G. McKinnell et al.), vol. VII, p. 212. Springer-Verlag, Berlin.

Friend, C., W. Scher, J.G. Holland, and T. Sato. 1971. Hemoglobin synthesis in murine virus-induced leukemic cells *in vitro*: Stimulation of erythroid differentiation by dimethylsulfoxide. *Proc. Natl. Acad. Sci.* **68:** 378.

Fuchs, E. and H. Green. 1980. Changes in keratin gene expression during terminal differentiation of the keratinocyte. *Cell* **19:** 1033.

Gard, D.L. and E. Lazarides. 1980. The synthesis and distribution of desmin and vimentin during myogenesis *in vitro. Cell* **19:** 263.

———. 1982a. Cyclic AMP-modulated phosphorylation of intermediate filament proteins in cultured avian myogenic cells. *Mol. Cell. Biol.* **2:** 1104.

———. 1982b. Analysis of desmin and vimentin phosphopeptides in cultured myogenic cells and their modulation by 8-bromo-adenosine 3′,5′ chick monophosphate. *Proc. Natl. Acad. Sci.* **79:** 6912.

Gard, D.L., P.B. Bell, and E. Lazarides. 1979. Coexistence of desmin and the fibroblastic intermediate filament subunit in muscle and nonmuscle cells: Identification and comparative peptide analysis. *Proc. Natl. Acad. Sci.* **76:** 3894.

Geisler, N. and K. Weber. 1981. Comparison of the proteins of two immunologically distinct intermediate-sized filaments by amino acid sequence analysis: Desmin and vimentin. *Proc. Natl. Acad. Sci.* **78:** 4120.

———. 1982. The amino acid sequence of chicken muscle desmin provides a common structural model for intermediate filament proteins. *EMBO J.* **1:** 1649.

———. 1983. Amino acid sequence data on glial fibrillary acidic protein (GFA); implications for the subdivision of intermediate filaments into epithelial and non-epithelial members. *EMBO J.* **2:** 2059.

Geisler, N., U. Plessmann, and K. Weber. 1982. Related amino acid sequences in neurofilaments and non-neuronal intermediate filaments. *Nature* **296:** 448.

Granger, B.L. and E. Lazarides. 1978. The existence of an insoluble Z-disc scaffold in chicken skeletal muscle. *Cell* **15:** 1253.

———. 1979. Desmin and vimentin coexist at the periphery of the myofibril Z disc. *Cell* **18:** 1053.

———. 1982. Structural associations of synemin and vimentin filaments in avian erythrocytes revealed by immunoelectron microscopy. *Cell* **30:** 263.

———. 1983. Expression of the major neurofilament subunit in chicken erythrocytes. *Science* **221:** 553.

Granger, B.L., E.A. Repasky, and E. Lazarides. 1982. Synemin and vimentin are components of intermediate filaments in avian erythrocytes. *J. Cell Biol.* **92:** 299.

Gusella, J., R. Geller, B. Clarke, V. Weeks, and D. Housman. 1976. Commitment to erythroid differentiation by Friend erythroleukemia cells: A stochastic analysis. *Cell* **9:** 221.

Gusella, J.F., S.C. Weil, A.S. Tsiftsoglou, V. Vollouh, J.R. Neumann, C. Keys, and D.E. Housman. 1980. Hemin does not cause commitment of murine erythroleukemia (MEL) cells to terminal differentiation. *Blood* **56:** 481.

Hanukoglu, I. and E. Fuchs. 1982. The cDNA sequence of a human epidermis keratin: Divergence of sequence but conservation of structure among intermediate filament proteins. *Cell* **31:** 243.

———. 1983. The cDNA sequence of a type II cytoskeletal keratin reveals constant and variable structural domains among keratins. *Cell* **33:** 915.

Hofer, E. and J.E. Darnell, Jr. 1981. The primary transcription unit of the mouse β-major globin gene. *Cell* **23:** 585.

Hofer, E., R. Hofer-Warbinek, and J.E. Darnell, Jr. 1982. Globin RNA transcription: A possible termination site and demonstration of transcriptional control correlated with altered chromatin structure. *Cell* **29:** 887.

Hong, B.-S. and P.F. Davison. 1981. Isolation and characterization of a soluble immunoactive peptide of glial fibrillary acidic protein. *Biochim. Biophys. Acta* **670:** 139.

Hu, H.-Y., Y. Hu, J. Gardner, P. Aisen, and A.I. Skoultchi. 1977. Inducibility of transferrin receptors on Friend erythroleukemia cells. *Science* **197:** 559.

Ip, W., S.I. Danto, and D.A. Fischman. 1983. Detection of desmin-containing intermediate filaments in cultured muscle and nonmuscle cells by immunoelectron microscopy. *J. Cell Biol.* **96:** 401.

Jorcano, J.L., T.M. Magin, and W.W. Franke. 1984. Cell type-specific expression of bovine keratin genes as demonstrated by the use of complementary DNA clones. *J. Mol. Biol.* **176:** 21.

Lagacé, L., T. Chundra, S.L.C. Woo, and A.R. Means. 1983. Identification of multiple species of calmodulin messenger RNA using a full-length complementary DNA. *J. Biol. Chem.* **258:** 1684.

Lai, C.-J., R. Dhar, and G. Khoury. 1978. Mapping the spliced and unspliced late lytic SV40 RNAs. *Cell* **14:** 971.

Lazarides, E. 1980. Intermediate filaments as mechanical integrators of cellular space. *Nature* **283:** 249.

———. 1981. Intermediate filaments: Chemical heterogeneity in differentiation. *Cell* **23:** 649.

———. 1982. Intermediate filaments: A chemically heterogeneous, developmentally regulated class of proteins. *Ann. Rev. Biochem.* **51:** 219.

Lazarides, E. and D.L. Gard. 1982. Modulation of intermediate filament organization by cyclic nucleotide analogues and β-adrenergic agonists during myogenesis *in vitro*. *UCLA Symp. Mol. Cell. Biol.* **26:** 343.

Lewis, S.A., J.M. Balcavek, V. Krek, M. Shelanski, and N.J. Cowan. 1984. Sequence of a cDNA clone encoding mouse glial fibrillary acidic protein: Structural conservation of intermediate filaments. *Proc. Natl. Acad. Sci.* **81:** 2743.

MacDonald, M.E., M. Letarte, and A. Bernstein. 1978. Erythrocyte membrane antigen expression during Friend cell differentiation: Analysis of two non-inducible variants. *J. Cell. Physiol.* **96:** 291.

Mager, D. and A. Bernstein. 1978. Early transport changes during erythroid differentiation of Friend erythroleukemia cells. *J. Cell. Physiol.* **94:** 275.

Marks, P.A. and R.A. Rifkind. 1978. Erythroleukemic differentiation. *Annu. Rev. Biochem.* **47:** 419.

Mikawa, T., S. Takeda, T. Shimizu, and T. Kitaura. 1981. Gene expression of myofibrillar proteins in single muscle fiber of adult chicken: Micro two-dimensional gel electrophoretic analysis. *J. Biochem.* **89:** 1951.

Moon, R.T. and E. Lazarides. 1983. Synthesis and post-translational assembly of intermediate filaments of avian erythroid cells: Vimentin assembly limits the rate of synemin assembly. *Proc. Natl. Acad. Sci.* **80:** 5495.

Moss, M. and R. Schwartz. 1981. Regulation of tropomyosin gene expression during myogenesis. *Mol. Cell. Biol.* **1:** 289.

Nevins, J.R. and J.E. Darnell, Jr. 1978. Steps in the processing of Ad2 mRNA: Poly(A)$^+$ nuclear sequences are conserved and poly(A) addition precedes splicing. *Cell* **15:** 1477.

Nevins, J.R., J.-M. Blanchard, and J.E. Darnell, Jr. 1980. Transcription units of adenovirus type 2: Termination of transcription beyond the poly(A) addition site in early regions 2 and 4. *J. Mol. Biol.* **144:** 377.

Ngai, J., Y.G. Capetanaki, and E. Lazarides. 1984. Differentiation of murine erythroleukemia cells results in the rapid repression of vimentin gene expression. *J. Cell Biol.* **99:** 306.

Nudel, U., J. Salmon, E. Fibach, M. Terada, R. Rifkind, P.A. Marks, and A. Bank. 1977. Accumulation of α- and β-globin messenger RNAs in mouse erythroleukemia cells. *Cell* **12:** 463.

Orkin, S.H., F.I. Harosi, and P. Leder. 1975. Differentiation in erythroleukemic cells and their somatic hybrids. *Proc. Natl. Acad. Sci.* **72:** 98.

Osborn, M., J. Caselitz, and K. Weber. 1981. Heterogeneity of intermediate filament expression in vascular smooth muscle: A gradient of desmin positive cells from the rat aortic arch to the level of the arteria iliaca communis. *Differentiation* **20:** 196.

Parnes, J.R. and R.R. Robinson. 1983. Multiple mRNA species with distinct 3′ termini are transcribed from the β_2-microglobin gene. *Nature* **302:** 449.

Price, M.G. and E. Lazarides. 1983. Expression of intermediate filament-associated proteins paranemin and synemin in chicken development. *J. Cell Biol.* **97:** 1860.

Pruss, R.M., R. Mirsky, M.C. Raff, R. Thorpe, A.J. Dowding, and B.H. Anderson. 1981. All classes of intermediate filaments share a common antigenic determinant defined by a monoclonal antibody. *Cell* **27:** 419.

Quax, W., W.V. Egberts, W. Hendriks, Y. Quax-Jeuken, and H. Bloemendal. 1983. The structure of the vimentin gene. *Cell* **35:** 215.

Quax-Jeuken, Y., W. Quax, and H. Bloemendal. 1983. Primary and secondary structure of hamster vimentin predicted from the nucleotide sequence. *Proc. Natl. Acad. Sci.* **80:** 3548.

Quinlan, R.A. and W.W. Franke. 1982. Heteropolymer filaments of vimentin and desmin in vascular smooth muscle tissue and cultured baby hamster kidney cells demonstrated by chemical crosslinking. *Proc. Natl. Acad. Sci.* **79:** 3452.

———. 1983. Molecular interactions in intermediate sized filaments revealed by chemical cross-linking. Heteropolymers of vimentin and glial filament protein in cultured human glioma cells. *Eur. J. Biochem.* **132:** 477.

Reuben, R.C., R.L. Wife, R. Breslow, R.A. Rifkind, and P.A. Marks. 1976. A new group of potent inducers of differentiation in murial erythroleukemia cells. *Proc. Natl. Acad. Sci.* **73:** 862.

Ross, J. and D. Sautner. 1976. Induction of globin mRNA accumulation by hemin in cultured erythroleukemia cells. *Cell* **8:** 513.

Ross, J., Y. Ikawa, and P. Leder. 1972. Globin messenger-RNA induction during erythroid differentiation of cultured leukemia cells. *Proc. Natl. Acad. Sci.* **69:** 3620.
Sabban, E.L., D.D. Sabatini, V.T. Marchesi, and M. Adesnik. 1980. Biosynthesis of erythrocyte membrane protein Band 3 in DMSO-induced Friend erythroleukemia cells. *J. Cell. Physiol.* **104:** 261.
Salditt-Georgieff, M. and J.E. Darnell, Jr. 1983. A precise termination site in the mouse β-major globin transcription unit. *Proc. Natl. Acad. Sci.* **80:** 4694.
Sassa, S. 1976. Sequential induction of heme pathway enzymes during erythroid differentiation of mouse Friend leukemia virus-infected cells. *J. Exp. Med.* **143:** 305.
Schmid, E., M. Osborn, E. Rungger-Brundle, and W. Franke. 1982. Distribution of vimentin and desmin filaments in smooth muscle tissue of mammalian and avian aorta. *Exp. Cell Res.* **137:** 329.
Schnitzer, J., W.W. Franke, and M. Schachner. 1981. Immunocytochemical demonstration of vimentin in astrocytes and ependymal cells of developing and adult mouse nervous system. *J. Cell Biol.* **90:** 435.
Setzer, D.R., M. McGrogan, J.H. Numberg, and R.T. Schimke. 1980. Size heterogeneity in the 3′ end of dihydrofolate reductase messenger RNAs in mouse cells. *Cell* **22:** 361.
Shani, M., D. Zevin-Sonkin, O. Saxel, Y. Carmon, D. Katcoff, U. Nudel, and D. Yaffe. 1981. The correlation between the synthesis of skeletal muscle actin, myosin heavy chain, and myosin light chain and the accumulation of corresponding mRNA sequences during myogenesis. *Dev. Biol.* **86:** 483.
Smith, R.L., I.G. Macara, R. Levenson, D. Housman, and L. Cantley. 1982. Evidence that a Na^+/Ca^+ antiport system regulates murine erythroleukemia cell differentiation. *J. Biol. Chem.* **257:** 773.
Steinert, P.M., W.W. Idler, and R.D. Goldman. 1980. Intermediate filaments of baby hamster kidney (BHK-21) cells and bovine epidermal keratinocytes have similar ultrastructures and subunit domain structures. *Proc. Natl. Acad. Sci.* **77:** 4534.
Steinert, P.M., W.W. Idler, F. Cabrals, M.M. Gottesman, and R.D. Goldman. 1981. *In vitro* assembly of homopolymer and copolymer filaments from intermediate filament subunits of muscle and fibroblastic cells. *Proc. Natl. Acad. Sci.* **78:** 3692.
Tapscott, S.J., G.S. Bennett, and H. Holtzer. 1981a. Neuronal precursor cells in the chick neural tube express neurofilament proteins. *Nature* **292:** 836.
Tapscott, S.J., G.S. Bennett, Y. Toyama, F. Kleinbart, and H. Holtzer. 1981b. Intermediate filament proteins in the developing chick spinal cord. *Dev. Biol.* **86:** 40.
Tosi, M., R.A. Young, O. Hagenbüchle, and U. Schibler. 1981. Multiple polyadenylation sites in a mouse α-amylase gene. *Nucleic Acids Res.* **9:** 2313.
Virtanen, I., M. Kurkinen, and V.-P. Lehto. 1979. Nucleus-anchoring cytoskeleton in chicken red blood cells. *Cell Biol. Int. Rep.* **3:** 157.
Winter, H. and J. Schweizer. 1983. Keratin synthesis in normal mouse epithelia and in squamous cell carcinomas: Evidence in tumors for masked mRNA species coding for high molecular weight keratin polypeptides. *Proc. Natl. Acad. Sci.* **80:** 6480.
Woodcock, C.L.F. 1980. Nucleus-associated intermediate filaments from chicken erythrocytes. *J. Cell Biol.* **85:** 881.
Yen, S.-H. and K.L. Fields. 1981. Antibodies to neurofilament glial filament and fibroblast intermediate filament proteins bind to different cell types of the nervous system. *J. Cell Biol.* **88:** 115.
Zehner, Z.E. and B.M. Paterson. 1983a. Characterization of the chicken vimentin gene: Single copy gene producing multiple mRNAs. *Proc. Natl. Acad. Sci.* **80:** 911.
———. 1983b. Vimentin gene expression during myogenesis: Two functional transcripts from a single copy gene. *Nucleic Acids Res.* **11:** 8317.

Cell-Cell Interaction and Cell-shape-related Control of Intermediate Filament Protein Synthesis

Avri Ben-Ze'ev
Department of Genetics
Weizmann Institute of Science
Rehovot 76100, Israel

Recent studies suggest that important growth-related cellular activities are mediated through changes in cell shape (for review, see Ben-Ze'ev 1984). Folkman and Moscona (1978) have provided quantitative data that support the view that there is a direct correlation between DNA synthesis and the extent of cell spreading. The response of cells to serum growth factors also appears to be linked to cell shape. Tucker et al. (1981) have demonstrated that the growth of nonneoplastic cells is enhanced with increasing concentrations of serum as they become well-spread. The involvement of cell-shape changes, mediated by the composition of the extracellular matrix in differentiation and morphogenesis, was recently reviewed by Bissell et al. (1982). In addition, the expression of the malignant phenotype (Brouty-Boyé et al. 1980) and the metastatic capability of tumor cells (Raz and Ben-Ze'ev 1983) were also shown in certain experimental systems to be modulated by changes in cell shape. Our own studies of cell-configuration-related gene expression have led us to conclude that changes in cell shape bring about responses at the level of macromolecular metabolism (Benecke et al. 1978, 1980; Farmer et al. 1978; Ben-Ze'ev et al. 1980; Ben-Ze'ev and Raz 1981; Wittelsberger et al. 1981; Ben-Ze'ev 1983a). Since cell shape is determined to a large extent by the organization of the subcellular cytoskeleton, we decided to investigate the relationship between changes in cell shape and the expression of cytoskeletal protein genes. A link between cell morphology and the expression of a cytoskeletal protein gene was suggested to us by the observation that depolymerization of the microtubular network leads to a rapid decrease in the rate of tubulin synthesis accompanied by a parallel reduction in the translational activity of tubulin mRNA in vitro (Ben-Ze'ev et al. 1979). This feedback-type regulation of tubulin synthesis was further demonstrated in a variety of cell types using cloned cDNA probes for tubulin (Cleveland et al. 1981).

The systematic analysis of macromolecular metabolism in the suspension-reattachment culture system (Otsuka and Moskowitz 1976; Benecke et al. 1978, 1980; Farmer et al. 1978; Ben-Ze'ev et al. 1980) of anchorage-dependent fibroblasts enabled the investigation of actin synthesis under various conditions. In this system, cell shape was modulated between flat and spherical. In suspension cultures, actin synthesis was found to be inhibited by a posttranscriptional mechanism. However, when the cells were replated and actin synthesis was induced, its synthesis was found to be controlled by the level of mRNA production (Farmer et al. 1983). The changes in actin synthesis under suspension-reattachment conditions parallel the changes observed in actin synthesis during stimulation of cells into growth (Riddle et al. 1979), suggesting a link between cytoskeletal organization (as manifested by changes in cell shape), the expression of a cytoskeletal gene, and growth control. The results presented below summarize our studies on the control of vimentin synthesis in cells under conditions of varied cell shape. In addition, the coexpression of vimentin and cytokeratins in cultured kidney epithelial cells in response to cell-shape changes and the extent of cell-cell interaction are presented as a model system for the regulation of genes involved in morphological differentiation.

DISCUSSION

Changes in Vimentin Synthesis as a Function of Cell Shape

Cells may respond to extracellular stimuli that require alterations in their morphology by changing the balance between the concentrations of polymerized and unpolymerized cytoskeletal proteins. This is possible with tubulin-containing microtubules and actin-containing microfilaments, since in most cells, both actin and tubulin are found in an assembled form and in a disassembled form. However, vimentin, the intermediate filament protein, is found in the cells mainly in an assembled form. Alternatively, the cells' response to morphological changes could be the reorganization and rearrangement of the various cytoskeletal elements. In other instances, cells react to morphological changes imposed on them over longer periods of time by regulating the expression of the cytoskeletal genes that code for the proteins involved in constructing the cytoskeleton, as described above for actin and tubulin. The experiments described below demonstrate that the synthesis of the intermediate filament protein, vimentin, is also in response to change in cell shape.

Figure 1, lanes C through E, shows the reversible down-regulation of vimentin synthesis in a suspension culture system. The decrease of vimentin synthesis when cells are put in suspension cultures is gradual (Fig. 1F–H). However, when cells are returned to monolayer cultures, the level of vimentin recovers to the control level. This reversion is completed in about 6 hours (Fig. 1E,J). The cell-shape-related modulation of vimentin synthesis shown in Figure 1 was obtained with anchorage-independent B16 malignant melanoma cells (Ben-Ze'ev 1983b; Raz and Ben-Ze'ev 1983) as well as with anchorage-dependent cells such as the 3T3 mouse fibroblasts and the established BSC-1 monkey kidney epithelial cell

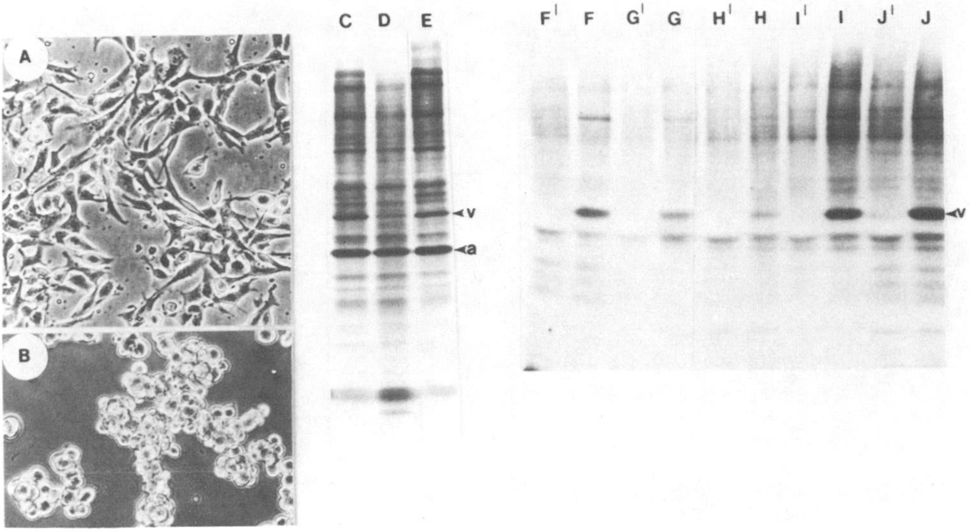

Figure 1
Reversible down-regulation of vimentin synthesis in suspension culture. [^{35}S]Methionine-labeled B16 melanoma cells cultivated in monolayer (*A,C,F*) or for 1 day (*G*) or 3 days in suspension culture (*B,D,H*) or for 6 hr (*E,I*) or 24 hr (*J*) of replating, following 3 days in suspension. (*C–E*) Total [^{35}S]methionine-labeled protein pattern. Immunoprecipitation with anti-vimentin serum (*F–J*) or nonimmune serum (*F′–J′*) from equal amounts of radioactive total cell protein.

line (Ben-Ze'ev 1983b). This suggests that inhibition of growth rate alone cannot account for this phenomenon. This is further substantiated by a double thymidine block experiment in which cell growth was arrested in the G_1 phase of the cell cycle and then the cells were stimulated into the S phase without an apparent change in the rate of vimentin synthesis. mRNA isolated from either monolayer or suspension cultures of B16 melanoma cells is equally active in the in vitro translation of vimentin, as measured by immunoprecipitation (Fig. 2D,F) or by two-dimensional gel electrophoresis of total in-vitro-translated proteins with mRNA from monolayer (Fig. 2J) and suspension cultures (Fig. 2I). Further analysis showed that poly(A)$^+$ RNAs from monolayer and suspension cultures of B16 melanoma contain similar amounts of vimentin mRNA sequences, as determined by RNA blot hybridization with nick-translated vimentin-cDNA-containing plasmids (Fig. 2G,H). This suggests a posttranscriptional control of vimentin synthesis in suspension cultures, as was also found for the regulation of actin synthesis under these conditions (Farmer et al. 1983). It would be of interest to investigate the mechanisms by which these translational controls regulate the synthesis of cytoskeletal proteins.

Cell-Cell Interaction and Cell Shape in the Control of Intermediate Filaments in Cultured Epithelial Cells

Intermediate filament proteins are tissue-type-specific (for review, see Lazarides 1982), but in culture, many epithelial cells express the vimentin-type mesenchy-

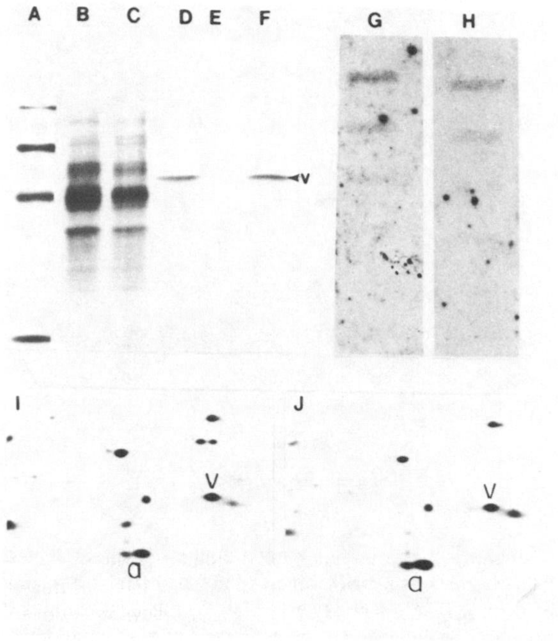

Figure 2
Posttranscriptional regulation of vimentin synthesis during suspension culture. Poly(A)+ cytoplasmic RNA (1 μg) from monolayer culture (*B,D,E,G,J*) or from 3-day suspension culture (*C,F,I,H*) was translated in vitro in a reticulocyte lysate (*B,C*), and the [^{35}S]methionine-labeled vimentin was immunoprecipitated with anti-vimentin (*D,F*) or nonimmune serum (*E*) or the total protein pattern was displayed on two-dimensional gels (*I,J*). (*G,H*) RNA blot hybridization with nick-translated vimentin cDNA. (*A*) Protein molecular-weight markers.

mal cytoskeleton in addition to the tissue-type-specific intermediate filament network (cf. Lane et al. 1982).

Since the synthesis of the vimentin-type mesenchymal intermediate filament protein was found to be related to changes in cell shape (see above), it was of interest to compare its synthesis with that of the cytokeratin-type intermediate filaments coexpressed in cultured epithelial kidney cells.

MDBK (Madin-Darby bovine kidney) epithelial cells, when cultivated either in monolayer (Fig. 3A) or in suspension, under conditions where tightly packed cell aggregates are formed (Fig. 3B), synthesized cytokeratins at comparable levels, as measured by analysis of [^{35}S]methionine-labeled Triton cytoskeletons (Fig. 3C,D). Thus, cytokeratin synthesis was not affected significantly by changes in cell shape, whereas vimentin synthesis was reduced extensively in suspended MDBK cells (Fig. 3D). Replating the cells for a few hours after 3 days in suspension culture induces a quick restoration of vimentin synthesis, as was also obtained with nonepithelial cells (Ben-Ze'ev 1983b). The cells maintaining the morphologies described in Figure 3, A and B, allowed the formation of extensive cell-cell interaction. These areas of cell-cell contact are characterized by epithelium-specific junctional complexes (Farquhar and Palade 1963) of the adherence and

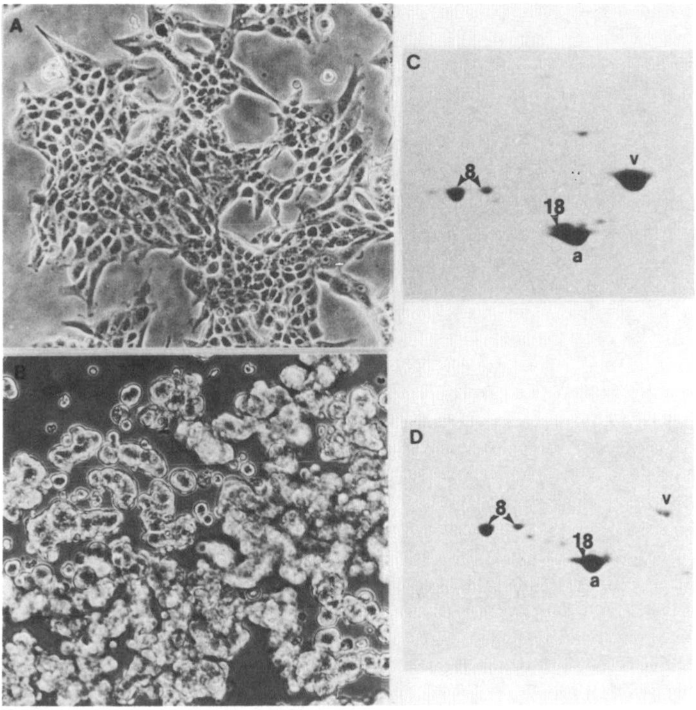

Figure 3
Cell-shape-dependent regulation of vimentin synthesis in cultured epithelial cells. MDBK cells in monolayer (*A,C*) or after 3-day suspension culture (*B,D*) were labeled with [^{35}S]methionine, and Triton cytoskeletons (*C,D*) were prepared from equal counts per minute of total cell protein and analyzed on two-dimensional gels. (a) Actin; (v) vimentin; (8,18) cytokeratins according to Moll et al. (1982).

the desmosomal types, as revealed by electron microscopy (results not shown). The formation of epithelium-specific intercellular junctional complexes is important for expressing the structural and functional polarization of kidney epithelium. Most relevant to the control of cytokeratin expression are the desmosome-type junctions, since the cytokeratin fibrils are anchored to the desmosomal plaque (for review, see Farquhar and Palade 1963), and it is suggested that they aid in its structural stabilization. The areas of desmosome-cytokeratin complex in MDBK cells are usually devoid of vimentin filaments. Therefore, the synthesis of cytokeratins and vimentin was also followed by sparse monolayer (Fig. 4B) and suspension cultures (Fig. 4C), where cell-cell interaction was minimal, and compared with very dense monolayer (Fig. 4A) and suspension cultures (Fig. 4D). In sparse monolayer cultures, the level of vimentin synthesis was maximal and that of cytokeratins was minimal (Fig. 5B,D). In dense monolayer cultures and in dense suspension cultures, the level of cytokeratin synthesis was high and that of vimentin was low (Figs. 5A,C and 3D). The ratio between cytokeratin synthesis and vimentin synthesis, determined by labeling cells with [^{35}S]methionine, was also reflected at the level of mRNA translation in vitro (Fig. 5E,F). Furthermore,

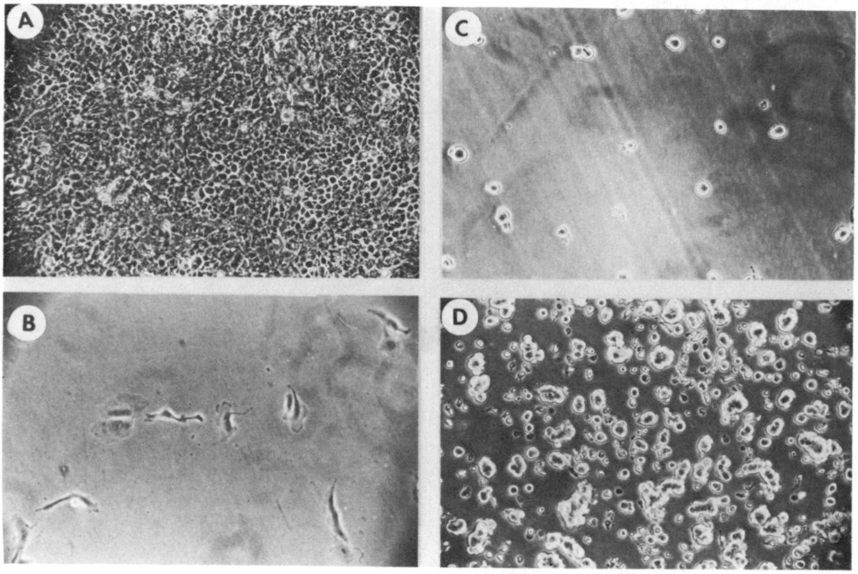

Figure 4
Modulation of cell shape and cell-cell interaction by varying the substrate adhesiveness and medium viscosity. MDBK cells after 3 days of culture either in dense monolayer culture (*A*) or in sparse monolayer culture (*B*), or in sparse suspension culture in medium containing 1.35% methylcellulose (*C*) or in dense suspension culture (*D*) on poly(HEMA)-coated plastic according to the method of Folkman and Moscona (1978).

in single-cell suspension cultures obtained by seeding MDBK cells in methylcellulose-containing medium (Fig. 4C), the level of cytokeratin synthesis was also low, as was found in sparse monolayer cultures (see Fig. 5B,D). It therefore appears that for optimal cytokeratin synthesis, extensive cell-cell interaction is required. The same conditions are known to induce the formation of desmosomes. Indeed, Geiger et al. (1983) have demonstrated, by double immunofluorescence antibody staining with antibody against cytokeratins and anti-desmosomal antibodies, that in MDBK cells, the cytokeratin fibrils terminate in the desmosomal plaques at the intercellular boundaries and that in sparse MDBK cells, the free edges of cells are devoid of desmosomes. The formation of desmosomes, the synthesis of desmosomal proteins, and the synthesis of cytokeratins also correlate with one another during embryogenesis (Jackson et al. 1980). The experiments described above suggest that in cells such as MDBK, where vimentin is coexpressed with cytokeratins, the two intermediate filament proteins are differentially regulated. Cytokeratin synthesis is relatively insensitive to changes in cell shape but requires extensive cell-cell interaction, whereas vimentin synthesis appears to respond more sensitively to changes in cell shape, as has also been shown in nonepithelial cells (Ben-Ze'ev 1983b). Thus, in sparse monolayer cultures, vimentin synthesis is maximal, but in very dense monolayer cultures (Fig. 4A) or in either sparse or dense suspension cultures, vimentin synthesis is minimal. The coexpression of vimentin and cytokeratins in vivo is rare, but recent

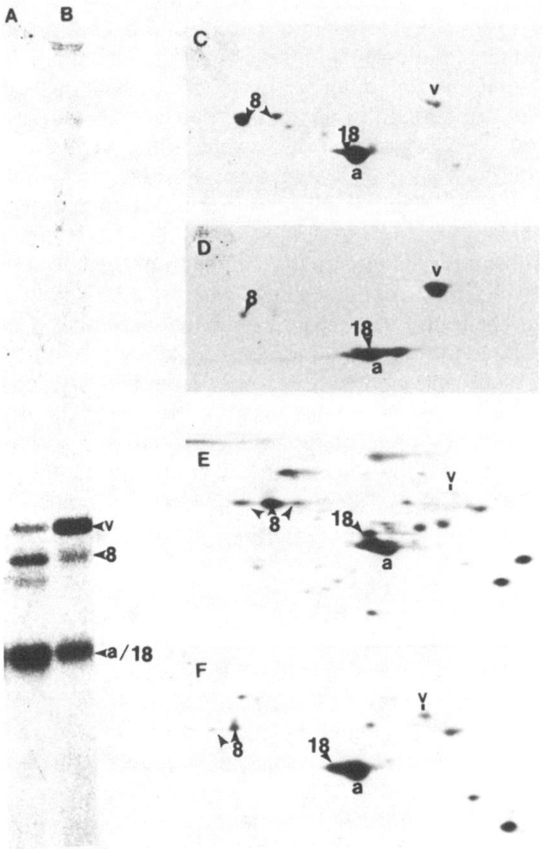

Figure 5

Cell-cell interaction-dependent regulation of cytokeratin synthesis in cultured MDBK cells. MDBK cells after 3 days in culture as a dense monolayer (*A,C,E*) or as a sparse monolayer culture (*B,D,F*) were labeled with [^{35}S]methionine, and Triton cytoskeletons from equal total counts per minute were analyzed on SDS gels (*A,B*) or two-dimensional gels (*C,D*). Equal amounts of poly(A)$^+$ cytoplasmic RNA were translated in vitro in a reticulocyte lysate (*E,F*).

studies have shown that in the parietal endoderm of the early mouse embryo, which constitutes an individual population of motile cells, such coexpression exists (Lane et al. 1983). In addition, Connell and Rheinwald (1983) have demonstrated cytokeratin and vimentin coexpression in mesothelial cells in vivo, and Ramaekers et al. (1983) have shown such coexpression in certain neoplastic cells of epithelial origin, when metastasizing to the ascites form. Thus, the expression of vimentin in epithelial cells in vivo could be related to reduced cell-cell contact and the independent existence of a cell, whereas the expression of cytokeratins appears to be dependent on extensive cell-cell interaction. These findings are compatible with our studies in which cell shape and the extent of cell-cell interactions were varied in vitro (see above).

Under the various culture conditions employed in our study, there were extensive differences in the growth rates of the cells, and the observed changes in the level of synthesis of the various intermediate filament proteins correlate partially with the changes in growth level. Most recently, using mesothelial cells that also coexpress vimentin and cytokeratins, Connell and Rheinwald (1983) have shown that vimentin synthesis correlates with a rapid growth level, whereas the synthesis of the cytokeratins parallels a slow growth level. To mimic changes in growth level, we employed various techniques such as the double thymidine block, which slows down growth almost completely, followed by the induction of the cells into S phase, or by completely blocking the synthesis of DNA with inhibitors such as cytosine arabinoside. We have also used the serum starvation–serum stimulation assay. Under none of the above-mentioned conditions could we change the ratio of vimentin synthesis to cytokeratin synthesis in either sparse or semiconfluent cell cultures. These studies, however, do not exclude the possibility that the rate of growth is an important factor in the control of the synthesis of various intermediate filament proteins. It is possible that, as in the control of actin synthesis (Farmer et al. 1983), changes in cell shape and cell-cell interaction in conjunction with changes in the growth level are important factors in the control of cytoskeletal protein gene expression.

CONCLUSION

The induction of cytokeratin synthesis and the assembly of cytokeratins into tonofilaments may be related to the assembly of the desmosomal proteins into desmosomes, which are located at the intercellular junctions in epithelia. In analogy to the intercellular junctions that contain desmoplakin-cytokeratin complexes, the cell-substrate attachment sites in fibroblasts or the adherence-type junctions in epithelia contain vinculin (Geiger 1979; Geiger et al. 1983) into which the actin-containing filaments terminate, thus forming actin-vinculin complexes. Since some of the adherence-type junction proteins and desmosome-type junction proteins are characterized and it is conceivable that the genes coding for these proteins will soon be cloned, the connection between these proteins at one end with the plasma membrane and at the other end with the cytoskeletal network that terminates at the nuclear lamina provides us with a useful model system for studying structurally linked skeletal elements that respond to cell-cell interaction and cell-shape changes in coupling with growth control. In addition, this system can be useful in the study of cytoarchitecture and differentiation-related gene expression.

ACKNOWLEDGMENT

This study was supported in part by a grant from the Israel-U.S.A. Binational Foundation, Jerusalem, Israel.

REFERENCES

Benecke, B.J., A. Ben-Ze'ev, and S. Penman. 1978. The control of mRNA production, translation and turnover in suspended and reattached anchorage-dependent fibroblasts. *Cell* **14:** 931.

———. 1980. The regulation of RNA metabolism in suspended and reattached anchorage-dependent 3T6 fibroblasts. *J. Cell. Physiol.* **103:** 247.
Ben-Ze'ev, A. 1983a. Virus replication in infected epithelial cells is coupled to cell shape-responsive metabolic controls. *J. Cell. Physiol.* **114:** 145.
———. 1983b. Cell configuration-related control of vimentin biosynthesis and phosphorylation in cultured mammalian cells. *J. Cell Biol.* **97:** 858.
———. 1984. Cell shape, the complex cellular networks and gene expression: Cytoskeletal protein genes as a model system. In *Cell and muscle motility, 6* (ed. J.W. Shay). Plenum Press, Dallas. (In press.)
Ben-Ze'ev, A. and A. Raz. 1981. Multinucleation and inhibition of cytokinesis in suspended cells: Reversal upon reattachment to a substrate. *Cell* **26:** 107.
Ben-Ze'ev, A., S.R. Farmer, and S. Penman. 1979. Mechanisms of regulating tubulin synthesis in cultured mammalian cells. *Cell* **17:** 319.
———. 1980. Protein synthesis requires cell-surface contact while nuclear events respond to cell shape in anchorage-dependent fibroblasts. *Cell* **21:** 365.
Bissell, M.J., G.H. Hall, and G. Parry. 1982. How does the extracellular matrix direct gene expression? *J. Theor. Biol.* **39:** 31.
Brouty-Boyé, D., R.W. Tucker, and J. Folkman. 1980. Transformed and neoplastic phenotype: Reversibility during culture by cell density and cell shape. *Int. J. Cancer* **26:** 501.
Cleveland, D.W., M.A. Lopata, P. Sherline, and M.W. Kirshner. 1981. Unpolymerized tubulin modulates the level of tubulin mRNAs. *Cell* **25:** 537.
Connell, N.D. and J.G. Rheinwald. 1983. Regulation of the cytoskeleton in mesothelial cells: Reversible loss of keratin and increase in vimentin during rapid growth in culture. *Cell* **34:** 245.
Farmer, S.R., A. Ben-Ze'ev, B.-J. Benecke, and S. Penman. 1978. Altered translatability of messenger RNA from suspended anchorage dependent fibroblasts: Reversal upon cell attachment to a surface. *Cell* **15:** 627.
Farmer, S.R., K.M. Wan, A. Ben-Ze'ev, and S. Penman. 1983. Regulation of actin mRNA levels and translation responds to changes in cell configuration. *Mol. Cell. Biol.* **3:** 182.
Farquhar, M.G. and G.E. Palade. 1963. Junctional complexes in various epithelia. *J. Cell Biol.* **17:** 375.
Folkman, J. and A. Moscona. 1978. Role of cell shape in growth control. *Nature* **273:** 345.
Geiger, B. 1979. A 130K protein from chicken gizzard: Its localization at the termini of microfilament bundles in cultured chicken cells. *Cell* **18:** 193.
Geiger, B., E. Schmid, and W.W. Franke. 1983. Spatial distribution of proteins specific for desmosomes and adhaerens junctions in epithelial cells demonstrated by double immunofluorescence microscopy. *Differentiation* **23:** 189.
Jackson, B.W., C. Grund, E. Schmid, K. Bürke, W.W. Franke, and K. Illmensee. 1980. Formation of cytoskeletal elements during mouse embryogenesis. *Differentiation* **17:** 161.
Lane, E.B., S.L. Goodman, and L.K. Trejdosiewicz. 1982. Disruption of the keratin filament network during epithelial cell division. *EMBO J.* **1:** 1365.
Lane, E.B., B.L.M. Hogan, M. Kurkinen, and J.I. Garrels. 1983. Coexpression of vimentin and cytokeratins in parietal endoderm cells of early mouse embryo. *Nature* **303:** 701.
Lazarides, E. 1982. Intermediate filaments: A chemically heterogeneous, developmentally regulated class of proteins. *Annu. Rev. Biochem.* **51:** 219.
Moll, R., W.W. Franke, D. Schiller, B. Geiger, and R. Krepler. 1982. The catalog of human cytokeratins: Patterns of expression in normal epithelia, tumors and cultured cells. *Cell* **31:** 11.
Otsuka, H. and M. Moskowitz. 1976. Arrest of 3T3 cells in G_1 phase in suspension culture. *J. Cell. Physiol.* **87:** 213.
Ramaekers, F.C.S., D. Haag, A. Kant, O. Moesker, P.H.K. Jap, and G.P. Vooijs. 1983. Coexpression of keratin- and vimentin-type intermediate filaments in human metastatic carcinoma cells. *Proc. Natl. Acad. Sci.* **80:** 2618.

Raz, A. and A. Ben-Ze'ev. 1983. Modulation of the metastatic capability in B16 melanoma by cell shape. *Science* **221:** 1307.

Riddle, V.G.H., R. Dubrow, and A.B. Pardee. 1979. Changes in the synthesis of actin and other cell proteins after stimulation of serum arrested cells. *Proc. Natl. Acad. Sci.* **76:** 1298.

Tucker, R.S., C.E. Butterfield, and J. Folkman. 1981. Interaction of serum and cell spreading affects the growth of neoplastic and nonneoplastic fibroblasts. *J. Supramol. Struct. Cell. Biochem.* **15:** 29.

Wittelsberger, S.C., K. Kleene, and S. Penman. 1981. Progressive loss of shape responsive metabolic controls in cells with increasingly transformed phenotype. *Cell* **24:** 859.

Organization and Sequence of the Genes for Desmin and Vimentin

Wim Quax, Yvonne Quax-Jeuken,
Richard van den Heuvel, Wilma Vree Egberts,
and Hans Bloemendal
Department of Biochemistry
University of Nijmegen
6525 EZ Nijmegen, The Netherlands

The intermediate filaments form a group of related cytoskeletal proteins that have obtained a quite different expression pattern during evolution. From biochemical and biophysical characterizations, it has become clear that long α-helical regions are common structures found in all intermediate filament molecules. Recent sequence data on the proteins (Geisler and Weber 1982, 1983; Geisler et al. 1983) and on cDNA clones (Hanukoglu and Fuchs 1982, 1983; Quax-Jeuken et al. 1983; Steinert et al. 1983) have allowed the precise localization of α-helices in different intermediate filament proteins. The data show that despite the heterogeneity in sequence, the secondary structures of all intermediate filament molecules studied so far share common α-helical domains, conserved in length and position, that are capable of forming intermolecular coiled-coil interactions. The main differences between the various intermediate filament molecules are found in the nonhelical terminal domains, which show large variations in sequence as well as in size.

In addition to facilitating the study of protein structure, cDNA clones are useful tools for studying the organization and expression of the genes for intermediate filaments. With the aid of cDNA probes, it has been shown that the epithelial keratins are encoded by a multigene family, as is also the case with the actins and tubulins. In contrast, vimentin is encoded by a single gene in mammals (Quax et al. 1983) as well as in birds (Capetanaki et al. 1983; Zehner and Paterson 1983).

Recently, we have isolated desmin-specific cDNA clones. In this paper, we compare the nucleotide sequence of desmin cDNA with that of vimentin cDNA. We also discuss the evolutionary conservation of the nontranslated 3′ nucleotides of vimentin mRNA and the structural organization of the vimentin and desmin genes.

RESULTS AND DISCUSSION

Nucleotide Sequences of Hamster Vimentin and Desmin cDNAs

Lens cells are a convenient source for the isolation of vimentin and vimentin mRNA. We have selected vimentin cDNA clones from a cDNA library constructed on hamster lens mRNA (Dodemont et al. 1982). The nucleotide sequence of these clones has revealed virtually the total primary structure of vimentin (Quax-Jeuken et al. 1983). We now use this vimentin clone as a hybridization probe to obtain desmin-specific cDNA clones. Screening of 5000 recombinant plasmids constructed from BHK-21 mRNA has yielded four desmin-specific clones. The combined sequence of these clones is shown in Figure 1. From the nucleotide sequence, the 298 carboxyterminal amino acids for hamster desmin can be derived. Downstream from the stop codon, there is a stretch of 677 nontranslated bases, followed by a poly(A) tail. The amino acid sequence shows over long regions the heptade regularity in the distribution of hydrophobic residues, which is typical for α-helices organized in coiled-coils. In Figure 1, the dots indicate the amino acids that form part of the hydrophobic backbone.

The primary structure deduced for hamster desmin shows a close homology (92%) with that of chicken desmin (Geisler and Weber 1982). It is remarkable that the nonhelical carboxyterminal domain of desmin is as well conserved as the helical region. This is in contrast to the great variations in sequence and size that are seen in the carboxyl terminus when the different intermediate filament types are compared.

In Figure 2 we compare the desmin nucleotide sequence of Figure 1 with the corresponding vimentin nucleotide sequence using a dot matrix computer program (Staden 1982). There is a very clear diagonal line demonstrating the homology of desmin and vimentin at the nucleotide level. The line stops when we go beyond the stop codon, and no homology can be detected between the nontranslated regions of desmin and vimentin cDNAs. In the part of the line that corresponds to the nonhelical carboxyl residues, there are several interruptions, showing that there is a relatively lower level of homology between these regions of the two different intermediate filament genes. On the other hand, the nonhelical sequence (200–350 bp) that separates the two long α-helices does not show lower homology. The absence of lines that run parallel to the central diagonal demonstrates that there are no intramolecular duplications within intermediate filament genes. Therefore, one can conclude that the two long α-helical domains of intermediate filament molecules have not arisen from a single ancestor sequence by means of intramolecular duplication.

The 3′ Nontranslated Region of Vimentin mRNA Is Evolutionarily Conserved

In an earlier report, we described several homologous stretches of sequences between the nontranslated regions of hamster and chicken vimentin (Quax et al.

```
1
ARG ALA ARG VAL ASP VAL GLU ARG ASP ASN LEU ILE ASP ASP LEU GLN ARG LEU LYS ALA
CGT GCC CGT GTC GAC GTG GAG CGC GAC AAC TTG ATC GAC GAC CTC CAG AGG CTC AAG GCC  (60)
21
LYS LEU GLN GLU GLU ILE GLN LEU ARG GLU GLU ALA GLU ASN ASN LEU ALA ALA PHE ARG
AAG CTA CAG GAG GAA ATC CAA CTG AGA GAA GAA GCA GAG AAC AAC CTG GCT GCC TTC CGA  (120)
41
ALA ASP VAL ASP ALA ALA THR LEU ALA ARG ILE ASP LEU GLU ARG ARG ILE GLU SER LEU
GCG GAC GTA GAT GCA GCC ACT CTG GCT CGC ATC GAC CTA GAG CGC AGA ATC GAA TCG CTC  (180)
61
ASN GLU GLU ILE ALA PHE LEU LYS LYS VAL HIS GLU GLU GLU ILE ARG GLU LEU GLN ALA
AAC GAG GAA ATC GCA TTC CTG AAG AAA GTG CAC GAA GAG GAG ATC CGT GAG CTT CAG GCT  (240)
81
GLN LEU GLN GLU GLN GLN VAL GLN VAL GLU MET ASP MET SER LYS PRO ASP LEU THR ALA
CAG CTT CAG GAA CAG CAG GTC CAG GTG GAG ATG GAC ATG TCC AAG CCA GAC CTC ACA GCG  (300)
101
ALA LEU ARG ASP ILE ARG ALA GLN TYR GLU THR ILE ALA ALA LYS ASN ILE SER GLU ALA
GCC CTC AGG GAC ATC CGG GCT CAG TAC GAG ACC ATT GCG GCT AAG AAC ATC TCT GAA GCT  (360)
121
GLU GLU TRP TYR LYS SER LYS VAL SER ASP LEU THR GLN ALA ALA ASN LYS ASN ASN ASP
GAG GAG TGG TAC AAG TCC AAG GTT TCA GAC TTG ACC CAG GCA GCC AAT AAG AAC AAT GAT  (420)
141
ALA LEU ARG GLN ALA LYS GLN GLU MET MET GLU TYR ARG HIS GLN ILE GLN SER TYR THR
GCC CTG CGC CAG GCC AAG CAG GAG ATG ATG GAG TAC CGA CAC CAG ATC CAG TCC TAC ACC  (480)
161
CYS GLU ILE ASP ALA LEU LYS GLY THR ASN ASP SER LEU MET ARG GLN MET ARG GLU LEU
TGC GAG ATT GAT GCC CTC AAG GGC ACC AAT GAC TCC CTG ATG AGG CAG ATG AGA GAG CTG  (540)
181
GLU ASP ARG PHE ALA SER GLU ALA SER GLY TYR GLN ASP ASN ILE ALA ARG LEU GLU GLU
GAG GAT CGC TTT GCC AGC GAG GCC AGT GGC TAT CAG GAT AAC ATT GCA CGC CTG GAG GAG  (600)
201
GLU ILE ARG HIS LEU LYS ASP GLU MET ALA ARG HIS LEU ARG GLU TYR GLN ASP LEU LEU
GAG ATC CGG CAC CTG AAG GAT GAG ATG GCC CGC CAC CTG CGG GAG TAC CAA GAC CTG CTC  (660)
221
ASN VAL LYS MET ALA LEU ASP VAL GLU ILE ALA THR TYR ARG LYS LEU LEU GLU GLY GLU
AAT GTG AAG ATG GCC TTG GAT GTG GAG ATT GCC ACC TAC CGC AAG CTG CTG GAG GGC GAG  (720)
241
GLU SER ARG ILE ASN LEU PRO ILE GLN THR PHE SER ALA LEU ASN PHE ARG GLU THR SER
GAG AGC CGG ATC AAC CTT CCC ATC CAG ACC TTC TCT GCT CTC AAC TTC CGA GAA ACC AGC  (780)
261
PRO GLU GLN ARG GLY SER GLU VAL HIS THR LYS LYS THR VAL MET ILE LYS THR ILE GLU
CCT GAA CAA AGG GGT TCT GAA GTC CAC ACC AAA AAG ACG GTG ATG ATC AAG ACC ATC GAG  (840)
281                                                                 298  "STOPCODON"
THR ARG ASP GLY GLU VAL VAL SER GLU ALA THR GLN GLN GLN HIS GLU VAL LEU
ACC CGG GAT GGA GAG GTC GTC AGC GAG GCC ACA CAG CAA CAA CAC GAG GTG CTC TAA GCC  (900)

AGACACTGTCCTGGTCCCCGTGGTCACTGCCTCCTGAAGCCAGCCTCTTCCACTCTCGGATGTCACACCCAGCCACTTTCCTTCACTCACAGAATCTGACCCTTCCTCACCGATCACCCC  (1020)
TTTGTGGTCTTCATGCTGCCCAGGAAACACCCCAGCACCTCTGCAGACCTTACCATGAGTCCTGGCTGTCGGCAGTCGCAAGCCTGGCTCTTCAGATAGAACCTAGTTCAAGTCATGGCC  (1140)
CTTTCCCTCCCACCTTTGTAACCTCAGGCTCTACGCTTTGGCTTTGGAGATGGTACCAGAGAAGGTGTTGGGATCTGTAGGGTCAGGACAGAGCTTTATAGACACCCTCACATTCGACCC  (1260)
CCAGCCTGGGTCAGAGACAGAGTGAAGCCTCTCAGCTGAGGTGGGGAGGGGCTGAAAAAATGTCCTTGCGTCCCCTCTCTTTCCCATCCCAGCCCAGGATGGGTTAGAAAAGCTGGGGC  (1380)
TGTAAGAGGGAACCTGAAGGTGCTGGATGTGGGAGCAGGAGATTCAGAAGGAGAGCGGGTGGGTGAGAAGCTGGAGGGAAAGAAGAGAGAGGCAGAGAGTGGGCCCAGGCTGGTGGGAGG  (1500)
                                                "POLY-A"
GCCCCACCTCTCACGCCTGCCCCTCCCACTGCAGGGGCCCTGGACAGAAACAATAAAGAGCCAAGCACAAACCTAAAAAAAAAAAAAAAAAAAAAAAA  (1590)
```

Figure 1

Nucleotide sequence and derived primary structure for hamster desmin. The nucleotide sequences of four independently isolated desmin clones were determined by the M13 dideoxy method. In the overlapping segments, the clones are found to be identical and therefore they have been combined into a single sequence. The dots above the amino acids indicate the a and d positions of the heptade repeat, which mainly harbor hydrophobic residues. At the 5′ end, the information for about one third of the primary structure is missing. The 3′-untranslated region carrying 677 bases is complete, since a residual poly(A) track is present at the end.

1983). We have now extended the sequence at the 3′ end of the hamster gene, and in Figure 3 we compare this sequence to the 3′ end of the chicken gene

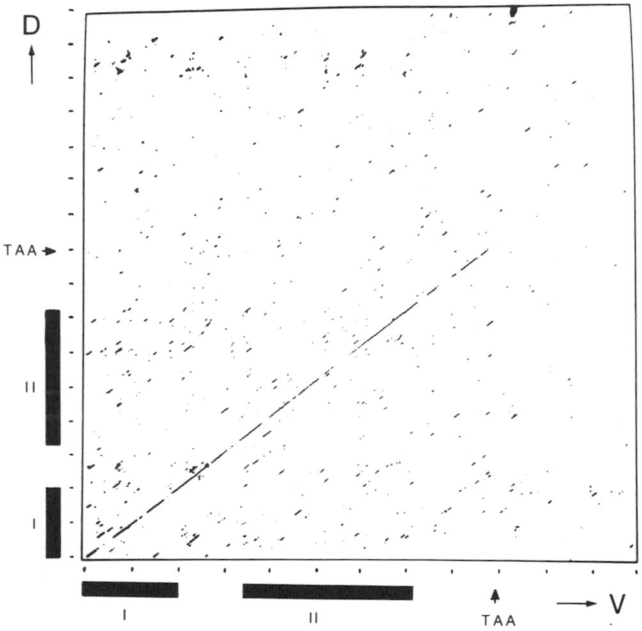

Figure 2
Dot matrix comparison of the desmin and vimentin nucleotide sequences. A dot matrix comparison of nucleotide sequences is a powerful tool in searching for homologies between DNAs. For this analysis, one set of sequences is indicated along the abscissa and the other set, along the ordinate; the computer places a dot at every position where defined stretches of the two DNAs show a certain level of homology (Staden 1982). The span length for the compared stretches was 17; of these, 11 had to be identical to result in a dot in the matrix. The desmin sequence (D) of Fig. 1 is indicated along the ordinate, and the vimentin sequence (V) is indicated along the abscissa. From the latter sequence (Quax-Jeuken et al. 1983), the codons for the first third of the polypeptide have been omitted, since that information is also lacking in the desmin cDNA. The black bars represent the regions that code for the α-helical domains. TAA marks the end of the translated region.

(Zehner and Paterson 1983) using the dot matrix comparison described in Figure 2. Homology lines are found from the stop codon up to 40 bases downstream from the single polyadenylation site of the hamster gene. The 250 additional 3′-end nucleotides that are only present in the 2.3-kb chicken mRNA do not show homology with the hamster gene. The line shows several divergences from the perfect diagonal, indicating that some insertion or deletion events have occurred. Since there are no common open reading frames in the hamster and chicken nontranslated regions, it is unlikely that the conserved regions encode a kind of regulatory polypeptide, as has been assumed (Zehner and Paterson 1983). It seems more likely that the secondary structure of the RNA or protein–nucleic acid interactions demand the conservation of the 3′ noncoding region. In this context, it is interesting to note that interspecies conservation of 3′-

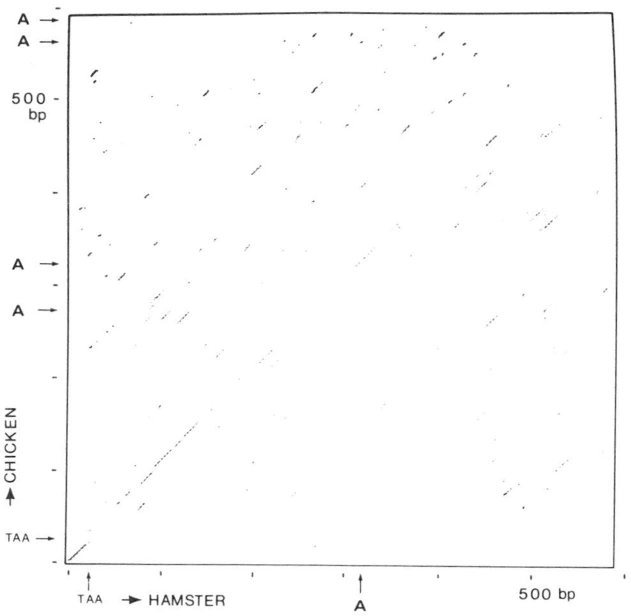

Figure 3
Homology between the 3′-untranslated regions of the hamster and chicken vimentin genes. The dot matrix program of Fig. 2 was used with the same span length of 17 and the same percentage score of 11. At the abscissa are indicated the 7 carboxyterminal codons plus 568 nucleotides following the stop codon of the hamster vimentin gene. At the ordinate are indicated the corresponding chicken codons plus the same number of 3′ downstream nucleotides. TAA marks the beginning of the untranslated region. A indicates the position of a consensus polyadenylation signal, AATAAA. The chicken vimentin gene contains four of these signals, which give rise to two different size classes of mRNA (Zehner and Paterson 1983). We see that there is a clear diagonal homology line between the hamster and the chicken sequences. The region that begins about 40 nucleotides beyond the single hamster polyadenylation signal, however, does not show homology, indicating that only transcribed 3′-untranslated nucleotides are conserved.

untranslated regions has recently been reported for several other cytoskeletal mRNAs (Ponte et al. 1983; Cowan et al. 1983).

Organization and Structure of the Vimentin and Desmin Genes

The cloned vimentin and desmin cDNAs were used to determine the copy number of the corresponding genes in the hamster genome. Southern blotting analysis with 5′-specific and 3′-specific probes shows that there is a single vimentin gene and a single desmin gene present per haploid genome. Screening of a Syrian gold hamster genomic library yielded several vimentin phages, all originating from the same chromosomal locus. The structure of the vimentin gene has been extensively studied using restriction enzyme mapping, S1-nuclease mapping, and sequence determination techniques. It was found that the hamster vimentin gene is interrupted by eight intervening sequences (Fig. 4). The total length of the gene is about 9 kb; 1848 bases of these are coding. At the 5′ end there are a consensus "TATA" box sequence and a consensus "CCAAT" sequence that are found in many other eukaryotic genes. All of the exon-intron junctions show the consensus bordering intron nucleotides GT and AG. The second intron (with 2.6 kb, the largest in the vimentin gene) shows at the 3′ end an alternating CA sequence of 44 residues (Fig. 5). An interesting feature of this sequence is that it can form Z-DNA (Zimmer et al. 1982). These $(CA)_n$ sequences appear to be widely dispersed throughout the genomes of a number of eukaryotic species (Hamada et al. 1982). Whether these elements can influence gene expression or whether they play a role in gene recombination is not yet clear (Rogers 1983).

Knowledge of the precise exon-intron pattern of an intermediate filament gene opens the possibility of searching for a possible correlation between the gene structure and the protein domain structure (Gilbert 1978). In Figure 6, we have correlated the intron positions with the model that we have proposed for the vimentin molecule in a previous report (Quax-Jeuken et al. 1983). At first glance, there is no clear correlation visible between the proposed helical and nonhelical domain borders and the intron pattern. Upon closer inspection, one can see that intron 3 and intron 6 may be located between an α-helical region and a nonhelical domain. Most other intron positions, however, do not correlate with the proposed protein domain borders and therefore are not in accord with the hypothesis made by Gilbert (1978) and Blake (1978). To determine whether these intron positions are unique for the vimentin gene, a structural analysis of other intermediate filament genes is necessary. We have now begun to analyze a cloned desmin gene isolated from the above-mentioned hamster gene library. From this analysis, it is now becoming clear that at least some intron positions are conserved among

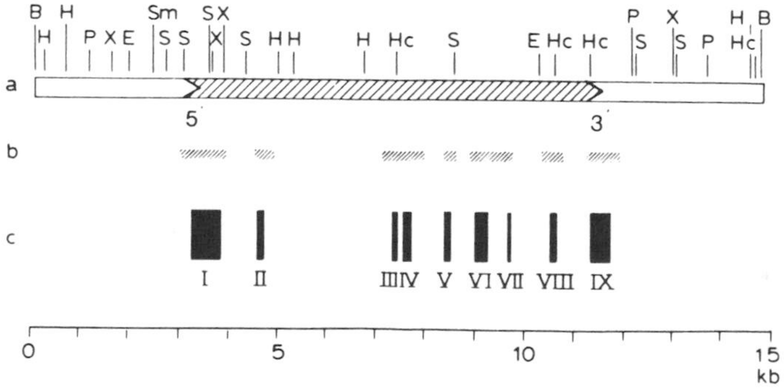

Figure 4
Structure of the vimentin gene. (*a*) Restriction pattern of a cloned hamster DNA fragment that contains the vimentin gene. Hatched area represents the vimentin transcription unit comprising 9 kb of DNA. (*b*) Hatched boxes represent the areas that have been sequenced to detect all the vimentin exons. (*c*) Black boxes show the sizes and positions of the nine exons of the vimentin gene. (Reprinted, with permission, from Quax et al. 1983.)

```
                                          "Z-DNA"
GATCTGACTGCCTTCTTCTGGCTTTCATGGTCAC[CACACACACACACACACACACACACACACACACACACACACACA]GAGTGCACTTAAATACGTGCAGGCAAAATAGCCACATAAAAT

                                                                        INTRON 2    207
                                                                                     D  V  D  N  A  S  L  A  R  L  D  L
AGAATGAAAAATAATTAGTATAAATAGATAAGATAAAGAATAGACTTTAGTACCGCCACATCATCTTTTCTCTCTTTAAAACAGGATGTTGACAATGCCTCTCTGGCACGCCTCGACCTT
```

Figure 5
Potential Z-DNA in an intron. Boxed segment is a stretch of 44 nucleotides of alternating C and A residues present in the second intervening sequence of the hamster vimentin gene. The Z-DNA-forming potential of this sequence has been shown in vitro (Zimmer et al. 1982).

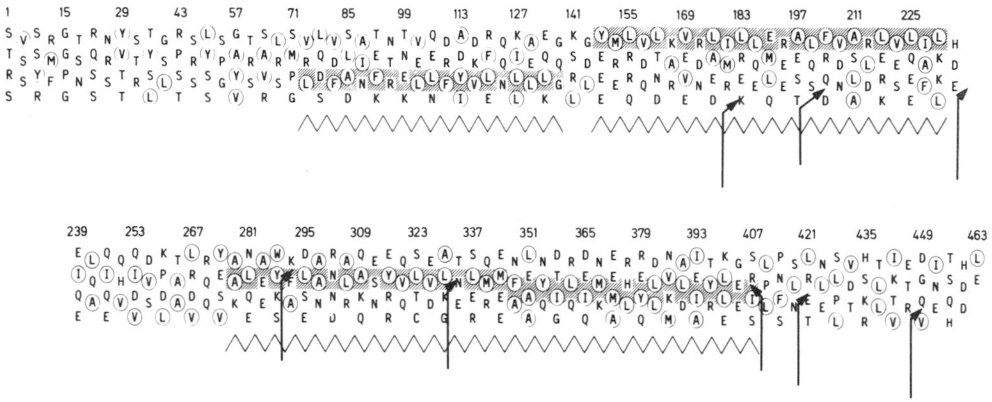

Figure 6
Correlation of vimentin gene structure and protein structure. To test whether the Gilbert hypothesis (Blake 1978; Gilbert 1978) holds true for intermediate filament proteins, we have translated the positions of the introns (arrows) to the corresponding positions in the model that we have proposed for vimentin (Quax-Jeuken et al. 1983). The model is a heptade drawing of the complete primary structure of vimentin in which the hydrophobic backbone of the α-helix is shaded. We see that apart from introns 3 and 6, no arrow matches with a border between different domains.

intermediate filament genes (W. Quax et al., in prep.). All intervening sequences found in the desmin gene thus far are located at positions that fully correspond to vimentin introns. If this conservation of intron positions is true for all intermediate filament genes, then it will be difficult to determine whether exon shuffling has played a role during intermediate filament gene evolution. An investigation of possible intermediate filament genes of lower eukaryotic species could reveal more details of intermediate filament evolution. In this context, it is interesting to note that vimentin and desmin cDNA probes hybridize to DNAs from representatives of all vertebrate classes. Both probes recognize different restriction fragments in all DNAs tested. We conclude from this observation that the gene duplication event that most likely has been on the basis of the origin of separate desmin and vimentin genes took place before the evolution of the vertebrate species ($\pm$550 million years ago; Romer 1966).

CONCLUSIONS

Insight into the organization and detailed structure of genes coding for intermediate filament proteins is an important step toward the understanding of the molecular regulation of the tissue-specific expression of intermediate filament subunits. A thorough comparison of the detailed sequences surrounding intermediate filament genes, combined with expression experiments of cloned genes in heterologous cells, should make it possible to identify which elements in or around the genes are responsible for their regulated expression. Future gene-transfer experiments with mutated and recombined genes can also provide an answer to the question of which elements of the primary structure of intermediate filament molecules are necessary for the formation of 10-nm filaments.

ACKNOWLEDGMENTS

We thank Wiljan Hendriks and Leon van den Broek for assistance with some aspects of the experiments described. The present investigations have been carried out in part under the auspicies of the Netherlands Foundation for Chemical Research (SON) and with financial aid from the Netherlands Organization for the Advancement of Pure Research (ZWO).

REFERENCES

Blake, C.C.F. 1978. Do genes-in-piece imply proteins-in-piece? *Nature* **273:** 267.

Capetanaki, Y.G., J. Ngai, C.N. Flytzanis, and E. Lazarides. 1983. Tissue-specific expression of two mRNA species transcribed from a single vimentin gene. *Cell* **35:** 411.

Cowan, N.J., P.R. Dobner, E.V. Fuchs, and D.W. Cleveland. 1983. Expression of human α-tubulin genes: Interspecies conservation of 3′ untranslated regions. *Mol. Cell. Biol.* **3:** 1738.

Dodemont, H., P. Soriano, W. Quax, F. Ramaekers, J. Lenstra, M. Groenen, G. Bernardi, and H. Bloemendal. 1982. The genes coding for the cytoskeletal proteins actin and vimentin in warm-blooded vertebrates. *EMBO J.* **1:** 167.

Geisler, N. and K. Weber. 1982. The amino acid sequence of chicken muscle desmin provides a common structural model for intermediate filament proteins. *EMBO J.* **1:** 1649.

———. 1983. Amino acid sequence data on glial fibrillary acidic protein (GFA): Implications for the subdivision of intermediate filaments into epithelial and non-epithelial members. *EMBO J.* **2:** 2059.

Geisler, N., E. Kaufmann, S. Fisher, U. Plessmann, and K. Weber. 1983. Neurofilament architecture combines structural principles of intermediate filaments with carboxyterminal extensions increasing in size between triplet proteins. *EMBO J.* **2:** 1295.

Gilbert, W. 1978. Why genes in pieces? *Nature* **271:** 501.

Hamada, H., M. Petrino, and T. Kakunaga. 1982. A novel repeated element with Z-DNA forming potential is widely found in evolutionary diverse eukaryotic genomes. *Proc. Natl. Acad. Sci.* **79:** 6465.

Hanukoglu, I. and E. Fuchs. 1982. The cDNA sequence of a human epidermal keratin: Divergence of sequence but conservation of structure among intermediate filament proteins. *Cell* **31:** 243.

———. 1983. The cDNA sequence of a type II cytoskeletal keratin reveals constant and variable structural domains among keratins. *Cell* **33:** 915.

Ponte, P., P. Gunning, H. Blau, and L. Kedes. 1983. Human actin genes are single copy for α-skeletal and α-cardiac actin but multicopy for β- and γ-cytoskeletal genes: 3′ Untranslated regions are isotype specific but are conserved in evolution. *Mol. Cell. Biol.* **3:** 1783.

Quax, W., W. Vree Egberts, W. Hendriks, Y. Quax-Jeuken, and H. Bloemendal. 1983. The structure of the vimentin gene. *Cell* **35:** 215.

Quax-Jeuken, Y., W. Quax, and H. Bloemendal. 1983. Primary and secondary structure of hamster vimentin predicted from the nucleotide sequence. *Proc. Natl. Acad. Sci.* **80:** 3548.

Rogers, J. 1983. CACA sequences—The ends and the means? *Nature* **305:** 101.

Romer, A.S., ed. 1966. *Vertebrate paleontology.* 3rd edition. University of Chicago Press, Illinois.

Staden, R. 1982. An interactive graphics program for comparing and aligning nucleic acid and amino acid sequences. *Nucleic Acids Res.* **10:** 2951.

Steinert, P.M., R.H. Rice, D.B. Roop, B.L. Trus, and A.C. Steven. 1983. Complete amino acid sequence of a mouse epidermal keratin subunit and implications for the structure and intermediate filaments. *Nature* **302:** 794.

Zehner, Z.E. and B.M. Paterson. 1983. Characterization of the chicken vimentin gene: Single copy gene producing multiple mRNAs. *Proc. Natl. Acad. Sci.* **80:** 911.

Zimmer, C., S. Tymen, C. Marck, and W. Guschlbauer. 1982. Conformational transition of poly(dA-dC)•poly(dG-dT) induced by high salt or in ethanolic solution. *Nucleic Acids Res.* **10:** 1081.

Structure of Mouse Glial Fibrillary Acidic Protein and Its Expression by In Situ Hybridization Using a Cloned cDNA Probe

Sally A. Lewis, Joanna M. Balcarek, and Nicholas J. Cowan
Department of Biochemistry
New York University School of Medicine
New York, New York 10016

In spite of their relative abundance in the central nervous system, the role of astroglial cells is not well understood. Possible functions for this cell type include their involvement in the nutritional and structural support of neurons (Orkand 1975), regulation of neurotransmitters and ion exchange (Henn and Hamberger 1971; Henn et al. 1972), and neuronal migration during development (Hatten and Liem 1981). There is also evidence that astroglia may take part in the response of the central nervous system to injury and disease: For example, fibrous gliosis—a proliferation and hypertrophy of astrocytes—occurs following spinal cord injury and in a range of pathological conditions including multiple sclerosis and Alzheimer's disease (Bignami et al. 1980; Eng and DeArmond 1981). Neurological mutants of the mouse such as staggerer and jimpy show similar changes (Caviness and Rakic 1978) and may therefore provide useful experimental models.

All classes of astrocytes contain bundles of filaments known as glial filaments that extend from the cytoplasm into the processes characteristic of this cell type (Bignami and Dahl 1974). Because of their size (7–11 nm), glial filaments are classified as intermediate filaments and are assembled from an immunologically unique protein named glial fibrillary acidic protein (GFAP). The molecular weight of GFAP is 49,000–54,000, depending on the species of origin (Eng 1980). Although immunocytochemical studies using fluorescent anti-intermediate filament antisera have been useful in localizing specific intermediate filament proteins (including GFAP), such studies cannot address questions relating to gene expression at the transcriptional level. In addition, although a great deal of structural information on intermediate filament proteins has been obtained by sequence analysis of the proteins themselves, this approach is laborious in comparison with the sequence determination of a specific cloned cDNA. Furthermore, the construction of cloned intermediate-filament-specific probes provides the starting point for an exploration of the genes themselves and the factors that regulate their expression.

RESULTS AND DISCUSSION

Construction and Screening of a cDNA Expression Library from Mouse Brain mRNA

With the above considerations in mind, we sought the isolation of a GFAP-specific cDNA clone. Mouse was selected as the species because of (1) the extensive corpus of genetic information for this species, (2) its fully documented embryological and developmental history, and (3) the availability of neurological mutants. The strategem for the isolation of a GFAP cDNA clone was influenced by two factors. First, the abundance of GFAP in the brains of higher vertebrates (including mouse) is only on the order of 0.2%. Second, the size of GFAP is about 50,000 daltons, suggesting that the mRNA encoding GFAP might be close to the average size for eukaryotic mRNA species. We therefore avoided any attempt at physicochemical fractionation of brain mRNA. Instead, we exploited the availability of a GFAP-specific antiserum. Two possible methods were considered: (1) immunoprecipitation of GFAP-specific polysomes and (2) immunological screening of a cDNA library designed to express the cloned inserts as fusion proteins. The latter procedure was chosen because of the potential for screening large numbers of recombinants and also because such a library can be permanently stored and screened for other sequences of interest.

Library construction was carried out according to the method of Helfman et al. (1983). The procedure depends on the generation of random cDNA clones, using unfractionated poly(A)$^+$ mouse brain mRNA as the starting material. Double-stranded cDNA molecules are then inserted into a vector designed to express each cloned sequence as a fusion protein in a suitable bacterial host. Given that the cloned insert is in the correct orientation and reading frame, the fusion protein is potentially detectable with antisera to the protein of interest. Very briefly, the method is as follows: (1) synthesis of total mouse brain cDNA by primed extension with reverse transcriptase; (2) hydrolysis of the template strand and synthesis of a second strand by self-primed extension, also using reverse transcriptase; (3) "blunt-ending" the second strand by reaction with DNA polymerase I (Klenow fragment); (4) coupling of a synthetic oligonucleotide linker (e.g., an *Eco*RI linker) to the open (blunt) end of the double-stranded cDNA; (5) cleavage of the "hairpin" (closed) end of the cDNA with S1 nuclease; (6) coupling of a second synthetic oligonucleotide linker (e.g., *Sal*I) to the newly opened end; (7) cut-back of linkers using the appropriate restriction enzymes (e.g., *Eco*RI and *Sal*I); (8) insertion of the restriction-site-flanked cDNA molecules into the plasmid vectors pUC8 and pUC9. These vectors contain a selectable marker (Amp^r) and the *Escherichia coli lacZ* gene promoter (to facilitate expression of the inserted cDNA) located directly adjacent to the cloning site. The two vectors, pUC8 and pUC9, differ only in the orientation of their cloning sites; thus, by constructing libraries in each, the possible complication that cDNA molecules might contain restriction sites identical with those contained in the added oligonucleotide linkers can be taken into account. (9) Transformation of bacterial hosts; (10) plating out and screening of replicas using primary antibody; and (11) detection of positive clones using ^{125}I-labeled *Staphylococcus aureus* protein A.

Identification and Sequence of a GFAP-specific Clone

A single clone giving a positive signal with bovine anti-GFAP antiserum (Liem et al. 1978) was identified following a screen of about 40,000 clones. Hybrid-selection experiments were then performed using the purified plasmid DNA to confirm that the cloned sequences indeed encoded GFAP (Fig. 1). The filter-bound plasmid DNA selected mRNA from mouse brain (but not from mouse liver) that translated in a cell-free system to yield a polypeptide with a molecular weight of 49,000. This product was immunoprecipitable with the anti-GFAP antiserum used to screen the cDNA library. On the basis of this evidence, the strong amino acid homology between GFAP and other intermediate filament proteins predicted by the cDNA sequence (see below), and the known specificity of the anti-GFAP antibody (Liem et al. 1978), we conclude that the isolated cDNA clone encodes mouse GFAP.

The complete sequence of the GFAP cDNA clone is shown in Figure 2. The

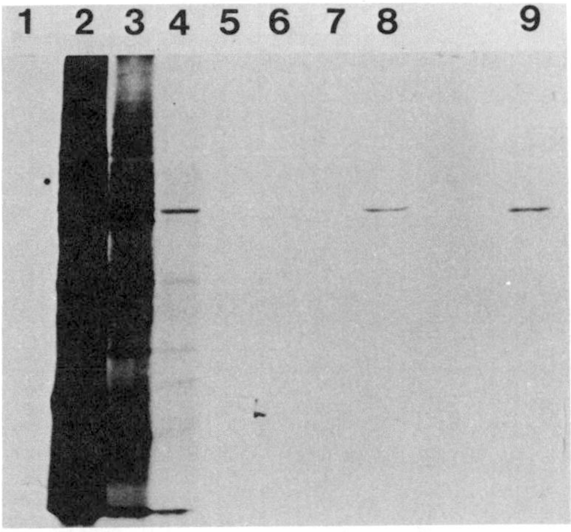

Figure 1
Translation of mRNA selected by a cDNA clone expressing mouse GFAP. Purified cDNA was linearized, denatured, and spotted onto a nitrocellulose filter. The filter was prehybridized and hybridized with 40 μg of mouse brain poly(A)$^+$. Selected RNA was eluted and translated in a rabbit reticulocyte lysate (Pelham and Jackson 1976) containing [^{35}S]-methionine in parallel with total mouse liver poly(A)$^+$ mRNA, total mouse brain poly(A)$^+$ mRNA, and a negative (H_2O) control. Two thirds of each translation mixture was immunoprecipitated with anti-GFAP antiserum (Liem et al. 1978). Total translation mixtures and immunoprecipitates were electrophoresed in a 9% SDS-polyacrylamide gel that was autoradiographed following fluorography. (*1–4*) Total translation products; (*5–8*) immunoprecipitated translation products; (*1,5*) no mRNA added; (*2,6*) mouse liver poly(A)$^+$; (*3,7*) mouse brain poly(A)$^+$; (*4,8*) GFAP cDNA-selected mRNA; (*9*) immunoprecipitation of GFAP fusion protein from an extract of [^{14}C]amino-acid-labeled bacteria expressing the cloned GFAP cDNA.

1 · 10 · 20 · 30
Leu Gly Thr Met Pro Arg Phe Ser Leu Ser Arg Met Thr Pro Pro Leu Pro Ala Arg Val Asp Phe Ser Leu Ala Gly Ala Leu Asn Ala
CTG GGT ACC ATG CCA CGT TTC TCC TTG TCT CGA ATG ACT CCT CCA CTC CCT GCC AGG▲GTG GAC TTC TCC CTG GCC GGG GCG CTC AAT GCT

40 · 50 · 60
Gly Phe Lys Glu Thr Arg Ala Ser Glu Arg Ala Glu Met Met Glu Leu Asn Asp Arg Phe Ala Ser Tyr Ile Glu Lys Val Arg Phe Leu
GGC TTC AAG GAG ACA CGG GCG AGC GAG CGT GCA GAG ATG ATG GAG CTC AAT GAC CGC TTT GCT AGC TAC ATC GAG AAG GTC CGC TTC CTG

70 · 80 · 90
Glu Gln Gln Asn Lys Ala Leu Ala Ala Glu Leu Asn Gln Leu Arg Ala Lys Glu Pro Thr Lys Leu Ala Asp Val Tyr Gln Ala Glu Leu
GAA CAG CAA AAC AAG GCG CTG GCA GCT GAA CTG AAC CAG CTT CGA GCC AAG GAG CCC ACC AAA CTG GCT GAT GTC TAC CAG GCG GAG CTT

100 · 110 · 120
Arg Glu Leu Arg Leu Arg Leu Asp Gln Leu Thr Ala Asn Ser Ala Arg Leu Glu Val Glu Arg Asp Asn Phe Ala Gln Asp Leu Gly Thr
CGG GAG CTG CGG CTG CGG CTG GAC CAG CTT ACG GCC AAC AGT GCC CGG CTG GAG GTG GAG AGG GAC AAC TTT GCA CAG GAC CTC GGC ACC

130 · 140 · 150
Leu Arg Gln Lys Leu Gln Asp Glu Thr Asn Leu Arg Leu Glu Ala Glu Asn Asn Leu Ala Ala Tyr Arg Gln Glu Ala Asp Glu Ala Thr
CTG AGG CAG AAG CTC CAA GAT GAA ACC AAC CTG AGG CTG GAG GCA GAG AAC AAC CTG GCT GCG TAT AGA CAG GAG GCA CAT GAA GCC ACC

160 · 170 · 180
Leu Ala Arg Val Asp Leu Glu Arg Lys Val Glu Ser Leu Glu Glu Glu Ile Gln Phe Leu Arg Lys Ile Tyr Glu Glu Glu Val Arg Asp
CTG GCT CGT GTG GAT TTG GAG AGA AAG GTT GAA TCG CTG GAG GAG GAG ATC CAG TTC TTA AGG AAG ATC TAT GAG GAG GAA GTT CGA GAT

190 · 200 · 210
Leu Arg Glu Gln Leu Ala Gln Gln Gln Val His Val Glu Met Asp Val Ala Lys Pro Asp Leu Thr Ala Ala Leu Arg Glu Ile Arg Thr
CTC CGG GAG CAG CTG GCC CAA CAG CAG GTC CAC GTG GAG ATG GAT GTG GCC AAG CCA GAC CTC ACA GCG GCC CTG AGA GAG ATT CGC ACT

220 · 230 · 240
Gln Tyr Glu Ala Val Ala Thr Ser Asn Met Gln Glu Thr Glu Glu Trp Tyr Arg Ser Lys Phe Ala Asp Leu Thr Asp Ala Ala Ser Arg
CAA TAC GAG GCA GTG GCC ACC AGT AAC ATG CAA GAG ACA GAG GAG TGG TAT CGG TCT AAG TTT GCA GAC CTC ACA GAC GCT GCG TCC CGC

250 · 260 · 270
Asn Ala Glu Leu Leu Arg Gln Ala Lys His Glu Ala Asn Asp Tyr Arg Arg Gln Leu Gln Ala Leu Thr Cys Asp Leu Glu Ser Leu Arg
AAC GCA GAG CTC CTC CGC CAA GCC AAA CAC GAA GCT AAC GAC TAT CGC CGC CAA CTG CAG GCC TTG ACC TGC GAT CTG GAG TCC CTC CGC

280 · 290 · 300
Gly Thr Asn Glu Ser Leu Glu Arg Gln Met Arg Glu Gln Glu Glu Arg His Ala Arg Glu Ser Ala Ser Tyr Gln Glu Ala Leu Ala Arg
GGC ACG AAC GAG TCC CTA GAG CGG CAA ATG CGC GAA CAG GAA GAG CGC CAT GCG CGG GAG TCG GCC AGT TAC CAG GAG GCA CTT GCT CGG

```
                                   310                                                 320                                                 330
Leu Glu Glu Glu Gly Gln Ser Leu Lys Glu Glu Met Ala Arg His Leu Gln Glu Tyr Gln Asp Leu Leu Asn Val Lys Leu Ala Leu Asp
CTG GAG GAG GAG GGC CAA AGC CTC AAG GAG GAG ATG GCC CGC CAC CTG CAG GAG TAC CAG GAT CTA CTC AAC GTT AAG CTA GCC CTG GAC

                                   340                                                 350                                                 360
Ile Glu Ile Ala Thr Tyr Arg Lys Leu Leu Glu Gly Glu Glu Asn Arg Ile Thr Ile Pro Val Gln Thr Phe Ser Asn Leu Gln Ile Arg
ATC GAG ATC GCC ACC TAC AGG AAA TTG CTG GAG GGC GAA GAA AAC CGC ATC ACC ATT CCT GTA CAG ACT TTC TCC AAC CTC CAG ATC CGA

                                   370                                                 380                                                 390
Glu Thr Ser Leu Asp Thr Lys Ser Val Ser Glu Gly His Leu Lys Arg Asn Ile Val Val Lys Thr Val Glu Met Arg Asp Gly Glu Val
GAA ACC AGC CTG GAC ACC AAA TCC GTG TCA GAA GGC CAC CTC AAG AGG AAC ATC GTG GTA AAG ACT GTG GAG ATG CGG GAT GGT GAG GTC

                                   400
Ile Lys Asp Ser Lys Gln Glu His Lys Asp Val Val Met Op
ATT AAG GAC TCG AAG CAG GAG CAC AAG GAC GTG GTG ATG TGA GGTGTGCCACCTGGTGGCCCTTGCCATGCAGTGTGAGGGCCCAAAGCTTAGCCTCAAATAGGC

CTGTTTGCCAGGCTCAGTTCCCACCCACACCAGCACTTCCCTTCCTTCCCTGGTTTTCTGCCTGTGTGCTGCCCAAGGCGTCAATCAGGTCATAAGCTTCATAGATGGCATAGATGGCA

TATACCCTTCACCTTCAACTAACAGGATACTCACCCCAAAGGCGCAGTCAGGAGGGGAGGGAACCCCAGCTGGTTAGAATTGGAAGGGAAGAGGAAAGATGAGCAGAGTAGAGAGATTT

AACAAATCACTTCCTTCATCCTTGTTGTTATGGAAACCGTTGCCAGAGCTGGAAGTTTCCACAGGCTGCTGGAGCTAGACAACAATTCAGACAGAAAGGGAAAGTCCCTGAGGCAAAGT

CTCTCTAGCCAGAGACCTATGCATCCCGAATGCCACTAAGGCAGTCCTGAAGGGCCCTCCAGGTGATGACTCCAGTGTGTCAGCCCCACTGAGCAGCTATGCAGGTTGACTGCCCACAG

GCATGTGAAACTTGGTTCTCAGCACTTGGCAGGGATCTATGGCATAAGTGGAGAGGGAAGGTGTACTGGACGGCGGAGAGGAGGGCTCCCTGGCCCCTAAGTGTGGATGCAGAGAGGTG

GAGCCCAGGAAGGGTCTCTGCTTAGGCTGCAGGGTACCGAATGGCAGAGGCACTGGTAGAGATCATTTGGACACTCGGAGTTGAAAGTTACAGGCAATCTGTTACACTTGGCTCTGAAT

CCTATCAATCAAGGAAATAACCCGTTCTCTGGAAGACACTGAAACAGGAGAGCAGGACTTCCGTCCACTGGGCAGGGTACAGATGTGTCTCAGTTGTGAAGGTCTATTCCTGGCTGCAC

AGTCCCCATCCGCTCAGGTCATCTTACCCCTGTGACTGCTCTCAGCCCTGGAAGAATCCACAACCATCCTTCCAAGGTTGTCCATCCCCACAATGACTAGCTGTTGCTCTCCAAGCTAA

GGGACCATTCCCTGCTCTATGCATATACGTAATGTCACCTATTTAGTATCATCCTATTTGAGAGTTTGAGGAACTGAAACACTGTTCTGTTCAAGCACCTGGTGCTATGCCTTCATATT

AGAGCACCTTCTCTGAGGCTGATTGGTGGGCAGGTAGGGAAGACATTGAGCAGACAGTGTCCGCTCAGTTGTCCTTCCCTCCCTTCCAAGGTCCCTCCCTCTTTCCAGGACATCGCCCC

CCCACCCCACCCCTCCTTTCCACCTCCGCTAACCTCCAGACGCACT
```

Figure 2

Nucleotide and predicted amino acid sequence of mouse GFAP. Dashed line indicates regions of homology with both hamster vimentin (Quax-Jeuken et al. 1983) and chicken gizzard desmin (Geisler and Weber 1982); dotted line indicates homology with only one of these proteins. Helical regions predicted by the intermediate filament model (Geisler and Weber 1982) are boxed. (▲) Demarcation between head and rod domains.

cloned insert is close to 2.5 kb, encodes 403 amino acids followed by a stop codon, and contains 1.4 kb of 3′-untranslated region. The amino acid sequence of GFAP predicted from the cDNA sequence reveals considerable homology with other intermediate filament proteins, especially desmin (Geisler and Weber 1982) and vimentin (Quax-Jeuken et al. 1983). The homology is particularly evident in the three α-helical domains postulated in structural models of intermediate filament proteins (Geisler and Weber 1982; Steinert et al. 1983; Quax-Jeuken et al. 1983). Even where amino acids in the α-helical regions differ, the charge or nonpolarity at that position is usually preserved. As in the case of desmin, charged and uncharged residues within the helix fall into clusters with a 28-residue (40A) repeat (Geisler and Weber 1982).

Evolutionary Conservation of GFAP Genes

The number of mouse genomic GFAP sequences and the ability of the mouse GFAP cDNA probe to cross-hybridize with genomic DNA from heterologous species were determined in DNA blot-transfer experiments (Fig. 3). Not more than two bands are evident in digests of mouse DNA probed with the GFAP cDNA clone, suggesting that there are at most two genes containing sequences ho-

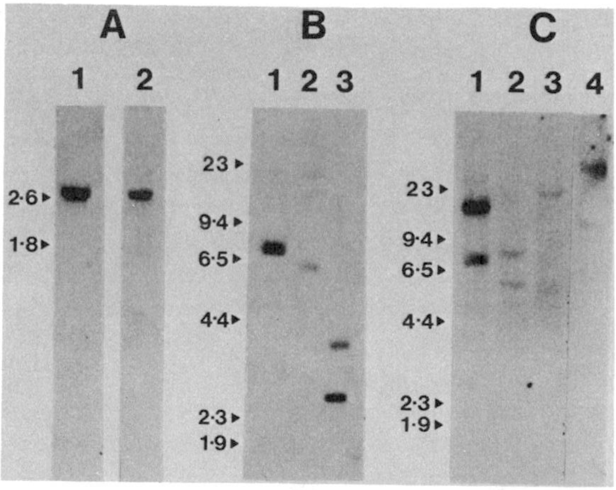

Figure 3

(*A*) RNA blot transfer of mouse brain poly(A)$^+$ mRNA. Mouse brain poly(A)$^+$ mRNA (10 μg) was resolved on a formaldehyde/agarose gel, transferred to nitrocellulose, and hybridized with 5×10^7 cpm of ^{32}P-labeled cloned GFAP-specific cDNA (lane *1*) or 2×10^7 cpm of ^{32}P-labeled subcloned fragment derived from the 3′-untranslated region of the GFAP-specific cDNA clone (lane *2*). (*B*) DNA blot transfer of mouse DNA. Aliquots (10 μg) of mouse DNA were digested with various restriction endonucleases, resolved in an agarose gel, transferred to nitrocellulose, and hybridized with 5×10^7 cpm of ^{32}P-labeled cloned GFAP-specific cDNA. (*1*) *Bam*HI; (*2*) *Hin*dIII; (*3*) *Sac*I. (*C*) Blot transfer of DNA from various vertebrate species. DNA from mouse (10 μg, lane *1*), rat (5 μg, lane *2*), human (10 μg, lane *3*), and chicken (10 μg, lane *4*) was digested with *Eco*RI, resolved on an agarose gel, and transferred to nitrocellulose. The blot was hybridized with 5×10^7 cpm of ^{32}P-labeled cloned GFAP-specific cDNA. Final wash conditions following hybridization were 0.3 M NaCl/0.02 M Na citrate/0.1% SDS at 68°C.

mologous to GFAP (Fig. 3). Analogous experiments performed on *Eco*RI digests of DNA from rat, human, and chicken reveal some cross-hybridization even at high stringency (Fig. 3). Thus, there appears to be significant interspecies conservation of GFAP sequences.

Expression of GFAP mRNA in Astrocytes Visualized by In Situ Hybridization

The size of GFAP mRNA was determined in an RNA blot-transfer experiment using mouse brain poly(A)$^+$ mRNA (Fig. 3). A single band of about 2.7 kb is evident. Thus, allowing for approximately 100 3′ dA residues, the 2.5-kb cloned GFAP cDNA must be almost full length.

Although RNA blot-transfer analysis is useful in defining the size and abundance of different RNA species, the technique is limited in the sense that the RNA to be analyzed must be purified from an organ or tissue that almost without exception consists of a heterogeneous population of cell types. A more informative approach would involve the visualization of distinct mRNA species within a histologically recognizable cell type. Although in situ hybridization potentially provides the means to accomplish this, experimental success has been hampered by problems of specificity (in terms of high background) and the lengthy exposure times required using ^3H probes. To circumvent these difficulties, we have performed experiments using mouse brain cryostat sections using the GFAP cDNA probe labeled by nick translation with [^{35}S]dNTP 5′ -[α-thio]triphosphates. The hybridization procedure, which is essentially that of Gee and Roberts (1983) and Hafen et al. (1983), yields signals that are highly specific and essentially free of background. Furthermore, because the emission energy of ^{35}S is tenfold greater than that of ^3H, and because of the higher specific activity obtainable with ^{35}S-labeled nucleotides, readily detectable signals are generated in 24–48 hours. An illustration of the method is shown in Figure 4, in which the GFAP probe uniquely identifies the pial cell layer in a shallow sagittal section of a 12-day-old mouse brain. The pial layer consists of astocytes that are clearly expressing GFAP mRNA, in contrast to other surrounding cells. The method is currently being applied to study the developmental regulation of tubulin gene expression in the mouse, using subcloned gene-specific 3′-untranslated region probes.

CONCLUDING REMARKS

The existence of regions of conserved amino acid sequence within a large variety of intermediate filament proteins points to a common evolutionary origin. From both protein and cDNA sequence data, a structural model for intermediate filament subunits is emerging (Geisler and Weber 1982; Steinert et al. 1983). Each intermediate filament protein contains a central α-helical domain that is significantly homologous among all intermediate filaments, as well as a nonhelical head and tail region that is unique to each class. The amino acid sequence of GFAP derived from the cDNA sequence (Fig. 2) is consistent with this model. The homology with other intermediate filament proteins, particularly desmin and vimentin, is evident. In particular, there is conspicuous homology within the three α-helical regions postulated in the intermediate filament protein model (Geisler and

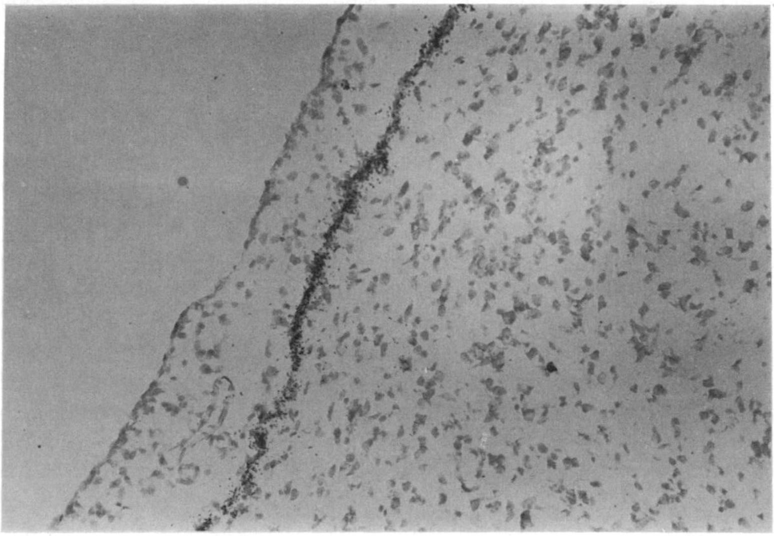

Figure 4
Detection of mouse GFAP mRNA by in situ hybridization. Sections (6 μm) of 12-day-old mouse brain were prepared using a cryostat microtome and mounted on gelatin-coated slides. Mounted sections were treated briefly with proteinase K, prehybridized, and hybridized as described previously (Gee and Robert 1983; Hafen et al. 1983) using the GFAP cDNA probe labeled by nick translation with [^{35}S]deoxyadenosine 5′-[α-thio]triphosphate.

Weber 1982; Steinert et al. 1983). The GFAP sequence also contains the heptade repeat found in other intermediate filament proteins, with a resulting helix of 3–5 residues per turn and a hydrophobic backbone. This backbone may be responsible for the aggregation of monomer subunits into coiled coils, which then polymerize into filaments.

Although there is extensive homology within the α-helical regions of intermediate filament proteins, structural data suggest that the amino- and carboxyterminal regions of these peptides are less conserved. The amino- and carboxyterminal domains of mouse GFAP are also unique. These nonhomologous regions may be responsible for the structural, functional, and antigenic individuality of different intermediate filament proteins. Strong antigenic determinants within nonconserved domains could explain the generation of non-cross-reacting antisera that show specificity for a given intermediate filament protein.

Although structural data point to a large multigene family encoding all intermediate filament proteins, the extent of evolutionary divergence has been such that no cross-hybridization is detectable at the nucleic acid level between different intermediate filament gene sequences at high stringency. However, within a given class, some cross-hybridization is detectable: The mouse GFAP probe detects sequences at high stringency in DNA from several vertebrate species. In each case, only one or two bands are visible per restriction digest, and it seems likely that a single functionally expressed gene encodes GFAP. The development of a rapid and sensitive in situ assay for the detection of specific mRNA sequences promises to yield data on the developmental regulation of this and other eukaryotic gene sequences.

REFERENCES

Bignami, A. and D. Dahl. 1974. Astrocyte-specific protein and neuroglial differentiation: An immunofluorescence study with antibodies to the glial fibrillary acidic protein. *J. Comp. Neurol.* **153:** 27.

Bignami, A., D. Dahl, and D.C. Reuger. 1980. Glial fibrillary acidic protein in normal neural cells and in pathological conditions. *Adv. Cell. Neurobiol.* **1:** 285.

Caviness, V. and P. Rakic. 1978. Mechanisms of cortical development: A view from mutations in mice. *Annu. Rev. Neurosci.* **1:** 297.

Eng, L.F. 1980. The glial fibrillary acidic (GFA) protein. In *Proteins of the nervous system* (ed. R.A. Bradshaw and D.M. Schneider), p. 85. Raven Press, New York.

Eng, L.F. and S.J. DeArmond. 1981. Glial fibrillary acidic protein immunocytochemistry in development and neuropathology. In *11th Congress of Anatomy: Glial and Neuronal Cell Biology*, p. 85. A.R. Liss, New York.

Gee, C.E. and J.L. Roberts. 1983. *In situ* hybridization histochemistry: A technique for the study of gene expression in single cells. *DNA* **2:** 159.

Geisler, N. and K. Weber. 1982. The amino acid sequence of chicken muscle desmin provides a common structural model for intermediate filament proteins. *EMBO J.* **1:** 1649.

Hafen, E., M. Levine, R.L. Garber, and W.J. Gehring. 1983. Improved *in situ* hybridization method for the detection of cellular RNAs in *Drosophila* tissue sections and its application for visualizing transcripts of homeotic *Antennapedia* gene complex. *EMBO J.* **2:** 617.

Hatten, M.E. and R.K.H. Liem. 1981. Astroglial cells provide a template for the positioning of developing cerebellar neurons *in vitro. J. Cell Biol.* **90:** 622.

Helfman, D.M., J.R. Feramisco, J.C. Fiddes, P. Thomas, and S.H. Hughes. 1983. Identification of clones that encode chicken tropomyosin by direct immunological screening of a cDNA expression library. *Proc. Natl. Acad. Sci.* **80:** 31.

Henn, F.A. and A. Hamberger. 1971. Glial cell function: Uptake of transmitter substances. *Proc. Natl. Acad. Sci.* **68:** 2686.

Henn, F.A., J. Haljame, and A. Hamberger. 1972. Glial cell function: Active control of extracellular K^+ concentration. *Brain Res.* **43:** 437.

Liem, R.K.H., S.H. Yen, G.D. Salomen, and M.L. Shelanski. 1978. Intermediate filaments in nervous tissues. *J. Cell Biol.* **79:** 637.

Orkand, R.K. 1975. Glial Cells. In *Handbook of physiology*, p. 855. Academic Press, New York.

Pelham, H. and R.T. Jackson. 1976. An efficient mRNA dependent translation system from reticulocyte lysates. *Eur. J. Biochem.* **67:** 247.

Quax-Jeuken, Y.E.F.M., W.J. Quax, and H. Bloemental. 1983. Primary and secondary structure of hamster vimentin predicted from the nucleotide sequence. *Proc. Natl. Acad. Sci.* **80:** 3548.

Steinert, P.M., R.H. Rich, D.R. Roop, B.L. Rus, and A.C. Steven. 1983. Complete amino acid sequence of a mouse epidermal keratin subunit and implications for the structure of intermediate filaments. *Nature* **302:** 794.

Characterization of the Chicken Actin Multigene Family

Robert J. Schwartz, Kun Sang Chang, Warren E. Zimmer, Jr., and Derk J. Bergsma

Department of Cell Biology
Baylor College of Medicine
Houston, Texas 77030

The actins are a family of highly conserved contractile proteins found in all eukaryotic cells. Multiple actin isoforms have been identified previously (Garrels and Gibson 1976; Vandekerckhove and Weber 1978), and some of these proteins are associated with diverse functions such as muscle contraction, cytoskeletal structure, cell motility, and chromosome movement (Pollard and Weihing 1974). Recombinant DNA techniques have shown that actin isoforms are represented by multigene families in a broad host of eukaryotic genomes (Kindle and Firtel 1978; Cleveland et al. 1980; Fyrberg et al. 1980; Schwartz and Rothblum 1980; Scheller et al. 1981; Engel et al. 1982).

From previous studies, the chicken genome was estimated to contain between 6 and 11 actin genes (Cleveland et al. 1980; Schwartz and Rothblum 1980). In this preliminary paper, we describe the screening of a chicken genomic library to obtain six different genes that represent a major portion of the known vertebrate actin isoforms. A more detailed report will be published elsewhere (Chang et al. 1984). Nucleic acid sequencing allowed us to identify α-skeletal, α-cardiac, α-smooth-muscle, and β-actin genes. A fragment of one gene was identified as a nonmuscle type that appeared to be expressed in the gizzard. Another chicken gene was identified as a nonmuscle actin that closely resembled the amphibian type-5 actin (Vandekerckhove et al. 1981). Each actin gene contains nonoverlapping DNA sequences and is represented by only a single copy in the chicken genome.

DISCUSSION

Separation of Actin Genomic Clones by Sequence Homology

For a number of multigene families, differences at the nucleic acid level have provided sufficient sequence divergence to discriminate between closely related genes. Studies of globin (Benz et al. 1977), vitellogenin (Wahli et al. 1979), and chorion (Sim et al. 1979) gene families have shown that a ΔT_m of 5°C or better would help to discriminate between related nucleic acid sequences. Previously, we showed that a ΔT_m of 10–13°C exists between chicken α-skeletal actin cDNA and that of β, γ, and smooth-muscle actin mRNAs (Schwartz et al. 1980; Schwartz and Rothblum 1981). Although these differences in nucleic acid homology allowed us to develop stringent hybridization conditions to quantitate actin mRNA species during myogenesis in culture (Schwartz and Rothblum 1981), it also provided a rationale in the present study for grouping homologous actin genes. From multiple screenings of a chicken genomic library (Dodgson et al. 1979), we obtained 30 clones that contained actin-coding sequences. Actin cDNA clones and poly(A)-containing RNA isolated from tissues enriched in different actin isoforms were used as probes for hybridization to actin genomic clones. Figure 1 illustrates the results of an experiment in which multiple samples of DNA purified from actin genomic clones (λAC) were spotted onto filters as dots and hybridized with various ^{32}P-labeled probes under conditions of 50% formamide and 0.5 M NaCl at 37°C. Since our primary interest was to obtain clones that contain homologous sequences, the actin genomic clones were evaluated by washes at 60–65°C followed by autoradiography.

When pA1, a cloned β-actin cDNA (Cleveland et al. 1980), was used as a probe, λAC9, λAC11, and λAC29 were found to share the closest homology with β-actin cDNA. Similarly, when pAC269 was used as a probe, both λAC5 and λAC25 showed hybridization of the highest stringency (Fig. 1). Even though these findings suggest that λAC5 and λAC25 were homologous to α-skeletal actin, it was possible that these clones might represent the closely related α-cardiac actin genes. To differentiate between these two possibilities, [^{32}P]cDNA was synthesized from cardiac poly(A)-containing RNA and then hybridized to the actin genomic clones. The presumed α-skeletal actin clones were not among the eight cardiac actin clones that survived stringent thermal washes (Fig. 1).

Brain tissue serves as a rich source of β- and γ-actin mRNAs (Cleveland et al. 1980). Since brain tissue is highly vascularized, we assume that it might also serve as a source of α-smooth-muscle actin mRNA. Hybridization probes were prepared by random cleavage of actin mRNA with alkaline Tris-HCl and were subsequently end-labeled with [γ-^{32}P]ATP and polynucleotide kinase. Hybrids to actin genomic clones λAC3, λAC8, λAC9, λAC11, λAC12, λAC17, λAC23, and λAC24 were resistant to stringent washes at 65°C. Since clones λAC9 and λAC11 were shown in Figure 1 to correspond to the β-actin gene clones, then λAC3, λAC8, λAC12, λAC17, λAC23, and λAC24 could represent the γ-actin and/or the vascular smooth-muscle actin genes. Hybridization probes were then developed from actin mRNAs isolated from gizzard and aorta tissues to help identify the smooth-muscle actin genes. Actin mRNA homologous to λAC2 was found primarily in the gizzard, whereas clones λAC3 and λAC17 correspond to actin mRNA found in the aorta and vascularized tissues. Therefore, comparison of chimeric

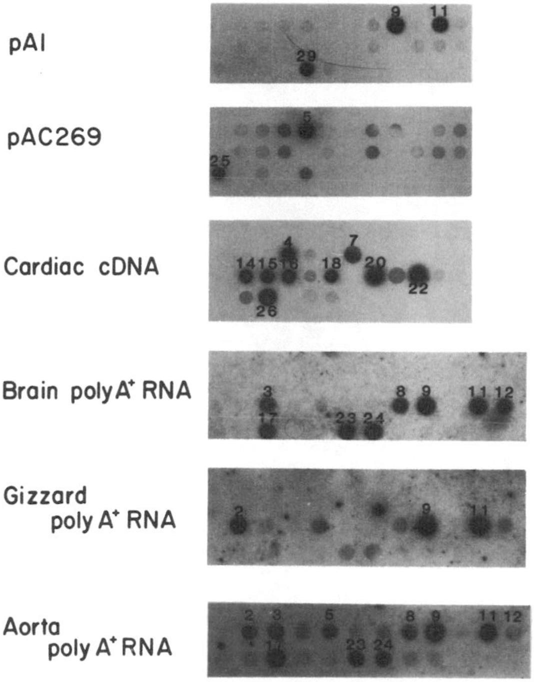

Figure 1
Identification of chicken actin genes screened from a λ genomic library were boiled in 0.3 M NaOH for 5 min, neutralized with 2 M ammonium acetate, and spotted onto nitrocellulose filters with the aid of a filtration manifold. Filters were hybridized with ^{32}P-labeled probes prepared from the following sources: pA1, β-actin cDNA clone; pAC269, α-skeletal actin cDNA clone; cardiac cDNA; poly(A)$^+$ gizzard RNA; poly (A)$^+$ aorta RNA. In the last three filters, clones λAC15, λAC16, λAC18, λAC20, λAC22, λAC25, and λAC29 were not spotted since they were known to be duplicates of other clones. Filters were washed under standard conditions and then with 50% formamide and 3 × SSC at 65°C for 15 min. Filters were autoradiographed on X-ray film. (Reprinted, with permission, from Chang et al. 1984.)

phase DNAs by their thermal stability allowed us to assign preliminarily the actin genes into six different groups: α-skeletal (λAC5 and λAC25), α-cardiac (λAC4, λAC7, λAC14, λAC15, λAC16, λAC18, λAC22, and λAC26), β (λAC9, λAC11, and λAC29), vascular smooth-muscle-enriched (λAC3 and λAC17), gizzard-enriched (λAC2), and nonmuscle-type (λAC8, λAC12, λAC23, and λAC24) actins.

*Hae*III Digests of Actin Genomic Clones

Members of each group were further analyzed by the electrophoretic pattern of restriction fragments generated by digestion with *Hae*III. The separated fragments were transferred to a nitrocellulose filter and hybridized to labeled pAC269. The autoradiograph in Figure 2 shows that the coding region of the actin genes was dissected into several heterogeneously labeled fragments. These autoradi-

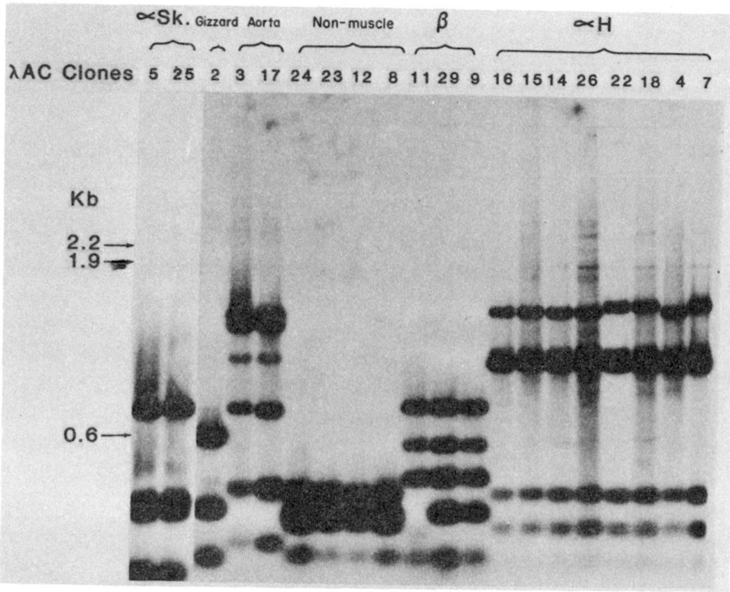

Figure 2
*Hae*III digestion of chicken actin genomic DNA clones. Purified bacteriophage λ DNA (2 μg) was digested to completion with 5 units of *Hae*III. The reaction mixtures were heated at 68°C for 5 min before loading and electrophoresis on a 1.5% agarose gel. DNA fragments were blotted onto a nitrocellulose filter. The filter was probed with ^{32}P-labeled nick-translated pAC269 DNA. Numbers at the top of each lane refer to the λ-actin genomic clones. (Reprinted, with permission, from Chang et al. 1984.)

ographs reveal a distinct restriction pattern for each group of actin genomic clones. For example, all of the eight cardiac actin genomic clones shared the same set of restriction fragments. The largest of the five *Hae*III fragments (~1.7 kb) varied in size because it was contiguous to one of the λ Charon 4A arms. The α-skeletal actin gene represented by the two overlapping clones contained at least three identically sized radioactive fragments. Even though cardiac and skeletal α-actins are closely related at the protein level, their genes did not have the same DNA fragment patterns. The dissimilar *Hae*III restriction pattern of the other four groups of genomic clones demonstrated the presence of different genes. These restriction digests confirmed our preliminary assignment of the actin genomic clones into six distinct groups.

Restriction Maps of the Actin Genomic Clones

The actin-coding sequences and genomic flanking regions were located on chimeric phage DNA by mapping with infrequent-cutting restriction enzymes (Fig. 3). Individual phage were singularly digested with at least *Eco*RI, *Bam*HI, *Xho*I, and *Xba*I, and then double digested with all possible combinations. The actin structural gene was positioned by transferring the restriction fragments onto ni-

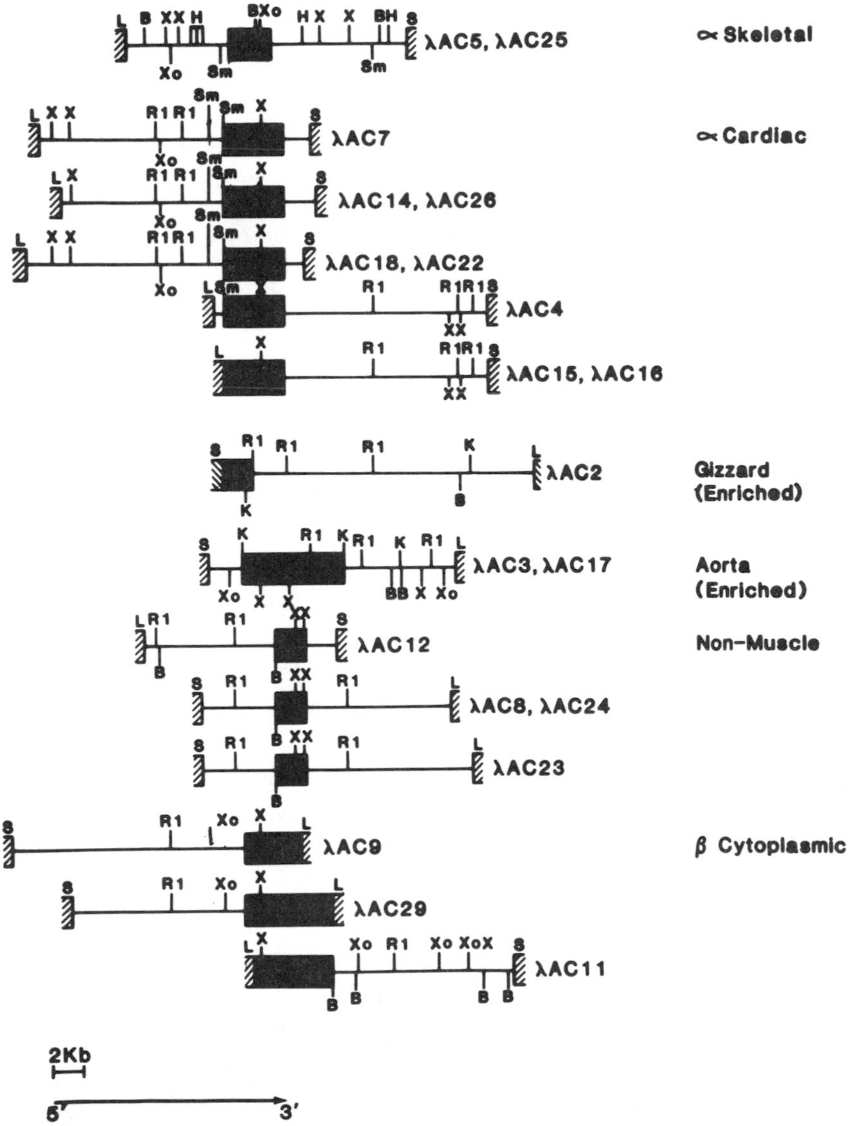

Figure 3
Restriction maps of isolated chicken actin genomic clones. The locations of actin-coding regions were demonstrated by hybridization of the restriction fragments to nick-translated pAC269 and are shown as solid black bars. L and S represent long and short arms of λ Charon 4A DNA. All of the clones were mapped with the following restriction enzymes: *Bam*HI (B); *Eco*RI (R1); *Xba*I (X); *Xho*I (Xo). In addition, a few clones were mapped with *Hin*dIII (H), *Kpn*I (K), and *Sma*I (Sm). Their exclusion on other maps does not suggest the absence of cutting sites. (Reprinted, with permission, from Chang et al. 1984.)

trocellulose filters and hybridization with ^{32}P-labeled pAC269. Figure 4 diagrams and summarizes the restriction endonuclease maps of chimeric phage DNA.

These restriction maps provided the following salient points about the genomic actin clones. First, actin clones grouped by thermal analysis (Figs. 1 and 2) and *Hae*III digestion (Fig. 2) were found to contain overlapping restriction maps to the same identical portion of genomic DNA. Thus, each group of actin clones represents a separate nonalleic actin gene. Second, a unique restriction enzyme digestion pattern was also mapped for the single clone λAC2. The absence of genomic flanking sequences in λAC2, presumably removed from this gene during the construction of the genomic library, precludes any statement about its immediate linkage to the other actin genes. Finally, restriction maps showed that the other five actin genes were derived from different and nonoverlapping regions of genomic DNA.

Identification of the Six Isolated Actin Genes

To identify these isolated genes as specific actin isoforms, it was necessary to sequence portions of their DNA coding regions. The limits of the coding regions allowed us to determine the first 17 amino acids at the amino terminus and the amino acids toward the carboxyl terminus of the isolated actin genes as shown in Figure 4. The amino acid sequences at amino termini allow for definitive identification of each actin isoform, whereas the carboxyterminal polypeptide sequences provide information that separates actins into either striated or nonmuscle isoforms. The gene sequences shown to be homologous to pAC269 were represented by two overlapping clones λAC5 and λAC25. Their aminoterminal sequence contains Asp-Glu-Asp-Glu-Thr-Thr, Cys, Leu, and Val at amino acid positions 1–6, 10, 16, and 17, which identifies this gene as the α-skeletal actin isoform, in agreement with the preliminary dot blots. The actin gene that hybridized selectively to ^{32}P-labeled cardiac cDNA was represented by the largest number of overlapping clones and was found to contain amino acids diagnostic for a striated actin. The Asp and Glu at positions 2 and 3 as well as the Leu and Ser at positions 298 and 357 prove that this gene codes for the α-cardiac isoform (Vandekerckhove and Weber 1978, 1979). One group of actin genomic DNAs (λAC3 and λAC17) that hybridized selectively to aorta and brain RNAs was found to contain amino acids Glu-Glu-Glu-Asp-Ser-Thr-, Cys, Leu, and Cys at positions 1–6, 10, 16, and 17. This gene represents the striated α-smooth-muscle actin isoform (Vandekerckhove and Weber 1979). Each of these three different α-actin genes contain initiator methionine-cysteine codons thought to be processed from primary translation products (Gunning et al. 1983).

We initially identified λAC2 as an actin gene whose product is enriched in gizzard smooth-muscle tissue. Sequencing revealed that the 5′ portion of this gene was truncated at the 42nd amino acid, which precluded an exact identification of this actin. Nevertheless, the amino acid sequences near the carboxyl terminus showed that λAC2 cannot be categorized as a striated isoform, but instead must encode a nonmuscle actin. The actin clones that showed a high degree of specificity to the β-actin cDNA clone, pA1, were found to contain amino acids Asp-Asp-Asp-Ile-Ala, Val, Met, and Cys at positions 2–5, 10, 16, and 17. These amino

```
                        NH2 Terminal Amino Acid Sequence

1               5                   10                  15
Met Cys Asp Glu Asp Glu Thr Thr Ala Leu Val Cys Asp Asn Gly Ser Gly Leu Val  α-Skeletal
ATG TGT GAC GAG GAC GAG ACC ACC GCG CTC GTG TGC GAC AAC GGC TCC GGC CTG GTG
        --- --- --- --- --- ---             ---                     --- ---
Met Cys Asp Asp Glu Glu Thr Thr Ala Leu Val Cys Asp Asn Gly Ser Gly Leu Val  α-Cardiac
ATG TGT GAC GAC GAG GAG ACC ACC GCG CTG GTG TGC GAC AAC GGC TCG GGG CTG GTC
        --- *** *** --- --- ---             ---                     --- ---
Met Cys Glu Glu Glu Asp Ser Thr Ala Leu Val Cys Asp Asn Gly Ser Gly Leu Cys  α-Smooth
ATG TGT GAG GAG GAG GAC AGC ACT GCC CTT GTT TGT GAC AAT GGC TCA GGG CTC TGT
        +++ +++ +++ +++ +++ ---             ---                     --- +++
        Met Asp Asp Asp Ile Ala Ala Leu Val Val Asp Asn Gly Ser Gly Met Cys  β
        ATG GAT GAT GAT ATT GCT GCG CTC GTT GTT GAC AAT GGC TCC GGT ATG TGC
        === === === === ===                 ===                     === ===
    Met Ala Asp Glu Glu Ile Ala Ala Leu Val Val Asp Asn Gly Ser Gly Met Cys  Type 5
    ATG GCA GAT GAG GAA ATT GCT GCC CTG GTG GTG GAC AAT GGC TCT GGT ATG TGC
            === === === === ===                 ===                     === ===

                        COOH Terminal Amino Acid Sequence

                        271 278 286 296 298 357 364
                        Ala Tyr Ile Asn Met Thr Ala  α-Skeletal
                        GCT TAC ATT AAC ATG ACA GCC

                        Ala Tyr Ile Asn Leu Ser Ala  α-Cardiac
                        GCT TAC ATT AAC CTC AGC GCT
                                        *** ***
                                            Ser Ala  α-Smooth
                                            AGC GCT

                        Cys Phe Val Thr Leu Ser Ser  β
                        TGT TTC GTG ACA CTG AGC TCC

                        Cys Phe Val Thr Leu Ser Ser  Type 5
                        TGC TTC GTG ACC CTG AGC TCT

                        Cys Phe Val         Ser Ser  Non-Muscle
                        TGC TTC GTC         AGC TCC
```

Figure 4
Identification of the actin genomic clones by partial DNA sequencing. Partial nucleic acid sequences obtained from the amino- and carboxyterminal regions of the subcloned DNA fragments are shown above. Aminoterminal sequences underlined with the following symbols represent the diagnostic amino acids in striated skeletal (— — —), cardiac (* * *), α-smooth-muscle (+ + +), and nonmuscle (= = =) actins. The diagnostic amino acids toward the carboxyl terminus show the partition into muscle and nonmuscle actin types. (Reprinted, with permission, from Chang et al. 1984.)

acids clearly identify this gene as the nonmuscle β-actin isoform (Vandekerckhove and Weber 1978). The third nonmuscle actin expressed in brain and vascular tissues is represented by the overlapping clones λAC8, λAC12, λAC23, and λAC24. Unexpectantly, this nonmuscle actin type is not identical to either β- or γ-actins found in vertebrate tissues, nor has this type been previously identified in birds or mammals. Instead, the amino acids Asp-Glu-Glu at positions 2–4 show a striking similarity to the aminoterminal region of the type-5 nonmuscle actin found in amphibians (Vandekerckhove et al. 1981) and to the Dma 2 actin gene found in *Drosophila* (Fyrberg et al. 1981). This chicken nonmuscle actin gene also contains an initiation methionine and an alanine codon at amino acid position 1, which is missing from the mature *Xenopus* protein. This chicken actin gene might represent a primitive actin isoform that is still expressed in higher vertebrates.

The Chicken Actin Multigene Family Is Represented by at least Eight to Ten Different Genes

Even though six different actin genes were isolated and identified, it is still not clear how many chicken actin genomic loci actually exist. To demonstrate how many different actin-coding loci are represented in the chicken genome, we digested nuclear DNA with a variety of infrequent-cutting restriction enzymes and Southern blotted. The filters were then probed with a small DNA-coding fragment (0.46 kb *Xba*I) that overlaps amino acids 221–374 of the nonmuscle type-5-like actin gene. Our rationale for using a short probe was that these restriction enzymes would rarely cut within a 460-nucleotide fragment; thus, each labeled DNA fragment observed on the genomic blot might represent a single coding locus. We found eight *Bam*HI and ten *Bgl*I fragments as shown in Figure 5 (left panel). In addition, two genomic DNA blots were hybridized with ^{32}P-labeled probes prepared from either the full-length α-skeletal or the β-actin cDNA clones using standard conditions. The autoradiograph in Figure 5 (right panel) shows that the same major *Eco*RI fragments of 32, 22, 16, 13, 9.8, 6.8, 5.8, 4.8, 4.5, 3.3, and 3.1 kb were labeled by hybridization to both actin cDNA probes. Therefore, the chicken genome is probably represented by a minimum of eight to ten actin genes, since other actin sequences that are more divergent than the coding probes used here might not be detected by this assay.

The identities of the predominantly labeled genomic bands were then determined by hybridization to blots with DNA probes developed from each of the cloned actin genes (Fig. 3). Blots were then washed at increased thermal stringency to eliminate most cross-hybridizations and retain hybrids to the highly homologous sequences. The results shown in Figure 5 support the conclusion that the chicken genome contains actin-coding sequences on at least 11 *Eco*RI fragments. Seven of these fragments are represented by six distinct actin genes, whereas the 5.8-, 4.8-, 4.5-, and 3.1-kb *Eco*RI fragments remain to be identified.

Actin Genes Are Surrounded by Repetitive DNA

Interspersed repeated sequence elements in eukaryotes have been proposed to regulate gene expression at the level of transcription (Britten and Davidson 1969) or RNA processing (Davidson and Britten 1979). Only recently, however, has it been possible to make a precise analysis of the organization and transcription of repetitious DNA elements. We applied a routine test to each of the actin genes to determine the presence of repeated sequences within the cloned DNA fragments (Shen and Maniatis 1980). This involved hybridization of total genomic DNA labeled wtih ^{32}P by nick translation to actin genomic clones spotted onto nitrocellulose filters. Autoradiographs of the upper left panel in Figure 6 showed that the α-skeletal (λAC5 and λAC29) and the β-actin (λAC9, λAC11, and λAC29) genes contain high levels of repetitious DNA, whereas lower levels were detected for the α-cardiac and the type-5-like nonmuscle actin genes. Neither the α-smooth-muscle actin gene nor the genomic fragment of the unidentified nonmuscle actin appeared to contain repeated DNA sequences. To determine the organization of repeated DNA sequences within actin genomic DNA, we probed nitrocellulose blots of restriction-enzyme-digested actin clones with ^{32}P-labeled chicken DNA.

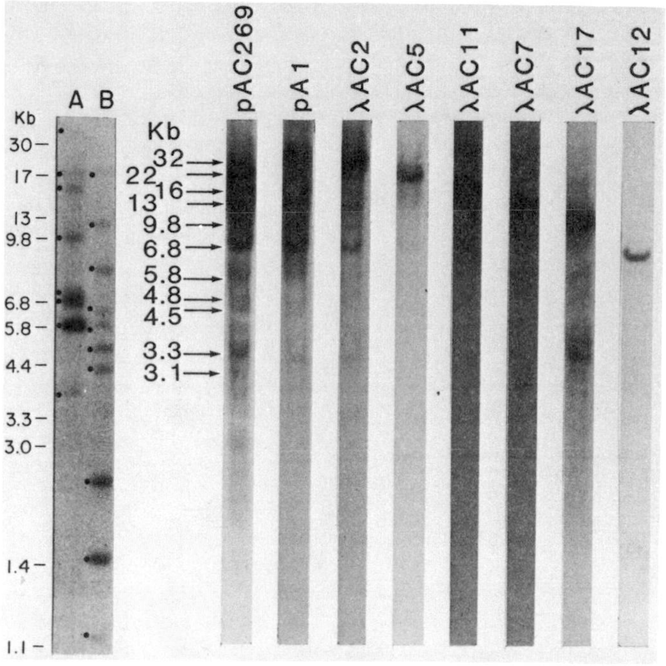

Figure 5

Restriction-endonuclease-digested chicken genomic DNA hybridized with actin natural gene fragments. (*Left*) Purified chicken genomic DNA (10 μg/lane) was digested separately with 100 units of *Bam*HI (slot *A*) and *Bgl*I (slot *B*) at a DNA concentration of 0.1 mg/ml for 8 hr. Digested DNA was electrophoresed in a 0.8% agarose gel and transferred to a nitrocellulose filter. Filters were probed with a ^{32}P-labeled nick-translated 460-bp *Xba*I fragment taken from the coding region of the type-5-like actin gene (λAC12). Since the intensity of the autoradiographic bands was highly variable, we placed an ink dot next to each band that was observed on exposed X-ray film. (*Right*) Chicken genomic DNA was digested with *Eco*RI and blotted as described above. Each lane was hybridized with one of the following ^{32}P-labeled probes: pAC269, pA1, λAC2 (1.5-kb *Eco*RI fragment), λAC5 (2.0-kb *Bam*HI-*Sma*I fragment), λAC11 (4.3-kb *Bam*HI-*Xba*I fragment), λAC7 (2.7-kb *Xba*I-*Sma*I fragment), λAC17 (5.9-kb *Kpn*I fragment), and λAC12 (6.6-kb *Eco*RI fragment). Filters probed with *Xba* fragment pAC269 were washed with 1 × SSC and 0.25% SDS. The other filters were washed with the same solution and then once in 0.5 × SSC and 0.15% SDS for 15 min at 65°C. (Reprinted, with permission, from Chang et al. 1984.)

Figure 6 (lower left panel) shows the hybridization pattern and the location of repetitive DNA summarized on restriction maps (right panel). This methodology demonstrates that the α-skeletal, β-, and type-5-like actin genes are closely surrounded by repetitive DNA elements, whereas repeated sequences flank only the 3′ end of the α-cardiac actin gene. In all cases, repeated DNA sequences failed to hybridize to the actin-coding region. Therefore, under these hybridization conditions, sequences that are repeated more than ten times in the genome were detected. We found by hybridization analysis that the repeated DNA se-

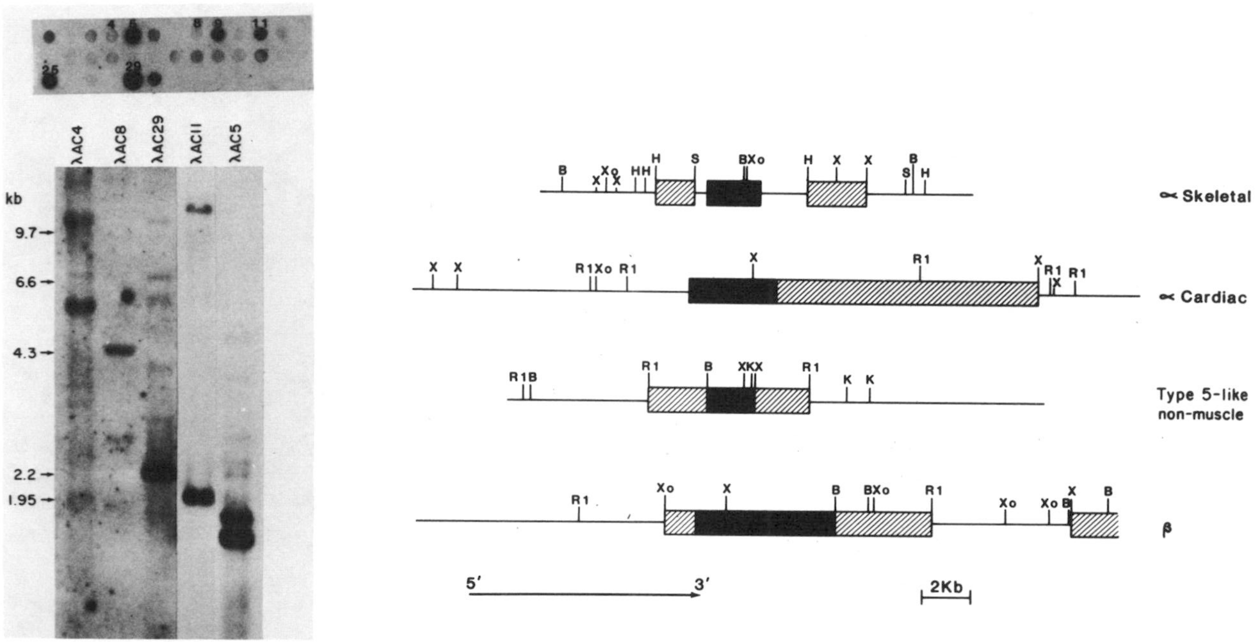

Figure 6

Localization of repeated DNA sequences on actin genomic DNA. Repeated DNA sequences in actin genomic clones were localized according to the method of Shen and Maniatis (1980). Purified DNA obtained from chicken livers was ^{32}P-labeled (10^{-8} cpm/μg DNA) and hybridized to a dot filter (*upper left*), which contains λ-actin genomic clones (see Fig. 1 and text), and to a DNA blot filter of electrophoresed restriction fragments (*lower left*). DNAs were digested with the following restriction enzymes: λAC4 (*Eco*RI-*Xba*I); λAC8 (*Eco*RI-*Kpn*I); λAC29 (*Eco*RI-*Xho*I-*Xba*I); λAC11 (*Eco*RI-*Xba*I); and λAC5 (*Xba*I-*Hin*dIII-*Sma*I). Sizes of the probed actin genomic DNA are as follows: λAC4 (4.5 kb and 7.5 kb); λAC8 (4 kb and 2.4 kb); λAC29 (2.2 kb); λAC11 (1.9 kb); λAC5 (1.6 kb and 1.2 kb). Restriction map at the right indicates the location of repetitious DNA (hatched area) and actin-coding regions (black area). (Reprinted, with permission, from Chang et al. 1984.)

quences that flank the α-skeletal actin-coding regions were not related to repetitious elements located on the other actin genes.

Actin Gene Switching during Myogenesis

We utilized the six unique actin genes as probes to examine the appearance and disappearance of actin mRNA species during chick embryonic leg-muscle development. The λ-actin clones were spotted onto filters in dots of uniform diameter and DNA content. Recombinant DNA clones λAC5, λAC4, λAC9, λAC17, λAC24, and λAC2 represented the α-skeletal, α-cardiac, β-cytoplasmic, α-smooth-muscle, type-5-like, and an unidentified nonmuscle actin genes. Each filter was then hybridized with ^{32}P-end-labeled poly(A) RNA isolated from chick embryonic muscle from one of four successive days of development. The filters were washed under highly stringent conditions (65°C; 0.4 M Na^{++} and 50% formamide) so that only the most homologous sequences would be retained by the genomic actin clone. As shown in Figure 7, the results indicate that the β-actin mRNA is the predominant actin transcript (95%; 4000 copies/cell nucleus) at 10 and 12 days of chick embryonic development. The thigh is mostly composed of replicating myoblasts (day 10) and fusing myoblasts (day 12) at these developmental time points. RNA products representing the α-cardiac actin gene appear to be expressed to a low degree in replicating myoblasts (5% of the actin RNA), whereas transcripts from the α-skeletal actin gene were not detected. However, by 16 days of chick leg development, the α-skeletal actin mRNA becomes the prevalent actin mRNA species (95%; 20,000 copies/cell nucleus) and the level of β-actin mRNA decreases. At this time of development, the thigh is composed of myotubes that are innervated by motor neurons and responsive to electrical stimulation (McLennan 1983). These results show that the expression of the α-skeletal actin within the actin multigene family is switched during myogenesis through a strict developmental pattern in the intact embryo.

SUMMARY

Genes representing six different actin isoforms were isolated from a chicken genomic library. Cloned actin cDNAs as well as tissue-specific mRNAs enriched in different actin species were used as hybridization probes to group individual ac-

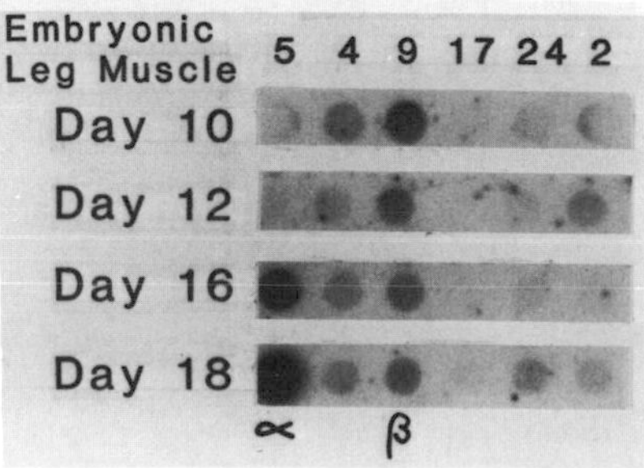

Figure 7
Expression of chicken actin genes during in vivo myogenesis. Genomic clones representing α-skeletal (λAC5), α-cardiac (λAC4), β-(λAC9), α-smooth-muscle (λAC17), type-5 (λAC24), and nonmuscle (λAC2) actin were spotted onto nitrocellulose filers and hybridized to ^{32}P-end-labeled RNA isolated from embryonic leg muscle. Filters were washed under highly stringent conditions (0.5 M Na^{+}, 50% formamide at 65°C) and then autoradiographed onto X-ray film.

tin genomic clones by their relative thermal stability. This methodology allowed us to isolate six distinct actin genes. Restriction maps showed that these actin genes were derived from separate and nonoverlapping regions of genomic DNA. Five of the six isolated genes included sequences from both the 5′ and 3′ ends of the actin-coding area. Amino acid sequence analysis from both the amino- and carboxyterminal regions provided for their unequivocal identification. The striated isoforms were represented by the isolated α-skeletal, α-cardiac, and α-smooth-muscle actin genes. The nonmuscle isoforms included the β-cytoplasmic actin gene and an actin gene fragment that lacked the 5′-coding and -flanking sequence; presumably, this region of DNA was removed from this gene during construction of the genomic library. Unexpectantly, a third nonmuscle chicken actin gene was found that resembled the amphibian type-5 actin isoform (Vandekerckhove et al. 1981). This nonmuscle actin type has not been previously detected in warm-blooded vertebrates. Analysis of genomic DNA blots showed that the chicken actin multigene family is represented by eight to ten separate coding loci. The 6 isolated actin genes correspond to 7 of 11 genomic *Eco*RI fragments. Only the α-smooth-muscle actin gene was shown to be split by an *Eco*RI site. Although we have not yet isolated the γ-smooth-muscle and γ-cytoplasmic actin genes, it appears that each actin isoform is encoded by a single gene in the chicken genome. Finally, the six different actin genes were used as probes to examine chicken myogenesis. The α-skeletal actin was found to be induced during late muscle development in the intact embryonic leg. We are currently developing specific 3′ noncoding actin DNA probes to quantitate transcription rates of various actin genes during muscle development, as well as examine tissue-specific expression during embryogenesis with in situ hybridization probes.

ACKNOWLEDGMENTS

We thank Paul Hobus for excellent technical assistance. This work was supported by U.S. Public Health Service grants NS-15050 and NS-16494 and a grant from the Muscular Dystrophy Association to R.J.S. R.J.S. is a recipient of a U.S. Public Health Service Career Development Award.

REFERENCES

Benz, E.J., C.E. Geist, A.W. Steggles, J.E. Barker, and A.W. Neinhus. 1977. Hemoglobin switching in sheep and goats: Preparation and characterization of complementary DNAs specific for the α, β, and γ-globin mRNAs of sheep. *J. Biol. Chem.* **252:** 1908.

Britten, R.J. and E.H. Davidson. 1969. Gene regulation for higher cells: A theory. *Science* **165:** 349.

Chang, K.S., W.E. Zimmer, Jr., D.J. Bergsma, J.B. Dodgson, and R.J. Schwartz. 1984. Isolation and characterization of six different chicken actin genes. *Mol. Cell. Biol.* (in press).

Cleveland, D.W., M.A. Lopata, R.J. McDonald, N.J. Cowan, W.J. Rutter, and M.W. Kirschner. 1980. Number and evolutionary conservation of α and β tubulin and cytoplasmic β- and γ-actin genes using specific cloned cDNA probes. *Cell* **20:** 95.

Davidson, E.H. and R.J. Britten. 1979. Regulation of gene expression: Possible role of repetitive sequences. *Science* **204:** 1052.

Dodgson, J.B., J. Strommer, and J.D. Engel. 1979. Isolation of the chicken β-globin gene and a linked embryonic β-like globin gene from a chicken DNA recombinant library. *Cell* **17:** 879.

Engel, J., P. Gunning, and L. Kedes. 1982. Human cytoplasmic actin proteins are encoded by a multigene family. *Mol. Cell. Biol.* **2:** 674.

Fyrberg, E.A., K.L. Kindle, N. Davidson, and A. Sodja. 1980. The actin genes of *Drosophila*: A dispersed multigene family. *Cell* **19:** 365.

Fyrberg, E.A., B.J. Bond, N.D. Hershey, K.S. Mixter, and N. Davidson. 1981. The actin genes of *Drosophila*: Protein coding regions are highly conserved but intron positions are not. *Cell* **24**: 107.

Garrels, J.E. and W. Gibson. 1976. Identification and characterization of multiple forms of actin. *Cell* **9:** 793.

Gunning, P., P. Ponte, H. Okayama, J. Engel, H. Blau, and L. Kedes. 1983. Isolation and characterization of full-length cDNA clones for human α-, β-, and γ-actin mRNA: Skeletal but not cytoplasmic actins have an amino terminal cysteine that is subsequently removed. *Mol. Cell. Biol.* **3:** 787.

Kindle, K.L. and R.A. Firtel. 1978. Identification and analysis of *Dictyostelium* actin genes: A family of moderately repeated genes. *Cell* **15:** 763.

McLennan, I.S. 1983. Differentiation of muscle fiber types in the chicken hindlimb. *Dev. Biol.* **97:** 222.

Pollard, T. and R.R. Weihing. 1974. Actin and myosin and cell movement. *Crit. Rev. Biochem.* **2:** 1.

Scheller, R.H., L.B. McAllister, W.R. Crain, D.S. Durica, J. Posakong, T.L. Thomas, R.J. Britten, and E.H. Davidson. 1981. Organization and expression of multiple actin genes in the sea urchin. *Mol. Cell. Biol.* **1:** 609.

Schwartz, R.J. and K. Rothblum. 1980. Regulation of muscle differentiation: Isolation and purification of chick actin messenger ribonucleic acid and quantitation with complementary deoxyribonucleic acid probes. *Biochemistry* **19:** 2506.

———. 1981. Gene switching in myogenesis: Differential expression of the chicken actin multigene family. *Biochemistry* **20:** 4122.

Schwartz, R.J., J.A. Haron, K.N. Rothblum, and A. Dugaizcyk. 1980. Regulation of muscle differentiation: Cloning of sequences from α-actin messenger ribonucleic acid. *Biochemistry* **19:** 5883.

Shen, J.C.K. and T. Maniatis. 1980. Identification of repetitive DNA. *Cell* **19:** 379.

Sim, G.K., F.C. Kafatos, C.W. Jones, M.D. Koehler, A. Efstratiadis, and T. Maniatis. 1979. Use of a cDNA library for studies on evolution and developmental expression of chorion multigene family. *Cell* **18:** 1303.

Vandekerckhove, J. and K. Weber. 1978. Mammalian cytoplasmic actins are the products of at least two genes and differ in primary structure in at least 25 identified positions from skeletal muscle actins. *Proc. Natl. Acad. Sci.* **75:** 1106.

———. 1979. The complete amino acid sequence of actins from bovine aorta, bovine heart, bovine fast skeletal muscle and rabbit slow skeletal muscle. *Differentiation* **14:** 123.

Vandekerckhove, J., W.W. Franke, and K. Weber. 1981. Diversity of expression of non-muscle actin in amphibia. *J. Mol. Biol.* **152:** 413.

Wahli, W., I.B. Dawid, T. Wyler, R.B. Jaggi, R. Weber, and G.U. Ryffel. 1979. Vitellogenin in *Xenopus laevis* is encoded in a small family of genes. *Cell* **16:** 535.

Isolation and Characterization of cDNA Clones Encoding Smooth-muscle and Nonmuscle Tropomyosins

David M. Helfman, Yuriko Yamawaki-Kataoka, William M. Ricci, and James R. Feramisco
Cold Spring Harbor Laboratory
Cold Spring Harbor, New York 11724

Stephen H. Hughes
Frederick Cancer Research Facility
Frederick, Maryland 21701

Tropomyosins are a family of highly related proteins present in muscle (skeletal, cardiac, and smooth) and nonmuscle cells, although different forms of the protein are characteristic of specific cell types. For example, on the basis of migration in one- and two-dimensional acrylamide gels, chicken skeletal muscle contains at least two forms of tropomyosin, called α and β, with apparent molecular weights of 34,000 and 36,000, respectively (Montarras et al. 1981). Chicken cardiac muscle contains a single form, called α, with an apparent molecular weight of 33,500 (Montarras et al. 1981), and smooth muscle contains at least two forms, called α and β, with apparent molecular weights of 35,000 and 43,000, respectively (Montarras et al. 1981). Nonmuscle cells also contain multiple forms of tropomyosin (Cohen and Cohen 1972; Fine et al. 1973; Fine and Blitz 1975; Lin et al. 1984). Chicken embryonic fibroblasts have been reported to contain as many as seven distinct species of tropomyosin (Lin et al. 1984).

Neither the structural basis nor the functional significance of all the various forms of tropomyosin is clear. Skeletal- and cardiac-muscle tropomyosins, in association with troponin, regulate the calcium-sensitive interaction of actin and myosin (Ebashi and Endo 1968). In contrast, smooth-muscle and nonmuscle cells are devoid of troponin. In these tissues, the phosphorylation of myosin by the enzyme myosin light-chain kinase appears to be the major calcium-sensitive regulatory mechanism controlling the interaction of actin and myosin (Adelstein et al. 1982). These differences in the regulation of the contractile apparatus in various cell types may require functionally distinct forms of tropomyosin. Biochemical differences in tropomyosins isolated from skeletal-muscle, smooth-muscle, and nonmuscle cells (brain and platelets) have been reported previously (Woods

1969; Fine and Blitz 1975; Sobieszek and Small 1981; Pearlstone and Smillie 1982). For example, tropomyosins from smooth and skeletal muscles, brain, and platelets have different effects on the ATPase activity of actomyosin (Sobieszek and Small 1981) and different binding affinities to fragments of troponin T (Pearlstone and Smillie 1981).

A first step in understanding the molecular basis for differences in various forms of tropomyosin will be a determination of their primary structure. The amino acid sequences of rabbit skeletal-muscle α- and β-tropomyosins (Mak et al. 1980), rabbit cardiac-muscle tropomyosin (Lewis and Smillie 1980), chicken skeletal-muscle α-tropomyosin (MacLeod 1982), and equine platelet β-tropomyosin (Lewis et al. 1983) have been published. Recently, we reported the isolation and characterization of a cDNA clone containing the entire coding region for smooth-muscle α-tropomyosin (Helfman et al. 1984). In this paper, we compare the derived amino acid sequence of smooth-muscle α-tropomyosin with the sequences of other tropomyosins. In addition, we have isolated and partially sequenced cDNA clones for two tropomyosins from rat embryonic fibroblasts. The relationships of the various tropomyosin sequences to function are discussed.

RESULTS AND DISCUSSION

Identification of cDNA Clones Encoding Smooth-muscle and Nonmuscle Tropomyosins

Two cDNA expression libraries of approximately 100,000 members each were constructed from embryonic chicken smooth-muscle mRNA (stomach and gizzard) and rat embryonic fibroblast mRNA using the plasmid expression vectors pUC8 and pUC9 (Helfman et al. 1983, 1984). Two approaches were used to identify recombinant clones encoding tropomyosin from the smooth-muscle cDNA expression library. First, the library was screened with a ^{32}P-labeled DNA probe made from a 900-bp cDNA encoding a chicken smooth-muscle tropomyosin, called pSMT-6, that we had identified previously (Helfman et al. 1983). pSMT-6 contains about 80% of the translated region of tropomyosin, beginning at approximately amino acid 50, and a portion of the 3′-untranslated region (Helfman et al. 1983). Second, the library was screened for bacterial colonies producing proteins antigenically related to tropomyosin by a procedure recently developed in our laboratory (Helfman et al. 1983). Fifty-two colonies were found to hybridize to a ^{32}P-labeled probe made from pSMT-6; 39 of these were in the pUC8 library and 13 were in the pUC9 library. The immunological screening procedure demonstrated that 13 of the 39 colonies in the pUC8 library and 2 of the 13 colonies in the pUC9 library were producing products detectable by the anti-tropomyosin serum. The 13 of the 39 in pUC8 agrees with the predicted 1 in 3 that would be in the right translational reading frame for expression. However, the lower percentage of expressors in the pUC9 library (2 of 13) would suggest some bias against expression of α-tropomyosin cDNAs in pUC9.

We believe that there is a *Sal*I restriction site in the 3′-untranslated region of the chicken smooth-muscle α-tropomyosin because we have identified a variety of clones that terminate at the same position at their 3′ ends. An internal *Sal*I restriction site would result in a high percentage of cDNAs unclonable in pUC9 due to the formation of cDNAs containing *Sal*I restriction sites on both their 5′ and 3′ ends. The relatively low number of total cDNAs encoding α-tropomyosin in the pUC9 library compared with the pUC8 library supports this notion. We previously suggested the use of both the pUC8 and the pUC9 systems in order to overcome problems due to internal restriction sites (Helfman et al. 1983). Since some cDNAs will contain restriction sites that are the same as the linkers, these will give either an unclonable DNA insert or an insert that is always cloned in the same, possibly incorrect, translational reading frame. These results demonstrate the advantage of using both vectors.

Colonies hybridizing to pSMT-6 were isolated, and the sizes of the cDNA inserts were determined (Helfman et al. 1984). Insert size ranged from 500 to 1100 bp. To characterize the bacterial fusion proteins, colonies producing proteins antigenically related to tropomyosin were grown overnight in liquid culture, and their proteins were analyzed on SDS-polyacrylamide gels (Fig. 1). As shown in Figure 1, fusion proteins with molecular weights ranging from 17,000 to 35,000 were present in the bacteria. The largest fusion proteins were found in bacteria containing plasmids with the largest cDNA inserts. pSMT-10 and pSMT-13 contained inserts of approximately 1100 bp each. The fusion proteins made in *Esch-*

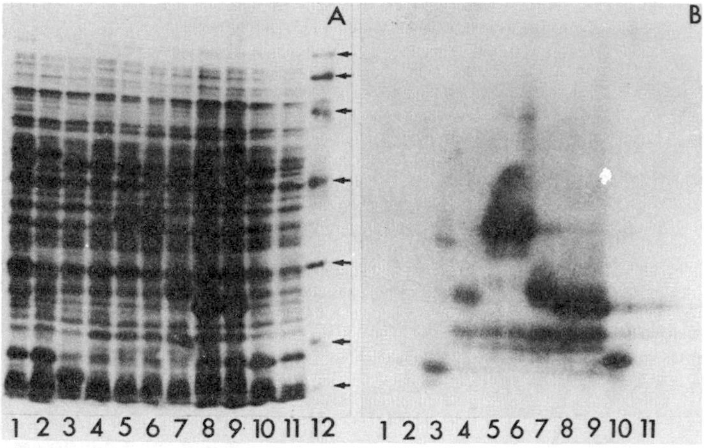

Figure 1

E. coli fusion proteins containing portions of tropomyosin. The proteins in bacterial lysates were analyzed on 12.5% SDS-polyacrylamide gels and electrophoretically transferred from the polyacrylamide gel to nitrocellulose according to the method of Towbin et al. (1979). The proteins were visualized by staining with Amido Black (*A*) or by reaction with anti-tropomyosin antibodies (*B*). (*1*) Parental strain DH-1 containing the pUC9; (*2*) DH-1 containing pUC8; (*3*) DH-1 containing pSMT-35; (*4*) DH-1 containing pSMT-34; (*5*) DH-1 containing pSMT-13; (*6*) DH-1 containing pSMT-10; (*7*) DH-1 containing pSMT-9; (*8*) DH-1 containing pSMT-8; (*9*) DH-1 containing pSMT-2; (*10*) DH-1 containing pSMT-3; (*11*) parental strain DH-1 without plasmid. (*12*) Molecular-weight markers from top to bottom: 116,000 (β-galactosidase), 94,000 (phosphorylase b), 68,000 (bovine serum albumin), 43,000 (ovalbumin), 30,000 (carbonic anhydrase), 21,000 (soybean trypsin inhibitor), and 14,000 (lysozyme). (Reprinted, with permission, from Helfman et al. 1984.)

erichia coli containing pSMT-10 and pSMT-13 had a molecular weight of 35,000, approximately the size of smooth-muscle α-tropomyosin. To show that these proteins were tropomyosin-related, the protein products were electrophoretically transferred to nitrocellulose (Fig. 1) and examined for immunological cross-reactivity. As shown in Figure 1, these proteins were recognized by anti-tropomyosin antibody. No obvious proteins related to tropomyosin were present in the parental bacterial DH-1 or the DH-1 containing plasmids pUC8 or pUC9. In bacteria expressing some piece of tropomyosin, protein bands that react with anti-tropomyosin antibodies can be readily identified. Presumably in these extracts, there is some degradation of the tropomyosin fusion proteins made in the bacteria that gives rise to the lower-molecular-weight bands detected by the antiserum. However, it is not clear whether the putative degradation takes place in vivo or is a result of the preparation of the bacterial lysates.

To determine whether pSMT-10 encoded sequences for the α-tropomyosin (M_{app} = 35,000) or the β-tropomyosin (M_{app} = 43,000), hybrid-selection translation was carried out using poly(A)$^+$ RNA from 11–13-day embryonic chicken smooth muscle (Helfman et al. 1984). The in vitro translation products of mRNA hybridized to pSMT-10 yielded a single protein product (M_{app} = 35,000 as determined by one-dimensional gel electrophoresis) that could be immunoprecipitated with anti-tropomyosin antibody (data not shown). Further examination of the in vitro translation product by two-dimensional gel electrophoresis verified that the translation product comigrated with the α-tropomyosin (data not shown). This is intriguing because the derived amino acid sequence of the protein based on the DNA sequence of pSMT-10 is clearly more homologous to the β-tropomyosin from rabbit skeletal muscle than to either the α-tropomyosin from rabbit skeletal muscle or the α-tropomyosin from chicken skeletal muscle (see below).

Two clones in the smooth-muscle pUC8 cDNA expression library were identified that produced proteins immunologically related to tropomyosin (i.e., detected by antisera) but did not contain sequences that cross-hybridized to a ^{32}P-labeled probe made from pSMT-6. One of these clones (pSMT-1) was isolated and further characterized. Plasmid pSMT-1 contained a cDNA insert of about 600 bp and was found to hybrid-select mRNA for β-tropomyosin from smooth muscle. DNA sequence analysis indicated that it contained sequences from approximately amino acid 20 to amino acid 200. These results would tend to suggest that immunological screening of cDNA expression libraries may be advantageous in identifying members of a gene family where there exists strong conservation at the protein level (i.e., antibody cross-reactivity) but little homology at the nucleic acid level.

Since each muscle and nonmuscle tissue has distinct tropomyosins, we investigated the tissue specificity of the two tropomyosin cDNA clones obtained from smooth muscle. Clones pSMT-10 and pSMT-1 were used to probe RNA obtained from nonmuscle cells (fibroblasts) and from different striated muscle types (skeletal and cardiac) and smooth muscle (stomach and gizzard) isolated from chick embryos at day 13 in development. Figure 2 shows that each clone hybridized to a different species of mRNA from each tissue. Whether or not each species of tropomyosin mRNA is encoded by a separate gene or perhaps arises from a single gene via differential processing of the message is unknown at present. It is worth noting that similar results were obtained using a probe made from the 3′-untranslated region of pSMT-10, suggesting that the different messages may

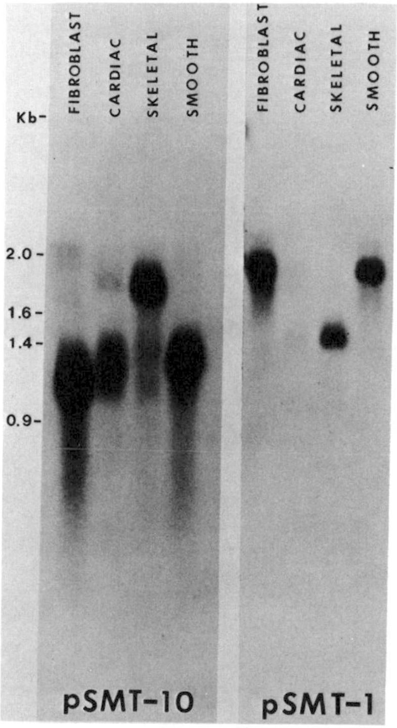

Figure 2
Northern blot analysis of mRNA from various tissues. Poly(A)$^+$ RNA was prepared from skeletal, cardiac, and smooth (stomach and gizzard) muscles obtained from chick embryos at day 13 in development and chick embryonic fibroblasts; 1 μg of each was then separated in a 1.2% agarose-formaldehyde gel. The RNA was then transferred to nitrocellulose and hybridized with ^{32}P-labeled probes prepared from pSMT-10 or pSMT-1. Filters were washed in 0.1 × SSC plus 0.1% SDS at 50°C and autoradiographed for 10 days with Dupont Cronex Lightning-plus X-ray intensifying screens.

be derived from a single gene (data not shown). We are currently analyzing genomic clones in order to address this problem.

Rat embryonic fibroblasts contain three major tropomyosins with apparent molecular weights of 40,000, 36,500, and 32,400 and two relatively minor tropomyosins with apparent molecular weights of 35,000 and 32,000 (Matsumura et al. 1983). To identify clones encoding nonmuscle tropomyosins, we screened the rat embryonic fibroblast cDNA expression library with antisera against tropomyosin and with ^{32}P-labeled probes made from plasmids pSMT-10 and pSMT-1. We have identified and isolated at least three different classes of clones (Y. Yamawaki-Kataoka and D.M. Helfman, unpubl.). One set of clones has been identified by using hybrid-selection translation to contain sequences complementary to tropomyosin 1 (M_{app} = 40,000) and another set of clones was found to contain sequences complementary to tropomyosin 4 (M_{app} = 32,400). We have begun to characterize these clones (see below).

Sequence Analysis of Smooth-muscle and Nonmuscle Tropomyosins

The complete nucleotide sequence of pSMT-10, along with the predicted amino acid sequence, is shown in Figure 3. The insert consists of approximately 1100 nucleotides derived from a smooth-muscle α-tropomyosin mRNA, of which 852 nucleotides encode the entire translated region of the protein; 22 are from the 5′-

```
                                                10
                         Met Glu Ala Ile Lys Lys Lys Met Gln Met Leu Lys Leu Asp Lys Glu Asn Ala Ile
CCGCGCCGCCGCCCCGCCGCACC  ATG GAG GCC ATC AAG AAG AAG ATG CAG ATG CTC AAA CTG GAC AAG GAG AAC GCC ATC

20                                     30
Asp Arg Ala Glu Gln Ala Glu Ala Asp Lys Lys Gln Ala Glu Asp Arg Cys Lys Gln Leu Glu Glu Glu Gln Gln Gly Leu
GAC CGT GCC GAG CAG GCG GAG GCT GAC AAG AAG CAG GCG GAG GAT CGC TGC AAG CAG CTG GAG GAG GAG CAG CAG GGC CTG

            50                                  60                                      70
Gln Lys Lys Leu Lys Gly Thr Glu Asp Glu Val Glu Lys Tyr Ser Glu Ser Val Lys Glu Ala Gln Glu Lys Leu Glu Gln
CAG AAG AAG CTG AAG GGC ACA GAA GAT GAG GTG GAG AAG TAC TCT GAA TCC GTC AAG GAG GCC CAG GAG AAG CTG GAG CAG

                        80                                      90                                      100
Ala Glu Lys Lys Ala Thr Asp Ala Glu Ala Glu Val Ala Ser Leu Asn Arg Arg Ile Gln Leu Val Glu Glu Glu Leu Asp
GCG GAG AAG AAA GCC ACG GAC GCC GAG GCC GAG GTG GCT TCT CTG AAC CGC CGC ATC CAG CTG GTG GAG GAG GAG CTG GAC

                                    110                                     120
Arg Ala Gln Glu Arg Leu Ala Thr Ala Leu Gln Lys Leu Glu Glu Ala Glu Lys Ala Ala Asp Glu Ser Glu Arg Gly Met
CGG GCC CAG GAG CGC CTG GCC ACC GCC CTG CAG AAG CTG GAA GAG GCT GAG AAG GCG GCG GAT GAG AGC GAG AGA GGC ATG

        130                                     140                                     150
Lys Val Ile Glu Asn Arg Ala Met Lys Asp Glu Glu Lys Met Glu Leu Gln Glu Met Gln Leu Lys Glu Ala Lys His Ile
AAG GTC ATT GAG AAC AGG GCC ATG AAG GAT GAG GAG AAA ATG GAA CTC CAG GAA ATG CAA CTG AAG GAG GCC AAA CAC ATA

                    160                                     170                                     180
Ala Glu Glu Ala Asp Arg Lys Tyr Glu Glu Val Ala Arg Lys Leu Val Val Leu Glu Gly Glu Leu Glu Arg Ser Glu Glu
GCG GAA GAG GCC GAC CGC AAA TAT GAG GAG GTG GCC CGC AAG CTG GTG GTC CTT GAG GGA GAG CTG GAG CGC TCA GAG GAG

                                190                                     200
Arg Ala Glu Val Ala Glu Ser Arg Val Arg Gln Leu Glu Glu Glu Leu Arg Thr Met Asp Gln Ser Leu Lys Ser Leu Ile
CGA GCA GAG GTC GCG GAG AGC CGA GTG AGA CAG CTG GAG GAG GAG CTG CGG ACC ATG GAC CAG AGC CTC AAA TCC CTC ATT

    210                                     220                                     230
Ala Ser Glu Glu Glu Tyr Ser Thr Lys Glu Asp Lys Tyr Glu Glu Glu Ile Lys Leu Leu Gly Glu Lys Leu Lys Glu Ala
GCC TCA GAG GAA GAG TAT TCC ACC AAG GAG GAC AAG TAC GAG GAG GAA ATC AAG CTT CTA GGG GAG AAA CTG AAG GAG GCT

                240                                     250                                     260
Glu Thr Arg Ala Glu Phe Ala Glu Arg Ser Val Ala Lys Leu Glu Lys Thr Ile Asp Asp Leu Glu Glu Ser Leu Ala Ser
GAG ACC AGG GCA GAG TTT GCT GAG CGG TCT GTG GCA AAG CTG GAG AAA ACC ATT GAT GAC TTA GAA GAG AGT CTG GCC AGT

                        270                                     280
Ala Lys Glu Glu Asn Val Gly Ile His Gln Val Leu Asp Gln Thr Leu Leu Glu Leu Asn Asn Leu
GCC AAA GAG GAG AAT GTG GGG ATA CAC CAG GTC CTG GAC CAG ACC TTG CTG GAG CTG AAC AAC CTC TGAGCTGCCTGGGGAG

CCCTCATCCCCCTACCCTACCCCACACATGCACCCCCTGGCACGCTCAGGACGTCACCAGCTAAACTGCATCTTGGCCAAATCTACACATCCATTGAGAAGGGA
AACCCTGCAGCCTTAAACCATGGCTCGGGGCGTCTTGTCTGTGCACAGCCTGAATGGCTTTGTCCTGTTTTCATGTCCAAATATTAAAGCAGTTTAACAAGA
```

Figure 3 (*See facing page for legend.*)

untranslated region and 223 are from the 3′-untranslated region. The absence of a poly(A) sequence suggests that the entire 3′-untranslated region is not present. Also absent from the 3′-untranslated region is the sequence AATAAA or ATAAA, usually found near the site of addition of the poly(A) tail in eukaryotic mRNA sequences (Proudfoot and Brownlee 1976). It is worth noting that Talbot and MacLeod (1983) recently published the carboxyterminal sequence of a chicken tropomyosin cDNA, termed pTM-7, isolated from a skeletal-muscle cDNA library. This clone, coding for amino acids 207–284, is identical in sequence to pSMT-10. pSMT-10 was found to encode a portion of the 5′-untranslated leader sequence of tropomyosin (Fig. 3). However, the bacteria containing this plasmid synthesize a fusion protein with a molecular weight of approximately 35,000. It is not clear whether the bacteria are using the AUG start codon of β-galactosidase or of tropomyosin, since both products would be of approximately the same size.

Comparing the amino acid sequences of the chicken smooth-muscle α-tropomyosin (Helfman et al. 1984) with those of chicken skeletal-muscle α-tropomyosin (MacLeod 1982), rabbit skeletal-muscle α- and β-tropomyosins (Mak et al. 1980), and equine platelet β-tropomyosin (Lewis et al. 1983) revealed extensive homology among the various proteins (Fig. 4). The majority of differences in amino acids represented conservative amino acid substitutions. Two regions where the greatest sequence divergence is evident are between amino acids 190 and 213 and amino acids 259 and 284. When the chicken smooth-muscle sequence is compared with those of the other tropomyosins, nonconservative amino acid substitutions are found at positions 190–192, 201, 208, 211, and 213. In addition, the chicken smooth-muscle α-tropomyosin is substantially different from tropomyosins isolated from skeletal muscle between amino acids 259 and 284, with nonconservative amino acid substitutions at positions 259, 261, 266–268, and 271–273. Interestingly, the carboxyterminal end of the equine platelet β-tropomyosin is nearly identical with that of the chicken smooth-muscle α-tropomyosin between amino acids 259 and 284, differing in only 6 of the last 25 amino acids; of these, only two, at positions 259 and 279, are nonconservative substitutions. In contrast, the platelet tropomyosin is more similar to the skeletal-muscle tropomyosins from amino acid 190 to 213, exhibiting little homology with the smooth-muscle form of the protein. Recently, we have identified and sequenced cDNA clones encoding tropomyosin 1 (M_{app} = 40,000) and tropomyosin 4 (M_{app} = 32,400) from rat embryonic fibroblasts (Y. Yamawaki-Kataoka and D.M. Helfman, unpubl.). The derived amino acid sequences of the carboxyterminal ends of these nonmuscle tropomyosins are compared in Figure 5 with those from smooth, skeletal, and cardiac muscles. As indicated, nonmuscle and smooth-muscle tropomyosins share striking sequence homology at their carboxyterminal ends. In contrast, the carboxyterminal ends of those found for the various striated muscle tropomyosins are almost identical to each other.

Figure 3

Nucleotide sequence and deduced protein sequence of smooth-muscle α-tropomyosin. The deduced protein sequence is numbered according to the known amino acid sequence of rabbit skeletal-muscle α-tropomyosin. The coding strand of clone pSMT-10 specifying the entire 284 amino acids is shown, as well as the 5′- and 3′-untranslated regions. (Reprinted, with permission, from Helfman et al. 1984.)

```
                       1                          10                            20
CSM - α                MetGluAlaIleLysLysLysMetGlnMetLeuLysLeuAspLysAlaAsnAlaIleAsp
CSK - α
RSK - α                   Asp                                                Leu
RSK - β                   Asp
EPL - β AlaGlyLeuAsnSerLeu      Val   Arg   Ile   Ala   GlnGlnGlnAlaAspGlu   Glu

                                   30                            40
CSM - α ArgAlaGluGlnAlaGluAlaAspLysLysGlnAlaGluAspArgCysLysGlnLeuGluGluGluGlnGlnGly
CSK - α                               Ala      Glu   Ser            Asp   LeuValAla
RSK - α                               Ala            Ser            Asp   LeuValSer
RSK - β                                                                         Ala
EPL - β       -  -  -  -  -  -  -  -  -  -  -  -  -  -  -  -  -  -  -  -  -

                    50                            60                            70
CSM - α LeuGlnLysLysLeuLysGlyThrGluAspGluValGluLysTyrSerGluSerValLysGluAlaGlnGluLys
CSK - α                                  LeuAsp               Leu   Asp
RSK - α                                  LeuAsp            AlaLeu   Asp
RSK - β                                                             Asp
EPL - β       ArgGlu   Asp   GluArgGluArgArg         -  -  -  -  -  -  -  -  -  -

                                   80                            90
CSM - α LeuGluGlnAlaGluLysLysAlaThrAspAlaGluAlaGluValAlaSerLeuAsnArgArgIleGlnLeuVal
CSK - α       Leu   Asp                     Ser
RSK - α       Leu                              Asp
RSK - β                                        Asp
EPL - β -  -  -  -  -  -  -  -  -  -        GlyAsp      Ala

                    100                           110                           120
CSM - α GluGluGluLeuAspArgAlaGlnGluArgLeuAlaThrAlaLeuGlnLysLeuGluGluAlaGluLysAlaAla
CSK - α
RSK - α
RSK - β
EPL - β

                                   130                           140
CSM - α AspGluSerGluArgGlyMetLysValIleGluAsnArgAlaMetLysAspGluGluLysMetGluLeuGlnGlu
CSK - α                                           Gln                     Ile
RSK - α                                  Ser      Gln                     Ile
RSK - β
EPL - β                                                                   Ile

                    150                           160                           170
CSM - α MetGlnLeuLysGluAlaLysHisIleAlaGluGluAlaAspArgLysTyrGluGluValAlaArgLysLeuVal
CSK - α Ile
RSK - α Ile                              Asp
RSK - β                                  AspSer
EPL - β
```

```
                                         180                             190
CSM-α   ValLeuGluGlyGluLeuGluArgSerGluGluArgAlaGluValAlaGluSerArgValArgGlnLeuGluGlu
CSK-α   IleIle      Asp         Ala               LeuSer      LysCysAlaGlu
RSK-α   IleIle   SerAsp         Ala               LeuSer   GlyLysCysAlaGlu
RSK-β   Ile                                                   LysCysGlyAsp
EPL-β   Ile                     Ala                  Ser   LeuLysCysGlyAsp


                    200                           210                           220
CSM-α   GluLeuArgThrMetAspGlnSerLeuLysSerLeuIleAlaSerGluGluTyrSerThrLysGluAspLys
CSK-α         Lys   ValThrAsnAsn            Glu   GlnAla   Lys      Gln
RSK-α         Lys   ValThrAsnAsn            Glu   GlnAla   Lys      Gln
RSK-β         LysIleValThrAsnAsn            Glu   GlnAlaAspLys
EPL-β         LysAsnValThrAsnAsn            Glu   AlaSer   Lys      Glu


                                   230                           240
CSM-α   TyrGluGluGluIleLysLeuLeuGlyGluLysLeuLysGluAlaGluThrArgAlaGluPheAlaGluArgSer
CSK-α                     Val   ThrAsp
RSK-α                     Val   SerAsp
RSK-β                           Glu
EPL-β                           SerAsp                                          Thr


                    250                           260                           270
CSM-α   ValAlaLysLeuGluLysThrIleAspAspLeuGluGluSerLeuAlaSerAlaLysGluGluAsnValGlyIle
CSK-α      Thr            Ser               AspGlu   TyrAlaGln   LeuLysTyrLysAla
RSK-α      Thr            Ser               AspGlu   TyrAlaGln   LeuLysTyrLysAla
RSK-β                                       AspGluValTyrAlaGln   MetLysTyrLysAla
EPL-β                                          Lys      Gln                     Leu


                                   280
CSM-α   HisGlnValLeuAspGlnThrLeuLeuGluLeuAsnAsnLeu
CSK-α   SerGluGlu      HisAla   AsnAspMetThrSerIle
RSK-α   SerGluGlu      HisAla   AsnAspMetThrSerIle
RSK-β   SerGluGlu      AsnAla   AsnAspIleThrSer
EPL-β         Thr               Asp         Cys
```

Figure 4
Comparison of the amino acid sequences of various tropomyosins. The amino acid sequences of chicken smooth-muscle α-tropomyosin (CSM-α; Helfman et al. 1984), chicken skeletal-muscle α-tropomyosin (CSK-α; MacLeod 1982), rabbit skeletal-muscle α- and β-tropomyosins (RSK-α, RSK-β; Mak et al. 1980), and equine platelet β-tropomyosin (EPL-β; Lewis et al. 1983) are shown. Where the various sequences are identical with chicken smooth-muscle α-tropomyosin, only the smooth-muscle sequence is shown while amino acids of nonidentity are indicated. Dashed lines indicate putative deletions in the equine platelet tropomyosin sequence. The available chicken skeletal-muscle α-tropomyosin sequence is from amino acid 24 to 284. (Reprinted, with permission, from Helfman et al. 1984.)

```
              265            270            275            280
REF-1   AlaLysGluGluAsnLeuGluIleHisGlnThrLeuAspGlnThrLeuLeuGluLeuAsnAsnLeu
REF-4                  ValGlyLeu                        Asn      LeuCysIle
EPL-β                  ValGlyLeu                        Asp         Cys
CSM-α                  ValGly         Val
CSK-α   Gln   LeuLysTyrLysAla   SerGluGlu      HisAla   AsnAspMetThrSerIle
RSK-α   Gln   LeuLysTyrLysAla   SerGluGlu      HisAla   AsnAspMetThrSerIle
RSK-β   Gln   MetLysTyrLysAla   SerGluGlu      AsnAla   AsnAspIleThrSer
RCM-α   Gln   LeuLysTyrLysAla   SerGluGlu      HisAla   AsnAspMetThrSerIle
```

Figure 5
Carboxyterminal ends of nonmuscle and muscle tropomyosins. The amino acid sequences of rat embryonic fibroblast tropomyosin 1 and 4 (REF-1, REF-4; Y. Yamawaki-Kataoka and D.M. Helfman, unpubl.), equine platelet β-tropomyosin (EPL-β; Lewis et al. 1983), chicken smooth-muscle α-tropomyosin (CSM-α; Helfman et al. 1984), chicken skeletal-muscle α-tropomyosin (CSK-α; MacLeod 1982), rabbit skeletal-muscle α- and β-tropomyosins (RSK-α, RSK-β; Mak et al. 1980), and rabbit cardiac-muscle α-tropomyosin (RCM-α; Lewis and Smillie 1980) are shown. Where the various sequences are identical with rat embryonic fibroblast tropomyosin 1, only the fibroblast sequence is shown while amino acids of non-identity are indicated. The sequences have been numbered according to those for skeletal-muscle tropomyosin.

The structural and functional significance of the differences in the carboxyterminal region of the various tropomyosins is not clear at present but may have important implications with respect to their interactions with the members of the troponin complex. Biochemical studies have indicated that the site of troponin binding is close to or at the carboxyterminal end of skeletal-muscle tropomyosin (Chong and Hodges 1972; Cohen et al. 1972, 1973; Hitchcock et al. 1973; Margossian and Cohen 1973; Ohtsuki 1974, 1979; Yamaguchi et al. 1974; McLachlan and Stewart 1976; Stewart and McLachlan 1976; Mak and Smillie 1981; Pearlstone and Smillie 1981). It is believed that there are at least two regions on both skeletal-muscle tropomyosin and troponin T involved in their mutual interaction (Hitchcock et al. 1973; Ohtsuki 1979: Mak and Smillie 1981; Pearlstone and Smillie 1981, 1982). Using defined fragments of troponin T, it has been shown that the T1 (residues 1–158) and CB1 (residues 1–151) fragments of troponin interact with the carboxyterminal end of tropomyosin (Pearlstone and Smillie 1982). The T2 fragment (residues 159–259) of troponin T interacts with tropomyosin in the region around the cysteine at position 190 in tropomyosin. Studies of the binding of rabbit skeletal-muscle troponin T to various types of tropomyosin have indicated biochemical differences in the tropomyosins (Pearlstone and Smillie 1982). The smooth-muscle tropomyosin and platelet tropomyosin bind troponin T and its T1 and CB1 fragments less well than the skeletal-muscle protein binds troponin (Pearlstone and Smillie 1982). This is intriguing because the smooth-muscle and platelet proteins are quite similar at their carboxyterminal ends (amino acids 260–284), the region in which they differ substantially from the skeletal-muscle proteins.The skeletal-muscle form is known to be phosphorylated at serine 283, whereas the smooth-muscle and platelet forms are not (Montarras et al. 1981). This serine is absent from both the smooth-muscle and the platelet tropomyosins, where it is replaced by asparagine and cysteine, respectively. The functional significance of the phosphorylation of the skeletal-muscle tropomyosin is not clear, but it may have a role in the regulation of these proteins.

The T2 fragment (residues 159–259) of troponin T interacts equally well with smooth-muscle, platelet, and skeletal-muscle tropomyosins (Pearlstone and Smillie 1982). These results suggested that the binding region for T2 would be similar among these tropomyosins (Pearlstone and Smillie 1982). As predicted, a comparison of platelet and skeletal-muscle tropomyosins in the region of cysteine 190 reveals extensive sequence homology. The smooth-muscle protein, in contrast, is substantially different from either the platelet or skeletal-muscle tropomyosins around cysteine 190. Thus, although there are differences in the primary structure of these proteins in the region of T2 binding (cysteine – 190), this is not reflected in their ability to bind the T2 fragment of troponin T. Whether there is a conservation of the gross three-dimensional structure of the protein in this region or a certain amount of flexibility in the structural requirements for the binding of troponin to tropomyosin in this region is not clear at present. Alternatively, the actual binding site for the T2 fragment of troponin T on tropomyosin may be to the aminoterminal side of cysteine 190, where the proteins show extensive homology.

SUMMARY

It is clear that different forms of tropomyosin exist in various cell types. Whether each protein species is functionally distinct remains to be determined. For example, the physiological significance of differences in troponin binding to various tropomyosins in vitro is unknown. Smooth-muscle and nonmuscle cells do not contain a troponin complex for the regulation of actomyosin (Adelstein et al. 1982). Instead, regulation of actomyosin in these cells seems to occur by phosphorylation of the light chains of myosin by the enzyme myosin light-chain kinase (Adelstein et al. 1982). It would appear that smooth-muscle and nonmuscle cells may require functionally, as well as structurally, different forms of tropomyosin compared with the skeletal-muscle proteins. In this respect, tropomyosins isolated from smooth-muscle and nonmuscle cells have been found to stimulate the ATPase activity of reconstituted skeletal-muscle actomyosin (Sobieszek and Small 1981). In contrast, under the same conditions, skeletal-muscle tropomyosin had no effect (Sobieszek and Small 1981). Whether these studies are of any significance in vivo remains unanswered. It is clear from the primary sequence data that tropomyosins are highly conserved proteins. However, substantial sequence differences exist near the carboxyterminal region of the protein. The obvious implication is either that the carboxyterminal region in some cell types is not essential for tropomyosin function or that such altered sequence confers different functional properties to the protein. The latter alternative may be more likely, but the problem remains unresolved. The conservation of structure among the nonmuscle and smooth-muscle tropomyosins would tend to favor a structure-function relationship. Finally, in addition to the problems concerning tropomyosins at the protein and cellular levels, the structure and organization of the genes encoding tropomyosin remain to be elucidated. Many questions concerning this gene family remain to be answered. For instance, do all forms of tropomyosin arise from separate but related genes or do some arise from a single gene via differential processing of the message and/or protein? How is expression of various tropomyosins in different cell and tissue types controlled? Analysis with gene-specific probes should help answer some of these questions.

ACKNOWLEDGMENTS

This work was supported by grants from the National Institutes of Health, the American Cancer Society, and the Muscular Dystrophy Association of America. We are also grateful to Madeline Szadkowski, Phil Renna, and Marilyn Goodwin for preparation of the manuscript.

REFERENCES

Adelstein, R.S., M.D. Pato, J.R. Sellers, P. DeLanerolle, and M.A. Conti. 1982. Regulation of actin-myosin interaction by reversible phosphorylation of myosin and myosin kinase. *Cold Spring Harbor Symp. Quant. Biol.* **46:** 921.

Chong, P.C.S. and R.S. Hodges. 1982. Photochemical cross-linking between rabbit skeletal troponin and α-tropomyosin. *J. Biol. Chem.* **257:** 9152.

Cohen, C., D.L.D. Caspar, D.A.D. Parry, and R.M. Lucas. 1972. Tropomyosin crystal dynamics. *Cold Spring Harbor Symp. Quant. Biol.* **36:** 205.

Cohen, C., D.L.D. Caspar, J.P. Johnson, K. Nauss, S.S. Margossian, and D.A.D. Parry. 1973. Tropomyosin-troponin assembly. *Cold Spring Harbor Symp. Quant. Biol.* **37:** 287.

Cohen, I. and C. Cohen. 1972. A tropomyosin-like protein from human platelets. *J. Mol. Biol.* **68:** 383.

Ebashi, S. and M. Endo. 1968. Calcium ion and muscle contraction. In *Progress in biophysics and molecular biology* (ed. J.A.V. Butler and D. Noble), vol. 18, p. 125. Pergamon Press, Oxford, England.

Fine, R.E. and A.L. Blitz. 1975. A chemical comparison of tropomyosins from muscle and non-muscle tissues. *J. Mol. Biol.* **95:** 447.

Fine, R.E., A.L. Blitz, S.E. Hitchcock, and B. Kaminer. 1973. Tropomyosin in brain and growing neurons. *Nat. New Biol.* **245:** 182.

Helfman, D.M., J.R. Feramisco, W.M. Ricci, and S.H. Hughes. 1984. Isolation and sequence of a cDNA clone that contains the entire coding region for chicken smooth muscle α-tropomyosin. *J. Biol. Chem.* (in press).

Helfman, D.M., J.R. Feramisco, J.C. Fiddes, G.P. Thomas, and S.H. Hughes. 1983. Identification of clones that encode chicken tropomyosin by direct immunological screening of a cDNA expression library. *Proc. Natl. Acad. Sci.* **80:** 31.

Hitchcock, S.E., H.E. Huxley, and A.G. Szent-Gyorgyi. 1973. Calcium sensitive binding of troponin to actin-tropomyosin: A two-site model for troponin action. *J. Mol. Biol.* **80:** 825.

Lewis, W.G. and L.B. Smillie. 1980. The amino acid sequence of rabbit cardiac tropomyosin. *J. Biol. Chem.* **255:** 6854.

Lewis, W.G., G.P. Cote, A.S. Mak, and L.B. Smillie. 1983. Amino acid sequence of equine platelet tropomyosin. Correlation with interaction properties. *FEBS Lett.* **156:** 269.

Lin, J.J.-C., F. Matsumura, and S. Yamashiro-Matsumura. 1984. Tropomyosin-enriched and α-actinin-enriched microfilaments isolated from chicken embryo fibroblasts by monoclonal antibodies. *J. Cell Biol.* **98:** 116.

MacLeod, A.R. 1982. Distinct α-tropomyosin mRNA sequences in chicken skeletal muscle. *Eur. J. Biochem.* **126:** 293.

Mak, A.S. and L.B. Smillie. 1981. Structural interpretation of the two-site binding of troponin on the muscle thin filament. *J. Mol. Biol.* **149:** 541.

Mak, A.S., L.B. Smillie, and G.R. Stewart. 1980. A comparison of the amino acid sequences of rabbit skeletal muscle α- and β-tropomyosins. *J. Biol. Chem.* **255:** 3647.

Margossian, S.S. and C. Cohen. 1973. Troponin subunit interactions. *J. Mol. Biol.* **81:** 409.

Matsumura, F., S. Yamashiro-Matsumura, and J.J.-C. Lin. 1983. Isolation and characterization of tropomyosin-containing microfilaments from cultured cells. *J. Biol. Chem.* **258:** 6636.

McLachlan, A.D. and M. Stewart. 1976. The troponin binding region of tropomyosin. Evidence for a site near residues 197 to 217. *J. Mol. Biol.* **106:** 1017.

Montarras, D., M.Y. Fiszman, and F. Gros. 1981. Characterization of the tropomyosin present in various chick embryo muscle types and in muscle cells differentiated in vitro. *J. Biol. Chem.* **256:** 4081.

Ohtsuki, I. 1974. Localization of troponin in thin filament and tropomyosin paracrystal. *J. Biochem.* **75:** 753.

———. 1979. Molecular arrangement of troponin-T in the thin filament. *J. Biochem.* **86:** 491.

Pearlstone, J.R. and L.B. Smillie. 1981. Identification of a second binding region of rabbit skeletal troponin-T for α-tropomyosin. *FEBS Lett.* **128:** 119.

———. 1982. Binding of troponin-T fragments to several types of tropomyosin. *J. Biol. Chem.* **257:** 10587.

Proudfoot, N.J. and G.G. Brownlee. 1976. 3′ Non-coding region sequences in eukaryotic messenger RNA. *Nature* **263:** 211.

Sobieszek, A. and J.V. Small. 1981. Effect of muscle and non-muscle tropomyosins in reconstituted skeletal muscle actomyosin. *Eur. J. Biochem.* **118:** 533.

Stewart, M. and A.D. McLachlan. 1976. Structure of magnesium paracrystals of α-tropomyosin. *J. Mol. Biol.* **103:** 251.

Talbot, K. and A.R. MacLeod. 1983. Novel form of non-muscle tropomyosin in human fibroblasts. *J. Mol. Biol.* **164:** 159.

Towbin, H., T. Staehelin, and J. Gordon. 1979. Electrophoretic transfer of proteins from polyacrylamide gels to nitrocellulose sheets: Procedure and some applications. *Proc. Natl. Acad. Sci.* **76:** 4350.

Woods, E.F. 1969. Comparative physicochemical studies on vertebrate tropomyosins. *Biochemistry* **11:** 4336.

Yamaguchi, M., M.L. Greaser, and R.G. Cassens. 1974. Interaction of troponin subunits with different forms of tropomyosin. *J. Ultrastruct. Res.* **48:** 33.

Antisense Gene Transcription as a Tool for Molecular Analysis

Jonathan G. Izant
Department of Genetics
Hutchinson Cancer Research Center
Seattle, Washington 98104

Several papers in this volume illustrate the power of genetic analysis for identifying the molecular components of cellular structures and for determining their in vivo activities. Naturally occurring and experimentally induced gene variants in phage, bacteria, yeast, fungi, algae, and even metazoans such as *Drosophila* and *Caenorhabditis* are valuable tools for distinguishing the gene products involved in a particular function. Because such assessments are performed in situ, in a living cell (or organism), mutational analysis is a critical complement to in vitro experimentation. But even genetic approaches have limitations. Mutations in fundamental structural genes such as cytoskeletal components or regulatory genes are often recessive or lethal and thus less informative. Conditional mutations in such genes, when available, do not always have the same phenotype. Many higher eukaryotic organisms are not readily amenable to classical genetic manipulations. Mutagenesis in mammals, for example, ranges from slow and tedious to logistically intractable. It is difficult to study the molecular details of the dynamic activity of cellular components such as the cytoskeleton in living vertebrate cells by genetic methods even though cell architecture makes important contributions to cellular structure, proliferation, and morphogenesis. The majority of examinations have been limited to in vitro experiments. With this problem in mind, we have been developing alternative modes of introducing specific "monkey wrenches" into living cells to attempt to mimic dominant gene mutations that are relatively inaccessible by conventional approaches.

Experimental Design

It is possible to alter the activity of specific gene products in living cells by treating them with drugs or by microinjecting specific antibodies into them. But the mechanism of drug action frequently is not simple, and pharmacological inhibitors do not exist for every process in a cell. The production of useful immunoglobulin monkey wrenches requires significant preparation and characterization of both the antigen and the antibodies, and some proteins are very inefficient immunogens. As a complementary approach to methodologies that interfere with the final gene product, we are studying the potential of nucleic acid oligomers, complementary in sequence to that of a predetermined RNA transcript, to disrupt normal gene expression by neutralizing the target transcript in living cells.

Hybrid-arrested translation is a standard in vitro analytical technique (Paterson et al. 1977), and we set out to determine whether it could work efficiently in the complex chemical environment of a living cell. The first attempt involved microinjecting single-stranded DNA copies of target genes (cloned into the bacteriophage M13) into the nuclei of cells. No inhibition of gene activity was detected because single-stranded DNA in eukaryotic cells is readily repaired by the cell to form duplex DNA (Cortese et al. 1980), which is subsequently transcribed to enhance (rather than reduce) the level of gene product (J.G. Izant and R. Harland, unpubl.).

Another approach is to use complementary sequence RNA to neutralize RNA transcripts. Microinjecting RNA synthesized in vitro would provide a significant but steadily decreasing concentration of inhibitor, and RNA is relatively unstable in vitro as well as in vivo (although as discussed below, this is a viable option). Therefore, we have tried introducing into cells double-stranded recombinant DNA constructions that direct transcription of the noncoding or antisense DNA strand of the gene of interest. An RNA transcript of the complementary, antisense DNA strand will be complementary to the normal mRNA transcript and will be capable of forming an RNA duplex. Furthermore, the cell itself will synthesize the antimessage RNA, and one would anticipate that the concentration of inhibitor would reach a steady-state level.

Plasmid DNAs that direct transcription of the noncoding strand of a cloned gene are simple to engineer. Since RNA polymerization is vectorial, only proceeding in a 5′ to 3′ direction, and since DNA has dyad symmetry, antisense transcription can be produced by manipulations that cause RNA polymerase to move in the opposite direction from normal. Reverse-direction DNA transcription can be accomplished by excising the protein-coding domain of a cloned gene and then reinserting it into the plasmid in reverse orientation. As diagramed in Figure 1, this will result in the normal promoter directing transcription of the antisense DNA strand. Furthermore, if transcription proceeds through the entire gene, it should be polyadenylated normally so that the fate of the complementary sequence antisense transcript should mimic that of normal transcripts and allow maximum opportunity for interaction.

Experiments using flipped gene constructions containing the coding regions of several different genes and containing various different promoters indicate that antisense RNA transcription is a viable method for phenocopying genetic mutations in mammalian cells.

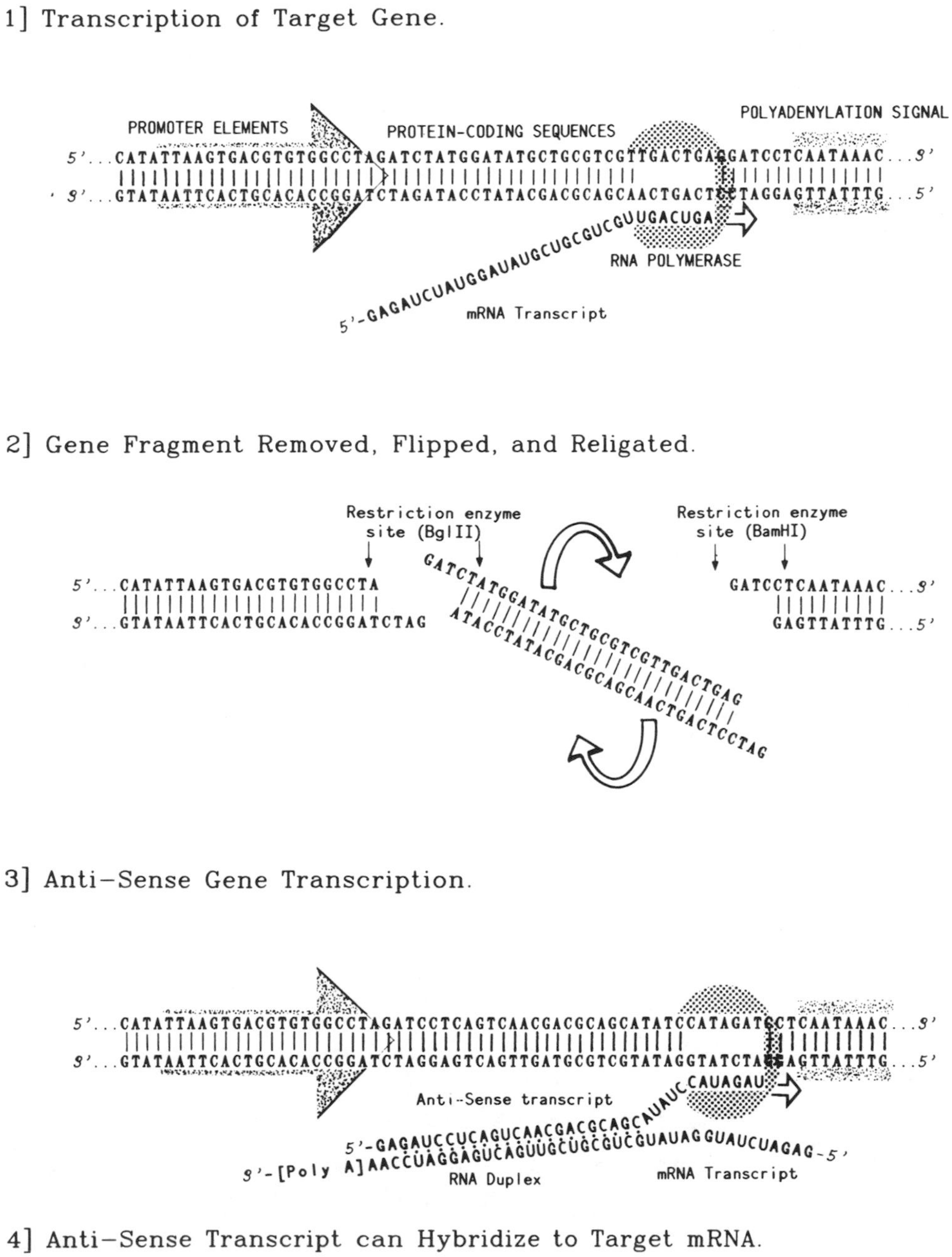

Figure 1

Antisense inhibition by flipped gene plasmids. Reverse orientation transcription of a gene produces an antimessage transcript that can neutralize the normal mRNA by duplex formation.

The thymidine kinase (*tk*) gene was chosen as a model system because *tk* expression can be readily analyzed cytologically in single cells by autoradiography and by solution assays in populations of cells. One can select for or against the growth of cells expressing *tk* by growth in the appropriate culture medium. Both the herpes simplex type I (HSV) *tk* gene and the chicken gene have been cloned and characterized, and such clones are efficiently transcribed when introduced into mammalian cells (Maitland and McDougall 1977; Wigler et al. 1977; Anderson et al. 1980; Capecchi 1980; Perucho et al. 1980; McKnight and Kingsbury 1982; Kwoh et al. 1983).

The plasmid DNAs shown in Figure 2 were constructed by standard recombinant DNA procedures (Maniatis et al. 1982) and are detailed elsewhere (Izant and Weintraub 1984). Briefly, the antisense plasmids contain promoter elements from the wild-type HSV *tk* gene (KF11) (McKnight 1982) or from the long terminal repeat of murine sarcoma virus (LTRS plasmids) (Van Beveren et al. 1981; B.J. Graves et al., in prep.). All of them contain polyadenylation signals from the SV40 VP1 transcription unit. The other antisense plasmids LTRS-chTK2, LTRS-5'TK2, and LTRS-βAc16 are identical to LTRS-TK3 with the exception that they have reverse orientation chicken genomic *tk*, 5' HSV *tk* (*Eco*RI to *Acc*I), and chicken β-actin cDNA (Cleveland et al. 1980) sequences, respectively, in place of the HSV *tk* sequences that are shown.

Antisense Transcription Inhibits Transient Gene Expression

Reasoning that the inhibition of gene expression may be more easily detected in the absence of pools of endogenous mRNA and enzyme, we first tested antisense Tk plasmids by trying to inhibit wild-type *tk* genes coinjected into Tk⁻ mouse

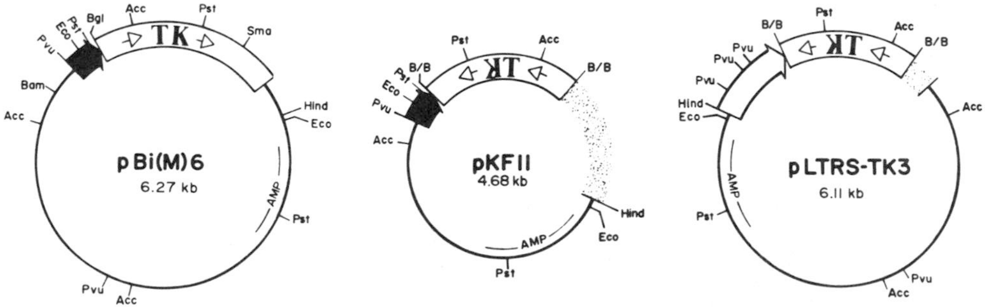

Figure 2
Diagrams of HSV thymidine kinase plasmid DNAs. Black arrows denote the location and orientation of the HSV *tk* promoter sequences. Open arrow represents the MSV LTR promoter and enhancer sequences. Stippled areas represent the SV40 sequences that contribute to the polyadenylation signals. The supercoiled plasmids contained HSV *tk* protein-coding sequences in the normal (Bi[M]6) or the antisense (KF11 and LTRS-TK3) orientations. LTRS-TK3 is a prototype antisense plasmid into which chicken *tk* and actin sequences have also been cloned. Restriction endonuclease site abbreviations: (Acc) *Acc*I; (Bam) *Bam*HI; (Bgl) *Bgl*II; (Eco) *Eco*RI; (Hind)*Hin*dIII; (Pst) *Pst*I; (Pvu) *Pvu*II; (B/B) *Bam*HI-*Bgl*II ligation sites.

L cells. Cloned wild-type *tk* genes (plasmid Bi[M]6) were introduced concomitantly with a large excess of flipped *tk*-gene constructions (KF11 and LTRS-TK3) or control plasmids by direct needle microinjection (Capecchi 1980; Mueller et al. 1981; Izant and Weintraub 1984). The cell cultures were incubated for 20–72 hours; 2 μCi/ml of [^{3}H]thymidine was added for the last 18 hours. The cells were washed in saline, fixed, dipped in photographic emulsion, and developed to assay the amount of labeled thymidine incorporated by each cell.

As summarized in Table 1, when an average of 5 copies of wild-type *tk* and 500–1000 copies of antisense plasmid per cell are coinjected, there is a four- to fivefold reduction in the frequency of Tk$^+$ cells after 20 or 72 hours. There is little effect of coinjecting an equimolar or even a five- or tenfold excess of the antisense plasmid KF11 with the HSV *tk* gene. There is no inhibition when antisense plasmids lacking promoter elements or lacking HSV *tk* sequences are coinjected at the same 200:1 ratio. The inhibition is sequence-specific. Antisense plasmids containing the non-cross-hybridizing chicken *tk* gene (LTRS-chTK2) do not significantly reduce the appearance of HSV *tk*-gene activity, but they do diminish the expression of chicken Tk plasmids in these cells; in a complementary fashion, antisense HSV *tk* does not reduce chicken *tk* expression.

Careful examination reveals that the cells coinjected with both normal and antisense *tk* (HSV and chicken) that are still Tk$^+$ show, on the average, a reduced accumulation of silver grains as compared with neighboring cells coinjected with Tk and control plasmids (Fig. 3). These observations suggest that antisense plasmids can reduce the efficiency of *tk*-gene expression in transient microinjection assays.

Integrated Antisense Plasmids Diminish Gene Expression

Antisense transcription can stably suppress the level of *tk* activity in L cells. As shown in Figure 4, there is a dramatic decrease in the transformation frequency

Table 1
Coinjection of Sense and Antisense Tk Plasmids

Plasmid DNA(s) injected	*Concentration*[a] *or ratio*	*Tk*$^+$ *cells*[b] *(%)*
HSV *tk* (Bi[M]6) alone	5 μg/ml	40
HSV *tk*:antisense HSV *tk* (KF11)	1:10	31
HSV *tk*:antisense HSV *tk* (KF11)	1:200	8.8
HSV *tk*:antisense HSV *tk* (LTRS-TK3)	1:200	5.8
HSV *tk*:antisense chicken *tk* (LTRS-chTK2)	1:200	32
HSV *tk*:promoterless antisense plasmid	1:200	34
HSV *tk*:promoter-only plasmid	1:200	51
Chicken Tk plasmid alone	2 μg/ml	44
Chicken *tk*:antisense chicken *tk* (LTRS-chTK2)	1:200	21

[a]Concentrations of Bi(M)6 and chicken *tk* were 2–5 μg/ml corresponding to approximately 10–25 copies per cell.

[b]Tk$^+$ phenotype was calculated from the number of cells incorporating [^{3}H]thymidine within 20 hr as a portion of the estimated number of cells surviving microinjection.

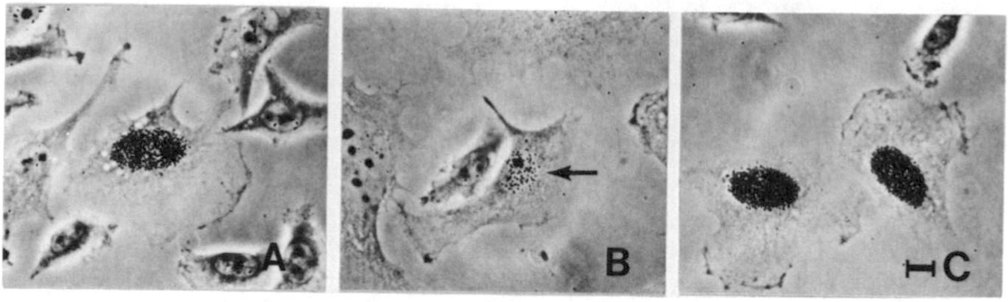

Figure 3
Autoradiography of L cells coinjected with wild-type and antisense DNA. Representative micrographs of mouse L cells injected with wild-type HSV *tk* (Bi[M]6) alone (*A*) or coinjected with wild-type *tk* and a 200-fold excess of antisense *tk* (KF11) (*B*) or a control promoterless antisense Tk plasmid (*C*). Cells coinjected with wild-type and antisense *tk* that are still Tk$^+$ (arrow) typically show a reduced autoradiographic grain density. Because all of the cells would not have passed through S phase during the labeling period and not every cell was injected, it is not anticipated that every cell in a field should be Tk$^+$. Bar, 10 μm.

of L cells to a Tk$^+$ phenotype (as measured by growth in aminopterin-containing culture medium) when excess antisense plasmid is included in calcium-phosphate-mediated DNA transformations (Wigler et al. 1977, 1979; Corsaro and Pearson 1981). No similar decrease is observed when equivalent concentrations of nonspecific plasmid or non-cross-hybridizing antisense plasmid are included.

DNA introduced into mammalian cells by injection or transfection undergoes homologous recombination at a significant rate (Capecchi 1980; Folger et al. 1982; Small and Scangos 1982; Miller and Temin 1983; Pomerantz et al. 1983). Single recombination events between sense and antisense plasmids within the Tk-coding region will inactivate both genes. Recombination does not occur, however, at a high enough frequency in this system to account for the observed inhibition (Izant and Weintraub 1984).

No general decrease in cell viability in cells containing the putative RNA duplex can be detected that could be responsible for the antisense inhibition observed. Interferon (and subsequent physiological stress) can be induced by the presence of RNA duplex (Lengyel 1982). However, we have not been able to detect interferon activity in any of our cultures.

Antisense plasmids and their target genes do not have to be introduced concomitantly in order to see antimessage inhibition. We have constructed several cell lines that produce anti-HSV *tk* RNA constitutively by cotransformation of the KF11 plasmid with other drug markers (Izant and Weintraub 1984). As shown in Table 2, these cells show a diminished capacity for transient expression of microinjected HSV *tk* genes; however, the non-cross-hybridizing chicken Tk plasmids are expressed normally. As with the transient assays, the average autoradiographic grain density of the cells that were still Tk$^+$ was reduced as compared with that of the parental L-cell line. Chicken *tk*-gene expression was similar in both the parental cell lines and the antisense HSV *tk* cell lines. These observations suggest that preloading a cell or an embryo with antisense transcript can

Table 2
Frequency of Expression of Tk Plasmids Microinjected into Antisense Cell Lines

		Tk⁺ cells (%)[a] *Tk plasmid microinjected*	
Cell line	*Genotype*	*HSV* tk	*chicken* tk
L/HA/2	*tk*⁻, *aprt*⁻ (wild type)	48	44
L/BB/3	KF11⁺, *aprt*⁺ (antisense HSV *tk*)	14	34
L/H1	KF11⁺, *aprt*⁺ (antisense HSV *tk*)	10	42

[a]Tk⁺ cells were calculated from the number of cells incorporating [^{3}H]thymidine within 20 hr as a proportion of the estimated number of cells surviving microinjection.

provide a measure of molecular "immunity" to subsequently activated or introduced genes.

Antisense Inhibition

Antisense transcription diminishes but does not eliminate *tk*-gene expression in most of these experiments. From 2% to 10% of the cells injected with HSV *tk* and an excess of antimessage plasmid still display a Tk⁺ phenotype. There are several factors that may contribute to the absence of an effect comparable to a null mutation. The Tk assay using [^{3}H]thymidine incorporation is especially sensitive and is nonlinear, since injection of five copies of HSV *tk* produces a Tk⁺ phenotype virtually indistinguishable from that produced by injection of 100 copies. This suggests that the Tk⁺ phenotype may be produced if only a few *tk* mRNAs escape antisense transcript hybridization. Thus, the assay could mask significant reductions of cellular *tk* mRNA activity and lead to an underestimate of the degree of inhibition that is actually occurring.

The molecular details of the mechanism of antisense inhibition are unknown. If an RNA duplex forms, it might block mRNA processing or transport in the nucleus or translation in the cytoplasm. To address this question and to try to construct more efficient antisense plasmids, plasmids containing fragments of cloned genes rather than the entire gene sequences have been constructed. Antisense Tk plasmids that are complementary to only the 5′-terminal 250 bases of the HSV transcript (LTRS-5′TK2) also inhibit *tk* expression in transient microinjection and cotransformation experiments. It is instructive to note that the predicted antisense LTRS-5′TK2 transcript duplex would cover the 5′ terminus of the target transcript as well as the initiator AUG codon. The original constructions (KF11 and LTRS-TK3) would cover only the protein-coding region including the initiator AUG codon. Remarkably, the LTRS-5′TK2 antisense plasmid virtually eliminates the expression of a chicken *tk*-gene construction driven by the HSV *tk* promoter even though it is complementary to only 54 bases at the 5′-untranslated terminus of the mRNA. This indicates that small antisense transcripts may possess significant inhibitory activity and, more significantly, that antisense genetics may be capable of discriminating between different isoforms of genes whose transcripts differ only in their untranslated domains. This also suggests that flipped gene antisense plasmids containing regions from the 5′ end of a gene will inhibit more efficiently than 3′ fragments.

The real goal is to inhibit the expression of endogenous genes in eukaryotic cells. Therefore, LTRS plasmids containing antisense actin genes (LTRS-βAc16) have been introduced into cultured cells to determine whether endogenous genes, coding for cytoarchitectural components, can be manipulated with antisense methodology. Such plasmids have little obvious effect on L cells 24–48 hours after microinjection. This is not surprising given the large size of actin mRNA and protein pools in mammalian cells. Antisense actin plasmids, however, are toxic when transformed into L cells. The transformation frequency of cells cotransfected with antisense actin plasmid and a drug marker is reduced compared with other control antisense plasmids and sense orientation actin plasmids (Fig. 4). The actin antisense colonies start to grow, but frequently stop proliferating and die after 1 week. The antisense actin plasmid appears to produce a dominant lethal phenotype, presumably by reducing actin synthesis. We are currently studying this effect in detail, but the initial results provide reasons for confidence that both endogenous genes and genes coding for cytoskeletal, in addition to metabolic, proteins can be inhibited by antimessage plasmids.

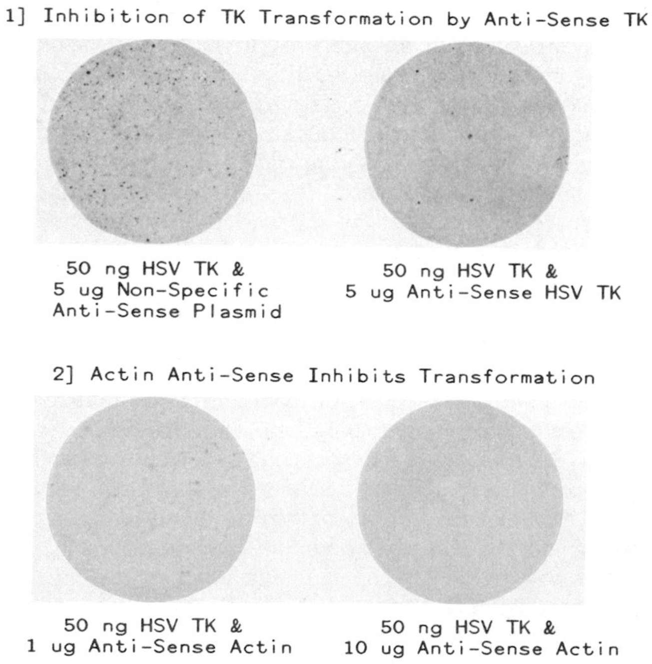

Figure 4

Antisense plasmids introduced into cells by transfection diminish transformation frequency. Tk⁻ L cells (5×10^5) were transfected with calcium-phosphate-precipitated plasmid DNA and carrier DNA. After growth in selective medium, the plates were fixed and stained to assay transformation frequency. Control experiments with nonspecific antisense plasmids showed no similar inhibition.

DISCUSSION

The results indicate that antisense RNA can be used to interfere with normal gene expression. The flipped gene antimessage plasmids can be constructed from genomic or cDNA clones of cellular and viral genes. Antisense plasmids are effective inhibitors of both endogenous and experimentally introduced genes. High ratios of antisense to target are necessary for both DNA-mediated and direct RNA antisense inhibition. The reasons such excesses are needed are not obvious, but it is possible that even in the complex cellular milieu, a significant concentration of inhibitor is required to obtain sufficient hybridization (R_0t) to inhibit gene function.

Antisense transcript regulation of gene activity is not without precedent. Bacteria utilize complementary RNA transcripts as one mode of regulating the expression and replication of DNA. Tomizawa's laboratory (Tomizawa et al. 1981; Tomizawa and Itoh 1982) has presented elegant evidence that short antisense transcripts form an RNA-RNA duplex with the 5′ end of the replication primer for the ColE1 plasmid. This interaction alters the tertiary structure of the primer and consequently its capacity to promote replication. Simons and Kleckner (1983) have suggested that a short antisense transcript regulates the expression of the transposase activity from the Tn*10* element in *Escherichia coli*. Mizuno et al. (1984) have shown that *E. coli* use a 170-base transcript with 70% homology with the 5′ end of the outer membrane protein OmpF mRNA to modulate the expression of that protein.

Bimolecular RNA hybridization may play an important role in the normal mechanism of gene expression in eukaryotes. Lerner et al. (1980), Rogers and Wall (1980), and Murray and Holliday (1979) have suggested that the RNA component in small, nuclear ribonucleoprotein particles contributes to mRNA splicing by base pairing with RNA splice junctions. These results have additional significance: Ribonucleoprotein complexes form rapidly on nascent RNA transcripts (Gall and Callan 1962; Miller 1965; Laird et al. 1976; McKnight and Miller 1976), which could limit the accessibility of both the target and the antisense transcript. Nevertheless, both the poly(A) tails (Kish and Pederson 1977) and the splice sites (Calvert and Pederson 1981; Calvert et al. 1982; Mount et al. 1983; Pederson 1983) on mRNA are available for base pairing. Therefore, it is a reasonable assumption that RNA transcripts in eukaryotes are at least transiently available for base pairing.

PROSPECTIVES

Antisense plasmids are one of many applications of molecular technology available for the study of cell structure and function. Sense (normal) orientation plasmids (often a serendipitous by-product of constructing antisense plasmids) can be used to examine the effect of overproducing a gene product or to rescue or complement some other deficiency in a cell or an organism. A "mutant" form of a sense orientation plasmid, such as one altered in vitro or a variant from another species or tissue, could be used to examine the consequences of altered protein structure in vivo.

Antisense RNA can be introduced directly into cells. Relatively large amounts of pure RNA can be transcribed from cloned genes using the *Salmonella* bacteriophage SP6 RNA polymerase (Butler and Chamberlin 1982; Green et al. 1983; Krieg and Melton 1984 and in prep.; D.A. Melton et al., in prep.). Microinjection of the appropriate antisense RNAs into *Xenopus* oocytes specifically diminishes or abolishes the translation of the HSV *tk* gene, the *E. coli* chloramphenicol acetyltransferase gene (R. Harland and H. Weintraub, pers. comm.), and *Xenopus* globin genes (Melton 1984). This approach has the merit of rapidly producing a sufficient concentration of the complementary inhibitor RNA and may have particular value for short-term studies. The direct injection of anti-message RNA has obvious advantages in transcriptionally quiescent cells and cells that tolerate the introduction of large amounts of RNA.

The antisense methodology represents a versatile tool for molecular analysis. It could be used to examine the activities of untranslated RNAs, such as small nuclear and heterogeneous nuclear RNAs. Antimessage plasmids could be used to analyze the cellular functions associated with a variety of gene sequences for which other functional probes are not available. Moreover, the methodology could be used to identify the cellular functions associated with unknown genomic or cDNA sequences. For while current technology allows the cloning of virtually any transcribed or nontranscribed DNA sequence, there are, unfortunately, very few ways to identify the cellular functions associated with such randomly cloned DNA.

The success of the antimessage plasmids in transformation studies indicates that they can be used to provide a specific molecular immunity to specific RNA transcripts that are produced after the introduction of the antisense plasmids. One fascinating application of this approach, which has research value and therapeutic potential, is the modulation of gene expression during viral or spontaneous transformation. Furthermore, with the use of appropriate promoter elements, this approach could be combined with germ-line transformation to produce antisense expression at specific times or in specific tissues of an embryo, which could be a valuable tool for gene therapy as well as for investigating cytodifferentiation and morphogenesis.

Our motivation for this investigation has been to provide some measure of the flexibility and power of genetic approaches available in such organisms as *Saccharomyces* to more complex multicellular systems such as mammals. Antisense RNA neutralization may not work efficiently for every gene, but it does represent an alternative strategy for the investigation of the molecular details of cellular systems, such as the cytoskeleton, that are otherwise not amenable to traditional genetic analysis.

ACKNOWLEDGMENTS

It is a pleasure to express my appreciation to Hal Weintraub for his stimulating interest, patient support, and invaluable participation in these studies as well as for his tolerance and lab facilities. Many colleagues at Hutchinson Cancer Research Center, especially Richard Harland and Steve McKnight, have provided crucial conversation, tutoring, and plasmids.

REFERENCES

Anderson, W.F., L. Killos, L. Sanders-Haigh, P.J. Kretschmer, and E.G. Diacumakos. 1980. Replication and expression of thymidine kinase and human globin genes microinjected into mouse fibroblasts. *Proc. Natl. Acad. Sci.* **77:** 5399.

Butler, E.T. and M.J. Chamberlin. 1982. Bacteriophage SP6-specific RNA polymerase: Isolation and characterization of the enzyme. *J. Biol. Chem.* **257:** 5772.

Calvert, J.P. and T. Pederson. 1981. Base-pairing interactions between small nuclear RNA precursors as revealed by psoralen cross-linking. *Cell* **26:** 363.

Calvert, J.P., L.M Meyer, and T. Pederson. 1982. Small nuclear RNA U2 is base-paired to heterogeneous nuclear RNA. *Science* **217:** 456.

Capecchi, M. 1980. High efficiency tranformation by direct microinjection of DNA in cultured mammalian cells. *Cell* **22:** 479.

Cleveland, D.W., M.A. Lopata, R.J. MacDonald, N.J. Cowan, W.J. Rutter, and M.W. Kirschner. 1980. Number and evolutionary conservation of α- and β-tubulin and cytoplasmic β- and γ-actin genes using specific cloned cDNA probes. *Cell* **20:** 95.

Corsaro, C.M. and M.L. Pearson. 1981. Enhancing the efficiency of DNA-mediated gene transfer in mammalian cells. *Somatic Cell Genet.* **7:** 603.

Cortese, R., R. Harland, and D. Melton. 1980. Transcription of tRNA genes *in vivo*: Single-stranded compared to double-stranded templates. *Proc. Natl. Acad. Sci.* **77:** 4147.

Folger, K.R., E.A. Wong, G. Wahl, and M. Capecchi. 1982. Patterns of integration of DNA microinjected into cultured mammalian cells: Evidence for homologous recombination between injected plasmid DNA molecules. *Mol. Cell. Biol.* **2:** 1372.

Gall, J.G. and H.G. Callan. 1962. ^{3}H-Uridine incorporation of lampbrush chromosomes. *Proc. Natl. Acad. Sci.* **48:** 562.

Green, M.R., T. Maniatis, and D.A. Melton. 1983. Human β-globin pre-mRNA synthesized *in vitro* is accurately spliced in *Xenopus* oocyte nuclei. *Cell* **32:** 681.

Izant, J.G. and H. Weintraub. 1984. Inhibition of thymidine kinase gene expression by antisense RNA: A molecular approach to genetic analysis. *Cell* **36:** 1007.

Kish, V.M. and T. Pederson. 1977. Heterogeneous nuclear RNA secondary structure: Oligo(U) sequences base paired with poly(A) and their possible role as binding sites for heterogeneous nuclear RNA-specific proteins. *Proc. Natl. Acad. Sci.* **74:** 1426.

Krieg, P.A. and D.A. Melton. 1984. Formation of the 3′ end of histone mRNA by post-transcriptional processing. *Nature* **308:** 203.

Kwoh, T.J., D. Zipser, and M. Wigler. 1983. Mutational analysis of the cloned chicken thymidine kinase gene. *J. Mol. Appl. Genet.* **2:** 191.

Laird, C.D., L.E. Wilkinson, V.E. Foe, and W.Y. Chooi. 1976. Analysis of chromatin-associated fiber arrays. *Chromosoma* **58:** 169.

Lengyel, P. 1982. Biochemistry of interferons and their actions. *Annu. Rev. Biochem.* **51:** 251.

Lerner, M.R., J.A. Boyle, S.M. Mount, S.L. Wolin, and J.A. Steitz. 1980. Are snRNPs involved in splicing? *Nature* **283:** 220.

Maitland, N.J. and J.K. McDougall. 1977. Biochemical transformation of mouse cells by fragments of herpes simplex virus DNA. *Cell* **11:** 233.

Maniatis, T., E.F. Fritsch, and J. Sambrook. 1982. *Molecular cloning: A laboratory manual.* Cold Spring Harbor Laboratory, Cold Spring Harbor, New York.

McKnight, S.L. 1982. Functional relationships between transcriptional control signals of the thymidine kinase gene of herpes simplex virus. *Cell* **31:** 355.

McKnight, S.L. and R. Kingsbury. 1982. Transcriptional control signals of a eukaryotic protein-coding gene. *Science* **217:** 316.

McKnight, S.L. and O.L. Miller. 1976. Ultrastructural patterns of RNA synthesis during early embryogenesis of *Drosophila melanogaster. Cell* **8:** 305.

Melton, D.A. 1984. Injected anti-sense RNAs specifically block messenger RNA translation *in vivo. Proc. Natl. Acad. Sci.* (in press).

Miller, C.K. and H.M. Temin. 1983. High-efficiency ligation and recombination of DNA fragments by vertebrate cells. *Science* **220:** 606.

Miller, O.L., Jr. 1965. Fine structure of lampbrush chromosomes. *Natl. Cancer Inst. Monogr.* **18:** 79.

Mizuno, T., M.-T. Chou, and M. Inouye. 1984. A unique mechanism regulating gene expression: Translational inhibition by a complementary RNA transcript (mic RNA). *Proc. Natl. Acad. Sci.* **81:** 1966.

Mount, S.M., I. Petterson, M. Hinterberger, A. Karmas, and J.A. Steitz. 1983. The U1 small nuclear RNA-protein complex selectively binds to a 5′ splice site *in vitro. Cell* **33:** 509.

Mueller, C., M. Graessmann, and A. Graessmann. 1981. A microinjection technique converting living cells into test tubes. *International cell biology,* 1980-1981 (ed. H.G. Schweiger), p. 119. Springer Verlag, Berlin.

Murray, V. and R. Holliday. 1979. Mechanism for RNA splicing of gene transcripts. *FEBS Lett.* **106:** 5.

Paterson, B., B. Roberts, and E. Kuff. 1977. Structural gene identification and mapping by DNA mRNA hybrid-arrested cell-free translation. *Proc. Natl. Acad. Sci.* **74:** 4370.

Pederson, T. 1983. Nuclear RNA-protein interactions and messenger RNA processing. *J. Cell Biol.* **97:** 1321.

Perucho, M., D. Hanahan, L. Lipsich, and M. Wigler. 1980. Isolation of the chicken thymidine kinase gene by plasmid rescue. *Nature* **285:** 207.

Pomerantz, B.J., M. Naujokas, and J.A. Hassell. 1983. Homologous recombination between transfected DNAs. *Mol. Cell. Biol.* **3:** 1680.

Rogers, J. and R. Wall. 1980. A mechanism for RNA splicing. *Proc. Natl. Acad. Sci.* **77:** 1877.

Simons, R.W. and N. Kleckner. 1983. Translational control of IS*10* transposition. *Cell* **34:** 683.

Small, J. and G. Scangos. 1982. Recombination during gene transfer into mouse cells can restore the function of deleted genes. *Science* **219:** 174.

Tomizawa, J. and T. Itoh. 1982. The importance of RNA secondary structure in ColE1 primer formation. *Cell* **31:** 575.

Tomizawa, J., T. Itoh, G. Selzer, and T. Som. 1981. Inhibition of ColE1 primer formation by a plasmid-specified small RNA. *Proc. Natl. Acad. Sci.* **78:** 1421.

Van Beveren, C., F. van Straaten, J.A. Galleshaw, and I.M. Verma. 1981. Nucleotide sequence of the genome of the murine saroma virus. *Cell* **27:** 97.

Wigler, M., A. Pellicer, S. Silverstein, R. Axel, G. Urlaub, and L. Chasin. 1979. DNA-mediated transfer of the adenine phosphoribosyltransferase locus into mammalian cells. *Proc. Natl. Acad. Sci.* **76:** 1373.

Wigler, M., S. Silverstein, L. Lee, A. Pellicer, Y. Cheng, and R. Axel. 1977. Transfer of purified herpes virus thymidine kinase gene to cultured mouse cells. *Cell* **11:** 223.

Author Index

Subject Index